Handbook of Navigation and Observation

Volume I

Handbook of Navigation and Observation
Volume I

Edited by **Stuart Cooper**

CLANRYE INTERNATIONAL

New Jersey

Published by Clanrye International,
55 Van Reypen Street,
Jersey City, NJ 07306, USA
www.clanryeinternational.com

Handbook of Navigation and Observation: Volume I
Edited by Stuart Cooper

International Standard Book Number: 978-1-63240-279-0 (Hardback)

Printed in the United States of America.

Contents

Permissions

List of Contributors

Preface

Navigation and observation is an area of study which focusses on procedural concepts related to controlling, monitoring, and the motion of air craft, land vehicle and water ships. In the European medieval period, navigation was included as an important component of the set of seven mechanical arts. The earliest known form of open ocean navigation, which was fascinatingly based on memory and observation, was Polynesian navigation. Also, it was in 1522 that the first circumnavigation of the earth was completed with the Magellan-Elcano expedition.

Navigation and observation fields can be categorised into four areas:

1. Land navigation,
2. Marine navigation,
3. Aeronautic navigation, and
4. Space navigation.

Most of the techniques in navigation systems are based on line of position. Some of the techniques in this area are as follows:

1. Dead reckoning
2. Pilotage
3. Celestial navigation
4. Radio navigation
5. Radar navigation
6. Satellite navigation

This process includes highly specific concepts of longitude, latitude, and loxodrome, to set the foundation of navigational path and connections between controller and objects in motion. Today, most modern navigation is based on positions determined on the basis of information received electronically from satellites. The future of navigation system planning is in electronic integrated bridge concepts.

This book discusses aspects of navigation and observation systems in different chapters. It is the outcome of contributions by many international experts in this field. I especially wish to acknowledge the contributing authors, without whom a work of this magnitude would clearly not have been realizable. I thank all the authors for allocating much of their scarce time to this project. This book is a true reflection of the vast research that is applicable to the area of navigation and observation. The chapters in this book cover most of the significant concepts and their development in this field of study. Finally, I hope that this book will achieve success, serving as a support to the readers and prove to be very useful to the scientific community.

Editor

GPS Vulnerability to Spoofing Threats and a Review of Antispoofing Techniques

Ali Jafarnia-Jahromi, Ali Broumandan, John Nielsen, and Gérard Lachapelle

Position Location and Navigation (PLAN) Group, Schulich School of Engineering, University of Calgary, 2500 University Drive, NW, Calgary, AB, Canada T2N 1N4

Correspondence should be addressed to Ali Jafarnia-Jahromi, ajafarni@ucalgary.ca

Academic Editor: Elena Lohan

GPS-dependent positioning, navigation, and timing synchronization procedures have a significant impact on everyday life. Therefore, such a widely used system increasingly becomes an attractive target for illicit exploitation by terrorists and hackers for various motives. As such, spoofing and antispoofing algorithms have become an important research topic within the GPS discipline. This paper will provide a review of recent research in the field of GPS spoofing/anti-spoofing. The vulnerability of GPS to a spoofing attack will be investigated and then different spoofing generation techniques will be discussed. After introducing spoofing signal model, a brief review of recently proposed anti-spoofing techniques and their performance in terms of spoofing detection and spoofing mitigation will be provided. Limitations of anti-spoofing algorithms will be discussed and some methods will be introduced to ameliorate these limitations. In addition, testing the spoofing/anti-spoofing methods is a challenging topic that encounters some limitations due to stringent emission regulations. This paper will also provide a review of different test scenarios that have been adopted for testing anti-spoofing techniques.

1. Introduction

GPS-dependent systems are ubiquitous in current positioning and navigation applications. There is an ever-increasing attention to safe and secure GPS applications such as air, marine, and ground transportations, police and rescue services, telecommunication systems, mobile phone location, and tracking the criminal offenders. Nowadays, most mobile phones as well as vehicles are equipped with positioning and navigation systems utilizing GPS. In addition, countless time tagging and synchronization systems in the telecom and electrical power grid industries rely primarily on GPS. As a consequence, such a widely used system is becoming an increasingly attractive target for illicit disruption by terrorists and hackers.

GPS signals are vulnerable to in-band interferences because of being extremely weak broadcasted signals over wireless channels. Therefore, even a low-power interference can easily jam or spoof GPS receivers within a radius˜of several kilometres. In addition, GPS is a backward compatible technology whose signal structure is in the public domain [1]. This makes GPS technology more susceptible to disruptive interfering methods. For example, spoofing attack could effectively mislead an activity monitoring GPS receiver mounted on a cargo transport or fishing vessel. Therefore, the GPS receiver will be logging a counterfeit trajectory with various consequences.

Spoofing and antispoofing mechanisms are emerging issues in modern GPS applications that will increasingly attract research in future [1]. Spoofing is a deliberate interference that aims to coerce GNSS receivers into generating false position/navigation solutions [2]. The spoofer attempts to mimic authentic GPS signals in order to mislead the target receiver. The spoofing attack is potentially significantly more menacing than jamming since the target receiver is not aware of this threat. Recently the implementation of sophisticated spoofers has become more feasible, flexible, and less costly due to rapid advances in software-defined radio (SDR) technology [3].

In recent years, research has been initiated on spoofing discrimination and mitigation [2–9]. This paper first provides a brief review of different spoofing generation techniques. Subsequently, the vulnerability of civilian GPS receivers to spoofing attacks will be investigated in different operational layers. Then, a brief review of current anti-spoofing techniques will be provided in terms of spoofing detection and spoofing mitigation. Furthermore, three test scenarios will be investigated that are useful for testing the spoofing/antispoofing algorithms in the real-world scenarios.

This paper is organized as follows: a brief discussion on different spoofing generation techniques is provided in Section 2. GPS vulnerability against spoofing attacks is investigated in Section 3, and then Section 4 demonstrates the received signal model for a GPS receiver under spoofing attack. Antispoofing techniques will be discussed in more detail in Section 5. In Section 6, the test scenarios are investigated in real spoofing environments. Concluding remarks are provided in Section 7.

2. Classification of Spoofing Generation Techniques

Spoofing generation can be divided into three main categories [2, 3, 7].

2.1. GPS Signal Simulator. In this category a GPS signal simulator concatenated with a RF front-end is employed to mimic authentic GPS signals. The signals generated by this kind of spoofer are not essentially synchronized to the real GPS signals. Therefore, the spoofing signals look like noise for a receiver operating in the tracking mode (even if the spoofer power is higher than the authentic signals). However, this type of spoofers can effectively mislead commercial GPS receivers especially if the spoofing signal power is higher than the authentic signals. A GPS signal simulator is the simplest GPS spoofer and it can be detected by different antispoofing techniques such as amplitude monitoring, consistency checks among different measurements, and consistency check with inertial measurement units (IMUs).

2.2. Receiver-Based Spoofers. A more advanced type of spoofer consists of a GPS receiver concatenated with a spoofing transmitter. This system first synchronizes to the current GPS signals and extracts the position, time, and satellite ephemeris, and then it generates the spoofing signal knowing the 3D pointing vector of its transmit antenna toward the target receiver antenna. This kind of spoofer is difficult to discriminate from the authentic signals and is more complicated than the first category. The main challenge toward realization of this kind of spoofer is projecting the spoofing signals to the intended victim receiver with the correct signal delay and strength. Note that the spoofing power should be slightly higher than the authentic signal power in order to successfully mislead the target receiver but it should not be much more than the typical power of GPS signals.

Aligning the carrier frequency and phase to the authentic GPS signals, minimizing the self-jamming effect and suppressing relative data bit latencies are other limitations that a receiver-based spoofer should deal with. Carrier phase

alignment to the authentic signals requires centimetre level knowledge of the 3D pointing vector from the spoofer transmit antenna phase centre toward the target receiver antenna phase centre. Therefore, it would be a great advantage for this kind of spoofers if the spoofer antenna were very close to the target receiver antenna. This type of spoofers is relatively hard to detect since they are synchronized to the real GPS satellites and can spoof receivers in tracking mode. Figure 1 shows a repeater-spoofer structure proposed by [3].

2.3. Sophisticated Receiver-Based Spoofers. This category is the most complex and effective type of the spoofing categories. This type is assumed to know the centimetre level position of the target receiver antenna phase centre to perfectly synchronize the spoofing signal code and carrier phase to those of authentic signals at the receiver [7]. This type of spoofer can take advantage of several transmit antennas in order to defeat direction of arrival antispoofing techniques. In this case the spoofer needs to synthesize an array manifold that is consistent with the array manifold of the authentic signal to defeat an angle of arrival (AOA) discriminating GPS receiver.

The complexity of constructing such a spoofer is much higher than the two previous categories discussed above. Compared to the previous spoofing categories, the effectiveness region of this type of spoofer is much more limited. The reason is that carrier phase alignment and array manifold synchronization might be achieved only for a very small region where target receiver antennas are located. In addition, there are some physical limitations regarding the spoofer antenna placement relative to the target receiver antenna(s). As such, the realization of this type of spoofers is very difficult and in many cases impossible due to the geometry and movement of the target receiver antenna(s).

3. GPS Vulnerability against the Spoofing Attack

The vulnerability of GPS to spoofing can be investigated in three operational layers of GPS receivers, namely, the signal processing, data bit, and position/navigation solution levels.

3.1. GPS Vulnerability in Signal Processing Level. The structure of GPS signals, including the modulation type, pseudorandom noise (PRN) signals, transmit frequency, signal bandwidth, Doppler range, and signal strength publicly known. Furthermore, GPS is a backward compatible technology whose L1 signal features do not significantly change through different generations of GPS satellites. Most of the commercial GPS receivers are equipped with automatic gain control (AGC) block that compensates the power variations in the received GPS signal. However, AGC can increase the vulnerability of GPS receivers against higher power spoofing signals since it automatically adjusts the receiver input gain according to the more powerful spoofing signals [8]. Therefore, knowing the general structure and operational basics of a civilian GPS receiver, a spoofing module can generate counterfeit signals that are arbitrarily similar to the

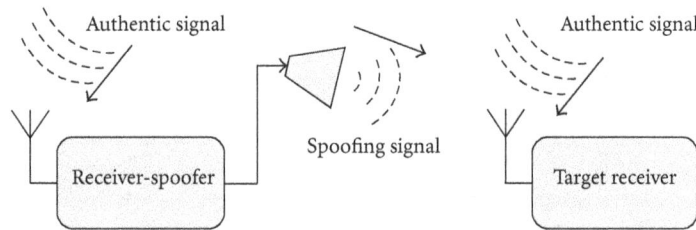

FIGURE 1: Repeater spoofer block diagram (modified after [3]).

authentic GPS signals such that they can effectively mislead GPS receivers.

3.2. GPS Vulnerability in Data Bit Level.
The framing structure of the GPS signals is publicly known. The navigation frame consists of different parts such as almanac, telemetry information, and satellite ephemeris. This information does not change rapidly during short time intervals; for example, the satellite ephemeris information can be acquired in less than 1 minute but it remains unchanged for 12.5 minutes [1]. Therefore, the spoofer can take advantage of this stability in order to regenerate the GPS data frame. In addition, the satellite health status bits can be manipulated by a spoofer in order to mislead the receiver toward rejecting the valid satellite signals [10].

3.3. GPS Vulnerability in Navigation and Position Solution Level.
The spoofer can inject counterfeit pseudorange measurements into the receiver, leading to a wrong position, velocity, and time (PVT) solution. The PVT error is proportional to the range residuals multiplied by a geometry related factor. In [11] the authors have developed a vulnerability index against spoofing (VIAS) that indicates the geometric relationship between GPS constellation and the spoofer position that results in receiver position solution deviations. It is assumed that the receiver is already running a receiver autonomous integrity monitoring (RAIM) procedure. Therefore, the VIAS coefficient is proposed for the case where the spoofing signal is not detected by the RAIM technique. It is shown that the VIAS changes over time and position and it has a higher value where the position dilution of precision (PDOP) value is high. The VIAS index can be used in the design and development of antispoofing methods.

In some applications, GPS receivers are strictly used for timing synchronization such as CDMA/GSM cell towers. In this case, the spoofing attack can highly disrupt the accuracy of the estimated timing, and this can seriously disturb the handoff processing between neighboring cells.

4. Received Signal Model

Antispoofing techniques can be generally investigated for two receiver categories, namely, single-antenna and multiple, antenna receivers. This section describes the received signal model for these receivers in the presence of spoofing attack.

4.1. Single Antenna Receiver.
Considering the GPS L1 C/A code, the received signal subjected to a spoofing attack can be modeled as

$$r(nT_s) = \sum_{m=1}^{N_{\text{Auth}}} \sqrt{p_m^a} F_m^a(nT_s) + \sum_{q=1}^{N_{\text{Spoof}}} \sqrt{p_q^s} F_q^s(nT_s) + \eta(nT_s),$$

$$(1)$$

where

$$F_m^a(nT_s) = h_m^a(nT_s - \tau_m^a) c_m^a(nT_s - \tau_m^a) e^{j\phi_m^a + j2\pi f_m^a nT_s},$$
$$F_q^s(nT_s) = h_q^s(nT_s - \tau_q^s) c_q^s(nT_s - \tau_q^s) e^{j\phi_q^s + j2\pi f_q^s nT_s},$$

$$(2)$$

and N_{Auth} and N_{Spoof} are the number of authentic and spoofing PRN signals, respectively. The superscripts s and a refer to the spoofing and authentic signals, respectively. T_s is the sampling interval, and ϕ, f, p, and τ are the carrier phase, Doppler frequency, signal power and code delay of the received signals, respectively. In this model, $h(nT_s)$ is the transmitted navigation data bit and $c(nT_s)$ is the PRN sequence ant time instant nT_s. The subscripts m and q correspond to the mth authentic signal and qth spoofing signal, respectively. η is the complex additive white Gaussian noise with variance σ^2 and j is the square root of -1.

4.2. Multiple-Antenna Receiver.
Assume an arbitrary N-element antenna array configuration. In this configuration, one antenna is chosen as the reference antenna. Without loss of generality assume that the reference coordinate system is located at the reference antenna (r_1) as shown in Figure 2. Here, it is assumed that the spoofer is a single-antenna transmitter that is transmitting several PRN signals from the same direction. Therefore, the complex baseband representation of N received spatial samples of authentic and spoofing signals impinging on the antenna array before de-spreading can be written in vector form as

$$\mathbf{r}(nT_s) = \begin{bmatrix} r_1(nT_s) \\ \vdots \\ r_N(nT_s) \end{bmatrix} = \sum_{m=1}^{N_{\text{Auth}}} \mathbf{a}_m \sqrt{p_m^a} F_m^a(nT_s)$$

$$+ \mathbf{b} \sum_{q=1}^{N_{\text{Spoof}}} \sqrt{p_q^s} F_q^s(nT_s) + \boldsymbol{\eta}(nT_s),$$

$$(3)$$

where $\boldsymbol{\eta}$ is the $N \times 1$ complex additive white Gaussian noise vector with covariance matrix $\sigma^2 \mathbf{I}$ and $\mathbf{I}$ represents a N by N

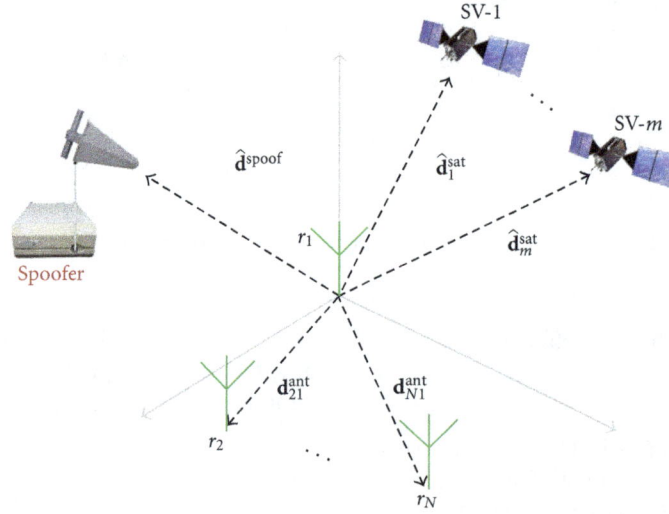

FIGURE 2: Multiple-antenna configuration.

identity matrix. $\mathbf{a}_m$ and $\mathbf{b}$ are steering vectors incorporating all spatial characteristics of the antenna array for authentic and spoofing signals, which can be written as

$$
\mathbf{b} = \begin{bmatrix} 1 \\ b_2 \\ \vdots \\ b_N \end{bmatrix} = \begin{bmatrix} e^{-j((2\pi \mathbf{d}_{11}^{ant} \cdot \hat{\mathbf{d}}^{spoof})/\lambda)} \\ e^{-j((2\pi \mathbf{d}_{21}^{ant} \cdot \hat{\mathbf{d}}^{spoof})/\lambda)} \\ \vdots \\ e^{-j((2\pi \mathbf{d}_{N1}^{ant} \cdot \hat{\mathbf{d}}^{spoof})/\lambda)} \end{bmatrix},
$$

$$
\mathbf{a}_m = \begin{bmatrix} 1 \\ (a_m)_2 \\ \vdots \\ (a_m)_N \end{bmatrix} = \begin{bmatrix} e^{-j((2\pi \mathbf{d}_{11}^{ant} \cdot \hat{\mathbf{d}}_m^{sat})/\lambda)} \\ e^{-j((2\pi \mathbf{d}_{21}^{ant} \cdot \hat{\mathbf{d}}_m^{sat})/\lambda)} \\ \vdots \\ e^{-j((2\pi \mathbf{d}_{N1}^{ant} \cdot \hat{\mathbf{d}}_m^{sat})/\lambda)} \end{bmatrix},
$$

(4)

where $\mathbf{d}_{i1}^{ant}$ represents a vector pointing from the origin (reference antenna phase centre) to the ith antenna phase centre. $\hat{\mathbf{d}}_m^{sat}$ and $\hat{\mathbf{d}}^{spoof}$ represent the pointing unit vectors from the origin to the mth authentic satellite and spoofing source respectively. λ represents the GPS carrier wavelength at L1 frequency.

5. Classification of Antispoofing Techniques

Several antispoofing techniques have been proposed in the open literature and can generally be classified into two main categories, namely *spoofing detection* and *spoofing mitigation*. Spoofing detection algorithms concentrate on discriminating the spoofing signals but they do not necessarily perform countermeasures against the spoofing attack, while spoofing mitigation techniques mainly concentrate on neutralizing the detected spoofing signals and help the victim receiver to retrieve its positioning and navigation abilities. In the following subsections a brief introduction is provided on different techniques proposed for each category.

5.1. Spoofing Detection

5.1.1. Methods Based on the Signal Power Monitoring

(a) C/N₀ Monitoring. Most GPS receivers employ C/N_0 measurements as a parameter that characterizes the received signal quality. In open sky conditions, only satellite movement and ionosphere variations can cause gradual smooth changes in the received signal power. However, when a higher power spoofer misleads a GPS receiver, the received C/N_0 may experience a sudden change that can indicate the presence of the spoofing signal. The antispoofing receiver can continuously monitor the C/N_0 and look for any unusual variation that can be a sign of spoofing attack. It is easy for a GPS receiver to store a time history of the signal received from each satellite.

Consider the correlator output for the lth authentic signal as the following equation:

$$
y_l^a(kNT_s) = \underbrace{\sqrt{p_l^a}e^{j\varphi_l^a}}_{S:\ \text{Desired Signal}} + \underbrace{\sum_{\substack{m=1 \\ m \neq l}}^{N_{Auth}} \sqrt{p_m^a}C_{ml}^a(kNT_s)}_{I_{Auth}:\ \text{Interference caused by} \atop \text{other authentic PRNs}}
$$

$$
+ \underbrace{\sum_{q=1}^{N_{Spoof}} \sqrt{p_q^s}C_{ql}^s(kNT_s)}_{I_{Spoof}:\ \text{Interference caused by} \atop \text{spoofer generated PRNs}} + \underbrace{\overline{\eta}(kNT_s)}_{\text{Gaussian Noise}},
$$

(5)

where

$$
C_{ml}^a(kNT_s) = \frac{1}{N}\sum_{n=(k-1)N+1}^{kN} F_m^a(nT_s)\hat{F}_l(nT_s),
$$

$$
C_{ql}^s(kNT_s) = \frac{1}{N}\sum_{n=(k-1)N+1}^{kN} F_q^s(nT_s)\,\hat{F}_l(nT_s),
$$

(6)

$$
\hat{F}_l(nT_s) = c_l(nT_s - \hat{\tau}_l)\,e^{-j2\pi \hat{f}_l nT_s},
$$

where N determines the coherent integration interval and kNT_s is the time instant at which the correlator output is updated. $C_{ml}^a(kNT_s)$ is the cross-correlation between $F_m^a(nT_s)$ and the lth locally generated PRN signal replica, $\hat{F}_l(nT_s)$, whose Doppler and code delay are $\hat{\tau}_l$ and $\hat{f}_l$, respectively. Herein, the effect of data bits has been neglected to simplify the notations. In (5) the second and third additive terms (I_{Auth} and I_{Spoof}) are interference terms caused by cross-correlation effect of other authentic and spoofing signals. $\bar{\eta}[kNT_s]$ is the filtered noise component with variance σ^2/N. The C/N_0 measurement for each GPS signal is proportional to the ratio between the despread signal power at the correlator output to the noise power plus other signal interferences. The postprocessing signal-to-noise ratio (SNR), which is linked to the C/N_0 value, can be shown as

$$\text{SNR}_l^a = \frac{p_l^a}{|I_{Auth}|^2 + |I_{Spoof}|^2 + (\sigma^2/N)}. \tag{7}$$

GPS signals are designed such that $|I_{Auth}|^2$ is negligible compared to the filtered Gaussian noise variance. However, $|I_{Spoof}|^2$ increases as the total spoofing power (TSP) increases. TSP is the sum of signal powers for different spoofing PRNs (i.e., $\text{TSP} = \sum_{q=1}^{N_{Spoof}} \sqrt{p_q^s}$). Therefore, an asynchronous spoofing source that is transmitting several PRNs with considerable power can effectively reduce the C/N_0 of the authentic signals. However, if a spoofing signal is despread, its corresponding C/N_0 measurement would be in the normal authentic C/N_0 range. In this case, the spoofer has generated higher power correlation peaks over an elevated noise floor. This procedure can effectively mislead those spoofing detection techniques that are based on C/N_0 monitoring. As a consequence, the receiver might be tracking the higher power spoofing correlation peaks while its C/N_0 measurement does not show any abnormalities [12].

Figure 3 illustrates the authentic and spoofing SNR values versus the TSP for the case of 10 equal power authentic signals and 10, 20, and 30 equal power spoofing signals. The power of each authentic signal is -158 dBW and the coherent integration time is $T_c = NT_s = 1$ ms. A typical detection SNR threshold has been depicted in this figure. It is observed that the SNR of the authentic signals decreases as the TSP increases, while, on the contrary, the SNR of the spoofing signals increases up to a certain level as the TSP increases. The maximum spoofing SNR level depends on the number of transmitted spoofing PRN signals and the distribution of TSP among them. The receiver noise floor estimate at the 1 ms integration time is also depicted on the right-hand side of the y-axis of Figure 3. This curve is useful for analyzing the noise floor increase at a certain TSP level.

(b) Absolute Power Monitoring. As the path loss between the spoofer and target receiver is highly variable, it is difficult for a spoofer to estimate the transmit power required to impose sufficient signal strength at the target receiver while not excessively exceeding the typical power level of the authentic GPS signals [8]. The maximum received power of the GPS signals at earth terminals is around -153 dBW at the L1 frequency [13]. Therefore, reception of a spoofing signal whose absolute power is considerably higher than the expected authentic GPS signal power is a simple direct means of detecting a spoofing attack.

Figure 4 provides a comparison between the spoofing vulnerability regions for a C/N_0 monitoring receiver and an absolute power monitoring receiver. It has been assumed that the absolute power monitoring receiver is able to discriminate the elevated noise floor as well as higher power PRN signals within a 2 dB accuracy range. In other words, this receiver discriminates those PRNs whose absolute power is 2 dB or higher than the maximum possible received power of GPS L1 C/A signal. Furthermore, this receiver is capable of detecting a 2 dB increase in noise floor from its desired value. On the other hand, the C/N_0 monitoring receiver is only able to discriminate the signals whose SNR is higher than the maximum possible SNR of the GPS L1 C/A signal (this value is assumed to be 21.8 dB for $T_c = 1$ ms and temperature = $300°$K).

Hence, as it is shown in Figure 4, the vulnerability region of the absolute power monitoring receiver is much smaller than the vulnerability region of the C/N_0 monitoring receiver. Furthermore, if the receiver is able to detect the absolute receiver power more accurately, it can considerably reduce the size of its vulnerability window in the presence of a spoofing attack [12].

Implementation of this power monitoring technique requires the receiver ability to measure the absolute amplitude of the received signal within a certain accuracy level. Hence, the hardware complexity slightly increases. In addition, the relatively high dynamic range of the GPS signal strength imposes another limitation to the performance of the amplitude discrimination techniques.

(c) Received Power Variations versus Receiver Movement. Based on the free space square law of propagation, the received power of a free space propagating signal is proportional to the inverse of the squared propagation distance. GPS satellites are around 20,000 kilometres away from the earth surface; therefore, if the receiver moves on the earth surface in low multipath open sky environments, no considerable change in the received power from authentic satellites should be observed other than the deterministic losses occurring at lower elevations. However, as discussed before, the spoofing signal is usually transmitted from a single directional antenna located much closer to the receiver compared to the GPS satellites. Therefore, the movement of the receiver relative to the spoofer antenna can considerably change the C/N_0 received from spoofing signals. Figure 5 illustrates the variations of spoofing and authentic received C/N_0 values versus the receiver distance from a spoofer antenna. It is observed that when the spoofer is very close to its target receiver, even a slight movement between spoofer and the target receiver can considerably affect the received spoofing signal C/N_0. It should be considered that all the spoofing signals are usually transmitted from the same antenna and therefore all experience the same propagation medium. As such, variations of all spoofing signals will be

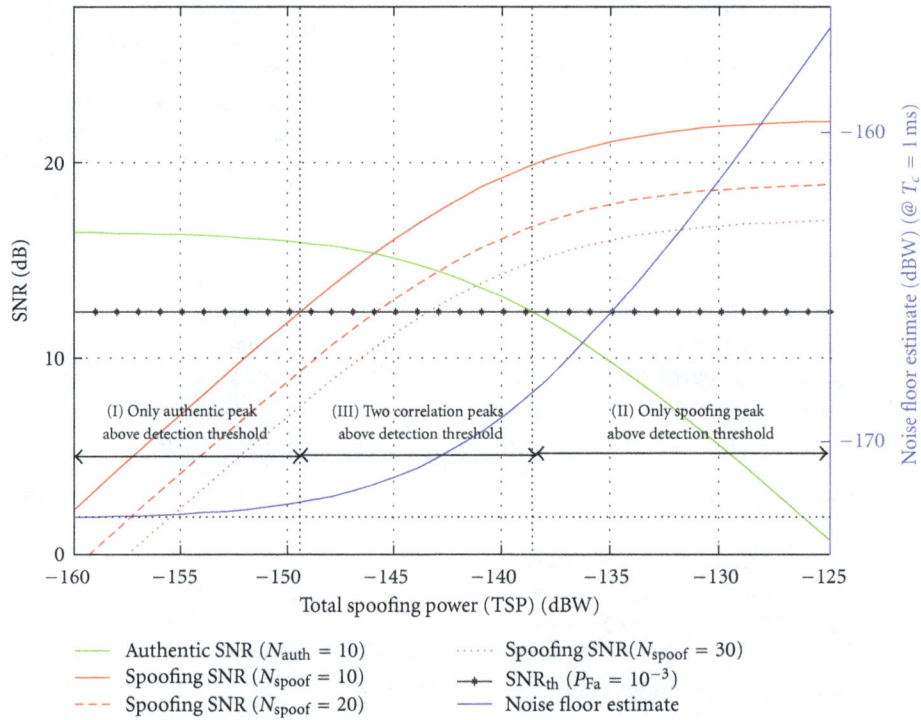

FIGURE 3: Received SNR versus TSP for authentic and spoofing correlation peaks [12].

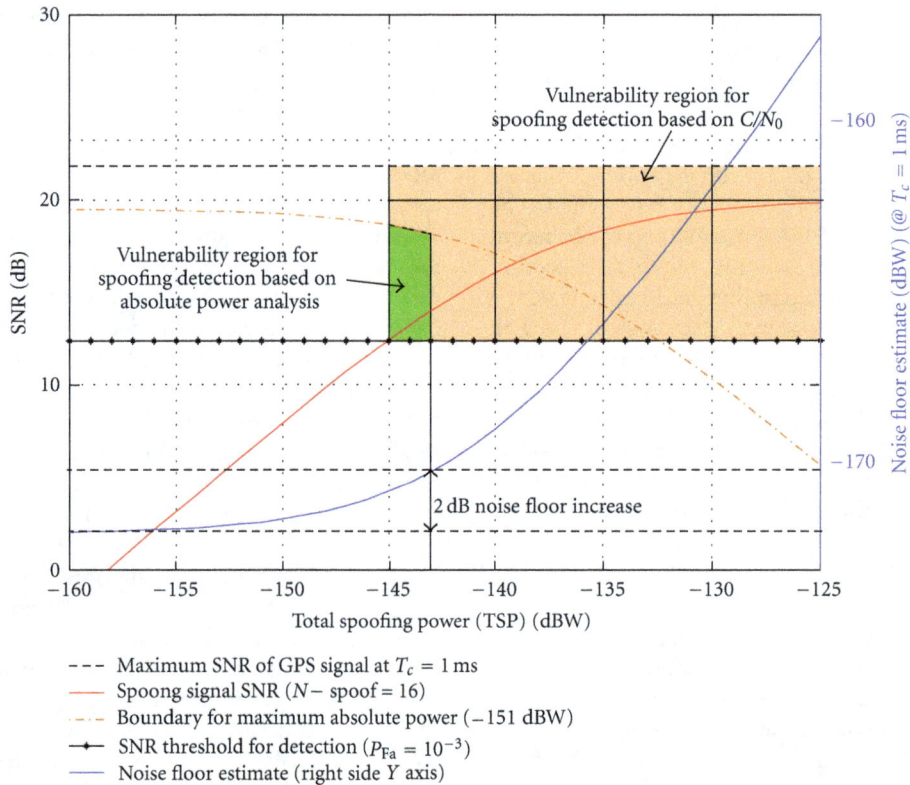

FIGURE 4: Vulnerability region comparison of C/N_0 versus absolute power monitoring techniques [12].

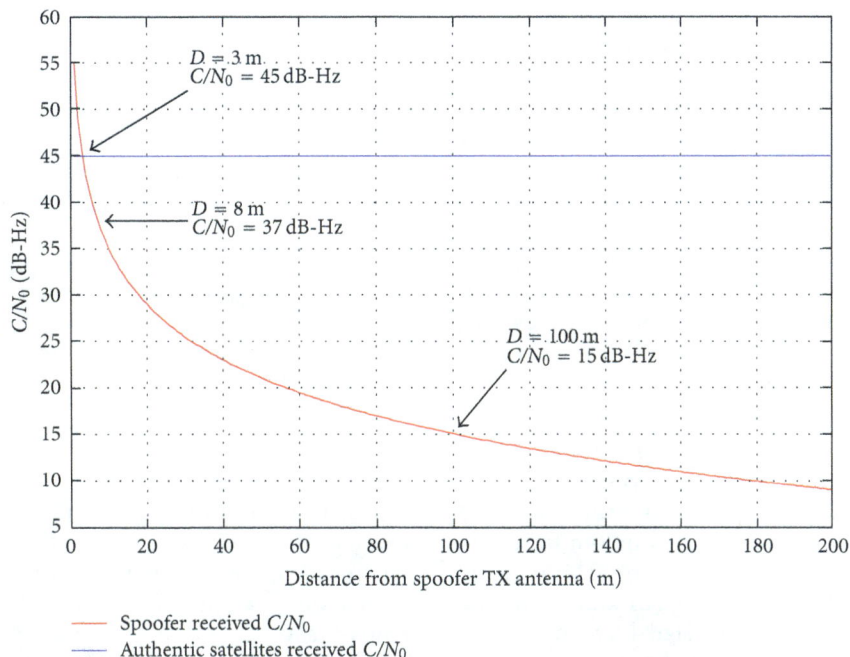

FIGURE 5: Variations of spoofing and authentic received C/N_0 versus receiver distance from spoofer transmitting antenna.

the same regardless of the receiver movement and multipath effects [11]. Here it is assumed that the spoofer does not differentially modulate the C/N_0 of the different PRN signals.

This method is a low-complexity spoofing discrimination technique that does not impose extensive hardware/software modifications to the GPS receiver. However, since the receiver does not necessarily know the position of the spoofer antenna and the distance variations with respect to the receiver antenna, there is no guarantee that the receiver movement considerably changes the received C/N_0 of the spoofer generated signals. For example, when both spoofing transmitter and GPS receiver are located in the same vehicle, the movement of vehicle does not cause variation in measure of spoofing signals C/N_0. Another disadvantage of this technique is that it cannot be employed for the case of static GPS receivers. Therefore, the effectiveness of this spoofing discrimination technique is limited to a few spoofing scenarios.

(d) L1/L2 Power Level Comparison. There is a predefined power level difference between GPS signals in different frequency bands [8] and many GPS receivers are able to monitor both L1 and L2 signals. However, a low-complexity spoofer may only generate L1 signal. Therefore, a large difference between L1 and L2 power levels or the absence of L2 signals can reveal the presence of a spoofing signal.

This method can successfully detect the single-band spoofers. However, most of the civil GPS receivers do not have the ability to monitor both L1 and L2 frequency bands and this discrimination technique imposes additional hardware complexity to the GPS receiver.

5.1.2. Spoofing Discrimination Using Spatial Processing. Due to logistical limitations, spoofing transmitters usually transmit several counterfeit signals from the same antenna while the authentic signals are transmitted from different satellites with different directions. Therefore, a spatial processing technique can be employed to estimate the spatial signature of received signals and discriminate those signals that are spatially correlated.

(a) Multiantenna Spoofing Discrimination. In [2] a spoofing detection technique is proposed which observes the phase difference between two fixed antennas for around one hour. Knowing the bearing of the antenna array and the satellites movement trajectory, the theoretical phase differences can be calculated and compared to the practical phase difference observed by the antenna array to discriminate the spoofing threat. The main drawback of the algorithm is that it takes a long time (about 1 hour) to discriminate the spoofing signals. In addition, this technique requires a calibrated antenna array with known array orientation in order to operate properly.

In [14] an antenna array structure is used to detect and mitigate spoofing signals based on their spatial correlation. The correlator output phase measurements for different PRN signals are mutually compared to discriminate the ones received from the same spatial sector. This technique can successfully detect spoofing signals and it does not need any array calibration or information regarding array orientation. This technique can effectively discriminate the spoofing scenarios that employ a single transmit antenna. In addition, the multipath propagation does not degrade the performance of this method since all the spoofing signals experience the same propagation channel characteristics.

However, this technique increases the hardware complexity of the GPS receiver as it necessitates the use of several antenna branches. Furthermore, applying this method increases the computational complexity of GPS receiver since the receiver needs to acquire and track both spoofing and authentic signals in order to be able to discriminate spoofing PRNs.

A multipleantenna spoofer might be able to defeat the multiple-antenna spoofing discrimination techniques depending on the number of transmit antennas, the number of receiver antennas, and the geometry of spoofer antennas with respect to the target receiver antennas. However, there are many practical limitations to realizing such a sophisticated spoofing scenario.

(b) Synthetic Array Spoofing Discrimination. In [6] a spoofing detection technique that employs a synthetic antenna array has been proposed. In this scenario a single-antenna handheld GPS receiver is moved along a random trajectory and forms a synthetic antenna array structure. This scenario is shown in Figure 6. The received signals amplitude and phase corresponding to different PRN signals are continually compared to each other using a correlation coefficient metric (ρ_{ij}). Therefore, after acquiring different PRN signals in the received signal set (both authentic and spoofing signals), spoofing signals are discriminated using the following normalized correlation coefficient:

$$\rho_{ij} = \left| \frac{E\left[(\mathbf{y})_i^H (\mathbf{y})_j \right]}{\sqrt{E\left[(\mathbf{y})_i^H (\mathbf{y})_j \right]} \sqrt{E\left[(\mathbf{y})_j^H (\mathbf{y})_j \right]}} \right|, \quad (8)$$

where $E[]$ represents the statistical expectation and the superscript H denotes the conjugate transpose. $(\mathbf{y})_i$ and $(\mathbf{y})_j$ represent the ith and jth columns of matrix $\mathbf{y}$, which is defined as follows:

$$\mathbf{y} = \begin{bmatrix} [\mathbf{y}^a[1], \mathbf{y}^s[1]] \\ [\mathbf{y}^a[2], \mathbf{y}^s[2]] \\ \vdots \\ [\mathbf{y}^a[M], \mathbf{y}^s[M]] \end{bmatrix}_{M \times L}, \quad (9)$$

$$\mathbf{y}^a[k] = \left[y_1^a(kNT_s), \ldots, y_{N_{\text{Auth}}}^a(kNT_s) \right],$$

$$\mathbf{y}^s[k] = \left[y_1^s(kNT_s), \ldots, y_{N_{\text{Spoof}}}^s(kNT_s) \right].$$

In (9), it is assumed that correlator outputs are monitored during M time instances and $\mathbf{y}$ is an $M \times L$ matrix where L is the number of acquired PRN signals ($L \leq N_{\text{Auth}} + N_{\text{Spoof}}$). $\mathbf{y}^a[k]$ is the set of correlator outputs for all acquired authentic signals at time instant kNT_s, whereas $\mathbf{y}^s[k]$ consists of all acquired spoofing peaks for that time instant. M is the number of equivalent spatial samples.

Figure 7 illustrates the normalized signal amplitude for acquired spoofing and authentic signals. During the data collection, the antenna was randomly moved. It is observed that the amplitude variations for spoofing signals are highly correlated (i.e., the plots representing the amplitudes of PRN-16, PRN-18, PRN-21, and PRN-29 are totally overlaid)

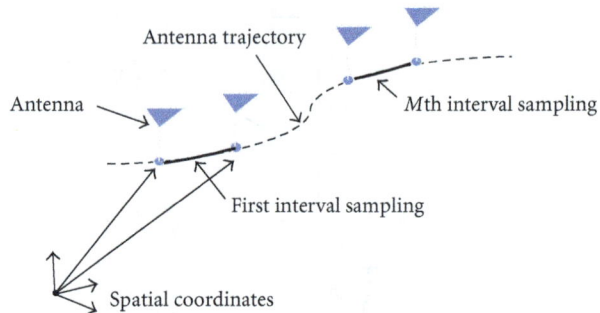

FIGURE 6: Spatial Sampling for a Moving Handheld GPS Receiver (modified after [6]).

while this correlation does not exist for the authentic signals (i.e., the amplitudes of PRN-22 and PRN-24 do not overlay). This technique works effectively even in multipath environments because all the spoofing signals experience the same fading path. Furthermore, since this method does not employ several receive antennas, its hardware complexity is much lower as compared to the techniques proposed in [2, 14]. However, in the case that spoofer differentially modulates the amplitude and/or phase of different PRN signals, some modifications should be applied to this method in order to successfully discriminate the counterfeit signals.

5.1.3. Time of Arrival (TOA) Discrimination

(a) PRN Code and Data Bit Latency. In the case that the receiver-based spoofer does not have any prior information regarding the navigation data bits, it should first decode the received GPS signals and then generate a processed replica as the spoofing signal. Hence, an unavoidable delay exists between the spoofing data bit boundaries with respect to the authentic ones [3, 15, 16]. Therefore, if the data bit transition happens at time instants with a spacing other than 20 ms, then a spoofing attack might be present.

This technique encounters some limitations because the GPS data frame structure is already known and it consists of different parts with different update frequencies. The update frequency of most parts of the GPS frame is very low. Therefore, the majority of the GPS data bits can be predicted by the spoofer if it has already acquired the GPS information before starting to transmit the fake spoofing signals.

(b) L1/L2 Signals Relative Delay. GPS satellites transmit encrypted P(Y) codes on both L1 and L2 frequencies. The signals received on these two frequencies have a relative delay/attenuation that is caused by the different frequency response of the ionosphere. Therefore, if a dual frequency GPS receiver correlates the L1 and L2 signals, it should observe only one correlation peak [8]. The propagation delay in L2 is larger than the L1 frequency; therefore the approximate relative delay of correlation peaks is already known to the GPS receiver. The spoofer should be able to generate signals on both frequencies in order to defeat this countermeasure.

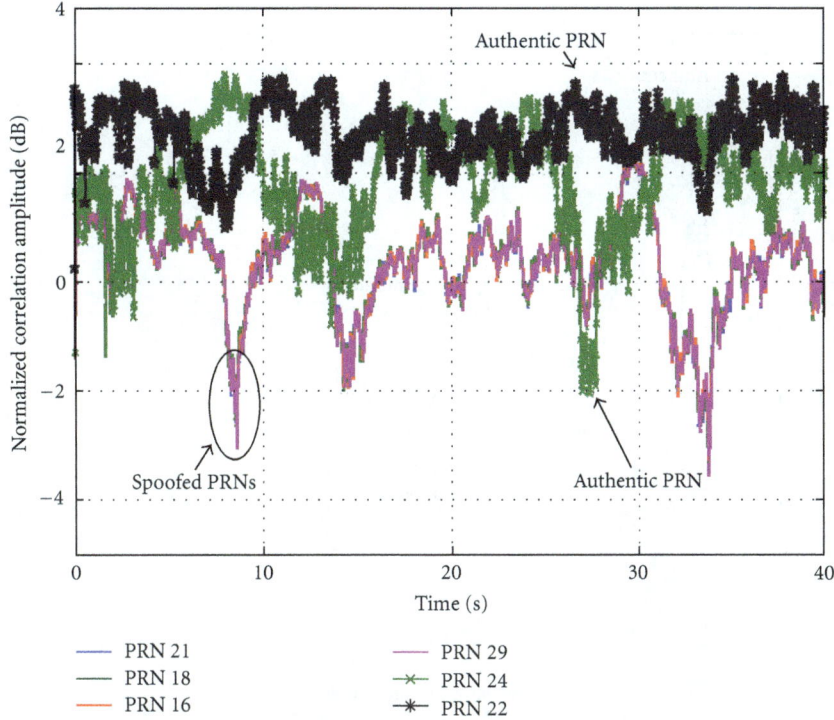

FIGURE 7: Correlation amplitude for spoofing and authentic PRN signals.

5.1.4. *Signal Quality Monitoring (SQM).* SQM techniques have been previously employed to monitor the GPS correlation peak quality in multipath fading environments [17]. Spoofing attacks on a tracking receiver can affect the correlator output in a way similar to that of multipath components [18]. Therefore, authors of [4, 5, 7] have extended the SQM techniques to detect the spoofing attack on tracking receivers that are working in line-of-sight condition. They have employed the *ratio* and *delta* SQM tests in order to detect any abnormal asymmetry and/or flatness of GPS correlation peaks that is imposed by the spoofing attack. It is assumed that the receiver has initially locked onto the authentic correlation peaks and a spoofing attack tries to deceive the receiver toward tracking its fake correlation peaks.

The SQM antispoofing techniques are powerful methods toward detecting the spoofing attack especially in the line-of-sight propagation environments. However, in the presence of multipath propagation, the SQM method might not be able to discriminate between spoofing signals and multipath reflections.

5.1.5. *Distribution Analysis of the Correlator Output.* In line-of-sight (LOS) conditions, the correlator output power for a tracking receiver approximately follows a Chi-squared (χ^2) distribution. For the case of a spoofing attack on a tracking receiver, the spoofing signal correlation peak should be located as close as possible to that of the authentic signal; therefore, the correlator output power is affected by the spoofing signals. As such, assuming that the receiver

is initially locked into tracking the authentic peak, the correlator output amplitude can be written as follows [19]:

$$
\begin{aligned}
y_l\left[\Delta f_l^{a,s}, \Delta \tau_l^{a,s}, kNT_s\right] &= \sqrt{p_l^a}\; e^{j\varphi_l^a} \\
&+ \sqrt{p_l^s} R\left(\Delta \tau_l^{a,s}\right) \frac{\sin\left(\pi \Delta f_l^{a,s} NT_s\right)}{N \sin\left(\pi \Delta f_l^{a,s} T_s\right)} \\
&\times e^{j\pi \Delta f_l^{a,s}(N-1)T_s + j\varphi_l^s} + \widetilde{\eta}_l[kNT_s],
\end{aligned}
\tag{10}
$$

where $\Delta \tau_l^{a,s}$ and $\Delta f_l^{a,s}$ are delay and frequency differences between the authentic and spoofing signals, respectively, and these parameters are generally functions of time. $R(\cdot)$ is the correlation function that is closely related to the choice of GPS signal subcarrier. This function is a triangle with a normalized height and two-chip base width for the GPS subcarrier. It is assumed that the spoofer smoothly changes the code delay and the Doppler frequency of its signal in order to lift off the tracking point of the target receiver. Therefore, the interaction between the authentic and spoofing signals leads to some fluctuations in the correlator output amplitude. These fluctuations cause the correlator output distribution to deviate from the expected χ^2 distribution. This feature can be used for detecting the presence of spoofing signals [19]. Figure 8 shows the correlator output distributions for different relative powers for authentic and spoofing signals. It is observed that the correlator output distributions are completely different in the presence and absence of spoofing attacks.

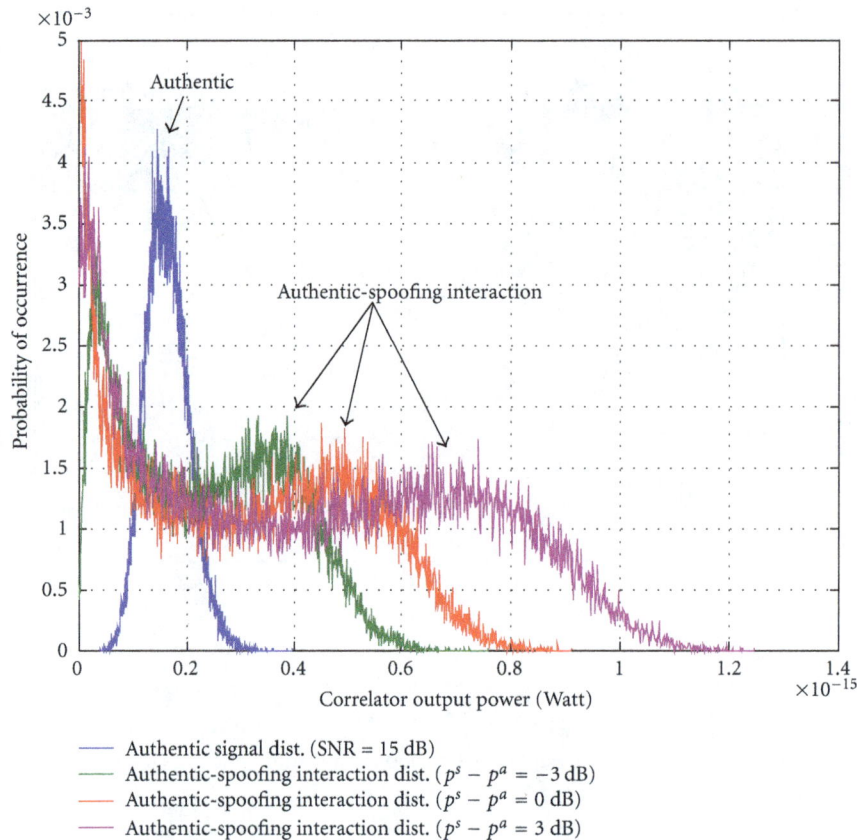

FIGURE 8: Distribution of prompt correlator output power for authentic signals and authentic-spoofing interaction for different spoofing powers [19].

This technique can successfully discriminate spoofing signals in the line-of-sight propagation environments. However, in presence of multipath propagation, the χ^2 distribution is not a valid assumption for the distribution of correlator output amplitude. Therefore, this method is of limited applicability in non-line of sight propagation environments.

5.1.6. Consistency Check with Other Navigation and Positioning Technologies. The augmenting data from auxiliary devices such as inertial measurement unit (IMU) can help the target receiver to discriminate the spoofing threat [2, 20]. In addition, the GPS receiver can compare the solution extracted by received GPS signals to the other position and navigation solutions obtained by mobile networks or WiFi stations. Therefore, if the confidence region of different solutions does not have intersection, there is a high likelihood of a spoofing condition.

Employing this spoofing detection technique increases the hardware and software complexity of GPS receiver. The IMU sensors require calibration before being employed for the positioning purposes [21]. In addition, alternative wireless location technologies such as cellular networks do not usually provide position solutions as accurate as GPS signals; therefore, they might not be very helpful if there is a small mismatch between spoofing solution and authentic position. Furthermore, there is a limited coverage of cellular

and WiFi networks which, in turn, limits the applicability of this spoofing discrimination technique.

5.1.7. Cryptographic Authentication. Authentication techniques can be employed to detect spoofing threat in both civil and military applications. This capability is considered in military version of GPS signals; however, some articles have discussed authentication techniques for current civil GPS receivers [10, 15, 22, 23]. Reference [22] has proposed authentication techniques for GPS L2C and L5 and wide area augmentation system (WAAS) signals.

Most of the cryptographic authentication techniques require some modifications in the GPS signal structure. Therefore, this method does not seem to be applicable to GPS in short term.

5.1.8. Code and Phase Rates Consistency Check. In the case of authentic signals, the Doppler frequency and the code delay rate are consistent because they are both affected by the relative movement between GPS satellite and receiver [24]. This consistency requires that

$$f_l^a = -f_{RF}\dot{\tau}_l^a, \qquad (11)$$

where f_{RF} is the RF frequency of L1 GPS signals ($f_{RF} = 1575.42$ MHz) and $\dot{\tau}_l^a$ is the code delay rate for the lth

TABLE 1: Summary of spoofing detection techniques.

Anti-Spoofing method	Spoofing feature	Complexity	Effectiveness	Receiver required capability	Spoofing scenario generality
C/N_0 monitoring	Higher C/N_0	Low	Medium	C/N_0 monitoring	Medium
Absolute power monitoring	Higher amplitude	Low	Medium	Absolute power monitoring	High
Power variation versus receiver movement	Higher power variations due to proximity	Low	Low	Antenna movement/C/N_0 monitoring	Low
L1/L2 power comparison	No L2 signal for spoofer	Medium	Low	L2 reception capability	Low
Direction of arrival comparison	Spoofing signals coming from the same direction	High	High	Multiple receiver antennas	High
Pairwise correlation in synthetic array	Spoofing signals coming from the same direction	Low	High	Measuring correlation coefficient	High
TOA discrimination	Inevitable delay of spoofing signal	Medium	Medium	TOA analysis	Low
Signal quality monitoring	Deviated shape of authentic correlation peak	Medium	Medium	Multiple correlators	Low
Distribution analysis of the correlator output	Perturbed amplitude distribution due to spoofing-authentic interaction	Low	Medium	Distribution analysis of correlator outputs	Medium
Consistency check with other solutions	Inconsistency of spoofing solution	High	High	Different navigation sensors	High
Cryptographic authentication	Not authenticated	High	High	Authentication	High
Code and phase rate consistency check	Mismatch between artificial code and phase rate	Low	Low	—	Low
GPS clock consistency check	Spoofing/authentic clock inconsistency	Low	Medium	—	Medium

authentic PRN signal. A low-quality spoofer might not keep this consistency between Doppler frequency and code delay rate [8]. As such, a spoofing aware receiver can successfully detect this type of spoofers if the loop filter output of phase locked loop (PLL) and delay locked loop (DLL) are not consistent. The PLL and DLL loop filter outputs are estimates of the phase and delay rates, respectively.

5.1.9. Received Ephemeris Consistency Check.
The navigation message of each satellite contains some ephemeris information corresponding to the position of other GPS satellites. Any inconsistency among these ephemeris data can alert an unsynchronized spoofing attack.

5.1.10. GPS Clock Consistency Check.
The navigation message of each PRN signal contains the GPS clock information. The GPS clock obtained from different satellites of GPS constellation should be consistent. However, the GPS time extracted from an unsynchronized spoofer might not be consistent with the GPS time extracted from other satellites and this can alert the presence of a spoofing attack.

Table 1 provides a summarized comparison among the previously discussed spoofing detection algorithms.

5.2. Spoofing Mitigation

5.2.1. Vestigial Signal Detection.
Suppressing the authentic signal is very hard for GPS spoofers because it requires precise knowledge of the victim antenna phase centre position relative to spoofer antenna phase centre. In most cases, after successful lift-off, a vestige of the authentic signal that can be used for spoofing detection and mitigation remains. In [3] the authors have proposed a vestigial detection technique in which the receiver employs the following software-defined technique. First, the receiver copies the incoming digitized front-end data into a buffer memory. Second, the receiver selects one of the GPS signals being tracked and removes the locally regenerated version of this signal from the buffered signal. Third, the receiver performs acquisition for the same PRN signal on the buffered data. This technique is very similar to the successive interference cancellation (SIC) used for removing strong signals in order to combat the near/far

problem in direct sequence code division multiple-access (DS-CDMA) networks [24].

The implementation of the vestigial signal detection increases the hardware and processing complexity of the receivers because this technique requires additional tracking channels to track both authentic and spoofing signals. In addition, in the presence of high power spoofing signals and limited bit resolution, the authentic vestige might not still be detectable since it might have been fallen under the sensitivity level of the GPS receiver quantizer.

5.2.2. Multiantenna Beam Forming and Null Steering. A multiantenna receiver can employ array processing techniques in order to shape its beam. As such, after detecting the direction of spoofing signal, this receiver can steer a null toward the spoofer source and suppress its harmful effect. Therefore, considering (3), spoofing signals can be mitigated if the received signal is multiplied by a complex ($N \times 1$) weighting vector (**f**) such that

$$\mathbf{f}^H \mathbf{b} = 0, \quad \text{constraint}: \|\mathbf{f}\| = 1. \quad (12)$$

The constraint avoids the trivial solution, which is $\mathbf{f} = \mathbf{0}$. Therefore, applying this gain vector to the sampled signal of (3), the following output signal will be achieved:

$$s(nT_s) = \mathbf{f}^H \mathbf{r}(nT_s) = \sum_{m=1}^{N_{\text{Auth}}} \mathbf{f}^H \mathbf{a}_m \sqrt{p_m^a} F_m^a(nT_s)$$

$$+ \underbrace{\mathbf{f}^H \mathbf{b}}_{=0} \sum_{q=1}^{N_{\text{Spoof}}} \sqrt{p_q^s} F_q^s(nT_s) + \mathbf{f}^H \boldsymbol{\eta}(nT_s). \quad (13)$$

Consequently, the spoofing signal is removed after proper combination of signals for different antenna branches [14, 25].

In [14] a spoofing mitigation technique is proposed that employs a multiantenna GPS receiver toward mitigating the spoofing attack. McDowell's method can effectively discard the spoofing signals after determining the spatial correlation between different received signal pairs. However, this method considerably increases the receiver hardware and processing complexity since the proper gain vector can be achieved after processing the despread version of all received authentic and spoofing GPS signals.

In [25] a very low computational complexity double-antenna spoofing mitigation method is proposed that is able to spatially filter out the spoofing signals. This method cross-correlates the received signals from different antennas and extracts the spatial signature of spoofing signals based on their spatial power dominancy. All these operations are performed on the raw samples before despreading the authentic and spoofing signals. Assuming that spoofer module transmits several PRN signals each of which having a power level comparable to authentic ones, the steering vector corresponding to spoofing signals (**b**) can be extracted because all spoofing signals are coming from the same direction. This method does not need array calibration or any prior information regarding antenna array orientation

and can be employed as an in-line stand-alone antenna combining block that mitigates the spoofing signals at before entering the conventional GPS receivers.

Figure 9 shows the average SNR of the authentic and spoofing signals as a function of the average input spoofing power for both the single-antenna and the proposed double-antenna receivers. For the case of single-antenna receiver, the authentic SNR decreases as the input spoofing power increases. However, it is observed that after proper combining of the signals of both antennas, the SNR of the authentic signal almost remains constant while the spoofing SNR is always far below the detection threshold for different input spoofing powers.

The spoofing mitigation technique proposed in [25] successfully mitigates the spoofing signals as long as their TSP is considerably higher than the average power of authentic signals. Nevertheless, in some cases it might unintentionally reduce the power of some authentic signals due to the inherent cone of ambiguity in the double-antenna beam pattern. This problem can be solved by employing larger antenna arrays because the ambiguity region of antenna beam pattern considerably decreases as the number of array elements increases [26]. This spoofing mitigation technique might not perform well in the case of multiple-antenna spoofing transmission.

5.2.3. Receiver Autonomous Integrity Monitoring (RAIM). Spoofing signals effectively inject counterfeit pseudoranges into the receiver measurements. These measurements might not be consistent and consequently do not lead to a reasonable position solution. Most of the GPS receivers perform measurements integrity monitoring in order to detect and reject the outlier measurements; this technique is known as receiver autonomous integrity monitoring [27]. In [7] the authors propose an extended RAIM technique that is able to detect and exclude the outlier measurements injected by the spoofing threat. In [11] a vulnerability index against spoofing (VIAS) is proposed that investigates the vulnerability of a GPS receiver that is protected by RAIM technique in the presence of misleading spoofing measurements. The author has shown that the maximum position deviation is the product of the RAIM level with the VIAS index.

RAIM techniques can be employed as useful antispoofing techniques at the position solution level. However, these methods are effective only in cases where only one or two spoofing measurements are present among several authentic pseudoranges; otherwise, if the spoofed pseudorange measurements are in majority, the RAIM technique might reject authentic measurements in order to decrease the residuals.

Table 2 provides a summarized comparison among the previously discussed spoofing mitigation algorithms.

5.3. Antispoofing Techniques from a Multilayer Perspective. From a multilayer perspective, the antispoofing techniques can be investigated in three different levels, namely, the signal processing, data bit and position solution, and navigation levels. Spoofing threat might be detected/mitigated at any

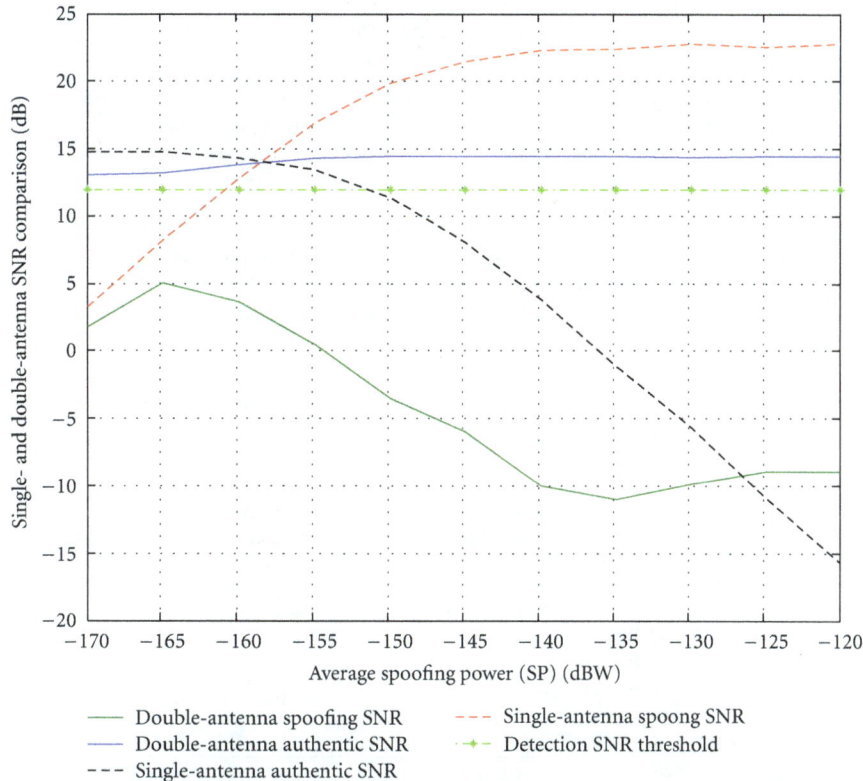

FIGURE 9: Authentic and spoofed SNR variations as a function of average spoofing power [25].

TABLE 2: Summary of spoofing mitigation techniques.

Anti-spoofing method	Spoofing feature	Complexity	Effectiveness	Receiver required capability	Spoofing scenario generality
Vestigial signal detection	The authentic signal is still present and can be detected	High	Medium	Multiple receive channels	Medium
Multi-antenna null steering	Spoofing signals coming from the Same direction	Medium	High	Multiple receiver antennas	High
RAIM	Higher residuals for spoofed measurements	Medium	Medium	—	Medium

of the above-mentioned levels. In other words, a successful spoofer should be able to overcome the antispoofing techniques implemented in different layers. In addition to the previously discussed antispoofing methods, cross-layer techniques can be developed to incorporate measurements from different operational levels in order to combat the harmful effect of spoofing signals. Figure 10 shows some of the previously discussed antispoofing techniques in a multilayer approach.

6. Spoofing/Antispoofing Test Scenarios

Testing a spoofer/antispoofer system is challenging since radio transmission regulations prohibit outdoors radio frequency (RF) power transmission in the GPS band. Therefore, special considerations should be taken into account in order to test a spoofing/antispoofing system in the presence of authentic satellites signals. This section presents some test scenarios that can be used for evaluating the performance of the antispoofing methods in real-world spoofing scenarios.

6.1. GPS Indoor Signal Retransmission. In [2] a rooftop GPS antenna has been used to receive authentic GPS signals. The received signals are amplified and then retransmitted indoors from a point source antenna. In this case, the spoofing transmission can take place indoors where it does not violate radio transmission regulations. This setup seems to be appropriate although it does not exactly represent real outdoor spoofing scenarios, especially for the case of multiantenna antispoofing techniques. In this case all authentic signals are also retransmitted from the single antenna (see Figure 11). Multipath propagation and relative spoofing and authentic signal powers are other issues that should be considered while employing indoor retransmission.

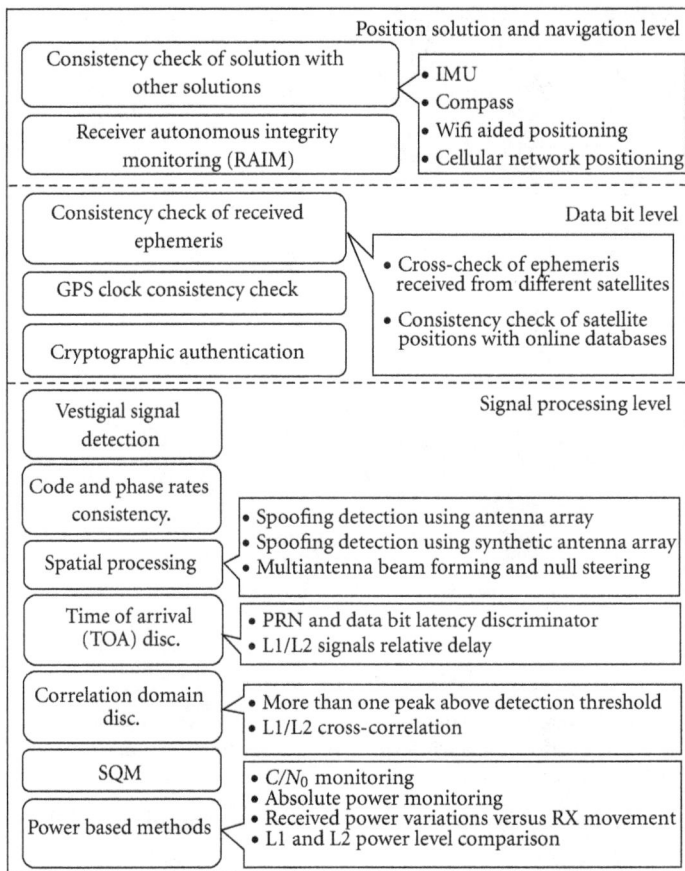

FIGURE 10: A multilayer approach to antispoofing techniques.

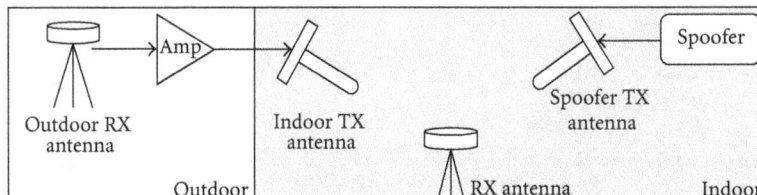

FIGURE 11: Spoofing test using GPS indoor signal retransmission.

6.2. Spoofing Using Recorded Data with No RF Transmission.
In this scenario, no real spoofer RF transmission takes place; instead, the intermediate frequency (IF) authentic GPS L1 signal is digitized and stored on a hard disk; then, the recorded data is fed to the GPS receiver-spoofer, which tracks present GPS signals and generates corresponding spoofing signals. These signals are combined into a quantized output bit stream. The output bit stream is then combined with the original data by interleaving, and the result of this process is fed to the target receiver [3]. Figure 12 depicts a block diagram of this test scenario.

6.3. Employing RF Combiners to Combine Authentic and Spoofing Signals. Authentic GPS signals can be combined with locally generated spoofing signals using RF power combiners. Spoofing signal power can be adjusted using a

cascaded setup of amplifier and variable attenuator. Figure 13 shows the block diagram of this test setup for validating the proper performance of a multiantenna antispoofing technique.

7. Conclusions

Spoofing attack on GPS receivers has been considered as a serious threat to safety of life applications; since there is enough motivation for illicit application of spoofers, the realization of spoofers is not prohibitively costly. As such, it is anticipated that many research activities will be conducted on increasing the security of GPS receivers against spoofing and jamming attacks. In this paper different spoofing/antispoofing scenarios were described and the vulnerabilities of GPS that can potentially be exploited

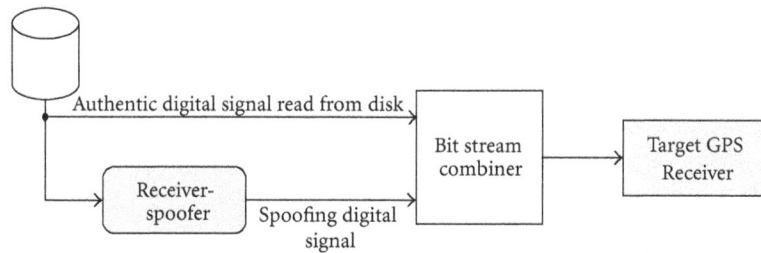

FIGURE 12: Spoofing test using recorded GPS data (modified after [3]).

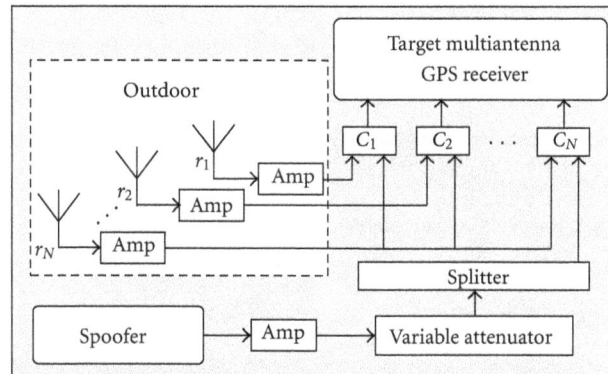

FIGURE 13: Spoofing test setup using RF combiners for a multiantenna GPS receiver.

by a spoofer were discussed in a multilayer GPS processing approach. It was shown that commercial GPS receivers are quite vulnerable to spoofing attacks generated by different spoofing scenarios. Nevertheless, by applying modest modifications, low-complexity spoofing detection and mitigation techniques can be employed in order to increase the robustness of commercial GPS receivers against spoofing attacks. Countermeasures to spoofing signals can be introduced in any (or all) of the processing levels of a GPS receiver. A powerful antispoofing technique should ideally be of low computational complexity and be effective for generic spoofing scenarios. Based on this paper, since most of the practical spoofing scenarios employ a single antenna to transmit counterfeit signals, the spatial characteristics of spoofing signals are different from those of authentic GPS signals. Therefore, spatial-processing-based antispoofing techniques can be employed as a generic and very effective countermeasure against most spoofing signals currently envisaged.

References

[1] X. J. Cheng, K. J. Cao, J. N. Xu, and B. Li, "Analysis on forgery patterns for GPS civil spoofing signals," in *Proceedings of the 4th International Conference on Computer Sciences and Convergence Information Technology (ICCIT '09)*, pp. 353–356, Seoul, Korea, November 2009.

[2] P. Y. Montgomery, T. E. Humphreys, and B. M. Ledvina, "Receiver-autonomous spoofing detection: experimental results of a multi-antenna receiver defense against a portable civil GPS spoofer," in *Proceedings of the Institute of Navigation—International Technical Meeting (ITM '09)*, pp. 124–130, Anaheim, Calif, USA, January 2009.

[3] T. E. Humphreys, B. M. Ledvina, M. L. Psiaki, B. W. O'Hanlon, and P. M. Kintner, "Assessing the spoofing threat: development of a portable gps civilian spoofer," in *Proceedings of the 21st International Technical Meeting of the Satellite Division of the Institute of Navigation (ION GNSS '08)*, pp. 2314–2325, Savannah, Ga, USA, September 2008.

[4] A. Cavaleri, B. Motella, M. Pini, and M. Fantino, "Detection of spoofed GPS signals at code and carrier tracking level," in *Proceedings of the 5th ESA Workshop on Satellite Navigation Technologies and European Workshop on GNSS Signals and Signal Processing (NAVITEC '10)*, pp. 1–6, December 2010.

[5] A. Cavaleri, M. Pini, L. Lo Presti, and M. Fantino, "Signal quality monitoring applied to spoofing detection," in *Proceedings of the 24th International Technical Meeting of The Satellite Division of the Institute of Navigation (ION GNSS '11)*, Portland, Ore, USA, September 2011.

[6] J. Nielsen, A. Broumandan, and G. Lachapelle, "Spoofing detection and mitigation with a moving handheld receiver," *GPS World*, vol. 21, no. 9, pp. 27–33, 2010.

[7] B. M. Ledvina, W. J. Bencze, B. Galusha, and I. Miller, "An in-line anti-spoofing device for legacy civil GPS receivers," in *Proceedings of the Institute of Navigation—International Technical Meeting (ITM '10)*, pp. 698–712, San Diego, Calif, USA, January 2010.

[8] H. Wen, P. Y. R. Huang, J. Dyer, A. Archinal, and J. Fagan, "Countermeasures for GPS signal spoofing," in *Proceedings of the 18th International Technical Meeting of the Satellite Division of The Institute of Navigation (ION GNSS '05)*, pp. 1285–1290, Long Beach, Calif, USA, September 2005.

[9] S. Savasta, L. Lo Presti, F. Dovis, and D. Margaria, "Trustworthiness GNSS signal validation by a time-frequency approach," in *Proceedings of the 22nd International Technical Meeting of the Satellite Division of the Institute of Navigation (ION GNSS '09)*, pp. 66–75, Savannah, Ga, USA, September 2009.

[10] X. J. Cheng, J. N. Xu, K. J. Cao, and W. Jie, "An authenticity verification scheme based on hidden messages for current civilian GPS signals," in *Proceedings of the 4th International Conference on Computer Sciences and Convergence Information Technology (ICCIT '09)*, pp. 345–352, Seoul, Korea, November 2009.

[11] J. C. Juang, "GNSS spoofing analysis by VIAS," in *Coordinates Magazine*, 2011.

[12] A. Jafarnia-Jahromi, A. Broumandan, J. Nielsen, and G. Lachapelle, "GPS spoofer countermeasure effectiveness based on signal strength, noise power and C/N0 observables," *International Journal of Satellite Communications and Networking*, vol. 30, no. 4, pp. 181–191, 2012.

[13] E. D. Kaplan and C. J. Hegarty, *Understanding GPS Principles and Applications*, Artech House, Boston, Mass, USA, 2nd edition, 2006.

[14] C. E. McDowell, "GPS Spoofer and Repeater Mitigation System using Digital Spatial Nulling—US Patent 7250903 B1," 2007.

[15] S. C. Lo and P. K. Enge, "Authenticating aviation augmentation system broadcasts," in *Proceedings of the IEEE/ION Position, Location and Navigation Symposium (PLANS '10)*, pp. 708–717, Indian Wells, Calif, USA, May 2010.

[16] S. Lo, D. De Lorenzo, P. Enge, D. Akos, and P. Bradley, "Signal Authentication, a secure civil GNSS for today," *GNSS magazine*, pp. 30–39, 2009.

[17] R. E. Phelts, *Multicorrelator techniques for robust mitigation of threats to GPS signal quality [Ph.D. thesis]*, Standford University, Palo Alto, Calif, USA, 2001.

[18] D. Shepard and T. Humphreys, "Characterization of receiver response to a spoofing attack," in *Proceedings of the 24th International Technical Meeting of The Satellite Division of the Institute of Navigation (ION GNSS '11)*, p. 2608, Portland, Ore, USA, September 2011.

[19] N. A. White, P. S. Maybeck, and S. L. DeVilbiss, "Detection of interference/jamming and spoofing in a DCPS-aided inertial system," *IEEE Transactions on Aerospace and Electronic Systems*, vol. 34, no. 4, pp. 1208–1217, 1998.

[20] A. Jafarnia-Jahromi, T. Lin, A. Broumandan, J. Nielsen, and G. Lachapelle, "Detection and mitigation of spoofing attack on a vector based tracking GPS receiver," in *Proceedings of the International Technical Meeting of The Institute of Navigation*, Newport Beach, Calif, USA, January 2012.

[21] M. G. Petovello, *Real-time integration of a tactical-grade IMU and GPS for high-accuracy positioning and navigation [Ph.D. thesis]*, Department of Geomatics Engineering, University of Calgary, Alberta, Canada.

[22] L. Scott, "Anti-Spoofing and Authenticated Signal Architectures for Civil Navigation Systems," in *Proceedings of the 16th International Technical Meeting of the Satellite Division of The Institute of Navigation (ION GPS/GNSS '03)*, Portland, Ore, USA, September 2003.

[23] G. W. Hein, F. Kneissl, J. A. Avila-Rodriguez, and S. Wallner, "Authenticating GNSS: Proofs Against Spoofs Part 2," *GNSS magazine*, pp. 58–63, 2007.

[24] S. Moshavi, "Multi-user detection for DS-CDMA communications," *IEEE Communications Magazine*, vol. 34, no. 10, pp. 124–135, 1996.

[25] S. Daneshmand, A. Jafarnia-Jahromi, A. Broumandan, and G. Lachapelle, "A low complexity gnss spoofing mitigation technique using a double antenna array," *GPS World Magazine*, vol. 22, no. 12, pp. 44–46, 2011.

[26] H. L. V. Trees, *Optimum Array Processing, Detection, Estimation, and Modulation Theory Part IV*, John Wiley & Sons, New York, NY, USA, 2002.

[27] H. Kuusniemi, A. Wieser, G. Lachapelle, and J. Takala, "User-level reliability monitoring in urban personal satellite-navigation," *IEEE Transactions on Aerospace and Electronic Systems*, vol. 43, no. 4, pp. 1305–1318, 2007.

On the Impact of Channel Cross-Correlations in High-Sensitivity Receivers for Galileo E1 OS and GPS L1C Signals

Davide Margaria,[1] Beatrice Motella,[1] and Fabio Dovis[2]

[1] Istituto Superiore Mario Boella, Via P.C. Boggio 61, 10138 Torino, Italy
[2] Department of Electronics and Telecommunications, Politecnico di Torino, Corso Duca degli Abruzzi 24, 10129 Torino, Italy

Correspondence should be addressed to Davide Margaria, margaria@ismb.it

Academic Editor: Shaojun Feng

One of the most promising features of the modernized global navigation satellite systems signals is the presence of pilot channels that, being data-transition free, allow for increasing the coherent integration time of the receivers. Generally speaking, the increased integration time allows to better average the thermal noise component, thus improving the postcorrelation SNR of the receiver in the acquisition phase. On the other hand, for a standalone receiver which is not aided or assisted, the acquisition architecture requires that only the pilot channel is processed, at least during the first steps of the procedure. The aim of this paper is to present a detailed investigation on the impact of the code cross-correlation properties in the reception of Galileo E1 Open Service and GPS L1C civil signals. Analytical and simulation results demonstrate that the S-curve of the code synchronization loop can be affected by a bias around the lock point. This effect depends on the code cross-correlation properties and on the receiver setup. Furthermore, in these cases, the sensitivity of the receiver to other error sources might increase, and the paper shows how in presence of an interfering signal the pseudorange bias can be magnified and lead to relevant performance degradation.

1. Introduction

In the context of Global Navigation Satellite Systems (GNSS) receivers, the interest on the new modulations that will be used for the modernized GPS L1C and Galileo E1 Open Service (OS) civil signals grew rapidly in past years. The definition of new signals structure results from an agreement between the European Commission and Unites States of America. A common Multiplexed Binary Offset Carrier Modulation (MBOC) signal baseline has been adopted, with the aim of assuring the compatibility and interoperability between GPS and Galileo systems [1]. For the GPS L1C signal, USA has chosen the Time Multiplexed BOC (TMBOC) solution that multiplexes a BOC(1,1) with a BOC(6,1) in time domain [2], while the composite BOC (CBOC) is the implementation selected for the Galileo E1 OS Signal In Space (SIS) [3].

One of the main features of the modernized civil and open access signals is the presence of the pilot channels. Pilot channel has been introduced to allow the receivers to perform coherent integration over a long time, without facing the issue of unpredictable data transitions. As a consequence, the receiver is able to acquire satellite signals at lower SNR than the nominal value. In order to deal with such a need in current GPS receiver, assistance data have been defined and standardized [4] in order to overcome the issues induced by the presence of data transition, which arms the entire correlation process. When dealing with standalone receivers, the presence of the pilot channel is of utmost importance since it allows to increase the integration time but avoiding the data wipe-off. In such a case, the correlation is performed with a local version of the code of the pilot channel, and the correlation value becomes the decision metrics for the detection of the satellite.

In this paper, the distortion of the discrimination function (S-curve) due to codes cross-correlation properties is assessed, considering the features of the modulation schemes adopted in Galileo E1 OS and GPS L1C civil signals and also investigating different receiver configurations (reception of data/pilot channels, variable correlators spacing).

This article is based on the preliminary results presented in [5]. A more detailed analysis has been carried out in order to assess and compare the auto- and cross-correlation properties of different families of codes and their impact on the receiver performance.

After this introduction, the paper is organized as follows: Section 2 discusses the features of Galileo E1 OS and GPS L1C modulations. Then, the impact of the codes cross-correlations and the receiver setup on the discrimination function is analysed (Sections 3 and 4). Simulation results are presented in Section 5 in order to show that, receiving a single channel (e.g., the pilot channel), the code cross-correlation distortion on the S-curve can be magnified by an inappropriate choice of the correlator spacing, leading to noticeable worsening in receiver performance when interfering signals are present. Section 6 outlines the conclusions.

2. Galileo E1 OS and GPS L1C Modulation Features

The MBOC signal baseline assures more power to the high-frequency spectral components if compared to the baseline BOC(1,1) and BPSK(1) modulations. This feature leads to a sharpener code correlation peak allowing to achieve improved tracking performance [6]. At the same time, since an amount of power is allocated to high frequencies, no benefits can be observed for receivers that use narrow front-end bandwidths. The MBOC(6,1,1/11) signal baseline has been recommended to obtain high degree of interoperability with receivers that might use narrowband front-ends. In such a case, since more than 90% of total power (10/11) remains available and the filtered MBOC(6,1,1/11) resembles a BOC(1,1) signal. Moreover, MBOC presents less susceptibility to narrow band interference at the worst case frequency. The TMBOC and CBOC modulations are designed to reduced side-lobe levels in the auto- and cross-correlation functions.

Both Galileo E1 OS and GPS L1C signals include two channels: the pilot signal, without any data message, that is spread by a ranging (pseudo-random noise—PRN) code; and the data channel, spread by a ranging code and modulated by a data message. At the receiver side, it is possible to consider only one of the two channels in order to exploit peculiar characteristics, as for example, if long integration times have to be used. The Galileo and the GPS signals differ on the modulation formats (CBOC versus TMBOC), on the data/pilot power allocations and on the code properties. Concerning the signals in space, it must be remarked that both the signals will be received with the same total power, that is −157 dBW [2, 3].

In addition, codes belonging to different families (memory codes for Galileo E1 OS signals, Weil codes for GPS L1C) will be used by the two systems in tiered code structures featuring different lengths, as summarized in Table 1.

The differences in terms of code properties, modulation formats (and consequent different spectral occupation), and relative power levels are then expected to lead to different system performance. As an example considering

the levels of interference robustness, it has been noticed that receiving only a single channel in case of continuous wave interference [7], better results have been obtained with the GPS L1C pilot channel and with the Galileo E1 data channel. Differences among modulations also depend on the setup of the correlator spacing and, as expected, tend to become negligible narrowing the correlators. More details on CBOC and TMBOC modulations are provided below, considering the correlation properties related to the data and pilot channels coexistence. Code properties will be discussed in the following sections.

2.1. Galileo E1 OS: CBOC(6,1,1/11). The main features of the E1 Open Service signal can be summarized as follows:

(i) 50% power split between data ($E1_B$) and pilot ($E1_C$) channels, for robust data demodulation;

(ii) optimized memory codes;

(iii) use of a tiered code structure including 4 ms primary and 100 ms secondary codes on the pilot channel.

Both pilot and data channel components take advantage of the CBOC(6,1,1/11) modulation: each PRN code chip is shaped by a weighted combination of BOC(1,1) and BOC(6,1) spreading symbols.

The chip shapes (normalized with unitary power) of the two Galileo E1 OS channels are reported in Figure 1, showing the CBOC(6,1,1/11) modulation. It must be pointed out the different sign in combining the BOC(1,1) and BOC(6,1) components between the data channel (denoted as CBOC$_+$ in Figure 1) and the pilot channel (denoted as CBOC$_-$), according to the Galileo OS Interface Control Document [3].

The theoretical autocorrelation functions computed on single chip of data (R_{CBOC+}) and pilot channels (R_{CBOC-}) are shown in Figure 2 where the data/pilot cross-correlation function ($R_{CBOC_\pm}$) is also represented.

In order to obtain unitary autocorrelation peaks, Figure 2 and following correlation plots have been obtained after normalizing the signal amplitudes.

Assuming now to demodulate the received signal (data and pilot channels) by using only the pilot component (local signal replica), the resulting correlation function $R(\tau)$ can be expressed by

$$R(\tau) = R_{CBOC_-}(\tau) + \alpha \cdot R_{CBOC_\pm}(\tau), \qquad (1)$$

where τ is the code delay in chips (omitted hereafter, in order to simplify the notation); R_{CBOC_-} is the (unitary) autocorrelation function of the pilot channel; $R_{CBOC_\pm}$ is the data/pilot cross-correlation function; α is a parameter representing the weight of the cross term, depending on both the data/pilot power splitting and on the code properties (relative amplitudes on the code auto- and cross-correlation functions).

The corresponding discrimination function is depicted in Figure 3, for different α, where the typical irregular trend on the slope due to the BOC(6,1) signal components can be observed. In Figure 3 an incoming signal composed by data and pilot channels is correlated with the pilot signal,

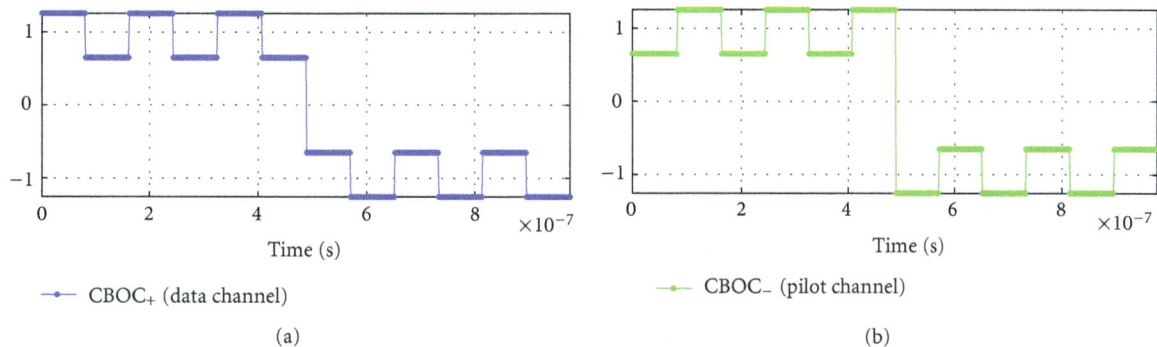

FIGURE 1: Chip shape of Galileo E1 OS data channel (in blue) and pilot channel (in green), CBOC modulation (BOC(1,1) ± BOC(6,1)).

TABLE 1: Galileo E1 OS and GPS L1C code lengths.

	Transmitted Channel	Code Length [chips]		Tiered Code Duration [ms]
		Primary	Secondary	
Galileo E1 OS Spreading Codes	E1-B (data)	4092	—	4
	E1-C (pilot)	4092	25	100
GPS L1C Spreading Codes	L1C$_D$ (data)	10230	—	10
	L1C$_P$ (pilot)	10230	1800	18000

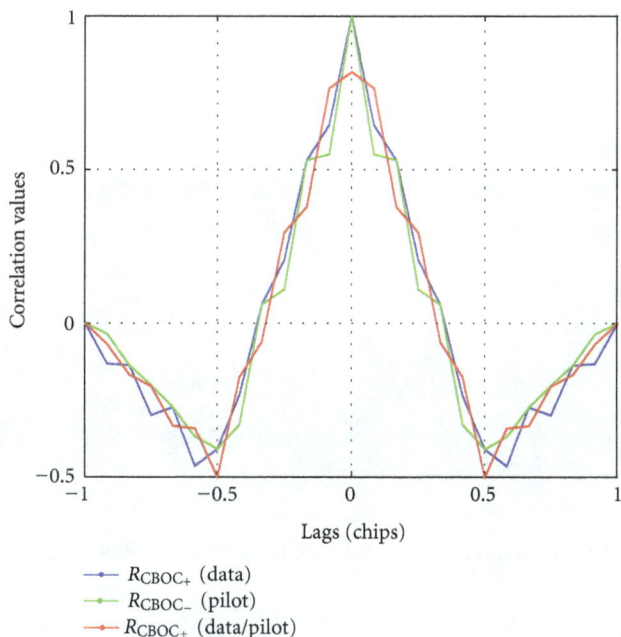

FIGURE 2: Theoretical CBOC circular correlation functions computed on single chip. Chip shape: BOC(1,1) ± BOC(6,1).

It is then possible to conclude that the intrinsic CBOC correlation and discrimination functions always appear symmetrical, regardless to the data/pilot relative power levels. Possible biases around the lock point will be due to other effects (codes cross-correlation impact), as it will be demonstrated in the following sections.

2.2. GPS L1C: TMBOC(6,1,4/33) and BOC(1,1).

The L1C signal, similarly to the Galileo E1 OS, consists of a data (L1C$_D$) channel and a pilot (L1C$_P$) channel [2]. Its main characteristics can be summarized as follows.

(i) 75% of power in the pilot component for enhanced signal tracking;

(ii) advanced Weil-based spreading codes;

(iii) use of a long overlay code (18 s) on the pilot channel.

The L1C MBOC implementation modulates the entire data component and 29 of every 33 code chips of the pilot channel with BOC(1,1), while 4 of every 33 pilot channel chips with a BOC(6,1) waveform. The code chip shapes for the BOC(1,1) data channel and the TMBOC(6,1,4/33) pilot channel are shown in Figure 4.

The optimized L1C signal has been designed to assure interoperability with Galileo E1 OS signal. The unequal power split improves the pilot tracking threshold by 1.87 dB compared with a 50% power split used in Galileo. It has been shown that a TMBOC pilot usage extends most of the advantages exploited by BOC(1,1) by more than 1 dB over BOC(1,1). The L1C modulation has been introduced to enhance the signal robustness in critical environments [8].

The TMBOC implementation assures a MBOC-like spectrum, but implies a slightly different correlation function with respect to the Galileo E1 CBOC case. Notice that, using

computing a coherent early-late discrimination function (early-late correlator spacing Δ of 1 chip) for different values of α. The variation of α impacts on the S-curve, affecting the slope in the linear region. In addition, it can be noticed that, even if α is different from zero, the discrimination function preserves symmetry and no tracking point biases are introduced.

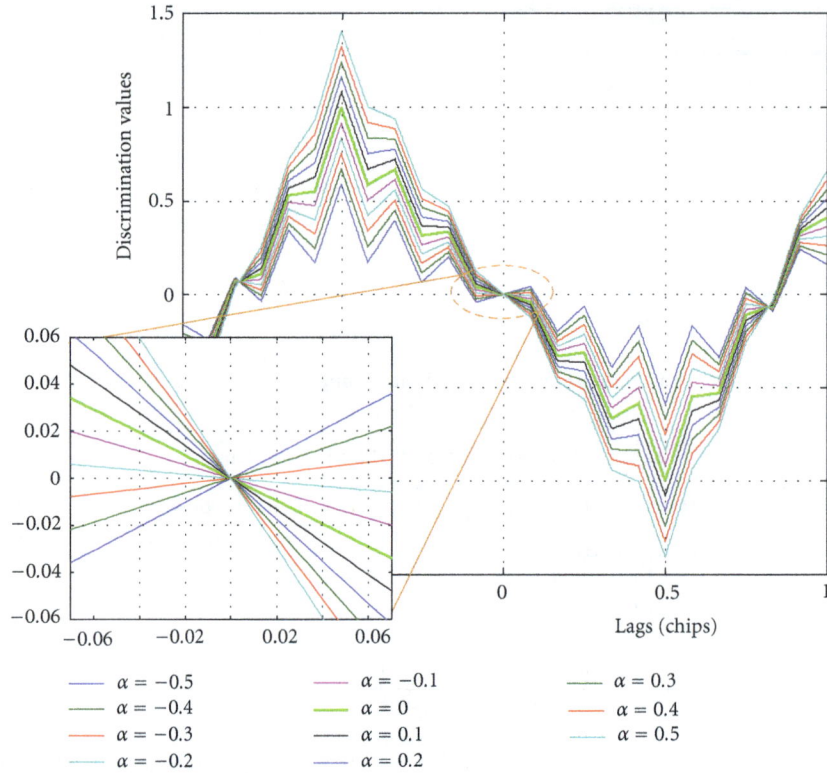

FIGURE 3: Discrimination function (coherent Early-Late, spacing $\Delta = 1$ chip) and its zoom, obtained using a single chip of the Galileo E1 OS signal and varying α. Received signal: both data and pilot channels, local signal: pilot channel only. PRN codes are neglected.

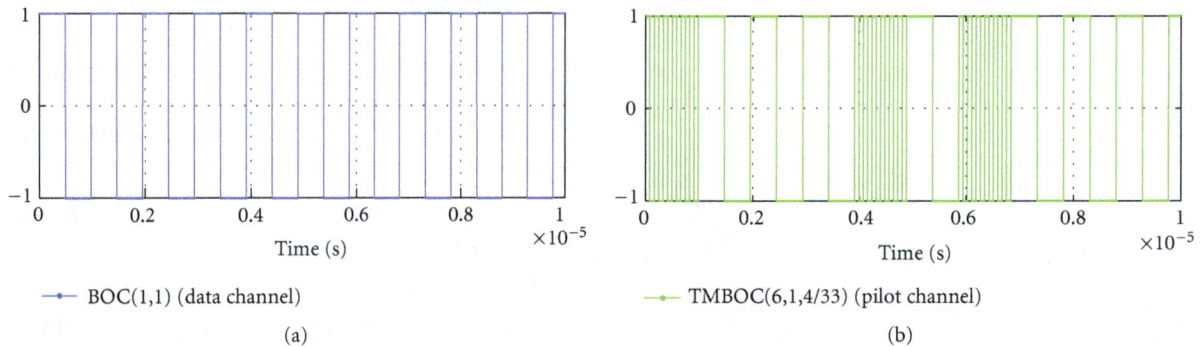

FIGURE 4: Code chip shape of GPS L1C data channel (in blue) and pilot channel (in green), due to the BOC(1,1) and TMBOC(6,1,4/33) modulations. PRN codes are neglected.

this kind of waveform, it is not possible to repeat the previous analysis concerning the theoretical S-curve for a single code chip: the TMBOC is in fact defined over a sequence of 33 chips. In order to obtain meaningful correlation and discrimination functions, it is necessary to use a whole code period.

3. Codes Cross-Correlation Impact on the S-Curve

Two are the signal elements that affect the discrimination function shape: the modulation and the code. While the former has already been investigated in the previous section, the impact of the code is the focus of the current one.

Both the Galileo E1 OS and the GPS L1C signals are taken into account, with specific analyses on how different code families can affect the shape of the discrimination function. In detail the memory codes (introduced for Galileo E1 OS) and the Weil codes (specific for GPS L1C) will be discussed.

3.1. Galileo E1 OS: Memory Codes. Memory codes are foreseen for the Galileo E1 OS signal [3]. 4092 chips 4 ms-long codes will be implemented as primary codes, in a tiered code structure (see Table 1).

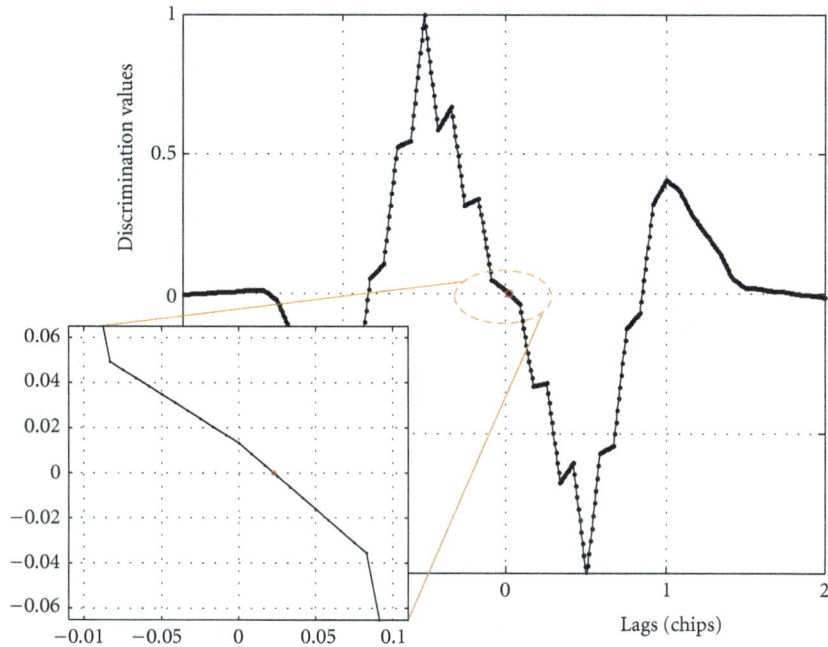

FIGURE 5: Discrimination function (coherent Early-Late, $\Delta = 1$ chip) and its zoom, obtained for a Galileo E1 OS signal using memory codes (4 ms primary codes, PRN 1). Received signal: both data and pilot channels, local signal: pilot channel only. The S-curve does not result symmetrical around the zero lag (bias = 6.61 m).

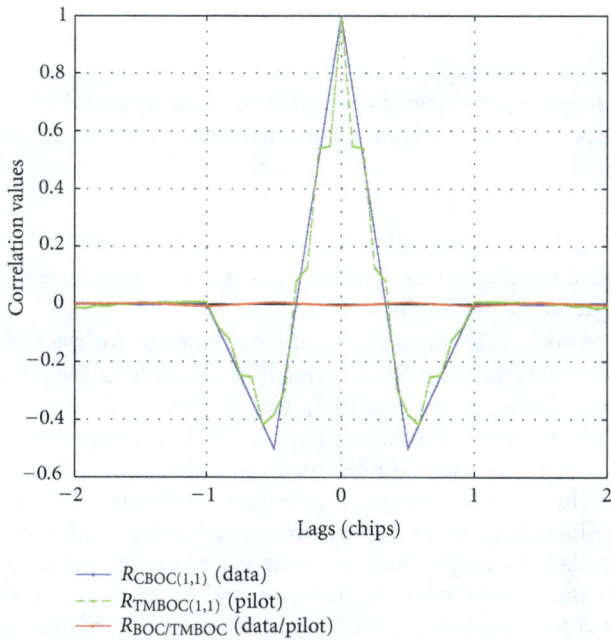

$\quad$ $R_{\text{CBOC}(1,1)}$ (data)
$\quad$ $R_{\text{TMBOC}(1,1)}$ (pilot)
$\quad$ $R_{\text{BOC/TMBOC}}$ (data/pilot)

FIGURE 6: Data (blue line) and pilot (green line) auto-correlation functions, data/pilot cross-correlation function (red line), obtained for a GPS L1C signal using Weil codes (10 ms, PRN 1 data and pilot channels).

It must be noted that, in spite of what happens for the Gold codes used by the GPS L1 current signal [9, 10], the

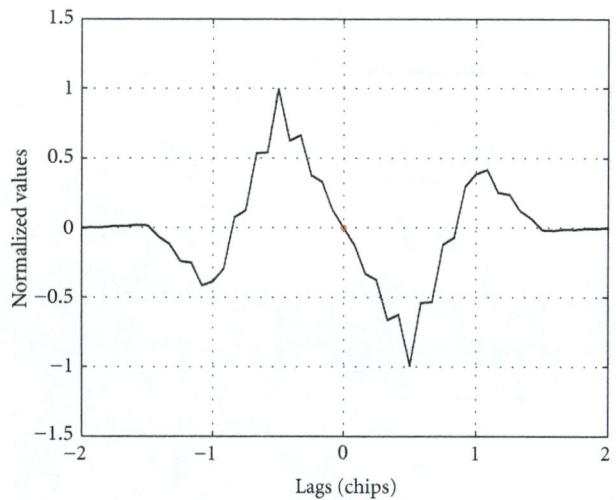

FIGURE 7: Discrimination function (coherent Early-Late, $\Delta = 1$ chip) obtained for a GPS L1C signal using Weil codes (10 ms, PRN 1). Received signal: both data and pilot channels, local signal: pilot channel only. The S-curve does not result symmetrical around the zero lag (bias = 0.09 m).

auto- and cross-correlation functions for memory codes take multiple values.

In addition, as explained in [11], the codes are designed to fulfill special properties, such as low autocorrelation side lobes. This guarantees that the autocorrelation values of every code correlate to zero with a replica of itself, delayed by one chip.

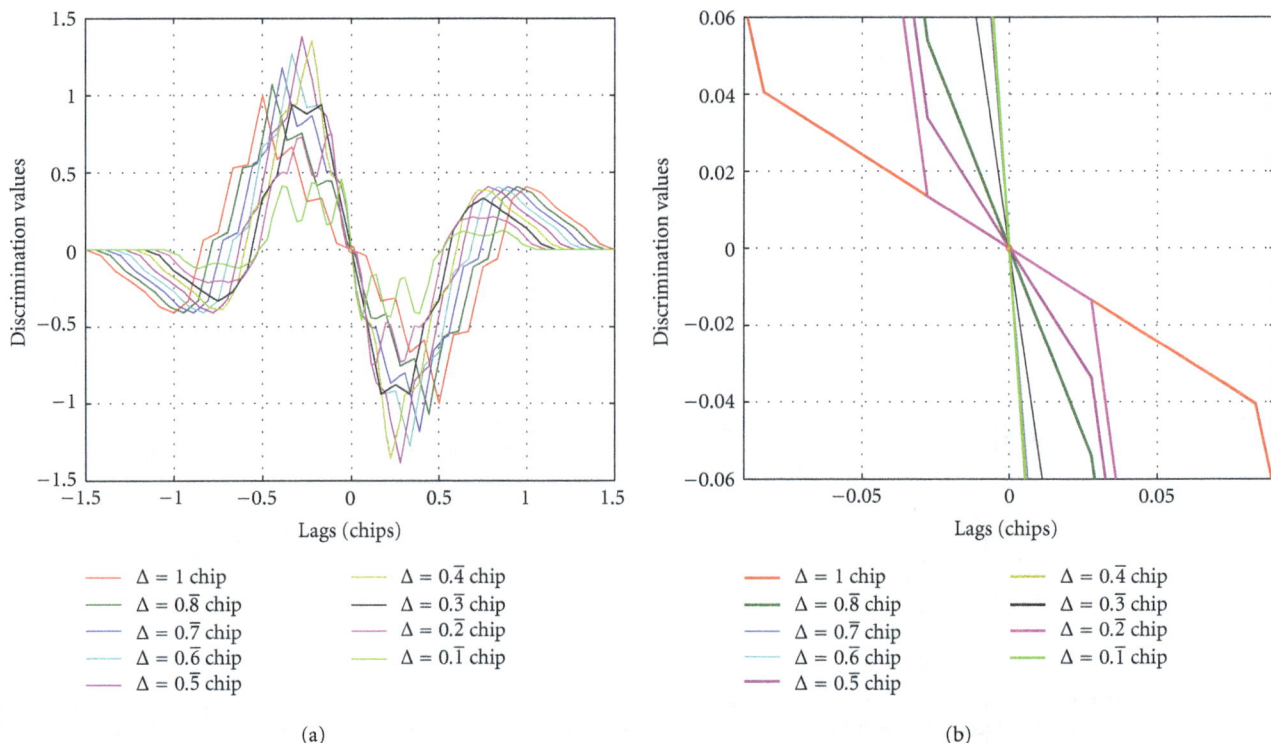

(a)   (b)

FIGURE 8: Theoretical coherent Early-Late discrimination function (a) and its zoom (b) varying the correlator spacing (Δ) and considering only the pilot channel of Galileo E1 OS (CBOC modulated chip without code cross-correlation effect, $\alpha = 0$).

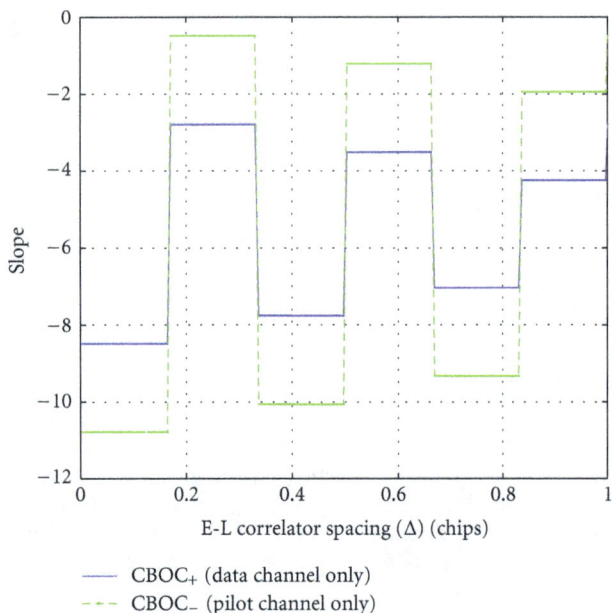

FIGURE 9: Theoretical S-curve slope varying the correlator spacing, considering both data and pilot channels of a Galileo E1 OS-like signal (CBOC modulated chip without code effect, $\alpha = 0$).

Proceeding with our analysis, the discrimination function of the Galileo E1 OS CBOC signal (obtained correlating only the pilot channel) might present an asymmetry and a change in the slope around the zero lag. An example is given in Figure 5, where an S-curve for the Galileo E1 OS signal (PRN 1) is depicted, showing a bias of 6.61 m around the origin.

3.2. GPS L1C. The codes foreseen for the GPS L1C TMBOC signal are Weil codes [2, 12]. They are based on Legendre sequences of 10230 chips (10 ms).

Also in this case, in spite of what happens for the Gold codes, the auto- and cross-correlation functions of the PRNs (L1C$_D$ Data channel and L1C$_P$ Pilot channel) take multiple values but have not been designed to fulfill special properties such as in the case of Galileo memory codes.

The effect of the Weil codes on the S-curve is again the introduction of an asymmetry caused by the cross-correlation contribution. In this case, due to the code length (10 ms, instead of 4 ms memory codes), the impact on the discrimination function results attenuated. This can be derived by observing Figure 6, where the data and pilot autocorrelation functions (with blue and green lines, resp.) are shown together with the data/pilot cross-correlation function (red line).

The cross-correlation term is not symmetrical, but the impact is this case is reduced. This is due to the fact that Weil codes are longer (10 ms) and consequently present better correlation properties. In fact, the longer the codes, the smaller the cross-correlation functions they have. This fact causes in a lighter effect on the S-curve asymmetry. The price

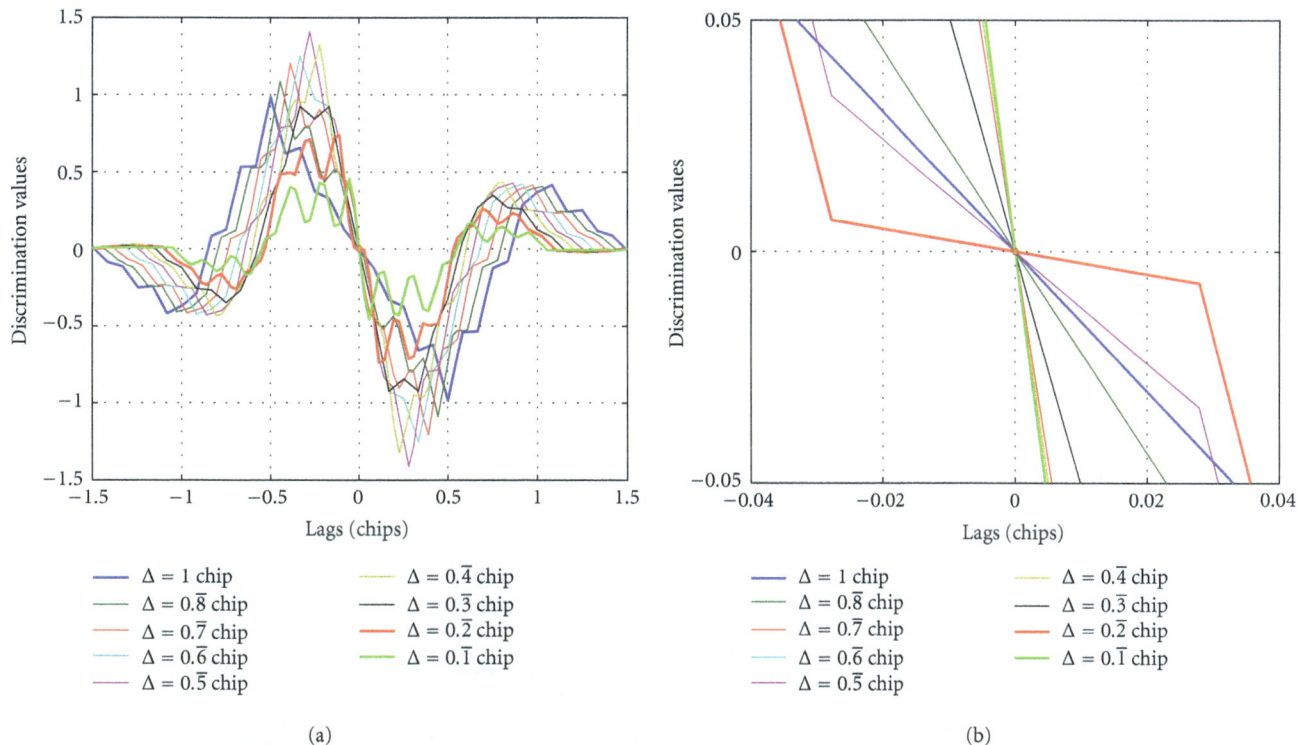

(a)

(b)

FIGURE 10: Theoretical coherent early-late discrimination function (a) and its zoom (b) varying the correlator spacing (Δ), considering only the pilot channel of a GPS L1C-like signal (TMBOC modulation, $\alpha = 0$) and using Galileo E1 4 ms primary codes (PRN 1).

to pay is that longer integration times are needed to align the local code with the incoming signal.

Simulating the GPS L1C signal with the PRN 1 Weil codes for data and pilot channels, the discrimination function in Figure 7 is affected by a bias of 0.09 m (smaller than that of the Galileo case).

4. S-Curve Analysis Varying the Receiver Setup

As previously outlined, in addition to the code and modulation features, the actual impact of cross-correlations on the discrimination function also depends on the receiver setup. Several parameters and architectural choices, including the correlator type and spacing, can lead to discrimination functions with different shapes and slopes in the lock point, affecting the receiver performance. A complete analysis of the shape and the slope of this discrimination function with MBOC signals has then been performed varying the correlator spacing and considering the well-known coherent early-late discriminator.

4.1. Galileo E1 OS. Simulation results obtained using a Galileo E1 OS-like signal are presented in Figure 8, where theoretical S-curves are plotted for different arbitrary spacing values (multiple of 1/9 chip) and considering only the pilot channel. In this case, the S-curves have been obtained starting from the theoretical autocorrelation functions R_{CBOC_+} and R_{CBOC_-} computed on a single chip (previously shown

in Figure 2) and neglecting the data/pilot cross-correlation effects ($R_{CBOC_\pm}$).

Observing the slope around the lock point in Figure 8(b), it is possible to notice that the steepest zero-crossing is obtained for $\Delta = 0.\overline{1}$ chip (green line). An important remark is that the trend of the slope is not directly related to the correlator spacing: a reduction of the spacing does not always lead to a steeper zero-crossing. The lowest slope, leading to the poorest receiver performance, is experienced using two spacings: $\Delta = 0.\overline{2}$ chip (purple line) and $\Delta = 1$ chip (red line).

A detailed analysis on the changes of the S-curve slope has been performed varying the spacing with a tiny step in the range (0, 1] chip. The results are shown in Figure 9 for both data and pilot channels of a Galileo E1 OS-like signal.

It is easy to observe from Figure 9 that the S-curve slope can assume only a finite number of values and the slope variations occur at multiple of 1/6 (0.$\overline{16}$) chip spacing. The range of possible slopes is larger receiving only the pilot channel and the steepest slope is obtained with a spacing smaller than 1/6 chip. Another remark is that the obtained slope values near to the correct lock point assume different magnitudes, but always the same sign (negative): in this case, an inversion on the S-curve slope, leading to false locks, has not been noticed varying the early-late correlator spacing.

4.2. GPS L1C. Previous analyses have been repeated also using a GPS L1C-like signal. The shape of the S-curve for the pilot channel (TMBOC modulation) using arbitrary

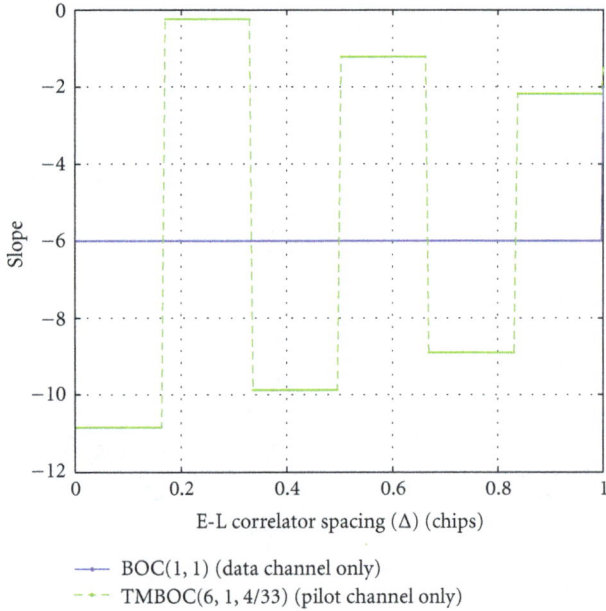

FIGURE 11: Theoretical S-curve slope varying the correlator spacing, considering the data and pilot channels of a GPS L1C-like signal (BOC and TMBOC modulations, $\alpha = 0$) and using Galileo E1 4 ms primary codes (PRN 1).

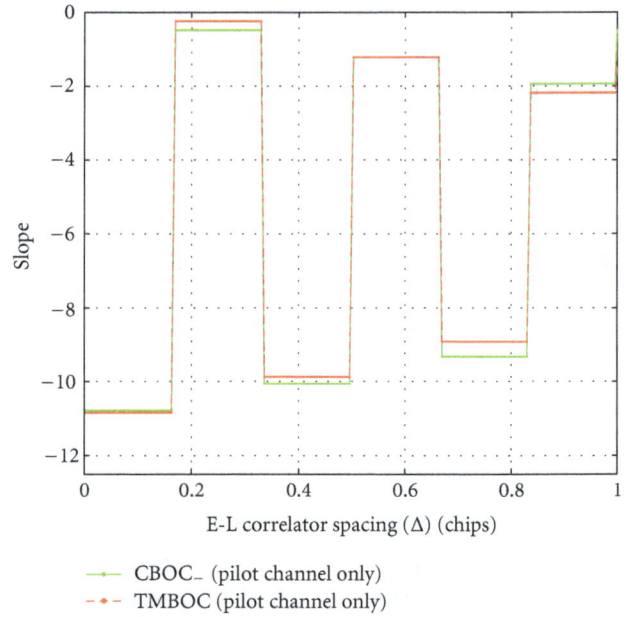

FIGURE 12: Comparison between theoretical slopes of the discrimination function varying the correlator spacing (Δ) and considering only the pilot channels of a GPS L1C-like signal (TMBOC modulation, $\alpha = 0$) and a Galileo E1 OS-like signal (CBOC modulation, $\alpha = 0$).

spacings is depicted in Figure 10. Comparing Figures 10 and 8, it can be noticed that in this case, the worst slope is obtained only in one case ($\Delta = 0.\overline{2}$ chip, red line). The spacing $\Delta = 1$ chip (blue line) using the TMBOC leads to an intermediate slope between the worst case ($\Delta = 0.\overline{2}$ chip) and the steepest slope ($\Delta = 0.\overline{1}$ chip).

A detailed analysis of the S-curve slope varying the spacing has been performed considering both data and pilot channels of GPS L1C, as reported in Figure 11. In this case, the slope for the L1C data channel is constant, due to the autocorrelation properties of the BOC(1,1) modulation.

On the other hand, the S-curve slope obtained using the pilot channel (TMBOC modulation) shows a similar behavior than the results with the Galileo pilot. In addition, as in the CBOC case, a variation on the early-late spacing does not lead to an inversion on the S-curve slope. These two cases are also compared in Figure 12, where small differences can be noticed.

The slopes in Figures 9, 11, and 12 have been computed assuming unitary autocorrelation functions. Taking into account also the relative power levels of data and pilot channels, it is demonstrated that the GPS L1C pilot channel leads to steeper discrimination functions than the Galileo E1 OS pilot channel (75% of the total power versus 50%). This does not prevent from drawing general conclusions about the relation between the receiver setup and the discrimination function. The correlator spacing impacts in a similar way on the shape and the slope of the S-curve using the pilot channels of the GPS L1C and Galileo E1 OS signals. In both cases, inappropriate choices of the spacing (e.g., $\Delta = 0.\overline{2}$ chip) can lead to a reduced slope, resulting

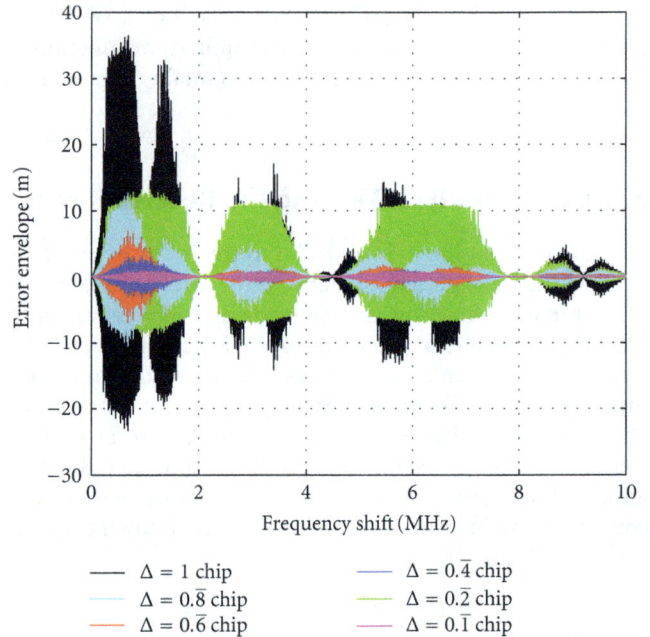

FIGURE 13: Interference Error Envelope comparison assuming different correlator spacings ($\Delta = 1 \div 0.\overline{1}$ chip) and receiving only the pilot channel of a Galileo E1 OS signal (CBOC modulation, PRN 1) in presence of CW interference (carrier to interference power ratio equal to 0 dB).

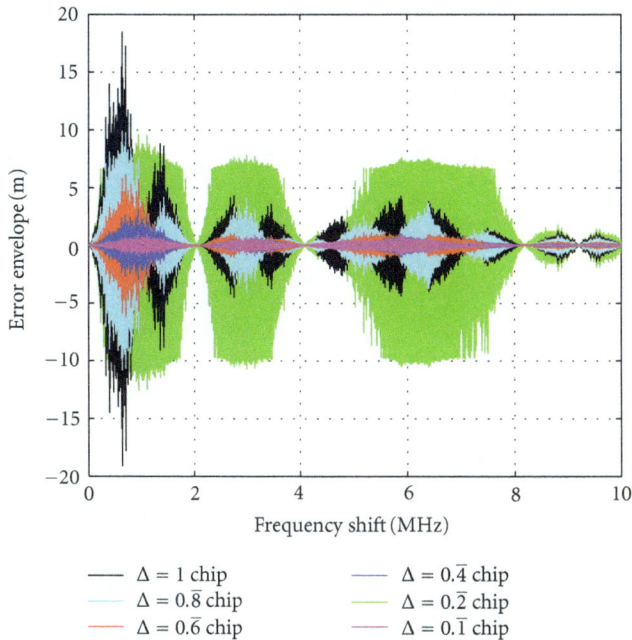

FIGURE 14: Interference Error Envelope comparison assuming different correlator spacings ($\Delta = 1 \div 0.\overline{1}$ chip) and receiving only the pilot channel of a GPS L1C-like signal (TMBOC modulation, using Galileo E1 PRN 1 primary code) in presence of CW interference (carrier to interference power ratio equal to 0 dB).

in a discrimination function more vulnerable to code cross-correlation distortions, as discussed in the following section.

5. Simulation Results in Presence of an Interfering Signal

The distortion induced by the cross-correlation of the channel not locally processed is not just threatening for the bias induced, but also the sensitivity to other error sources might increase. In this section, we show how the presence of an interfering source induces larger errors than expected. In order to compare the results, the interference error envelope (IEE) defined in [7, 13, 14] is used. The IEE curves measure the correlation distortion versus specific interferer characteristics (e.g., the carrier frequency for a CW interferer); the corresponding interference running average (IRA) curves can be easily derived averaging the interference error envelopes.

IEE result for Galileo E1 OS and GPS L1C signals in presence of CW interference and are then presented in Figures 13 and 14 respectively. In detail, the IEEs have been computed simulating the reception of Galileo E1 OS and GPS L1C-like signals using only local pilot codes and varying the early-late spacing.

Observing the results in Figure 13 for the Galileo E1 OS signal, it is easy to notice that the worst case ranging error strongly depends on the chosen correlator spacing. All error envelopes are nearly symmetrical, except the two cases

corresponding to the spacings $\Delta = 1$ chip and $\Delta = 0.\overline{2}$ chip: they lead to a clear asymmetry and to worse errors.

This effect can be explained taking into account previous remarks about code features and receiver setup. As previously shown in Figure 8, both the spacings $\Delta = 1$ chip and $\Delta = 0.\overline{2}$ chip lead to a low S-curve slope. This fact implies a noticeable vulnerability of the discriminator to possible distortions due to code cross-correlations and interference effects. In fact, assuming a low slope, the ranging errors in presence of an interfering signal result magnified. In addition, a noticeable asymmetry of the S-curve around the zero crossing (different slopes on each side) due to the data/pilot code cross-correlations leads to different magnitudes for positive and negative errors, as in Figure 13 (black and green envelopes).

Similar results have been obtained simulating a GPS L1C-like pilot channel in presence of a CW interference (see Figure 14).

In this case, the asymmetry on the envelope can be noticed only for the spacing $\Delta = 0.\overline{2}$ chip: in fact, as previously observed in Figure 10, a low S-curve slope is obtained only using such a spacing.

In conclusion, it must be remarked that both using Galileo E1 OS and GPS L1C signals the code cross-correlation distortion on the S-curve can be magnified by an inappropriate choice of the correlator spacings and it can lead to noticeable worsening in receiver performance in presence of an interfering signal. Such an effect can be noticed only in case of receiving a single channel (i.e., the pilot channel), whereas it is not present if the received signal is correlated with a coherent local replica including both data and pilot channels.

6. Conclusions

A comparative analysis of GPS L1C and Galileo E1 OS signals has been performed, pointing out how, when only the pilot channel is locally received in order to perform acquisition with long integration times, the residual cross-correlation due to the unprocessed channel cannot be neglected. Analytical and simulation results have been presented in order to demonstrate that in such a case the discrimination function can be affected by a bias around the lock point. The distortion can be noticed only in case of receiving a single channel (i.e., the pilot channel), whereas it is not present if the received signal is correlated with a coherent local replica including both data and pilot channels. It has also been shown how the distortion of the S-curve increases the sensitivity to other error sources, as for example to the presence of interfering signals. Moreover, the paper demonstrated that, in this case, inappropriate choices of the correlator spacing can lead to a discrimination function with reduced slope, thus enhancing the vulnerability of the receiver.

References

[1] M. Fantino, P. Mulassano, F. Dovis, and L. Lo Presti, "Performance of the proposed galileo CBOC modulation in heavy multipath environment," *Wireless Personal Communications*, vol. 44, no. 3, pp. 323–339, 2008.

[2] Global Positioning System Wing Systems Engineering & Integration, Interface Specification IS-GPS-800, Revision A. Navstar GPS Space Segment/User Segment L1C Interface, 2011, http://www.navcen.uscg.gov/pdf/gps/IS-GPS-800A_Final_08Jun10.pdf.

[3] European Union, European GNSS (Galileo) Open Service Signal In Space Interface Control Document, OS SIS ICD, Issue 1.1, 2010, http://ec.europa.eu/enterprise/policies/satnav/galileo/open-service/index_en.htm.

[4] F. van Diggelen, *Assisted GPS, GNSS, and SBAS*, Artech House, 2009.

[5] D. Margaria, S. Savasta, F. Dovis, and B. Motella, "Codes cross-correlation impact on the interference vulnerability of Galileo E1 OS and GPS L1C signals," in *Proceedings of the ION International Technical Meeting (ITM '10)*, pp. 1111–1121, San Diego, Calif, USA, January 2010.

[6] E. Rebeyrol, O. Julien, C. MacAbiau, L. Ries, A. Delatour, and L. Lestarquit, "Galileo civil signal modulations," *GPS Solutions*, vol. 11, no. 3, pp. 159–171, 2007.

[7] D. Margaria, S. Savasta, F. Dovis, and B. Motella, "Comparative interference vulnerability assessment of GPS TMBOC and Galileo CBOC signals," in *Proceedings of the 22nd International Technical Meeting of the Satellite Division of the Institute of Navigation (ION GNSS '09)*, pp. 38–48, Savannah, Ga, USA, September 2009.

[8] J. W. Betz et al., "Enhancing the future of civil GPS—overview of the L1C signal," *Inside GNSS*, pp. 42–49, 2007.

[9] R. Gold, "Optimal binary sequences for spread spectrum multiplexing," *IEEE Transactions on Information Theory*, vol. 13, no. 4, pp. 619–621, 1967.

[10] P. Misra and P. Enge, *Global Positioning System. Signal Measurements and Performance*, Ganga-Jamuna, Lincoln, Mass, USA, 2nd edition, 2006.

[11] G. W. Hein, "MBOC: the new optimized spreading modulation recommended for Galileo L1 OS and GPS L1C," *Inside GNSS*, pp. 57–66, 2006.

[12] G. W. Hein, J. A. Avila-Rodriguez, and S. Wallner, "The Galileo code and others," *Inside GNSS*, pp. 62–74, 2006.

[13] B. Motella, S. Savasta, D. Margaria, and F. Dovis, "Method for assessing the interference impact on GNSS receivers," *IEEE Transactions on Aerospace and Electronic Systems*, vol. 47, no. 2, pp. 1416–1432, 2011.

[14] B. Motella, S. Savasta, D. Margaria, and F. Dovis, "A method to assess robustness of GPS C/A code in presence of CW interferences," *Integrating Radio Positioning and Communications*, vol. 2010, Article ID 294525, 8 pages, 2010.

3

Two-Step Galileo E1 CBOC Tracking Algorithm: When Reliability and Robustness Are Keys!

Aleksandar Jovanovic,[1] Cécile Mongrédien,[2] Youssef Tawk,[1] Cyril Botteron,[1] and Pierre-André Farine[1]

[1] Electronics and Signal Processing Laboratory (ESPLAB), École Polytechnique Fédérale de Lausanne (EPFL), Rue A.-L. Breguet 2, 2000 Neuchâtel, Switzerland
[2] Fraunhofer Institute for Integrated Circuits IIS, Nordostpark 93, 90411 Nuernberg, Germany

Correspondence should be addressed to Aleksandar Jovanovic, aleksandar.jovanovic@epfl.ch

Academic Editor: Heidi Kuusniemi

The majority of 3G mobile phones have an integrated GPS chip enabling them to calculate a navigation solution. But to deliver continuous and accurate location information, the satellite tracking process has to be stable and reliable. This is still challenging, for example, in heavy multipath and non-line of sight (NLOS) environments. New families of Galileo and GPS navigation signals, such as Alternate Binary Offset Carrier (AltBOC), Composite Binary Offset Carrier (CBOC), and Time-Multiplex Binary Offset Carrier (TMBOC), will bring potential improvements in the pseudorange calculation, including more signal power, better multipath mitigation capabilities, and overall more robust navigation. However, GNSS signal tracking strategies have to be more advanced in order to profit from the enhanced properties of the new signals. In this paper, a tracking algorithm designed for Galileo E1 CBOC signal that consists of two steps, coarse and fine, with different tracking parameters in each step, is presented and analyzed with respect to tracking accuracy, sensitivity and robustness. The aim of this paper is therefore to provide a full theoretical analysis of the proposed two-step tracking algorithm for Galileo E1 CBOC signals, as well as to confirm the results through simulations as well as using real Galileo satellite data.

1. Introduction

New GPS and Galileo signals use new modulations, such as AltBOC, CBOC, and TMBOC that have the potential to improve navigation through advanced signal properties, such as more signal power, better multipath mitigation capabilities, and overall improved signal cross-correlation properties. Certainly, a major innovation brought by the new modulation schemes consists of the presence of two distinct components, namely, the data and pilot channels that carry two different pieces of information. The data channel contains the navigation message, whereas the pilot channel is dataless, allowing long coherent signal integration that, in turn, allows more precise determination of the ranging information. For carrier tracking, the presence of a pilot channel enables the combined use of pure PLL (Phase Lock Loop) discriminators and longer coherent integration time. Code tracking can be

organized as data/pilot collaborative tracking [1, 2], where two channels (data and pilot) are used in the estimation of the code error, decreasing the thermal noise error and improving overall tracking.

Several tracking algorithms proposed for Galileo E1 CBOC signals were derived from tracking schemes developed for BPSK (Binary Shift Keying) and BOC(1,1) signals. BOC(1,1) tracking faces the problem of biased tracking as already explained in [3]. Until now, the main algorithms that were proposed to specifically address the problem of biased tracking for BOC(1,1) and can be applied to CBOC tracking too are Single Side Lobe, bump and jump algorithm [4], ASPeCT (Autocorrelation Side-Peak Cancellation Technique) [3], and Double Estimator [5, 6]. The Single Side Lobe technique provides a robust solution to resolving BOC ambiguity, but it is suitable only for low-precision receivers. Note that the ASPeCT algorithm modifies the shape of the autocorrelation function and eliminates side peaks that can be points of false locking. However, it requires a different and more complex correlator architecture. Double Estimator assumes an additional loop, the Subcarrier Lock Loop (SLL), that tracks subcarrier delay. More correlators are needed, as well as more complex loop implementation. In addition, two tracking techniques have been proposed exclusively for Galileo E1 CBOC tracking: TM61 and Dual Correlator. The TM61 technique generates a one-bit local replica, either BOC(1,1) or BOC(6,1), simplifying the architecture, but also degrading tracking by at least 3 dB [7]. The Dual Correlator technique is based on the investigation of two parallel correlations: one between the incoming MBOC and a BOC(1,1) replica, and one between the incoming MBOC and a BOC(6,1) replica. Each correlation is weighted, and two outputs are linearly added, such that by changing the values of the weights the tracking can be easily modified [8]. These two E1 CBOC tracking techniques assume separate correlations, which degrades the tracking and brings additional complexity. In contrast, the two-step tracking algorithm considered in this paper is based on full four-level bit local replica generation and consists of two steps, coarse and fine, with different tracking parameters. The main requirements that were taken into consideration when defining this algorithm were (1) a relatively low implementation complexity suitable for a mass-market solution, (2) the shape of the E1 CBOC autocorrelation function, and (3) the correlator's structure since it conditions the tracking algorithm's properties. Regarding the second point, since the CBOC autocorrelation function has secondary peaks that can be potential false lock points, one of the main objectives was to design a tracking algorithm that can avoid such false locks or, at least, minimize their occurrence. Therefore, five complex correlators are used with several correlator spacing options (very early (VE) and a very late (VL) correlators have been added to the three conventional Early (E), Prompt (P), and Late (L) correlators). Regarding the third point, we wanted to design a tracking architecture that can minimize the tracking error caused by thermal noise and multipath. Moreover, DP (Dot-Product) and HRC (High Resolution Correlator) discriminators are used equally in the fine tracking step, depending on the tracking conditions.

This paper extends our previous results by fully analyzing the theoretical performance of the previously proposed two-step tracking schemes with respect to tracking accuracy, sensitivity, and robustness, as well as testing the algorithm using real satellite data. The paper is organized as follows. In Section 2, the Galileo E1 CBOC signal is described. Section 3 provides an overview of the two-step tracking algorithm with a thorough analysis of the discriminator curve outputs and discussion of its stability and linearity regions. Section 4 provides theoretical analysis of the tracking loops' performance, especially derivation and calculation of code tracking error and tracking thresholds for different discriminators. A multipath mitigation analysis is provided in Section 5, followed by simulation-based and realistic results provided in Section 6. Finally, conclusions and outlook are provided in Section 7.

2. Galileo E1 CBOC Signal Properties

New generations of Global Navigation Satellite Systems (Galileo, Glonass, Compass, etc.) are rapidly evolving, and old GPS system served localization purposes to users for more than twenty years in positioning that is currently undergoing a modernization process, and new European Galileo system will be available within a few years. MBOC modulation was chosen to be implemented on the Galileo E1 and GPS L1C signals [9]. The power density function of the MBOC modulation is a sum of the 10/11 normalized BOC(1,1) and the 1/11 normalized BOC(6,1) signal spectrum. MBOC is defined in the frequency domain, and two different implementations have been specified for Galileo and GPS: CBOC and TMBOC, respectively. Although having different time domain implementations, the power spectral densities in both configurations of the MBOC(6,1,1/11) are the same.

The GPS L1C signal has a pure BOC(1,1) data channel carrying 25% of the total signal power, while the pilot signal uses a TMBOC modulation with 75% of the total signal power. The Galileo E1 signal on the other hand shares its power equally between data and pilot channels, with both channels using a CBOC modulation based on a four-level subcarrier formed by the weighed sum of BOC(1,1) and BOC(6,1). To acquire 100% of the signal energy, a bandwidth of 16 MHz is needed. The only difference between data and pilot channels besides having different PRN (Pseudo-Random Noise) codes is in the sign of the weighted sum. The data channel is "in phase" (CBOC(6,1,"+")), and the pilot channel is in "antiphase" (CBOC(6,1,"−")). "Antiphase" configuration exhibits better tracking performance: autocorrelation peak is narrower. The analytical expression for the E1 CBOC signal can be represented as

$$s_{Gal}(t) = \sqrt{\frac{1}{2}}\Big[d_{Gal}(t)c_d(t)\text{CBOC}(6,1,p,``+")(t)$$
$$+c_p(t)\text{CBOC}(6,1,p,``-")(t)\Big],$$

$$\text{CBOC}(6,1,p,``+")(t)=\left(\sqrt{(1-p)}s_{BOC(1,1)}(t)+\sqrt{p}s_{BOC(6,1)}\right),$$

$$\text{CBOC}(6,1,p,``-")(t)=\left(\sqrt{(1-p)}s_{BOC(1,1)}(t)-\sqrt{p}s_{BOC(6,1)}\right),$$
$$(1)$$

and $p = 1/11$ for OS (Open Service), $c_d(t), c_p(t)$ are the data and pilot spreading codes, and CBOC(6, 1, p, "+") and CBOC(6, 1, p,"−") are the pilot and data subcarriers, respectively.

2.1. Galileo E1 CBOC Autocorrelation Function. The study of the autocorrelation function is important when trying to evaluate the tracking performance of the signal. The E1 CBOC autocorrelation function for the data and pilot channels, as well as for the combination of the two channels (for the combined tracking case), is shown in Figure 1 assuming an 18 MHz bandwidth. The CBOC autocorrelation function exhibits a very narrow main peak and two side-correlation peaks located at approximately 0.5 chips around

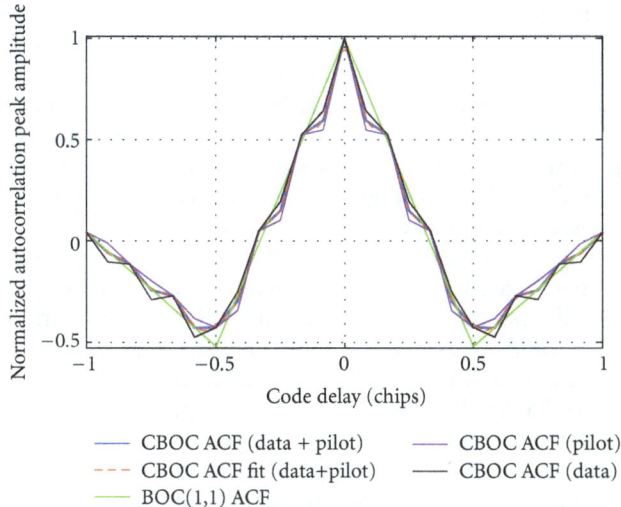

FIGURE 1: CBOC autocorrelation function for pilot, data, and averaged data/pilot channels and its fit by a sum of sinusoidal functions for bandwidth of 18 MHz.

the main peak. Dangerously, these secondary peaks can be potential false locking points, as explained earlier.

The CBOC autocorrelation function has a narrower main peak when compared to BOC(1,1), but it is not fully linear. The expression for the autocorrelation function can be approximated more simply, as it is done with the expression for the BOC(1,1) autocorrelation function inside the one chip width of the main autocorrelation peak [3]:

$$R_{BOC(1,1)}(\tau) = 1 - \alpha|\tau|, \qquad (2)$$

where α corresponds to the absolute value of the slope of the spreading sequence autocorrelation function's main peak. The value of α is equal to 3 for the BOC(1,1) signal, and it is larger than 3 for the CBOC data/pilot autocorrelation function for the narrow E-L spacing around the main correlation peak. Due to the existence of ripples in CBOC autocorrelation function, the value for the α is not constant and it depends on correlator spacing.

In the following, the first derivative of the autocorrelation function (α) is used as a parameter for the evaluation of the code tracking error and tracking threshold. α varies with the E-L correlator spacing, but for fixed E-L distances it can be considered constant. Analytical expressions for autocorrelation function should be able to provide both the properties of the CBOC modulation and the effects of front-end filtering on the autocorrelation peak. In order to compute α, the analytical expression for the E1 CBOC autocorrelation function is determined by fitting the autocorrelation curve with a sum of sinusoidal functions. By doing so, it is possible to compute α of the filtered autocorrelation function for different E-L correlator spacing. The resulting analytical expression for the CBOC autocorrelation function obtained

by fitting the autocorrelation curve and α can be written as [10]:

$$\widetilde{R}_F(t) = \sum_{i=1}^{8} a_i \cdot \sin(b_i \cdot t + c_i),$$

$$\alpha = \left(\frac{d\widetilde{R}_F(x)}{dx}\right)\bigg|_{x=-d/2}, \qquad (3)$$

where t is the distance from the main correlation peak (half of E-L spacing), and a_i, b_i, and c_i, where $i = 1, \ldots, 8$, are fitted constants that are different for the data and pilot channels. Since we assume later both data and pilot channels are used for tracking, and thus the averaged CBOC($+$) and CBOC($-$) autocorrelation is used for fitting. In this case, the infinite bandwidth is used so that 99.99% of the CBOC signal is received. Therefore, three sets of a_i, b_i, and c_i parameters were used. As can be observed from Figure 1, the analytical fit closely follows the averaged CBOC autocorrelation function over the considered range. α for different E-L spacings and three different combinations is provided in Table 1. If the correlator spacing decreases below $d = 1/12$ chips, the slope of the autocorrelation peak decreases as well, because of the rounding effect of the front-end filtering on the autocorrelation peak. Additionally, a smaller E-L spacing reduces the linear tracking region which, at some point, can make the tracking unreliable. Therefore, $d = 1/12$ was chosen as a good tradeoff between accuracy and reliability for the considered bandwidth of 18 MHz.

The existence of secondary code on the pilot channel on top of the primary PRN code of duration 100 ms or 25 chips additionally complicates the tracking and requires techniques for wiping off the secondary code before further process the signal. The main advantages of secondary code are increased resistance to narrow-band interference due to the additional spectrum line, better cross-correlation properties (which mostly help during acquisition to avoid the near-far effects), and more robust data bit synchronization than with the histogram method, although that does not apply to Galileo E1-B/C, since the data bit and PRN code have the same duration.

The algorithm used for secondary code extraction is based on a combination of serial and parallel searches. It is well described in [11, 12]. The search for the primary code phase is performed serially within one primary code length, and the secondary code phase is searched in parallel over the entire length of the secondary code. The algorithm takes into consideration the residual frequency offset, and it is shown to reduce the acquisition time compared to other methods.

3. Two-Step Tracking Scheme and Discriminator Curves Analysis

Achieving the favorable properties of the new signals is possible by increasing the complexity of the tracking loops design. There are many techniques proposed to provide unambiguous tracking, as already explained in the introduction. Most of them are listed with the corresponding tracking architectures in [13, 14] as well as the code tracking

TABLE 1: Slope of the autocorrelation function for different early-late spacing for data, pilot channels, and averaged data/pilot channel for infinite bandwidth.

Correlator spacing	Autocorrelation slope α		
	Data/pilot averaged	Pilot channel	Data channel
1/20	3.98	4.50	3.39
1/12	5.35	5.96	4.63
1/10	5.51	6.06	4.82
1/5	1.49	0.09	1.89

error comparison. Most of them need complex tracking architecture or modify the correlation function. In this section, we present the analysis about our proposed two-step tracking scheme, simple and with less complex tracking architecture. Carrier and code delay tracking are analyzed separately, putting more weight on code delay tracking.

3.1. Galileo E1 CBOC Carrier Tracking (PLL). The right choice of the PLL discriminator is the first step towards obtaining accurate phase error estimation. The choice of the PLL discriminator is dependent upon the parameters of the E1 CBOC signal structure. The pilot channel alone can be used for the phase error estimation, since there is no data bit on it, and, theoretically, coherent integration can be used for as long as needed. As a consequence, a discriminator that is insensitive to phase jumps can be used. Using pure PLL tracking on the pilot channel as well as longer coherent integration improves carrier tracking sensitivity. It also enhances the carrier tracking loop resistance to receiver dynamics. Although 3 dB are lost by ignoring the data channel for the carrier phase estimation, the noise sensitivity is improved by 3 dB. Therefore, the use of an extended arctangent discriminator has been selected (atan2), providing the widest linear tracking region. The discriminator output can be analytically expressed as

$$D_{\text{atan2}} = \arctan 2\left(\frac{Q}{I}\right). \tag{4}$$

I and Q are in-phase and quadrature-phase correlator outputs for the pilot channel. The extended *arctangent* (four-quadrant) discriminator has an operational range twice as large as the traditional *arctangent* discriminator ($[-\pi, \pi]$). It has good noise resistance performance for high C/N_0 (carrier-to-noise ratio). It can also track the phase modulo 2π, without a half cycle of ambiguity. Although pilot-only tracking is noisier than data/pilot combined tracking, the chance of losing lock is smaller, and the tracking is more stable and less complex.

3.2. Galileo E1 CBOC Code Delay Tracking (DLL). Code delay tracking is more stable than carrier tracking, since it provides the user with more robust measurements and initial estimates of the receiver position through pseudoranges. The

main idea behind the considered code tracking algorithm that we call the "two-step tracking technique" [1, 15] is to benefit from the narrowness of the CBOC autocorrelation peak while minimizing the risk of locking onto one of the secondary peaks of the autocorrelation function. This algorithm can also be applied to BOC(1,1) as well as CBOC(6,1,1/11) tracking and consists of two steps. The first step is a coarse tracking step that is used to ensure proper convergence towards the true lock point, since it relies on an unambiguous discriminator. This is achieved using an original unambiguous combination of all five correlators (E, L, P, VE, and VL), where VE and VL are always positioned on the secondary peaks. This ensures that the prompt correlator is indeed located on the main correlation peak.

However, once the code tracking has converged to the main peak, the VEMLP (Very Early Minus Late Power) discriminator that uses five correlators' outputs offers sub-optimal code tracking performance since it does not benefit from the narrow CBOC correlation peak. At this point, the tracking process switches to a fine tracking step, where a DP (or HRC) discriminator is used with a narrow correlator spacing. The description and parameters of the two-step tracking technique are provided in Table 2 [1]. Power in the name VEMLP comes from the fact that the discriminator function contains squared correlator outputs that relate to the power of the signal.

As shown in Table 2, the fine tracking step fully exploits the narrow autocorrelation peak by using (1) a DP (HRC) discriminator with narrow E-L spacing and (2) a reduced loop bandwidth (as enabled by the use of carrier aiding). We proposed an HRC discriminator [16] for the fine tracking step instead of a DP discriminator in order to improve multipath mitigation [1]. More specifically, a combination of traditional tracking (DP) and tracking with an HRC is used. In the case of strong multipath, an HRC correlator is used, since it shows good multipath mitigation, otherwise traditional tracking (DP) is assumed, since it has a wider stability tracking region. Details are provided in Section 5.2. Normalization of the discriminator outputs was also performed in order to provide an unbiased estimate of the small code errors. The use of a bump-jump algorithm [4] minimizes the false lock risk. It measures and compares the received power in the VE, P, and VL correlator outputs and jumps left or right if the VE or VL correlator output power is found to be consistently higher than that of the P. Note that the E-L correlator spacing should remain flexible, depending on the front-end bandwidth of the receiver. We used bandwidth of 18 MHz throughout the paper and generated the results based on this assumption. However, the parameters of the two-step tracking scheme, such as early-late spacing, width of the stability and linearity tracking regions, slope of the autocorrelation functions (as it is provided in Table 1), and tracking thresholds change depending on the available bandwidth. The scheme is designed to be flexible, for example, when decreasing the front-end bandwidth, early-late correlator spacing increases, and so forth.

In order to improve code tracking accuracy, both pilot and data channels are used for the code tracking delay estimation, making it more robust. Common data/pilot code

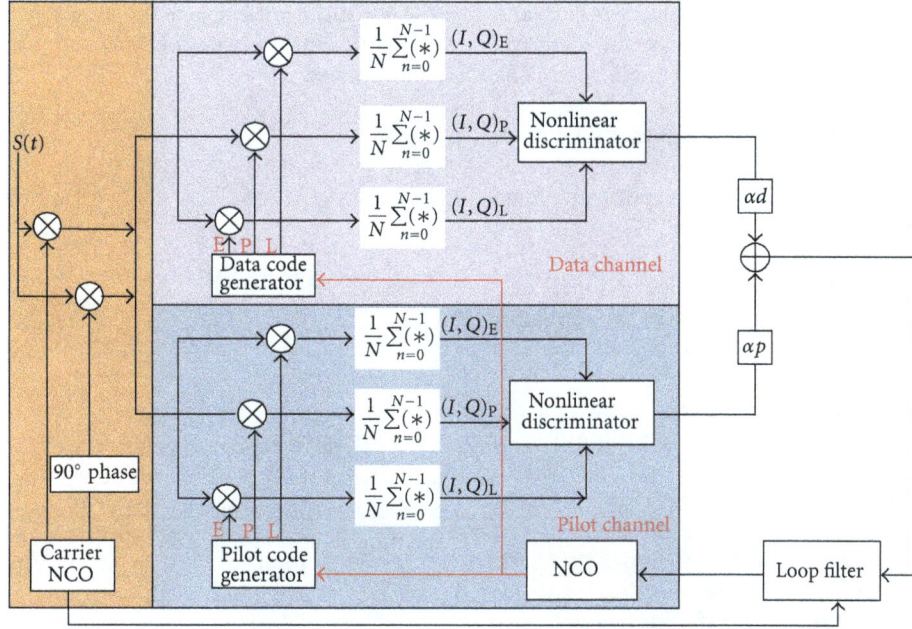

FIGURE 2: Noncoherent collaborative tracking architecture for the fine tracking step of two-step tracking algorithm.

TABLE 2: Two-step tracking scheme description for the front-end bandwidth of 18 MHz.

| | Two-step tracking technique | | |
Parameters	Coarse tracking	Fine tracking (pilot channel)	Fine tracking (data channel)
E-L spacing	$[-1/2; -1/4; 0; 1/4; 1/2]$	$[-1/2; -1/24; 0; 1/24; 1/2]$	$[-1/2; -1/24; 0; 1/24; 1/2]$
Discriminator type	VEMLP	DP/HRC	DP
Loop filter BW	2 Hz	1 Hz	1 Hz
False lock detector	Embedded	Bump-jump	Bump-jump
Carrier aiding	No	Yes	Yes
Advantage	Unambiguous	Accurate multipath gain	Accurate
Disadvantage	Noise and MP increased	Ambiguous	Ambiguous

tracking is performed by linearly combining them non-coherently. The incoming signal is separately combined with data and pilot PRN codes. The outputs are then combined in a noncoherent way, as shown in Figure 2. Coherent combining outperforms noncoherent combining from an accuracy standpoint, but it also requires data bit sign recovery for integration times greater than 4 ms, which increases the loop complexity and becomes unreliable at low C/N_0. Therefore, data and pilot discriminator outputs are averaged using a noncoherent combining method and fed into a unique loop filter that updates both pilot and data NCO (Numerically Controlled Oscillator). The discriminator output can then be written as

$$D_{\text{out}} = \alpha_p D_d + \alpha_d D_p, \tag{5}$$

where α_p and α_d are weights for the pilot and data channels, respectively, applied to the discriminator. The weights have to fulfill the following requirements:

$$\alpha_p + \alpha_d = 1,$$

$$\alpha_p = \frac{\sigma_p^2}{\sigma_p^2 + \sigma_d^2}, \tag{6}$$

$$\alpha_d = \frac{\sigma_d^2}{\sigma_p^2 + \sigma_d^2},$$

where σ_p and σ_d are the variances of the discriminator on the pilot and data channel, respectively. If the same modulation and code tracking loop update rate are used for the data and pilot channels, the same variance will be observed on both channels, reducing the channel combining to a simple averaging [2, 15]. However, since the data and pilot autocorrelation functions are not rigorously identical, the pilot channel having a slightly sharper main lobe peak that could provide slightly better tracking accuracy, it is therefore possible to further refine the weighting scheme so that the tracking relies more heavily on the pilot channel. Theoretical values for the weights for the pilot and data channels are shown to be approximately 0.6 and 0.4, respectively. They were obtained empirically, by computing the standard variation of the data and pilot channels and using a setup that consists of a Spirent GSS8000 simulator [17], connected with a wideband front-end [18], and the data was postprocessed in software.

The resulting collaborative tracking architecture for the fine tracking step of the two-step tracking algorithm is shown in Figure 2, highlighting the DLL part divided in separate tracking architectures for data and pilot channels. After wiping off the carrier using multiplication of the incoming signal with a local replica that is aligned using the information from the carrier NCO, the resulting in-phase and quadrature-phase components are multiplied with five code replicas: E, P, and L as well as VE and VL (not shown in Figure 2). Ideally, if the carrier phase is aligned with the carrier phase of the incoming signal, all the energy will be in the in-phase component. Estimation of the code delay error is performed in the DLL loop, using DLL discriminators. The output is then filtered by the DLL loop filter, and updates are provided for the code delay rate for the NCO. In the next section, analysis of the three different types of discriminators used in the tracking scheme described above is provided.

3.3. DLL Discriminators. As shown in Table 2, three types of discriminators are used in the two-step tracking scheme: VEMLP and DP, for the coarse and fine tracking steps, respectively, and HRC as an option when high multipath error is present in the fine tracking step. The VEMLP discriminator is noncoherent and needs two more correlators (VE and VL) with a fixed distance between them, to mark points that can be potential sources of false locks. It has a higher computational load than DP, which is quasicoherent. After converting a signal to *IF* frequency, *I* and *Q* baseband correlation outputs for the E, L, P, VE, and VL pilot channel can be written as

$$I_X = \sqrt{\frac{P}{2}} \widetilde{R}(\varepsilon_\tau + \beta) \frac{\sin\left(\pi \varepsilon_f T_i\right)}{\pi \varepsilon_f T_i} \cos\left(\varepsilon_{\phi_p}\right) + n_{IX},$$

$$Q_X = \sqrt{\frac{P}{2}} \widetilde{R}(\varepsilon_\tau + \beta) \frac{\sin\left(\pi \varepsilon_f T_i\right)}{\pi \varepsilon_f T_i} \sin\left(\varepsilon_{\phi_p}\right) + n_{QX}, \tag{7}$$

where $\widetilde{R}$ is the correlation of the local spreading code with the filtered incoming spreading code, ε_f is the frequency error, ε_τ is the code group delay error, and ε_ϕ is the carrier phase delay error. The noise components n_{IX}, n_{QX} are independent Gaussian noise components. X relates the correlator type (E, P, L, VE, VL), and β represents the spacing between correlators (for P $\beta = 0$, E $\beta = \delta/2$, for L $\beta = -\delta/2$, VE $\beta = \delta$, and VL $\beta = -\delta$). δ represents the E-L correlator

spacing and for the coarse tracking step ($\beta = 2\delta$). Therefore, the output for the types of the discriminators can be written analytically as

$$D_{\text{VEMLP}} = \sqrt{I_{\text{VE}}^2 + Q_{\text{VE}}^2 + I_{\text{E}}^2 + Q_{\text{E}}^2} - \sqrt{I_{\text{VL}}^2 + Q_{\text{VL}}^2 + I_{\text{L}}^2 + Q_{\text{L}}^2}, \tag{8}$$

$$D_{\text{DP}} = (I_{\text{E}} - I_{\text{L}})I_{\text{P}} + (Q_{\text{E}} - Q_{\text{L}})Q_{\text{P}}, \tag{9}$$

$$D_{\text{HRC}} = (I_{\text{E}} - I_{\text{L}}) + \frac{(I_{\text{VE}} - I_{\text{VL}})}{2}. \tag{10}$$

Analytical expressions for the VEMLP, DP, and HRC discriminators functions can be written using filtered auto-correlation function $(\widetilde{R})$ with the approximation that in tracking stage $\sin(\pi \varepsilon_f T_i)/\pi \varepsilon_f T_i \simeq 1$ and $\cos(\varepsilon_{\phi_d}) \simeq 1, \sin(\varepsilon_{\phi_d}) \simeq 0$ as

$$D_{\text{VEMLP}} = \sqrt{\frac{P}{2}} \left(\sqrt{\widetilde{R}^2\left(\tau + \frac{\delta}{2}\right) + \widetilde{R}^2(\tau + \delta)} \right.$$
$$\left. - \sqrt{\widetilde{R}^2\left(\tau - \frac{\delta}{2}\right) + \widetilde{R}^2(\tau - \delta)} \right), \tag{11}$$

$$D_{\text{DP}} = \sqrt{\frac{P}{2}} \left(\widetilde{R}\left(\tau + \frac{\delta}{2}\right) - \widetilde{R}\left(\tau - \frac{\delta}{2}\right) \right) \widetilde{R}(\tau), \tag{12}$$

$$D_{\text{HRC}} = \sqrt{\frac{P}{2}} \left(\widetilde{R}\left(\tau + \frac{\delta}{2}\right) - \widetilde{R}\left(\tau - \frac{\delta}{2}\right) - \frac{\widetilde{R}(\tau + \delta) - \widetilde{R}(\tau - \delta)}{2} \right). \tag{13}$$

The performance of the tracking loops in the presence of multipath and code tracking errors can be assessed by studying the discriminators curves. These curves present the discriminators' outputs as a function of code delays (inputs). Tracking parameters of interest that can be derived from the curves are the linear tracking region, defined as the region in which the discriminator responds without any bias, and the stability region (pull-in region), defined as the region where the discriminator reacts in the right direction [3].

A wider stability tracking region means more robust tracking. A wider linear tracking region corresponds to correct and unbiased discriminator response. As already shown in [3], the discriminator output does not directly estimate the input error. In order to obtain unbiased code delay error, the discriminator output has to be normalized by its gain, therefore achieving perfect normalization. Normalization removes amplitude sensitivity, improving performance under rapidly changing SNR (Signal-to-Noise Ratio) conditions and providing unbiased code delay error estimation. As normalization depends on the power of the signal, discriminator outputs should be first normalized by the estimated power of the signal. Therefore, using this approach, the normalization factors for three discriminator types can be written as

$$N_{VEMLP} = \sqrt{I_{VE}^2 + Q_{VE}^2 + I_E^2 + Q_E^2} + \sqrt{I_{VL}^2 + Q_{VL}^2 + I_L^2 + Q_L^2}, \quad (14)$$

$$N_{DP} = (I_E + I_L)I_P + (Q_E + Q_L)Q_P, \quad (15)$$

$$N_{HRC} = (I_E + I_L) + \frac{(I_{VE} + I_{VL})}{2}. \quad (16)$$

Expressions for the discriminator outputs can be simplified using the assumptions that the estimated code delay error is smaller than half of the correlator spacing, and the E and L correlator values correspond to the points on the autocorrelation function that are on the slope forming its main peak. Normalization provides unbiased estimation of the code delay error within $\pm\delta/2$ chips. Using the equations for the correlator outputs provided above and using the assumption that the autocorrelation function can be approximated by small linear parts with different slopes, the discriminator outputs can be written as

$$N_{VEMLP} = \sqrt{2P}\sqrt{2 - 2\alpha\varepsilon_\tau(\alpha\varepsilon_\tau - 2\delta) - \alpha\delta\left(\delta - 1 - \alpha - \alpha\frac{\delta}{4}\right)}, \quad (17)$$

$$N_{DP} = \sqrt{\frac{P}{2}}(2 - \alpha\delta)(1 - \alpha|\varepsilon_\tau|), \quad (18)$$

$$N_{HRC} = \sqrt{\frac{P}{2}}(2 - \alpha\delta)(1 - \alpha|\varepsilon_\tau|) + \sqrt{\frac{P}{2}}(1 - \alpha\delta). \quad (19)$$

Therefore, using the normalization expressions provided above, as well as the normalization factors that relate to discriminator gain, the expressions for the normalized VEMLP, HRC, and DP discriminators can be written as

$$D_{VEMLP_n} = \frac{2 - 3\alpha\delta + (5/4)\alpha^2\delta^2}{3\alpha\delta - 4}\frac{D_{VEMLP}}{N_{VEMLP}}, \quad (20)$$

$$D_{DP_n} = \frac{(2 - \alpha\delta)}{2\alpha}\frac{D_{DP}}{N_{DP}}, \quad (21)$$

$$D_{HRC_n} = \frac{2\alpha\delta - 3}{3\alpha}\frac{D_{HRC}}{N_{HRC}}. \quad (22)$$

These expressions are obtained from (18), (19), and (20) using the analytical expression for the autocorrelation function and obtaining the slope of the autocorrelation peak at $\alpha = -\delta/2$. Normalization factors were obtained analytically and tested empirically. They include normalization by the discriminator gain obtained by derivation of the discriminator output. Whereas normalization is straightforward for DP discriminator, for VEMLP and HRC discriminators, the following approximations should be used: $\alpha^2\varepsilon_\tau^2 \sim 0$ and $\alpha\varepsilon_\tau \sim 0$. These are valid for small code tracking errors, while the tracking is maintained.

These factors are constant for BOC(1,1) and BPSK signals, but for CBOC they vary depending on the E-L correlator distance. In both tracking steps, the correlators are

TABLE 3: Discriminator normalization factor for the two-step tracking scheme (E-L = 1/12, VE-VL = 1/2).

Discriminator normalization factors		
Discriminator type	Pilot channel	Data channel
DP	0.32	0.28
VEMLP	0.6	0.55
HRC	0.25	0.35

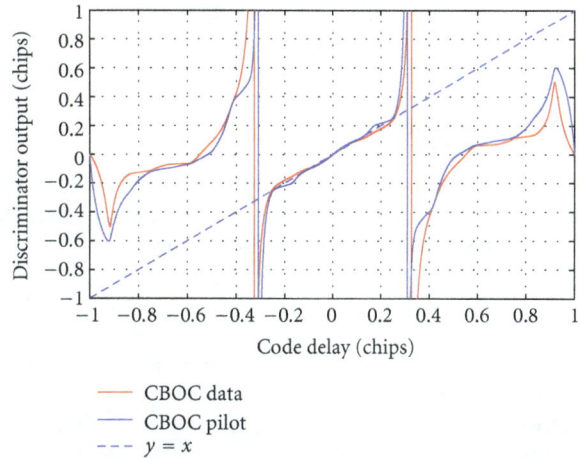

FIGURE 3: Normalized DP discriminator output for CBOC data and pilot channels and an E-L spacing of 1/12 chips.

located on the linear part of the discriminator curve, and the factors can be considered constant for fixed E-L spacing, as shown in Table 3. This table contains the discriminator normalization factors for the combined data/pilot tracking and the slope of the autocorrelation curve as provided in Table 1.

The discriminator outputs for DP and VEMLP discriminators are shown in Figures 3 and 4. The most important point that can be observed from these figures is that, by applying normalization, the discriminator curve's linear tracking region increases. Also, the discriminator curves for the pilot channel alone and the VEMLP discriminator (Figure 4) contain additional ripples. These ripples are due to high-frequency components introduced by BOC(6,1) modulation. As expected, the discriminators provide an unbiased response for input code errors below roughly 0.25 chips. For input code errors greater than that, the discriminators' output tends to flatten. This, in turn, should provide a smooth convergence phase. Since VEMLP offers only limited noise and multipath mitigation capabilities, it is therefore used directly after acquisition to ensure convergence of the code tracking loop to the main peak of the autocorrelation function. Upon convergence of both the code and carrier tracking loops, the tracking software switches to the fine tracking step.

As was the case for the VEMLP discriminator, the overall behavior of the DP discriminator is invariant across the two

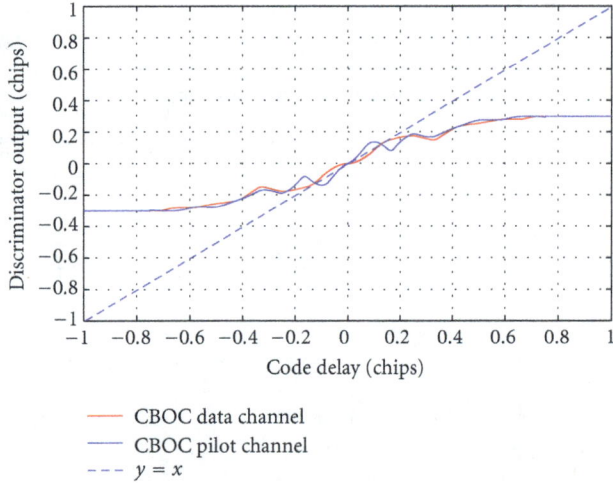

FIGURE 4: Normalized VEMLP discriminator output for CBOC data and pilot channels and an E-L spacing of 1/2 chips.

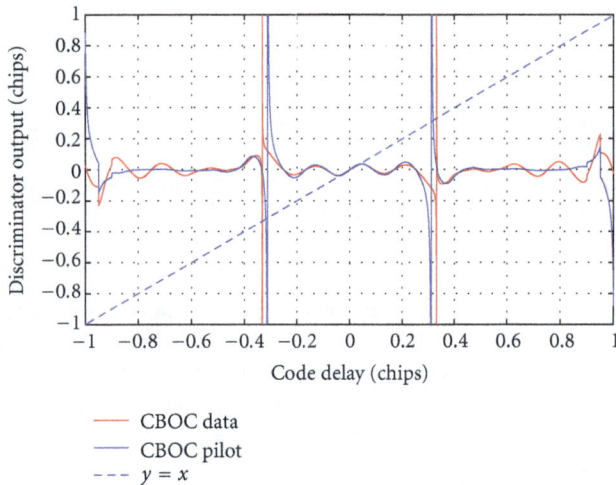

FIGURE 5: Normalized HRC discriminator output for CBOC data and pilot channels and an E-L spacing of 1/12 chips.

MBOC modulations apart from the additional ripples that can be observed on the two CBOC discriminators. Figure 3 also highlights the false-lock point issue inherent to all BOC and CBOC modulations (e.g., [3]). Indeed, it is obvious from this figure that the zero-crossing observed at approximately 0.5 chips would provide a stable lock point which would lead to a ranging bias of 150 m. The HRC normalized discriminator output is shown in Figure 5. It can be seen that

it has a much narrower linear tracking region in comparison to DP and VEMLP.

Again, normalization brings improvement by increasing the linearity region of the discriminator, making tracking more robust. However, the stability region remains similar to the unnormalized case. Normalization thus reveals the true code error present at the output of the discriminator. To conclude this section, the main advantage of using normalization is that the discriminator will perfectly react to a wide range of code delays. Both DP and VEMLP discriminators have similar widths of the linearity and stability regions before performing normalization. After normalization, the DP discriminator shows better stability and linearity (approximately 5% and more). Also, the normalized DP discriminator has vertical asymptotes. This means that the loop will overreact for large errors, and this can cause sudden tracking jumps that might lead to a false lock or degraded tracking.

So far, tracking loop and code discriminators have been described assuming no external disturbances. Since the DLL tracking loop can be affected by different sources of disturbances, such as thermal noise, multipath and signal dynamics, our next step is to investigate the influence of these error sources on the overall behavior of the tracking loop.

4. Code Tracking Error and Tracking Threshold

Code tracking error besides multipath and dynamic stress error is the dominant source of range errors in a GNSS receiver's DLL [19, 20]. Using carrier aiding, the dynamic stress error is negligible; therefore, only the error induced by thermal noise will be considered. The code tracking threshold can be derived using an analytical expression for the thermal noise-induced CBOC code tracking error variance. For the DP discriminator used in the fine tracking step, it is given in [3, 21], whereas, for the VEMLP discriminator, it is derived below. Analytical expressions for both discriminators can be written as

$$\sigma^2_{(DP,VEMLP,HRC)} = \frac{2B_L(1 - 0.5B_L T_I)S_N(0)}{K_{(DP,VEMLP,HRC)}}. \quad (23)$$

$S_N(0)$ is the noise power spectral density, and $K_{(DP,VEMLP,HRC)}$ represents the discriminator loop gain given by

$$K_{(DP,VEMLP,HRC)} = \left. \frac{dD_{(DP,VEMLP,HRC)}}{d\varepsilon_\tau} \right|_{\varepsilon_\tau=0}. \quad (24)$$

Combining this equation with the expressions for the correlator outputs provided in Section 3.3 and arranging them using the following equality for the filtered version of the autocorrelation function, we derived theoretical expressions for the code tracking errors for both DP and VEMLP discriminators for infinite front-end bandwidth:

$$\sigma^2_{VEMLP}$$
$$= \frac{B_L(1 - 0.5B_L T_I)\left(\left(\tilde{R}_F(0) - \tilde{R}_F(\delta)\right)\tilde{R}_F(\delta/2)\left(\tilde{R}_F(\delta/2) + \tilde{R}_F(\delta)\right) + \left(\tilde{R}_F(0) - \tilde{R}_F(2\delta)\right)\tilde{R}_F^2(\delta)\right)}{2(P/N_0)\left(\tilde{R}_F(\delta)\left(d\tilde{R}_F(x)/dx\right)\big|_{x=\delta} + \tilde{R}_F(\delta/2)\left(d\tilde{R}_F(x)/dx\right)\big|_{x=\delta/2}\right)^2} \cdot \theta,$$

$$\sigma_{\text{DP}}^2 = \frac{B_{\text{L}}(1 - 0.5B_{\text{L}}T_I)\left(\widetilde{R}_F(0) - \widetilde{R}_F(\delta)\right)}{2(C/N_0)\left\{d\widetilde{R}_F(x)/dx\right\}^2\Big|_{x=d/2}} \cdot \chi.$$

(25)

Expressions $\alpha_1 = (d\widetilde{R}_F(x)/dx)|_{x=\delta}$ and $\alpha_2 = (d\widetilde{R}_F(x)/dx)|_{x=\delta/2}$ represent the slopes of the autocorrelation function at E-L spacings δ and $\delta/2$. Expressions for θ and χ can be approximated as

$$\theta\left(\frac{\text{P}}{N_0}, T_I\right)$$
$$= 1 + \frac{3\left(\widetilde{R}_F^2(0) - \widetilde{R}_F^2(\delta)\right) + \widetilde{R}_F^2(0) - \widetilde{R}_F^2(2\delta)}{(2CT_I/N_0)\left(\left(\widetilde{R}_F(0) - \widetilde{R}_F(\delta)\right)\widetilde{R}_F(\delta/2)\left(\widetilde{R}_F(\delta/2) + \widetilde{R}_F(\delta)\right) + \left(\widetilde{R}_F(0) - \widetilde{R}_F(2\delta)\right)\widetilde{R}_F^2(\delta)\right)},$$

(26)

$$\chi\left(\frac{\text{P}}{N_0}, T_I\right) = 1 + \frac{1}{(CT_I/N_0)\widetilde{R}_F(0)}.$$

B_L is DLL loop bandwidth, T_I is coherent integration time, δ is the early-late spacing, C is the incoming power of the carrier, N_0 is the thermal noise, and $\widetilde{R}_F$ is the filtered correlation function of the incoming signal and $\theta, \chi = f(C/N_0, T_I)$ are factors that are a function of the integration time and carrier-to-noise ratio. The expression for the VEMLP discriminator can be simplified, using the following equality:

$$\widetilde{R}_F(0) - \widetilde{R}_F(\delta) \simeq \widetilde{R}_F\left(\frac{\delta}{2}\right) - \widetilde{R}_F\left(\frac{3\delta}{2}\right).$$

(27)

This simplification is valid only for wide E-L correlator spacings (E-L = 1/2) that are used in the coarse tracking step. Using the values for the slope of the autocorrelation function α from Table 1 for different correlator spacings, the code tracking error for the CBOC fine and coarse tracking steps can be computed. From our analysis, we observed that VEMLP discriminator provides very high code tracking noise error (33 m for C/N_0 of 25 dB-Hz) for the basic configuration (E-L = 1/2). High code tracking noise is the price one pays in the coarse tracking step. Decreasing the inner correlator spacing (E-L) while keeping the outer spacing (VE-VL) to one chip decreases the code tracking noise as well.

When increasing the correlator spacing, the DP discriminator code tracking error increases, but less dramatically pace than is observed for the VEMLP discriminator. The DP code tracking noise is very similar for different spacings since the E-L spacing is small (around 1.7 m for C/N_0 of 25 dB-Hz). The code tracking error for an E-L distance of 1/20 is higher than for E-L distance of 1/12. This happens due to front-end filtering, which rounds the autocorrelation peak, increasing the noise. DP outperforms VEMLP for large correlator spacings. Since the first stage of the two-step tracking scheme is short, the large code tracking noise does not greatly influence the overall tracking performance.

As was previously shown in [2], the code tracking error is approximately $\sqrt{2}$ lower for combined pilot and data channels tracking than for single channel tracking. Therefore, the tracking threshold decreases using combined tracking schemes. An analysis of the tracking threshold is provided below, starting from the code tracking noise derivation. Following the analysis provided in [19], the resulting analytical expression for the tracking threshold is obtained, and it is equal to

$$\left(\frac{C}{N_0}\right)_{\text{ThDP}} = \frac{18 \cdot W_D\left(1 + \sqrt{1 + \delta^2/9W_D T_I R_{\text{CBOC}}}\right)}{\delta^2}.$$

(28)

In the same way, the resulting tracking threshold for the VEMLP discriminator can be written as

$$\left(\frac{C}{N_0}\right)_{\text{ThVEMLP}} = \frac{18 \cdot W_T\left(1 + \sqrt{1 + \delta^2/9W_T T_I Y}\right)}{\delta^2}.$$

(29)

Parameters W_T and Y depend on the slope of the autocorrelation function, integration time, and DLL parameters and can be written as

$$W_T = \frac{B_{\text{L}}(1 - 0.5B_{\text{L}}T_I)\left(\left(\widetilde{R}_F(0) - \widetilde{R}_F(\delta)\right)\widetilde{R}_F(\delta/2)\left(\widetilde{R}_F(\delta/2) + \widetilde{R}_F(\delta)\right) + \left(\widetilde{R}_F(0) - \widetilde{R}_F(2\delta)\right)\widetilde{R}_F^2(\delta)\right)}{2\left(\widetilde{R}_F(\delta)\left(d\widetilde{R}_F(x)/dx\right)\Big|_{x=\delta} + \widetilde{R}_F(\delta/2)\left(d\widetilde{R}_F(x)/dx\right)\Big|_{x=\delta/2}\right)^2},$$

$$Y = \frac{2\left(\left(\widetilde{R}_F(0) - \widetilde{R}_F(\delta)\right)\widetilde{R}_F(\delta/2)\left(\widetilde{R}_F(\delta/2) + \widetilde{R}_F(\delta)\right) + \left(\widetilde{R}_F(0) - \widetilde{R}_F(2\delta)\right)\widetilde{R}_F^2(\delta)\right)}{3\left(\widetilde{R}_F^2(0) - \widetilde{R}_F^2(\delta)\right) + \widetilde{R}_F^2(0) - \widetilde{R}_F^2(2\delta)}.$$

(30)

Following our analysis that cannot be fully presented here, the tracking threshold decreases as the integration time increases and DLL bandwidth decreases. For very long integration durations, extremely low C/N_0 values can be tracked, such as 10 dB-Hz. What was observed is that the results for tracking threshold for two types of discriminators DP and VEMLP are very similar. For VEMLP, as the inner correlator spacing decreases, the tracking threshold decreases as well whereas, for coarse tracking step, does not drop below 20 dB-Hz. For DP, the trend is the same except the fact that for E-L spacing of 1/20 the threshold is increased, due to the rounding of autocorrelation peak.

5. Multipath Mitigation Analysis

Multipath represents a phenomenon that disturbs tracking causing phase offsets in code and carrier and should thus be properly mitigated. For automotive applications in urban environments, the multipath conditions will change continuously, making multipath one of the most disturbing problems of GNSS-based navigation systems in urban environments. As already analyzed and shown in [22], carrier multipath is not critical, causing maximum absolute multipath error of only 0.015 m for E1 CBOC. Code multipath represents a much more important issue and will be discussed here.

5.1. Code Multipath Analysis. Multipath is the main factor that affects the pseudoranges, and it does so through parameters such as the number of multipath signals, geometric path delay, phase, and relative power. Pseudorange measurement is performed by the code tracking loop, which aligns the locally generated PRN code with the PRN code of the incoming signal. The impact of multipath on code tracking accuracy is often represented as an error envelope representing the maximum error resulting from a single multipath with a certain phase, delay, and amplitude. This is called multipath error envelope (MEE). MEEs are computed for each discriminator type in the two-step tracking scheme, as already shown in [1], and for both steps and analyzed the optimal solution in order to mitigate or severely minimize the multipath error.

The first step, a coarse tracking step, uses a noncoherent VEMLP discriminator with wide correlator spacings. The resulting MMEs for the CBOC and BOC(1,1) signals are high MMEs, going up to 25 m as shown in [1]. Large multipath error is the price paid for reliable tracking in the coarse tracking step. As already explained in [1], the E-L correlator spacing should be as low as possible since the multipath error rapidly increases for spacings beyond 1/20 chips. For an E-L spacing of 1/12 chips, the MEE is close to that obtained for 1/20 chips but the linear region is wider which implies more robust tracking in the fine tracking step. For wider E-L spacings between the performances of the CBOC and BOC(1,1) are almost equal. As it it shown in [1], it is clear that CBOC offers a better resistance to long delay multipath than BOC(1,1) tracking. However, for short delays (<8 m), it leads to the same multipath envelope. An additional way to

mitigate multipath is to lower the loop bandwidth as much as possible (0.5 Hz). This way, the DLL will not be able to track the multipath-induced error, resulting in more accurate tracking.

5.2. Proposed Fine Tracking Step Implementation. The use of an HRC discriminator is proposed in the fine tracking step of two-step tracking algorithm. A potential issue with HRC is the limited stability tracking region, as shown in Section 3.3. This can be minimized through the implementation of carrier aiding which reduces the dynamics experienced by the discriminator. Since the HRC degrades the postcorrelation SNR and since the coherent integration time on the data channel is limited to 4 ms, it appears that implementing the HRC on the data channel would not provide a very robust solution. However, in the absence of unknown bit transition and with the help of carrier aiding, the coherent integration time can easily be extended to a couple hundred milliseconds on the pilot channel [3]. Consequently, implementation of the HRC on the pilot channel is the desired configuration.

Given the previous discussion, we proposed a CBOC tracking algorithm that combines a two-step tracking technique with the HRC on the pilot channel to lower the multipath error. Different correlator spacing configurations are used on the data and pilot channels. The data channel uses the configuration defined for the fine code tracking, that is, $[-1/2; -1/24; 0; 1/24; 1/2]$, and the pilot channel uses the HRC configuration defined as $[-1/12; -1/24; 0; 1/24; 1/12]$, where the spacings follow the pattern $[-2\delta_{HRC}; -\delta_{HRC}; 0; \delta_{HRC}; 2\delta_{HRC}]$ and δ_{HRC} is the E-L correlator spacing. The scheme then consists of using the data channel to perform false lock detection and, in case no false lock is detected, using the pilot channel to close the tracking loop, which may use HRC discriminator depending on the following two scenarios.

(1) *High Multipath Case.* Tracking using the HRC discriminator is desired. In this case, the use of the DP discriminator is not recommended as it would feed the combined discriminator with its multipath-induced error.

(2) *Reliably Critical Case.* Tracking using the DP discriminator is desired. In this case, the use of the HRC should be avoided so as to avoid integrating the potentially unreliable HRC discriminator output into the combined discriminator output.

The pilot channel should therefore implement an "either/or" discriminator combination that would alternately rely on the DP or the HRC discriminator. It was shown in [1] that the MEE does not exceed 0.01 chips for the considered E-L spacing. The discriminator output should not be greater than this if the tracking is correctly achieved. Therefore, to decide which discriminator to use, a test based on the DP correlator output is run. If the code tracking error is small (i.e., the DP discriminator output is in its linear tracking range), then the HRC is used. Otherwise, the tracking relies on the DP because of its wider stability range (0.3 chips in one direction). To summarize, the pilot discriminator output can be written as

$$D_{\text{pilot}} = D_{\text{DP}}, \quad \text{if } D_{\text{DP}} \geqslant \delta_{\text{HRC}},$$

$$D_{\text{pilot}} = D_{\text{HRC}}, \quad \text{if } D_{\text{DP}} \leqslant \delta_{\text{HRC}}.$$

$$(31)$$

The choice of threshold δ_{HRC} could be made to depend on C/N_0. This approach would bring robustness and will be investigated in the future. When higher C/N_0 is available, strong multipath mitigation could be achieved, while for lower C/N_0 the tracking would remain robust.

6. Results and Discussion

After providing the overview of the theoretical performance of the two-step tracking scheme, we present in this section further the simulation results in order analyze if they are consistent with the theoretically derived results. The data collection setup consists of a Spirent GSS8000 simulator [17] to emulate the Galileo E1 OS signal, a Fraunhofer triple-band front-end [18] with 18 MHz of bandwidth to filter, downconvert, and digitize the signal of interest (i.e., E1 CBOC signal), and a postprocessing architecture that contains a software receiver implemented in MATLAB, based on a modified version of the Kai Borre GPS L1 defined software radio [20].

6.1. Simulation-Based Tracking Results. To test the proposed tracking scheme, we created different scenarios. Using the Spirent simulator's build-in multipath simulation models, we tested two types of multipath models: first type of simple multipath models, that adds a multipath ray with 0.5 chips delay and 6 dB of attenuation to the direct signal, and in the second case, a multipath with a fixed delay of 100 m is added, since this represents the point at which multipath error is the highest. Third model adds three rays that are to the direct signal with delays of 0.5, 0.3, and 0.1 chips with the same attenuation of 6 dB. Fourth group of multipath models tested was based on Land Mobile Multipath (LMM) model, statistical model that emulates direct paths and reflected paths as Rician or Rayleigh, depending on the different type of environment: Urban, Suburban, and Urban Canyon scenarios were tested. Also, two different C/N_0 configurations were investigated: low (24 dB-Hz) and high (45 dB-Hz), as well as static and mobile receiver.

The outputs from the I correlator of the pilot channel for simple multipath model that assumes one reflected path with the delay of 100 m are shown in Figure 6 for all five correlators. The most important point to observe is that all outputs stabilize after approximately 300 code periods from the beginning of tracking. After this point, the tracking is stable. The transition from coarse to fine tracking happens around 1500 code periods, when it is assured that the main autocorrelation peak is being tracked. We could observe that late correlator has higher power at the beginning. The condition for transition from coarse to fine tracking step is related to the PLI (Phase Lock Indicator). If the PLI is less than 0.9 (0.7 in case of low C/N_0), the coarse tracking step

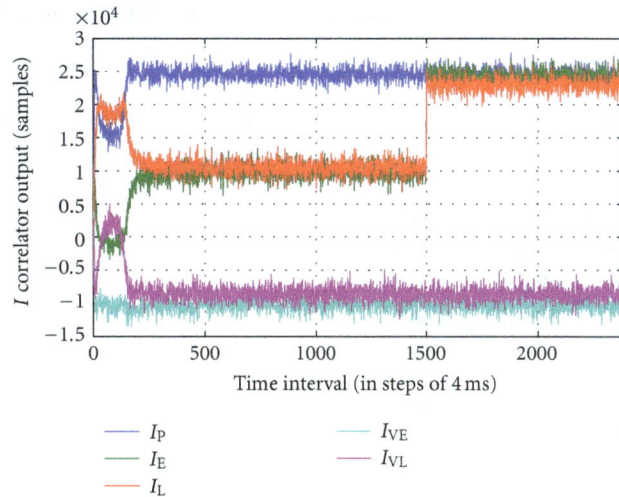

FIGURE 6: I correlation output for two-step tracking scheme with multipath model with fixed offset delay of 100 m.

proceeds. The transition from coarse to fine tracking step happens only when this value is exceeded. In the Figure 6, the threshold is achieved earlier, but the transition happens later in order to provide a better overview of the scheme.

The DLL outputs for the six different scenarios described above are shown in Figure 7. Note the different convergence times for some cases. Due to the deformation of the autocorrelation main peak, it takes more time for the DLL to converge when multipath is present, than when no multipath is present. Overall, approximately 500 periods are needed (or 2000 ms) to lock to the right correlation peak and then switch to the fine tracking step. The PLI indicators of the two multipath cases analyzed have multiple peaks lower than 0.5 until locked to the right correlation peak.

Once it is assured that the right correlation peak is being tracked, the fine tracking step turns on. Table 4 compares the fine and coarse tracking steps in terms of DLL output's standard deviation. It can be seen that the standard deviation drops by more than a factor of five when switching from coarse to fine tracking for all cases under investigation. It is also important to mention that the tracker succeeds to follow the right autocorrelation peak in all six cases. Peaks in the DLL output are present due to transition from coarse to fine step, as well as due to transition from HRC to DP in the fine tracking step. The PLL discriminator output is shown in Figure 8. It can be noted that variation of PLL error is similar for all cases considered, because carrier tracking is performed independently on the pilot channel only, and the frequency lock is well maintained. Only the secondary code was wiped off according to [12].

It has to be mentioned that in the fine tracking step a bump-jump algorithm was implemented, as previously proposed in [4]. Amplitude comparison was accomplished by a simple up/down counter mechanism. The absolute values of the VE, VL, and P correlators were compared and, if either VE or VL sample is the largest, the appropriate counter was increased and the other decreased. When the set

TABLE 4: Standard deviation of the DLL outputs for the two-step tracking scheme using six different scenarios.

| | DLL standard deviation | |
Scenario	Coarse tracking step	Fine tracking step
Low C/N_0 simple multipath model	0.0585	0.0124
Mobile receiver no multipath	0.0192	0.0035
One multipath ray of 0.5 chips delay	0.0169	0.0045
One multipath ray of 100 m delay	0.0180	0.0037
Land multipath model Urban Canyon	0.0154	0.0033
High C/N_0 no multipath	0.0134	0.0022

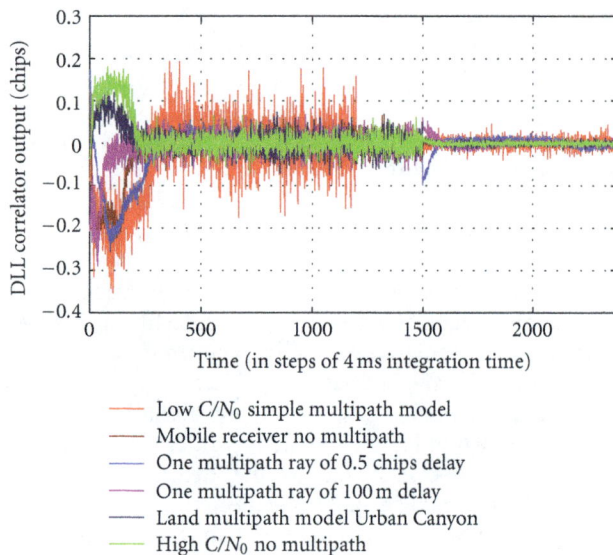

FIGURE 7: DLL loop output for the two-step tracking algorithm and six investigated scenarios.

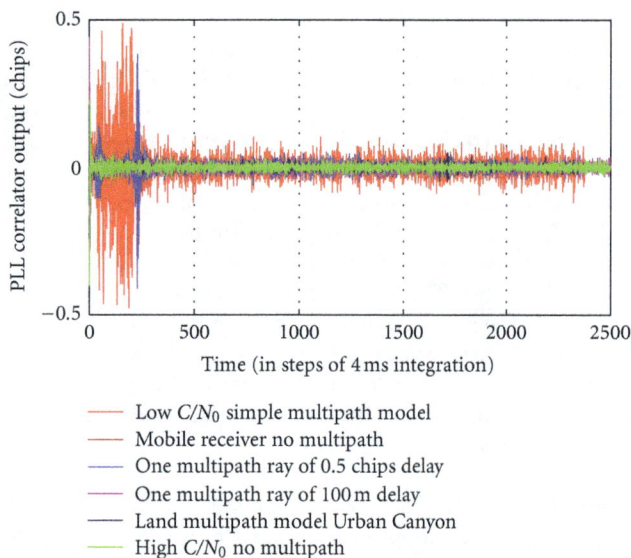

FIGURE 8: PLL loop output for the two-step tracking algorithm and different multipath scenarios.

threshold was exceeded, the jump to the new peak occurred. Here, we set the threshold to 50, which meant that if the VE or VL was larger than the P for 50 consecutive periods (or 200 ms), the tracking jumped to a new peak.

6.2. Galileo Real Data Tracking Results. On the 12th of December 2011, two Galileo in-orbit validation (IOV) satellites that were launched on 16th of October 2011, PFM and PM-FM2, started transmitting Galileo Open Service signals with pseudorandom code numbers SV11 and SV12 [9]. From the end of December 2011, the signals are available when the satellites are in the direct visibility domain of the receiver. Using the setup described above, we were able to receive signals from both satellites and successfully track them. We tested the two-step tracking scheme using real signals, and we show the first results here.

The setup differs from the simulation setup described above that differs only in the antenna. Instead of using the Spirent simulator output connected to the front-end receiver, we used a fixed, nondirectional rooftop wideband Antcomm antenna, connected via a two-stage amplifier to a front-end. The setup was located at our premises, in IMT Neuchatel, at latitude: 46.519617 degrees and longitude: 6.63221 degrees.

Digital *IF* (*IF* = 12.82 MHz) samples were stored in the memory and subsequently postprocessed, using the software receiver described earlier, on the evening of March 21st, 2012 when the satellites were in good view of the antenna. During recording, the elevation of the Galileo-PMF satellite (SV11) was 45.5 degrees and the elevation of the Galileo-FM2 satellite (SV12) 65.2 degrees.

The acquisition peak acquired in 3D space for satellite SV12 is shown in Figure 9. We can clearly observe the Galileo signal as the correlation peak rising above the noise level. After acquiring the signal, secondary code wipe-off was performed, and the values for the Doppler offset and code offset of the secondary code were passed to the tracking stage.

Correlator outputs for all five correlators are shown in Figure 10. The outputs are stable, and a clear transition from coarse to fine tracking is observed. The DLL discriminator output for both satellites is shown in Figure 11. The output is noisy in the coarse tracking step, but in the fine tracking step it is stable with a much smaller standard deviation and tracking is furthermore stable.

Once the receiver is in tracking mode, C/N_0 at the receiving antenna can be estimated. Approximately 20 code periods (around 100 ms) were used for the estimation of

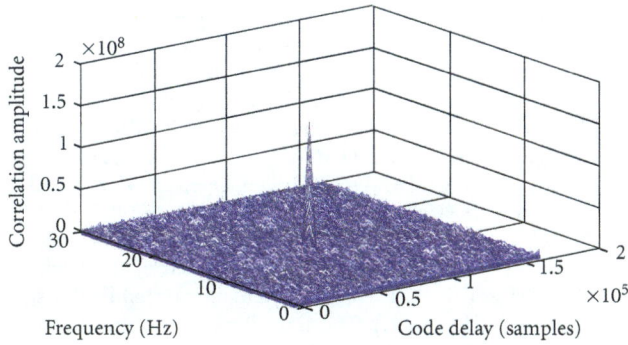

FIGURE 9: 3D acquisition peak acquired from the Galileo PFM satellite.

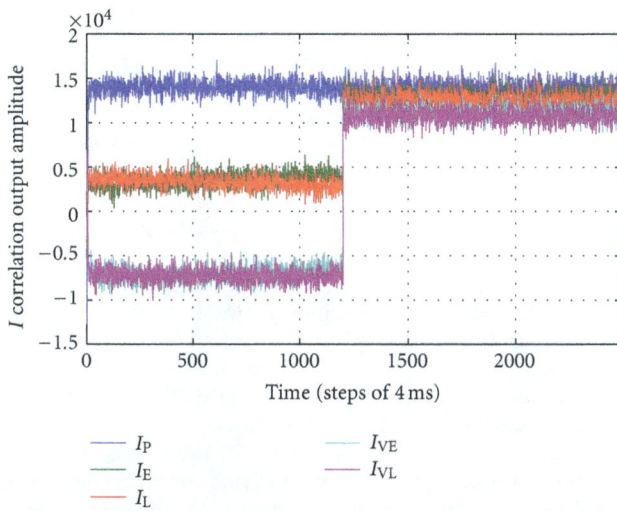

FIGURE 10: I correlator outputs for the two-step tracking algorithm using real data from Galileo FM2 and PFM satellites.

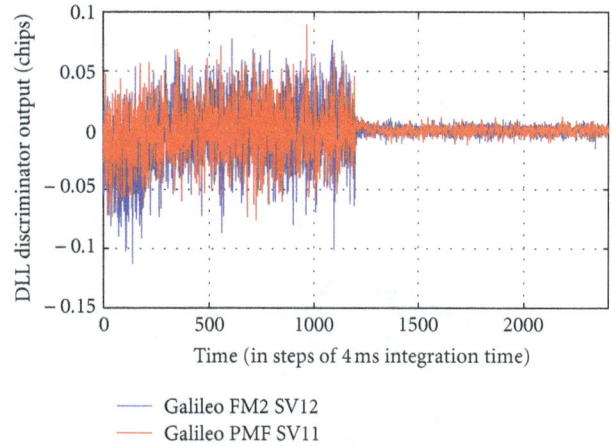

FIGURE 11: DLL loop output for the two-step tracking algorithm using real data from Galileo FM2 and PFM satellites.

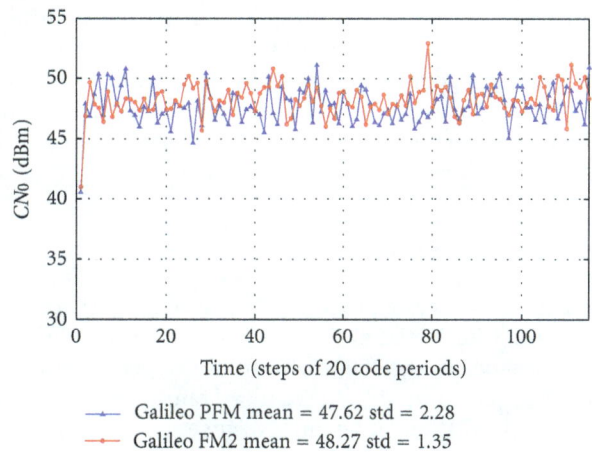

FIGURE 12: Estimated C/N_0 for the two-step tracking algorithm using real data from Galileo FM2 and PFM satellites.

the C/N_0. The estimated ratio is shown in Figure 12 for both satellites and is very similar C/N_0 profiles for both satellites. Note that the displayed C/N_0 relates to both the data and pilot channels. The single channel C/N_0 is around 3 dB lower. Measured C/N_0 is relatively high due to the clear view of the satellites and good position of the satellites.

7. Conclusions

This paper provided both theoretical and experimental study of the previously proposed two-step tracking algorithm for Galileo E1 CBOC tracking. The proposed algorithm has two independent tracking steps that make the algorithm flexible enough to be easily adaptable to changing signal environments. It was shown that this approach provides robust, accurate, and reliable tracking since the coarse step ensures convergence to the main autocorrelation peak while the fine step minimizes noise and multipath errors.

The analytical expression for the CBOC autocorrelation function was analyzed, as it is more complex than the BOC(1,1) autocorrelation function, and cannot be considered to be fully linear. Depending on the E-L correlator

spacing, the slope of the autocorrelation function varies in the range of 1–6, giving different code tracking noise behavior. Optimal correlator spacings were discussed (VE-VL = 1/2, E-L = 1/12), in order to minimize tracking errors. Still, scheme remains flexible, and, depending on the available front-end bandwidth, tracking parameters may vary.

The analytical expression for VEMLP code tracking error is derived, and this is one of the main contributions of this paper, since it was not provided before according to our best knowledge. Using this expression, it was shown that, although the code tracking error in the first step is very high ($\sim$35 m), stable lock is provided in the first tracking step, and then it decreases exponentially (down to 1.8 m) in the fine tracking step.

We also showed experimentally using theoretical derivations that the two-step tracking algorithm has good tracking sensitivity in both tracking steps and can be used to track CBOC signals with low C/N_0 ratio without losing lock. The tracking stability was analyzed as well in form of discriminator curves analysis. It was shown that the proposed scheme does not lose lock easily since the risk of biased tracking is circumvented. Since the two-step tracking algorithm assumes the usage of three discriminator types (VEMLP, DP, and HRC), each of them were analyzed separately and the normalization parameters optimized in order to improve the two-step tracking algorithm.

The main source of errors, including multipath and thermal noise, were evaluated, and optimal algorithm parameters (shown in Table 2) were obtained through this evaluation, but still staying flexible to different front-end bandwidths, correlator spacing, and so forth. High multipath error (up to 20 m) in the first tracking step is compensated in the fine tracking step using a combination of DP and HRC discriminators, reducing multipath error down to less than a meter. It was shown by simulations that the proposed scheme is able to well mitigate multipath in different scenarios, including both statistical and theoretical models.

Finally, the parameters of the algorithm were further optimized and tested in different configurations through simulations using a Spirent GSS8000 simulator. The two-step tracking algorithm was also implemented in a software receiver, and performance assessments were conducted using real data recorded from the recently launched Galileo PFM and FM2 satellites.

Acknowledgments

The authors are grateful for the financial support received from the Swiss National Science Foundation (http://www.snf.ch/) under Grant 200020 134766/1, as well as from the European Community's Seventh Framework Program (FP7/2007–2013) under Grant Agreement no. 228339.

References

[1] A. Jovanovic, C. Mongrédien, C. Botteron, Y. Tawk, G. Rohmer, and P. A. Farine, "Requirements and analysis for a robust E1 Galileo tracking algorithm in the scope of the GAMMA-A project," in *Proceedings of the Institute of Navigation—International Technical Meeting 2010 (ION ITM '10)*, pp. 973–985, San Diego, Calif, USA, January 2010.

[2] D. Borio, C. Mongrédien, and G. Lachapelle, "Collaborative code tracking of composite GNSS signals," *IEEE Journal on Selected Topics in Signal Processing*, vol. 3, no. 4, pp. 613–626, 2009.

[3] O. Julien, *Design of Galileo L1F receiver tracking loops [Thesis]*, University of Calgary, 2005.

[4] P. Fine and W. Wilson, "Tracking algorithm for GPS offset carrier signals," in *Proceedings of the US Institute of Navigation NTM Conference*, San Diego, Calif, USA, January 1999.

[5] M. S. Hodgart, P. D. Blunt, and M. Unwin, "Double estimator—a new receiver principle for tracking BOC signals," Inside GNSS, pp. 26–36, 2008.

[6] M. S. Hodgart, R. M. Weiler, and M. Unwin, "A triple estimating receiver of multiplexed binary offset carrier (MBOC) modulated signals," in *Proceedings of the 21st International Technical Meeting of the Satellite Division of the Institute of Navigation (ION GNSS '08)*, pp. 2295–2304, Savannah, Ga, USA, September 2008.

[7] O. Julien and C. Macabiau, "Two for one—tracking Galileo CBOC signal with TMBOC," Inside GNSS, 2007.

[8] O. Julien and C. Macabiau, "On potential CBOC/MBOC common receiver arcitectures," in *Proceedings of the International Technical Meeting of the Satellite Division of The Institute of Navigation (ION GNSS ITM '07)*, p. 1843, ENAC, 2007.

[9] Galileo Joint Undertaking, Galileo Open Service Signal in Space InterfaceControl Document, GAL OS SIS ICD, Draft 1, 2010.

[10] A. Jovanovic, C. Mongrédien, C. Botteron, Y. Tawk, G. Rohmer, and P. A. Farine, "Implementation and robustness analysis of the two-step CBOC tracking algorithm in the scope of the GAMMA-A project," in *Proceedings of the Proceedings of the ENC GNSS*, Braunscheig, Germany, October 2010.

[11] A. Jovanovic, C. Mongrédien, C. Botteron, Y. Tawk, G. Rohmer, and P. A. Farine, "Implementation and optimization of a Galileo E1 two-step tracking algorithm using data/pilot combining and extended integration time," in *Proceedings of the 24th International Technical Meeting of The Satellite Division of the Institute of Navigation (ION GNSS '11)*, Portland, Ore, USA, September 2011.

[12] Y. Tawk, A. Jovanovic, J. Leclere, C. Botteron, and P. A. Farine, "A new FFT-based algorithm for secondary code acquisition for Galileo signals," in *Proceedings of the IEEE Vehicular Technology Conference (VTC '11)*, San Francisco, Calif, USA, September 2011.

[13] P. B. Anantharamu, D. Borio, and G. Lachapelle, "Pre-filtering, side-peak rejection and mapping: several solutions for unambiguous BOC tracking," in *Proceedings of the 22nd International Technical Meeting of the Satellite Division of the Institute of Navigation 2009 (ION GNSS '09)*, pp. 3182–3195, Savannah, Ga, USA, September 2009.

[14] J. C. Juang and T. L. Kao, "Generalized discriminator and its applications in GNSS signal tracking," in *Proceedings of the 23rd International Technical Meeting of the Satellite Division of the Institute of Navigation 2010 (ION GNSS '10)*, pp. 3251–3257, Portland, Ore, USA, September 2010.

[15] C. Mongrédien, M. Overbeck, and G. Rohmer, "Development and integration of a robust signal tracking module for the triple-frequency dual-constellation GAMMA-A receiver," in *Proceedings of the 23rd International Technical Meeting of the Satellite Division of the Institute of Navigation 2010 (ION GNSS '10)*, pp. 2808–2819, Portland, Ore, USA, September 2010.

[16] M. S. Braasch and G. A. Mcgraw, "GNSS multipath mitigation using gated and high resolution correlator concepts," in *Proceedings of the National Technical Meeting of the Satellite Division of the Institute of Navigation (ION NTM '99)*, pp. 333–342, San Diego, Calif, USA, January 1999.

[17] http://www.spirent.com/Solutions-Directory/GSS8000.aspx.

[18] http://www.iis.fraunhofer.de/.

[19] E. D. Kaplan and C. J. Hegarty, *Understanding GPS—Principles and Applications*, Artech House, 1996.

[20] N. Bertelsen and K. Borre, *A Software Defined GPS and Galileo Receiver*, Birkhauser, Boston, Mass, USA, 2007.

[21] J. W. Betz and K. R. Kolodziejski, "Generalized theory of code tracking with an early-late discriminator part II: noncoherent processing and numerical results," *IEEE Transactions on Aerospace and Electronic Systems*, vol. 45, no. 4, article 1551, 2009.

[22] A. Jovanovic, Y. Tawk, C. Botteron, and P. A. Farine, "Multipath mitigation techniques for CBOC, TMBOC and AltBOC signals using advanced correlators architectures," in *Proceedings of the IEEE/ION Position, Location and Navigation Symposium (PLANS '10)*, pp. 1127–1136, Palm Springs, Calif, USA, May 2010.

4

A Comparison of Parametric and Sample-Based Message Representation in Cooperative Localization

Jaime Lien,[1] Ulric J. Ferner,[2] Warakorn Srichavengsup,[3] Henk Wymeersch,[4] and Moe Z. Win[5]

[1] Department of Electrical Engineering, Stanford University, Stanford, CA 94305, USA
[2] Research Laboratory of Electronics, Massachusetts Institute of Technology (MIT), Cambridge, MA 02139, USA
[3] Department of Computer Engineering, Thai-Nichi Institute of Technology, Bangkok 10250, Thailand
[4] Department of Signals and Systems, Chalmers University of Technology, Gothenburg 412 96, Sweden
[5] Laboratory for Information and Decision Systems, Massachusetts Institute of Technology (MIT), Cambridge, MA 02139, USA

Correspondence should be addressed to Jaime Lien, jlien@stanford.edu

Academic Editor: Elena Lohan

Location awareness is a key enabling feature and fundamental challenge in present and future wireless networks. Most existing localization methods rely on existing infrastructure and thus lack the flexibility and robustness necessary for large ad hoc networks. In this paper, we build upon SPAWN (sum-product algorithm over a wireless network), which determines node locations through iterative message passing, but does so at a high computational cost. We compare different message representations for SPAWN in terms of performance and complexity and investigate several types of cooperation based on censoring. Our results, based on experimental data with ultra-wideband (UWB) nodes, indicate that parametric message representation combined with simple censoring can give excellent performance at relatively low complexity.

1. Introduction

Location awareness has the potential to revolutionize a diverse array of present and future technologies. Accurate knowledge of a user's location is essential for a wide variety of commercial, military, and social applications, including next-generation cellular services [1, 2], sensor networks [3, 4], search-and-rescue [5, 6], military target tracking [7, 8], health care monitoring [9, 10], robotics [11, 12], data routing [13, 14], and logistics [15, 16]. Typically, only a small fraction of the nodes in the network, known as *anchors*, have prior knowledge about their location. The remaining nodes, known as *agents*, must determine their locations through a process of localization or positioning. The ad hoc and often dynamic nature of wireless networks requires distributed and autonomous localization methods. Moreover, location-aware wireless networks are frequently deployed in unknown environments and hence can rely only on minimal (if any) infrastructure, human maintenance, and a priori location information.

Cooperation is an emerging paradigm for localization in which agents take advantage of network connections and interagent measurements to improve their location estimates. Non-Bayesian cooperative localization in wireless sensor networks is discussed in [17]. Different variations of Bayesian cooperation have been considered, including Monte-Carlo sequential estimation [18] and nonparametric belief propagation in static networks [19]. For a comprehensive overview of Bayesian and non-Bayesian cooperative localization in wireless networks, we refer the reader to [20], which also introduces a distributed cooperative algorithm for large-scale mobile networks called SPAWN (sum-product algorithm over a wireless network). This message-passing algorithm achieves improved localization accuracy and coverage compared to other methods and will serve as the basic algorithm in this paper.

The complexity and cost associated with the SPAWN algorithm depend largely on how messages are represented for computation and transmission. As wireless networks typically operate under tight power and resource constraints, the choice of message representation heavily impacts the feasibility and ease of implementation of the algorithm. The method of message representation and ensuing tradeoff between communication cost and localization performance are thus of great practical importance in the deployment of realistic localization systems. Particle methods do not necessarily lend themselves well in practice to be exchanged wirelessly between devices, due to their high computational complexity and communication overhead [21]. Other message-passing methods have been developed that rely on parametric message representation, thus alleviating these drawbacks but limiting representational flexibility. In particular, in [22], the expectation propagation algorithm is considered with Gaussian messages, while in [23], variational message passing with parametric messages is shown to exhibit low complexity. A variation of SPAWN combining GPS and UWB was evaluated in [24], using a collection of parametric distributions with ellipsoidal, conic, and cylindrical shapes.

This paper addresses the need for accurate, resource-efficient localization with an in-depth comparison of various message representations for SPAWN. We describe and evaluate different parametric and nonparametric message representations in terms of complexity and accuracy. Additionally, we analyze the performance of various cooperative schemes and message representations in a simulated large-scale ultra-wide bandwidth (UWB) network using experimental UWB ranging data. UWB is an attractive choice for ranging and communication due to its ability to resolve multipath [25, 26], penetrate obstacles [27], and provide high resolution distance measurements [28, 29]. Recent research advances in UWB signal acquisition [30, 31], multiuser interference [32, 33], multipath channels [34, 35], non-line-of-sight (NLOS) propagation [28, 29], and time-of-arrival estimation [36] increase the potential for highly accurate UWB-based localization systems in harsh environments. Consequently, significant attention has been paid to both algorithm design [37–42] and fundamental limits of accuracy [43–46] for UWB localization. It is expected that UWB will be exploited in future location-aware systems that utilize coexisting networks of sensors, controllers, and peripheral devices [47, 48].

2. Problem Formulation

We consider a wireless network of N nodes in an environment $\mathcal{E}$. Time is slotted with nodes moving independently from time slot to time slot. The position of node i at time t is described by the random variable $\mathbf{x}_i^{(t)}$; the vector of all positions is denoted by $\mathbf{x}^{(t)}$. At each time t, node i may collect internal position-related measurements $z_{i,\text{self}}^{(t)}$, for example, from an inertial measurement unit. The set of all internal measurements is denoted by $\mathbf{z}_{\text{self}}^{(t)}$. Within the network, nodes communicate with each other via wireless transmissions. We denote the set of nodes from which node i can receive transmissions at time t by $\mathcal{S}_{\rightarrow i}^{(t)}$. Note that the communication link may not be bidirectional; that is, $j \in \mathcal{S}_{\rightarrow i}^{(t)}$ does not imply $i \in \mathcal{S}_{\rightarrow j}^{(t)}$. Using packets received from $j \in \mathcal{S}_{\rightarrow i}^{(t)}$, node i may collect a set of relative measurements, represented by the

vector $z_{j \rightarrow i}^{(t)}$, which we will limit to distance measurements. We denote the set of all relative measurements made in the network at time t by $\mathbf{z}_{\text{rel}}^{(t)}$. The full set of relative and internal measurements is denoted $\mathbf{z}^{(t)}$.

The objective of the localization problem is for each node i to determine the a posteriori distribution $p(\mathbf{x}_i^{(t)} \mid \mathbf{z}^{(1:t)})$ of its position $\mathbf{x}_i^{(t)}$ at each time t, given information up to and including t.

3. A Brief Introduction to SPAWN

In [20], we proposed a cooperative localization algorithm by factorizing the joint distribution $p(\mathbf{x}^{(0:T)} \mid \mathbf{z}^{(1:T)})$, formulating the problem as a factor graph with temporal and spatial constraints, and applying the sum-product algorithm. This leads to a distributed algorithm, known as SPAWN, presented in Algorithm 1. The aim of SPAWN is to compute a *belief* $b_i^{(t)}(\mathbf{x}_i^{(t)})$ available to node i at the end of any time slot t, which serves as an approximation of the marginal a posteriori distribution $p(\mathbf{x}_i^{(t)} \mid \mathbf{z}^{(1:t)})$. Note that each operation of SPAWN requires only information local to an individual node. Information is shared between nodes via physical transmissions. Each node can therefore perform the computations in Algorithm 1 using its local information and transmissions received from neighboring nodes.

Observe that Algorithm 1 contains a number of key steps.

(i) *Mobility update* (line 4), requiring knowledge of mobility models $p(\mathbf{x}_i^{(t)} \mid \mathbf{x}_i^{(t-1)})$ and self-measurement likelihood functions $p(z_{i,\text{self}}^{(t)} \mid \mathbf{x}_i^{(t-1)}, \mathbf{x}_i^{(t)})$.

(ii) *Message conversion* (line 10) of position information from neighboring devices to account for relative measurements, requiring knowledge of the neighbors and of relative measurement likelihood functions $p(z_{j \rightarrow i}^{(t)} \mid \mathbf{x}_i, \mathbf{x}_j)$.

(iii) *Belief update* (line 11), to fuse information from the mobility update with information from the current neighbors.

The first two operations can be interpreted as *message filtering*, while the latter operation is a *message multiplication*. How these operations can be implemented in practice will be the topic of Section 4.

4. Message Representation

4.1. Key Operations. In SPAWN, probabilistic information is exchanged and computed through *messages*. The manner in which these messages are represented for transmission between nodes and internal computation is closely related to the complexity and performance of the localization algorithm. In traditional communications problems, such as decoding, messages can be represented efficiently and exactly through, for instance, log-likelihood ratios [49]. In SPAWN, exact representation is impossible, so we must resort to different types of approximate message representations. Any representation must be able to capture the salient properties of the true message and must enable efficient computation

---

(1) Initialize belief $b^{(0)}(\mathbf{x}_i^{(0)}) = p(\mathbf{x}_i^{(0)})$, $\forall i$

(1) **for** $t = 1$ to T **do** {time index}

(2)   **for all** i **do** {mobility update}

(5)   Mobility update:
$$\widetilde{b}_i^{(t)}(\mathbf{x}_i^{(t)}) \propto$$
$$\int p(\mathbf{x}_i^{(t)}|\mathbf{x}_i^{(t-1)})p(z_{i,\text{self}}^{(t)}|\mathbf{x}_i^{(t-1)},\mathbf{x}_i^{(t)})$$
$$\times b_i^{(t-1)}(\mathbf{x}_i^{(t-1)})\,d\mathbf{x}_i^{(t-1)} \qquad \text{(A-1)}$$

(6)   **end for**

(7)   Initialize $b_i^{(t)}(\mathbf{x}_i^{(t)}) = \widetilde{b}_i^{(t)}(\mathbf{x}_i^{(t)})$, $\forall i$

(8)   **for** $l = 1$ to N_{it} **do** {iteration index; begin cooperative update}

(9)   **for all** i **do**

(10)   **for all** $j \in S_{\rightarrow i}^{(t)}$ **do**

(11)   Receive and convert $b_j^{(t)}(\mathbf{x}_j^{(t)})$ to a distribution $c_{j \rightarrow i}^{(l)}(\mathbf{x}_i^{(t)})$:
$$c_{j \rightarrow i}^{(l)}(\mathbf{x}_i^{(t)}) \propto$$
$$\int p(z_{j \rightarrow i}^{(t)}|\mathbf{x}_i^{(t)},\mathbf{x}_j^{(t)})b_j^{(t)}(\mathbf{x}_j^{(t)})d\mathbf{x}_j^{(t)} \qquad \text{(A-2)}$$

(12)   Update and broadcast $b_i^{(t)}(\mathbf{x}_i^{(t)})$:
$$b_i^{(t)}(\mathbf{x}_i^{(t)}) \propto \widetilde{b}_i^{(t)}(\mathbf{x}_i^{(t)}) \prod_{k \in S_{\rightarrow i}^{(t)}} c_{k \rightarrow i}^{(l)}(\mathbf{x}_i^{(t)}) \qquad \text{(A-3)}$$

(13)   **end for**

(14)   **end for**

(15)   **end for** {end cooperative update}

(16) **end for** {end current time step}

ALGORITHM 1: SPAWN.

of the key steps in SPAWN, namely, message *filtering* (A-1)-(A-2) and message *multiplication* (A-3). We consider three types of message representation: discretized, sample-based, and parametric.

For convenience, we will introduce a set of new notations. For the filtering operation, the incoming message is denoted by $p_{\mathbf{X}}(\mathbf{x})$, the filtering operation by $h(\mathbf{x}, \mathbf{y})$, and the outgoing message by $p_{\mathbf{Y}}(\mathbf{y})$, with

$$p_{\mathbf{Y}}(\mathbf{y}) \propto \int h(\mathbf{x}, \mathbf{y})p_{\mathbf{X}}(\mathbf{x})d\mathbf{x}. \qquad (1)$$

For the multiplication operation, we assume M incoming messages $p_{\mathbf{X}}^{(i)}(\mathbf{x})$ ($i = 1, \ldots, M$) over a single variable $\mathbf{X}$, and an outgoing message

$$\phi_{\mathbf{X}}(\mathbf{x}) \propto \prod_{i=1}^{M} p_{\mathbf{X}}^{(i)}(\mathbf{x}). \qquad (2)$$

Note that (1) maps to (A-1) through the following association:

$$\mathbf{x} \longrightarrow \mathbf{x}_i^{(t-1)},$$
$$\mathbf{y} \longrightarrow \mathbf{x}_i^{(t)},$$
$$h(\mathbf{x}, \mathbf{y}) \longrightarrow p(\mathbf{x}_i^{(t)} \mid \mathbf{x}_i^{(t-1)})p(z_{i,\text{self}}^{(t)} \mid \mathbf{x}_i^{(t-1)}, \mathbf{x}_i^{(t)}), \qquad (3)$$
$$p_{\mathbf{X}}(\mathbf{x}) \longrightarrow b_i^{(t-1)}(\mathbf{x}_i^{(t-1)}).$$

Similarly, (1) maps to (A-2) through the following association:

$$\mathbf{x} \longrightarrow \mathbf{x}_j^{(t)},$$
$$\mathbf{y} \longrightarrow \mathbf{x}_i^{(t)},$$
$$h(\mathbf{x}, \mathbf{y}) \longrightarrow p(z_{j \rightarrow i}^{(t)} \mid \mathbf{x}_i^{(t)}, \mathbf{x}_j^{(t)}), \qquad (4)$$
$$p_{\mathbf{X}}(\mathbf{x}) \longrightarrow b_i^{(t)}(\mathbf{x}_i^{(t)}).$$

4.2. Discretized Message Representation. A naive but simple approach to represent a continuous distribution $p_{\mathbf{X}}(\mathbf{x})$ is to uniformly discretize the domain of $\mathbf{X}$, yielding a set of quantization points $Q = \{\mathbf{x}_1, \ldots, \mathbf{x}_R\}$. The distribution is then approximated as a finite list of values, $\{p_{\mathbf{X}}(\mathbf{x}_k)\}_{k=1}^{R}$. The filtering operation then becomes

$$p_{\mathbf{Y}}(\mathbf{y}_k) \propto \sum_{l=1}^{R} h(\mathbf{x}_l, \mathbf{y}_k)p_{\mathbf{X}}(\mathbf{x}_l), \qquad (5)$$

requiring $\mathcal{O}(R^2)$ operations. The multiplication becomes

$$\phi_{\mathbf{X}}(\mathbf{x}_k) \propto \prod_{i=1}^{M} p_{\mathbf{X}}^{(i)}(\mathbf{x}_k), \qquad (6)$$

requiring $\mathcal{O}(RM)$ operations. Because R scales exponentially with the dimensionality of $\mathbf{X}$ and a large number of points are required in every dimension to capture fine features of the messages, discretization is impractical for SPAWN in UWB localization.

4.3. Sample-Based Message Representation. A sample-based message representation, as used in [19, 50], overcomes the drawback of discretization by representing messages as samples, concentrated where the messages have significant mass. Before describing the detailed implementation of the filtering and multiplication operations, we give a brief overview of generic sampling techniques (see also [51, 52]) and kernel density estimation (KDE).

4.3.1. Background: Sampling and Kernel Density Estimation. We say that a list of samples with associated weights $\{\mathbf{x}_k, w_k\}_{k=1}^R$ is a representation for a distribution $p_\mathbf{X}(\mathbf{x})$ if, for any integrable function $g(\mathbf{x})$, we have the following approximation:

$$I = \int g(\mathbf{x}) p_\mathbf{X}(\mathbf{x}) dx \approx \sum_{k=1}^R w_k g(\mathbf{x}_k). \qquad (7)$$

Popular methods for obtaining the list of weighted samples include (i) direct sampling, where we draw R i.i.d. samples from $p_\mathbf{X}(\mathbf{x})$, each with weight $1/R$; and (ii) importance sampling, where we draw R i.i.d. samples from a distribution $q_\mathbf{X}(\mathbf{x})$, with a support that includes the support of $p_\mathbf{X}(\mathbf{x})$, and set the weight corresponding to sample $\mathbf{x}_k$ as $w_k = p_\mathbf{X}(\mathbf{x}_k)/q_\mathbf{X}(\mathbf{x}_k)$. In both cases, it can easily be verified that the approximation is unbiased with mean I and variance that reduces with R (and that depends on $q_\mathbf{X}(\mathbf{x})$, for importance sampling). Most importantly, the variance does not depend on the dimensionality of $\mathbf{x}$.

A variation of importance sampling that is not unbiased but that often has smaller variance is obtained by setting the weights as follows: $w_k \propto p_\mathbf{X}(\mathbf{x}_k)/q_\mathbf{X}(\mathbf{x}_k)$, $\sum_k w_k = 1$. This approach has the additional benefit that it does not require knowledge of the normalization constants of $p_\mathbf{X}(\mathbf{x})$ or $q_\mathbf{X}(\mathbf{x})$. A list of R equally weighted samples can be obtained from $\{\mathbf{x}_k, w_k\}_{k=1}^R$ through resampling, that is, by drawing (with repetition) R samples from the probability mass function defined by $\{\mathbf{x}_k, w_k\}_{k=1}^R$.

For numerical stability reasons, weights are often computed and stored in the logarithmic domain, that is, $\lambda_k = \log p_\mathbf{X}(\mathbf{x}_k) - \log q_\mathbf{X}(\mathbf{x}_k)$. When the distributions involved contain exponentials or products, the log-domain representation is also computationally efficient. Operations such as additions can be evaluated efficiently in the log-domain as well, using the Jacobian logarithm [49, pages 90–94]. Once all R log-domain weights are computed, they are translated, exponentiated, and normalized: $w_k \propto \exp(\lambda_k - \max_l \lambda_l)$.

Given a sample representation $\{\mathbf{x}_k, w_k\}_{k=1}^R$ of a distribution $p_\mathbf{X}(\mathbf{x})$, we obtain a kernel density estimate of $p_\mathbf{X}(\mathbf{x})$ as

$$\hat{p}_\mathbf{X}(\mathbf{x}) = \sum_{k=1}^R w_k K_\sigma(\mathbf{x} - \mathbf{x}_k), \qquad (8)$$

where $K_\sigma(\mathbf{x})$ is the so-called *kernel* with bandwidth σ. The kernel is a symmetric distribution with a width parameter

that is tuned through σ. For instance, a two-dimensional Gaussian kernel is given by

$$K_\sigma(\mathbf{x}) = \frac{1}{2\pi\sigma^2} \exp\left(-\frac{\|\mathbf{x}\|^2}{2\sigma^2}\right). \qquad (9)$$

While the choice of kernel affects the performance of the estimate to some limited extent (e.g., in an MMSE sense, where the error is $\int |p_X(x) - \hat{p}_X(x)|^2 p_X(x) dx$), the crucial parameter is the bandwidth σ, which needs to be estimated from the samples $\{\mathbf{x}_k, w_k\}_{k=1}^R$. A large choice of σ makes $\hat{p}_\mathbf{X}(\mathbf{x})$ smooth, but it may no longer capture the interesting features of $p_\mathbf{X}(\mathbf{x})$. When σ is too small, $\hat{p}_\mathbf{X}(\mathbf{x})$ may exhibit artificial structure not present in $p_\mathbf{X}(\mathbf{x})$ [53].

With this background in sampling techniques and KDE, we return to the problem at hand: filtering and multiplication of messages.

4.3.2. Message Filtering. We assume a message representation of $p_\mathbf{X}(\mathbf{x})$ as $\{\mathbf{x}_k, w_k\}_{k=1}^R$ and wish to obtain a message representation of $p_\mathbf{Y}(\mathbf{y}) \propto \int h(\mathbf{x}, \mathbf{y}) p_\mathbf{X}(\mathbf{x}) dx$. Let us interpret $h(\mathbf{x}, \mathbf{y})$ as a conditional distribution $p_{\mathbf{Y}|\mathbf{X}}(\mathbf{y} \mid \mathbf{x})$, up to some arbitrary constant. Suppose we can draw samples $\{[\mathbf{x}_k, \mathbf{y}_k], w_k\}_{k=1}^R \sim p_{\mathbf{Y}|\mathbf{X}}(\mathbf{y} \mid \mathbf{x}) p_\mathbf{X}(\mathbf{x})$; then $\{\mathbf{y}_k, w_k\}_{k=1}^R$ will form a sample representation of $p_\mathbf{Y}(\mathbf{y})$. Now the problem reverts to drawing samples from $p_{\mathbf{Y}|\mathbf{X}}(\mathbf{y} \mid \mathbf{x}) p_\mathbf{X}(\mathbf{x})$. This can be accomplished as follows: first, for every sample $\mathbf{x}_k$, draw $\mathbf{y}_k \sim q_{\mathbf{Y}|\mathbf{X}}(\mathbf{y} \mid \mathbf{x}_k)$ from some distribution $q_{\mathbf{Y}|\mathbf{X}}(\mathbf{y} \mid \mathbf{x}_k)$. Second, set the weight of sample $[\mathbf{x}_k, \mathbf{y}_k]$ as

$$v_k = w_k \frac{p_{\mathbf{Y}|\mathbf{X}}(\mathbf{y}_k \mid \mathbf{x}_k)}{q_{\mathbf{Y}|\mathbf{X}}(\mathbf{y}_k \mid \mathbf{x}_k)}. \qquad (10)$$

Finally, renormalize the weights v_k to $v_k/\sum_l v_l$. The complexity of the filtering operation scales as $\mathcal{O}(R)$, a significant improvement from $\mathcal{O}(R^2)$ for discretization. In addition, R can generally be much smaller in a particle-based representation.

Let us consider some examples of the filtering operation in SPAWN.

(i) *Mobility update* (A-1): let $p_\mathbf{X}(\mathbf{x})$ be the belief before movement (represented by $\{\mathbf{x}_k, w_k\}_{k=1}^R$) and $p_\mathbf{Y}(\mathbf{y})$ the belief after movement. Assume that we are able to measure perfectly the distance traveled (given by z_self), but have no information regarding the direction, and furthermore that the direction is chosen uniformly in $(0, 2\pi)$. In that case,

$$h(\mathbf{x}, \mathbf{y}) \propto \delta(z_\text{self} - \|\mathbf{x} - \mathbf{y}\|), \qquad (11)$$

where δ is a Dirac delta function, so that $q_{\mathbf{Y}|\mathbf{X}}(\mathbf{y} \mid \mathbf{x}_k) = p_{\mathbf{Y}|\mathbf{X}}(\mathbf{y} \mid \mathbf{x}_k) \propto \delta(z_\text{self} - \|\mathbf{x}_k - \mathbf{y}\|)$ is a reasonable choice. For every $\mathbf{x}_k$, we can now draw values for $\mathbf{y} = \mathbf{x}_k + r \times [\cos\theta \sin\theta]^T$ by drawing $\theta \sim \mathcal{U}(0, 2\pi)$ and setting $r = z_\text{self}$, leading to $\mathbf{y}_k = \mathbf{x}_k + z_\text{self} \times [\cos\theta_k \sin\theta_k]^T$, with $v_k \propto w_k$.

(ii) *Ranging update* (A-2): let $p_\mathbf{X}(\mathbf{x})$ be a message (represented by $\{\mathbf{x}_k, w_k\}_{k=1}^R$) from a node with which we have performed ranging, resulting in a range estimate z. Let $h(\mathbf{x}, \mathbf{y}) = p_{Z|D}(z \mid d)$, where $d = \|\mathbf{x} - \mathbf{y}\|$. Note that $p_{Z|D}(z \mid d)$ is a likelihood function, since the measurement z is known. Assume that we have a model for the ranging performance in the form of distributions $p_{Z|D}(z \mid d)$ for any value of d. We then sample $p_Y(\mathbf{y})$ as follows: for every $\mathbf{x}_k$, draw $\mathbf{y} = \mathbf{x}_k + r \times [\cos\theta \sin\theta]^T$ by drawing $\theta \sim \mathcal{U}(0, 2\pi)$ and $r \sim q_{R|Z}(r \mid z)$, for some well-chosen $q_{R|Z}(r \mid z)$ (e.g., a Gaussian distribution with mean equal to the distance estimate, z, and a standard deviation that is sufficiently large with respect to the standard deviation of $p(z|d)$ for any d) . The weights are set as

$$v_k = w_k \frac{p_{Z|D}(z \mid r_k)}{q_{R|Z}(r_k z)}. \tag{12}$$

4.3.3. Message Multiplication. Here we assume message representations $\{\mathbf{x}_k^{(i)}, w_k^{(i)}\}_{k=1}^{R}$ for $p_X^{(i)}(\mathbf{x}_k)$, $i = 1, \ldots, M$. In contrast to the discretization approach, we cannot directly compute $\prod_{i=1}^{M} p_X^{(i)}(\mathbf{x})$ for arbitrary values of $\mathbf{x}$. Rather, for every message $p_X^{(i)}(\mathbf{x}_k)$, we create a KDE $\hat{p}_X^{(i)}(\mathbf{x}) = \sum_{k=1}^{R} w_k K_{\sigma^{(i)}}(\mathbf{x} - \mathbf{x}_k^{(i)})$ with a Gaussian kernel and a bandwidth estimated using the methods from [53]. Suppose we now draw R samples from a distribution $q_X(\mathbf{x})$; then the weights are

$$v_k \propto \frac{\prod_{i=1}^{M} \hat{p}_X^{(i)}(\mathbf{x}_k)}{q_X(\mathbf{x}_k)}, \tag{13}$$

which can be computed efficiently in the log-domain. A reasonable choice for $q_X(\mathbf{x})$ could be one of the incoming messages $p_X^{(i)}(\mathbf{x})$ (e.g., the one with the smallest entropy) or a mixture of the incoming messages. The computational complexity of the message multiplication operation scales as $\mathcal{O}(MR^2)$. This appears worse than the discretized case (complexity $\mathcal{O}(MR)$), but note that R is much smaller for sample-based representations than for discretization (e.g., $R = 10^3$ or $R = 10^4$ for the sample-based representation compared to $R = 10^8$ in the discretization).

4.4. Parametric Message Representation

4.4.1. Choosing a Suitable Parameterization. From the previous section, it is clear that the bottleneck of the sample-based message representation lies in the message multiplication, which scales quadratically with the number of samples. An alternative approach is to represent each message as a set of parameters (e.g., a Gaussian distribution characterized by a mean and covariance matrix). In contrast to the sample-based message representation, which can represent messages of any shape, parametric representations must be specially tailored to the problem at hand. For example, single two-dimensional Gaussian parametric messages are utilized in [22] for localization with both range and angle measurements. Our choice of parametric message is based on the following observations.

(i) For the filtering operation with a two-dimensional Gaussian input $p_X(\mathbf{x})$, the output $p_Y(\mathbf{y})$ can be approximated by a circular distribution with the same mean for both the mobility update (A-1) and the ranging update (A-2).

(ii) Multiplying Gaussian distributions yield a Gaussian distribution.

(iii) The multiplication of multiple circular distributions can be approximated by a Gaussian distribution or a mixture of Gaussian distributions.

We will use as a basic building block the following distribution in two dimensions:

$$\mathcal{D}(\mathbf{x}; m_1, m_2, \sigma^2, \rho)$$

$$= \frac{1}{C(\sigma^2, \rho)} \exp\left\{ -\frac{\left[\sqrt{(x_1 - m_1)^2 + (x_2 - m_2)^2} - \rho \right]^2}{2\sigma^2} \right\}, \tag{14}$$

where $[m_1, m_2]$ is the midpoint of the distribution, ρ is the radius, σ^2 is the variance, and $C(\sigma^2, \rho)$ is a normalization constant equal to

$$C(\sigma^2, \rho)$$

$$= 2\pi\sigma^2 \left[\exp\left(-\frac{\rho^2}{2\sigma^2} \right) + \frac{1}{2}\sqrt{\frac{2\pi\rho^2}{\sigma^2}} \left(1 + \mathrm{erf}\sqrt{\frac{\rho^2}{2\sigma^2}} \right) \right]. \tag{15}$$

As a special case, we note that, when $\rho = 0$, (14) reverts to a two-dimensional Gaussian. Moreover, we will represent all messages as a mixture of two distributions of the type (14), so that

$$p_X(\mathbf{x})$$

$$= \frac{1}{2}\mathcal{D}\left(\mathbf{x}; m_1^{(a)}, m_2^{(a)}, \sigma^2, \rho\right) + \frac{1}{2}\mathcal{D}\left(\mathbf{x}; m_1^{(b)}, m_2^{(b)}, \sigma^2, \rho\right), \tag{16}$$

which can be represented by the six-dimensional vector $[m_1^{(a)}, m_2^{(a)}, m_1^{(b)}, m_2^{(b)}, \rho, \sigma^2]$. We will denote the family of distributions of the form (16) by $\mathcal{D}_2$. Note that it is trivial to extend this distribution, which is designed for two-dimensional localization systems, for use in three-dimensional systems. Before we describe the message filtering and message multiplication operations, let us first show how the parameters of (14) can be estimated from a list of samples.

4.4.2. ML Estimation of the Parameters m_1, m_2, σ^2, ρ. Given a list of samples $\{\mathbf{x}_k\}_{k=1}^{R}$, we can estimate the parameters $[m_1, m_2, \sigma^2, \rho]$ as follows. The midpoint $\mathbf{m} = [m_1, m_2]$ is estimated by

$$\hat{\mathbf{m}} = \frac{1}{R}\sum_{k=1}^{R} \mathbf{x}_k. \tag{17}$$

To find the radius ρ and variance σ^2 of $\mathcal{D}$-distribution, we use maximum likelihood (ML) estimation, assuming the R samples are independent. Introducing $\boldsymbol{\alpha} = [\rho \sigma^2]^T$, we find that

$$\hat{\boldsymbol{\alpha}}_{\mathrm{ML}} = \arg \max_{\boldsymbol{\alpha}} \underbrace{\sum_{k=1}^{R} \log p_{\mathbf{X}}(\mathbf{x}_k; \hat{\mathbf{m}}, \boldsymbol{\alpha})}_{\Lambda(\boldsymbol{\alpha})}. \qquad (18)$$

Treating the log-likelihood function (LLF) $\Lambda(\boldsymbol{\alpha})$ as an objective function, we find its maximum through the gradient ascent algorithm

$$\hat{\boldsymbol{\alpha}}^{(n+1)} = \hat{\boldsymbol{\alpha}}^{(n)} - \varepsilon \nabla_{\boldsymbol{\alpha}} \Lambda(\boldsymbol{\alpha}) \big|_{\boldsymbol{\alpha} = \hat{\boldsymbol{\alpha}}^{(n)}}, \qquad (19)$$

where ε is a suitably small step size, and the gradient vector can be approximated using finite differences. To initialize (19), we consider two initial estimates for σ^2 and ρ: one assuming $\rho \ll \sigma$ and a second assuming $\rho \gg \sigma$. The LLF is evaluated for both preliminary solutions, and the one with the largest log-likelihood is used as the initial estimate in (19).

4.4.3. Message Filtering. To perform message filtering, we use the fact that sample-based message filtering is a low-complexity operation. We decompose $p_{\mathbf{X}}(\mathbf{x})$, represented in parametric form, into its two mixture components. From each component, we draw $R/2$ samples and perform sample-based message filtering, as outlined in Section 4.3.2. We can then estimate the new $\mathcal{D}$-parameters for each mixture component using the ML method described above. We thus have $p_{\mathbf{Y}}(\mathbf{y})$ in parametric form. The complexity of this operation scales as $\mathcal{O}(R)$.

4.4.4. Message Multiplication. The motivation for using the parametric message representation is to avoid the complexity associated with sample-based message multiplication. Given M distributions $p_{\mathbf{X}}^{(i)}(\mathbf{x}) \in \mathcal{D}_2$, our goal is to compute

$$\phi_{\mathbf{X}}(\mathbf{x}) = \prod_{i=1}^{M} p_{\mathbf{X}}^{(i)}(\mathbf{x}). \qquad (20)$$

Typically, $\phi_{\mathbf{X}}(\mathbf{x}) \notin \mathcal{D}_2$, so we will *approximate* $\phi_{\mathbf{X}}(\mathbf{x})$ by $q_{\mathbf{X}}^*(\mathbf{x}) \in \mathcal{D}_2$ by projecting $\phi_{\mathbf{X}}(\mathbf{x})$ onto the family $\mathcal{D}_2$:

$$q_{\mathbf{X}}^*(\cdot) = \arg \min_{q_{\mathbf{X}} \in \mathcal{D}_2} D_{\mathrm{KL}}(q_{\mathbf{X}} \| \phi_{\mathbf{X}}), \qquad (21)$$

where $D_{\mathrm{KL}}(\cdot \| \cdot)$ denotes the Kullback Leibler (KL) divergence, defined as

$$D_{\mathrm{KL}}(q_{\mathbf{X}} \| \phi_{\mathbf{X}}) = \int q_{\mathbf{X}}(\mathbf{x}) \log \frac{q_{\mathbf{X}}(\mathbf{x})}{\phi_{\mathbf{X}}(\mathbf{x})} d\mathbf{x}. \qquad (22)$$

Observe that all elements of $\mathcal{D}_2$ are characterized by the parameters $\mathbf{p} \triangleq [m_1^{(a)}, m_2^{(a)}, m_1^{(b)}, m_2^{(b)}, \rho, \sigma^2]^T$ and that the optimization (21) is therefore a six-dimensional problem over all possible $\mathbf{p}$. The divergence $D_{\mathrm{KL}}(q \| \phi)$ for an arbitrary

TABLE 1: Comparison of complexity of message representations, where R is the number samples taken from each distribution and M is the number of distributions or messages.

Approach	Operation	Complexity	Value of R
Discretized	Filtering	$\mathcal{O}(R^2)$	Large
Discretized	Multiplication	$\mathcal{O}(RM)$	Large
Sample-based	Filtering	$\mathcal{O}(R)$	Small
Sample-based	Multiplication	$\mathcal{O}(R^2M)$	Small
Parametric	Filtering	$\mathcal{O}(R)$	Small
Parametric	Multiplication	$\mathcal{O}(RM)$	Small

$\mathbf{p} \in \mathbb{R}^4 \times \mathbb{R}_+^2$ can be determined using Monte-Carlo integration as follows. We rewrite (22) as

$$D_{\mathrm{KL}}(q_{\mathbf{X}} \| \phi_{\mathbf{X}}) = \int q_{\mathbf{X}}(\mathbf{x}) f(\mathbf{x}) d\mathbf{x}, \qquad (23)$$

where $f(\mathbf{x}) = \log q_{\mathbf{X}}(\mathbf{x}) - \sum_{i=1}^{M} \log p_{\mathbf{X}}^{(i)}(\mathbf{x})$. By drawing R weighted samples $\{w_k, \mathbf{x}_k\}_{k=1}^{R}$ from $q_{\mathbf{X}}(\mathbf{x})$ (e.g., through importance sampling), we can approximate (23) by

$$D_{\mathrm{KL}}(q_{\mathbf{X}} \| \phi_{\mathbf{X}}) \approx \sum_{k=1}^{R} w_k f(\mathbf{x}_k). \qquad (24)$$

Using this approximation, the six-dimensional optimization problem (21) is solved through gradient descent, similar to (19). The complexity of this operation scales as $\mathcal{O}(RM)$. The initial estimate of $\mathbf{p}$ is obtained through a set of heuristics: we first decide whether $\phi_{\mathbf{X}}(\mathbf{x})$ can reasonably be represented by a distribution in $\mathcal{D}_2$. If not, the outgoing message is not computed. Otherwise, we use a geometric argument to find at most two midpoints. The initial estimates for ρ and σ^2 are set to a small constant value.

4.5. Comparison of Message Representations. The complexities of the discretized, sample-based, and parametric message representations are compared in Table 1.

5. Performance Analysis

In this section, we compare the performance of the SPAWN algorithm with sample-based versus parametric message representation in a simulated wireless network. We also analyze the use of different subsets of information in the algorithm and its effect on localization performance.

5.1. Simulation Setup and Performance Measures. We simulate a large-scale ultra-wide bandwidth (UWB) network in a $100\,\mathrm{m} \times 100\,\mathrm{m}$ homogeneous environment, with 100 uniformly distributed agents and 13 fixed anchors in a grid configuration. Each node is able to measure its range to other nodes within 20 meters. The simulated ranging measurements are independently drawn from the UWB ranging model developed in [54]. The model, based on data collected in a variety of indoor scenarios, consists of three component Gaussian densities, where the mean and variance of each component are experimentally determined functions of the true distance between the ranging nodes. To decouple the effect of mobility with the message representation, we consider a single time slot, where every agent has a uniform a priori distribution over the environment $\mathcal{E}$. SPAWN was run for $N_{it} = 20$ iterations, though convergence was generally achieved well before 10 iterations. For the sample-based representation, the number of samples is set to $R = 2048$ unless otherwise stated.

We quantify localization performance using the complementary cumulative distribution function (CCDF) of the localization error $e = \|\mathbf{x}_i - \hat{\mathbf{x}}_i\|$, where $\hat{\mathbf{x}}_i$ is the estimated location of node i, taken as the mean of the belief, similar to [18]. To estimate the CCDF, we consider 50 random network topologies and collect position estimates at every iteration for every agent. Note that a CCDF of 0.01 at an error of, say, $e = 1$ m means that 99% of the nodes have an error less than 1 meter.

5.2. Cooperation with Censoring.
In Section 3, we considered processing messages between all neighboring pairs of nodes. However, information from neighbors may not always be useful: (i) when the receiving node's belief is already very informative (e.g., concentrated around the mean); or (ii) when the transmitting node's belief is very uninformative. To better understand how much cooperative information is beneficial to localization, we will consider varying the subset of nodes that broadcast and update their location beliefs at each iteration. We distinguish between these subsets by the *level of cooperative information* they induce in the algorithm. The level of cooperative information indicates how each node utilizes information from its neighbors at each iteration.

We introduce the following terminology: a distribution is said to be "sufficiently informative" when 95% of the probability mass is located within 2 m of the mean; a node becomes a *virtual anchor* when its belief is sufficiently informative; a *virtual bianchor* is a node with a bimodal belief, with each mode being sufficiently informative; a node that is neither a virtual anchor nor a virtual bianchor will be called a *blind agent*. We are now ready to introduce four levels of cooperative information at each iteration.

(i) *Level 1 (L1)*: virtual anchors broadcast their beliefs, while all other nodes censor their belief broadcast. Virtual anchors do not update their beliefs.

(ii) *Level 2 (L2)*: virtual anchors and virtual bianchors broadcast their beliefs, while blind nodes censor their belief broadcast. Virtual anchors do not update their beliefs.

(iii) *Level 3 (L3)*: all nodes broadcast their beliefs. Virtual anchors do not update their beliefs.

(iv) *Level 4 (L4)*: all nodes broadcast their beliefs. All nodes update their beliefs.

In terms of cooperation, note that L4 utilizes more cooperative information than L3, L3 utilizes more cooperative information than L2, and L2 utilizes more cooperative

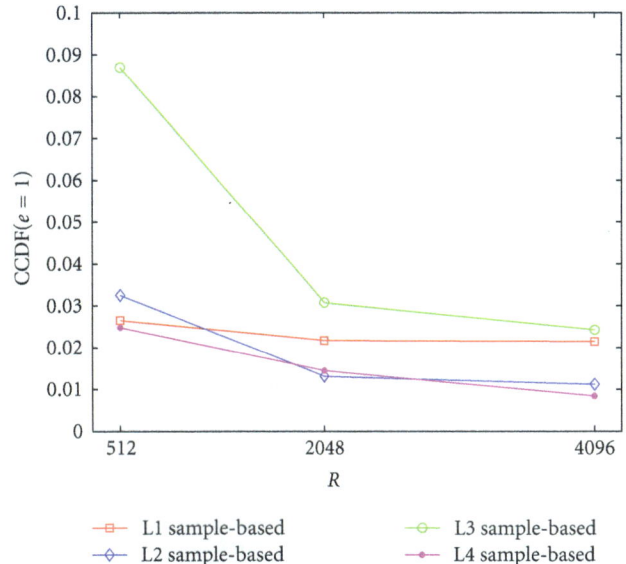

FIGURE 1: The effect of R and the level of cooperative information on the CCDF at $e = 1$ meter. All results are after $N_{it} = 10$ iterations.

information than L1. In this sense, the levels of cooperative information are strict subsets.

From previous sections, we know that the algorithm complexity scales linearly in M, the number of incoming messages in the multiplication operation. Hence, the level of cooperative information directly affects the algorithm's computational cost, with lower levels requiring less computation.

5.3. Numerical Results.
We now examine how localization performance varies with the algorithm parameters. In particular, numerical results show the effect of message representation (sample-based or parametric) and level of cooperative information (L1, L2, L3, or L4) on the CCDF of the localization error.

We first consider the localization performance as a function of the number of samples R and level of cooperative information. Figure 1 displays the CCDF at $e = 1$ m after 10 iterations. As expected, for any level of cooperative information, the CCDF decreases as the number of samples is increased. However, the decrease in CCDF comes with a cost in computation time; as R is increased, the per-node complexity increases quadratically. Figure 1 also shows that levels L1, L2, and L4 are not as sensitive to R as L3 and that each generally outperforms L3. This effect is particularly pronounced when R is small. L3 broadcasts more complex distributions than L2 and L1, and these elaborate distributions are not accurately represented with a small number of samples.

Secondly, we investigate level of cooperative information and its effect on localization performance, with numerical results represented in Figures 2 and 3, after $N_{it} = 20$ iterations. Note that each curve exhibits a "floor" because there is always some subset of nodes that have insufficient information to localize without ambiguity. This may be due to lack of connectivity or large flip ambiguities. Let us

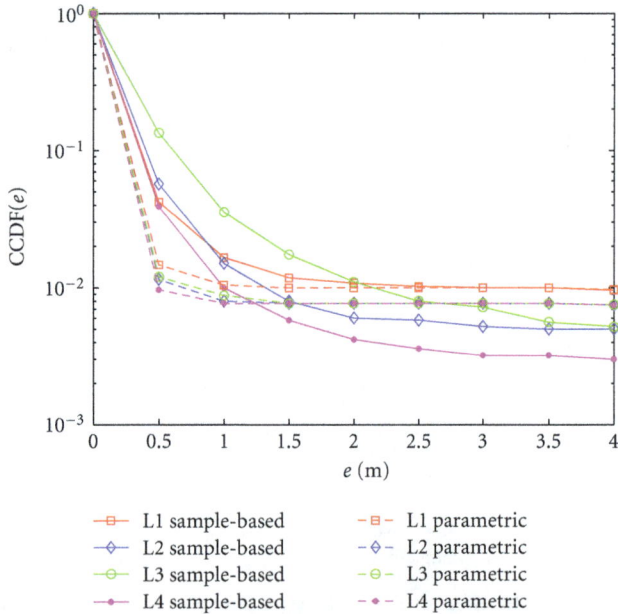

FIGURE 2: Effect of the level of cooperative information on localization accuracy for the distributed sample-based and parametric algorithms, when $N_{it} = 20$, and averaged over 50 networks.

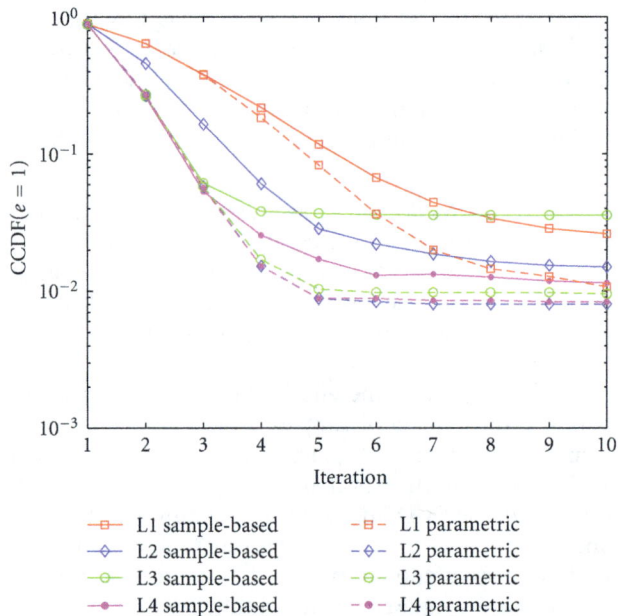

FIGURE 3: Effect of the level of cooperative information on the convergence speed for $e = 1$ m, for the distributed sample-based and parametric algorithms. Results are averaged over 50 networks.

focus on the sample-based representation in Figure 2 and consider the effect of the level of cooperative information on localization performance. In general, L4 has the best performance in terms of accuracy and floor. Intuitively, one might expect L3 to have the next best performance, followed by L2, and then L1. However, Figure 2 demonstrates that in some cases L3 has poorer accuracy than L2 and a similar

floor. This effect can be explained as follows. Agents that do not become virtual anchors within $N_{it} = 20$ tend to have large localization errors, creating a floor. Such agents comprise 1.7% of the total nodes for L1 and 0.3% for both L2 and L3. Since L2 and L3 have a similar fraction of agents that do not become virtual anchors, they have similar floors. In addition, the accuracy of beliefs belonging to agents that have become virtual anchors turns out to be highest for L1, followed by L2, and then L3. This is because L3 uses less reliable information than L2, which in turn is less reliable than L1. The final CCDF depends both on the fraction of virtual anchors (lowest for L1) and the accuracy of those virtual anchors (highest for L1). Note that we cannot compare L4 in this context, since there is no concept of a virtual anchor in L4.

We now move on to the parametric representation, still in Figure 2. We observe that L4 has the lowest overall CCDF for any e, for both types of message representation. For the parametric messages, the differences among different levels of cooperative information are smaller, and we generally obtain better performance (for $e < 1$ m) compared to sample-based messages.

Finally, in Figure 3, we evaluate the convergence speed of the different message representations and levels of cooperation, for a fixed error of 1 meter. We see that the parametric messages generally lead to faster convergence and lower CCDF than their sample-based counterparts. Levels L2, L3, and L4 all converge in around 5 iterations with a final CCDF at $e = 1$ m of around 0.01 for the parametric representation. Our results show that more cooperative information leads to faster improvement in terms of accuracy. The lowest level of cooperative information, L1, is consistently slower to converge and less accurate. However, higher levels of cooperative information also require the computation and representation of more complicated distributions. As a possible consequence, convergence issues may occur for levels L3 and L4. We also see that the parametric message representation performs approximately equal to or better than the sample-based messages in terms of both convergence and accuracy, while requiring much less execution time. Overall, parametric message representations yield a better performance/complexity tradeoff. This is due to the fact that the parametric distributions are well tailored to the localization problem and the homogeneous simulation environment.

6. Conclusions and Extensions

In this paper, we considered different message representations for Bayesian cooperative localization in wireless networks: a generic sample-based representation and a tailored parametric representation. We used experimentally derived UWB ranging models to evaluate the performance of SPAWN as a function of message representation and level of cooperative information. Our results show that the tradeoffs between message representation, cooperative information, localization accuracy, and algorithm convergence are not straightforward and should be tailored to the scenario.

Through large-scale network simulations, we demonstrated that more cooperative information may improve localization accuracy but also increase the complexity of messages. Higher levels of cooperative information do not always correspond to an improvement in localization accuracy or convergence rate. As complicated distributions associated with location-uncertain nodes are computed and transmitted, the resulting increases in computational complexity and signal interference can actually reduce localization performance. It may therefore be advantageous to broadcast only confident information in cooperative localization networks, especially considering the resources saved by a node censorship policy.

We also demonstrated that though parametric messages have less representational flexibility, they can outperform nonparametric message representation at a much lower computational cost. In our simulations, the parametric representation achieved a lower probability of outage for errors under 1 meter while converging in equal or fewer iterations than the sample-based representation. Clearly, a parametric representation well tailored to the localization scenario is desirable in terms of both resource efficiency and localization accuracy.

The use of parametric distributions for localization can be extended to (i) different ranging models; (ii) different types of measurements; (iii) more general scenarios. In terms of ranging models, the proposed distributions can be applied as long as typical distributions in SPAWN roughly resemble a distribution in the $\mathcal{D}_2$-family. Note that a Gaussian ranging error satisfies this criteria, as would many other, more realistic, models. Other models, such as those derived from received signal strength, will require different types of parametric distributions. The same comment applies to the use of different types of measurements. For instance, with angle-of-arrival measurements, the parametric distributions should include a collection of linear distributions. Finally, more general scenarios may require tailor-made distributions. With NLOS measurements that can be modeled as biased Gaussians [20], for example, mixtures of $\mathcal{D}_2$ distributions would easily accommodate LOS/NLOS propagation, without relying on explicit NLOS identification.

References

[1] J. J. Caffery and G. L. Stüber, "Overview of radiolocation in CDMA cellular systems," IEEE Communications Magazine, vol. 36, no. 4, pp. 38–45, 1998.

[2] A. H. Sayed, A. Tarighat, and N. Khajehnouri, "Network-based wireless location: challenges faced in developing techniques for accurate wireless location information," IEEE Signal Processing Magazine, vol. 22, no. 4, pp. 24–40, 2005.

[3] A. Mainwaring, D. Culler, J. Polastre, R. Szewczyk, and J. Anderson, "Wireless sensor networks for habitat monitoring," in Proceedings of the 1st ACM International Workshop on Wireless Sensor Networks and Applications (WSNA '02), pp. 88–97, ACM Press, September 2002.

[4] T. He, C. Huang, B. M. Blum, J. A. Stankovic, and T. Abdelzaher, "Range-free localization schemes for large scale sensor networks," in Proceedings of the 9th Annual International Conference on Mobile Computing and Networking (MobiCom '03), pp. 81–95, September 2003.

[5] K. Pahlavan, X. Li, and J. P. Mäkelä, "Indoor geolocation science and technology," IEEE Communications Magazine, vol. 40, no. 2, pp. 112–118, 2002.

[6] S. J. Ingram, D. Harmer, and M. Quinlan, "Ultrawideband indoor positioning systems and their use in emergencies," in Proceedings of the Position Location and Navigation Symposium (PLANS '04), pp. 706–715, April 2004.

[7] C.-Y. Chong and S. P. Kumar, "Sensor networks: evolution, opportunities, and challenges," Proceedings of the IEEE, vol. 91, no. 8, pp. 1247–1256, 2003.

[8] H. Yang and B. Sikdar, "A protocol for tracking mobile targets using sensor networks," in Proceedings of the 1st IEEE International Workshop on Sensor Network Protocols and Applications, pp. 71–81, May 2003.

[9] X. Ji and H. Zha, "Sensor positioning in wireless ad-hoc sensor networks using multidimensional scaling," in Proceedings of the 23rd Annual Joint Conference of the IEEE Computer and Communications Societies (INFOCOM '04), vol. 4, pp. 2652–2661, March 2004.

[10] R. A. Marjamaa, P. M. Torkki, M. I. Torkki, and O. A. Kirvelä, "Time accuracy of a radio frequency identification patient tracking system for recording operating room timestamps," Anesthesia & Analgesia, vol. 102, no. 4, pp. 1183–1186, 2006.

[11] D. Fox, W. Burgard, F. Dellaert, and S. Thrun, "Monte Carlo localization: efficient position estimation for mobile robots," in Proceedings of the 16th National Conference on Artificial Intelligence (AAAI '99), pp. 343–349, Orlando, Fla, USA, July 1999.

[12] J. J. Leonard and H. F. Durrant-Whyte, "Mobile robot localization by tracking geometric beacons," IEEE Transactions on Robotics and Automation, vol. 7, no. 3, pp. 376–382, 1991.

[13] R. Jain, A. Puri, and R. Sengupta, "Geographical routing using partial information for wireless ad hoc networks," IEEE Personal Communications, vol. 8, no. 1, pp. 48–57, 2001.

[14] H. Frey, "Scalable geographic routing algorithms for wireless ad hoc networks," IEEE Network, vol. 18, no. 4, pp. 18–22, 2004.

[15] R. J. Fontana and S. J. Gunderson, "Ultra-wideband precision asset location system," in Proceedings of IEEE Conference on Ultra Wideband Systems and Technologies (UWBST '02), vol. 21, no. 1, pp. 147–150, Baltimore, Md, USA, May 2002.

[16] W. C. Chung and D. Ha, "An accurate ultra wideband (UWB) ranging for precision asset location," in Proceedings of IEEE Conference on Ultra Wideband Systems and Technologies (UWBST '03), pp. 389–393, November 2003.

[17] N. Patwari, J. N. Ash, S. Kyperountas, A. O. Hero III, R. L. Moses, and N. S. Correal, "Locating the nodes: cooperative localization in wireless sensor networks," IEEE Signal Processing Magazine, vol. 22, no. 4, pp. 54–69, 2005.

[18] M. Castillo-Effen, W. A. Moreno, M. A. Labrador, and K. P. Valavanis, "Adapting sequential Monte-Carlo estimation to cooperative localization in wireless sensor networks," in Proceedings of IEEE International Conference on Mobile Ad Hoc and Sensor Sysetems (MASS '06), pp. 656–661, Vancouver, Canada, October 2006.

[19] A. T. Ihler, J. W. Fisher III, R. L. Moses, and A. S. Willsky, "Nonparametric belief propagation for self-localization of sensor networks," IEEE Journal on Selected Areas in Communications, vol. 23, no. 4, pp. 809–819, 2005.

[20] H. Wymeersch, J. Lien, and M. Z. Win, "Cooperative localization in wireless networks," Proceedings of the IEEE, vol. 97, no. 2, pp. 427–450, 2009.

[21] A. T. Ihler, J. W. Fisher, and A. S. Willsky, "Particle filtering under communications constraints," in *Proceedings of the 13th IEEE/SP Workshop on Statistical Signal Processing*, pp. 89–94, Bordeaux, France, July 2005.

[22] M. Welling and J. J. Lim, "A distributed message passing algorithm for sensor localization," in *Proceedings of the 17th International Conference on Artificial Neural Networks (ICANN '07)*, pp. 767–775, September 2007.

[23] C. Pedersen, T. Pedersen, and B. H. Fleury, "A variational message passing algorithm for sensor self-localization in wireless networks," in *Proceedings of IEEE International Symposium on Information Theory Proceedings (ISIT '11)*, pp. 2158–2162, Saint-Petersburg, Russia, July 2011.

[24] M. Caceres, F. Penna, H. Wymeersch, and R. Garello, "Hybrid cooperative positioning based on distributed belief propagation," *IEEE Journal on Selected Areas in Communications*, vol. 29, no. 10, pp. 1948–1958, 2011.

[25] M. Z. Win and R. A. Scholtz, "Characterization of ultra-wide bandwidth wireless indoor channels: a communication-theoretic view," *IEEE Journal on Selected Areas in Communications*, vol. 20, no. 9, pp. 1613–1627, 2002.

[26] M. Z. Win and R. A. Scholtz, "On the robustness of ultra-wide bandwidth signals in dense multipath environments," *IEEE Communications Letters*, vol. 2, no. 2, pp. 51–53, 1998.

[27] A. F. Molisch, J. R. Foerster, and M. Pendergrass, "Channel models for ultrawideband personal area networks," *IEEE Wireless Communications*, vol. 10, no. 6, pp. 14–21, 2003.

[28] J.-Y. Lee and R. A. Scholtz, "Ranging in a dense multipath environment using an UWB radio link," *IEEE Journal on Selected Areas in Communications*, vol. 20, no. 9, pp. 1677–1683, 2002.

[29] D. Dardari, A. Conti, U. Ferner, A. Giorgetti, and M. Z. Win, "Ranging with ultrawide bandwidth signals in multipath environments," *Proceedings of the IEEE*, vol. 97, no. 2, pp. 404–425, 2009.

[30] W. Suwansantisuk and M. Z. Win, "Multipath aided rapid acquisition: optimal search strategies," *IEEE Transactions on Information Theory*, vol. 53, no. 1, pp. 174–193, 2007.

[31] H. Xu and L. Yang, "Timing with dirty templates for low-resolution digital UWB receivers," *IEEE Transactions on Wireless Communications*, vol. 7, no. 1, pp. 54–59, 2008.

[32] M. Z. Win, P. C. Pinto, and L. A. Shepp, "A mathematical theory of network interference and its applications," *Proceedings of the IEEE*, vol. 97, no. 2, pp. 205–230, 2009.

[33] N. C. Beaulieu and D. J. Young, "Designing time-hopping ultrawide bandwidth receivers for multiuser interference environments," *Proceedings of the IEEE*, vol. 97, no. 2, pp. 255–284, 2009.

[34] M. Z. Win, G. Chrisikos, and A. F. Molisch, "Wideband diversity in multipath channels with nonuniform power dispersion profiles," *IEEE Transactions on Wireless Communications*, vol. 5, no. 5, pp. 1014–1022, 2006.

[35] M. Z. Win, G. Chrisikos, and N. R. Sollenberger, "Performance of Rake reception in dense multipath channels: implications of spreading bandwidth and selection diversity order," *IEEE Journal on Selected Areas in Communications*, vol. 18, no. 8, pp. 1516–1525, 2000.

[36] C. Falsi, D. Dardari, L. Mucchi, and M. Z. Win, "Time of arrival estimation for UWB localizers in realistic environments," *EURASIP Journal on Advances in Signal Processing*, vol. 2006, Article ID 32082, pp. 1–13, 2006.

[37] S. Venkatesh and R. M. Buehrer, "Non-line-of-sight identification in ultra-wideband systems based on received signal

[38] statistics," *IET Microwaves, Antennas & Propagation*, vol. 1, no. 6, pp. 1120–1130, 2007.

[38] D. B. Jourdan, J. J. Deyst Jr., M. Z. Win, and N. Roy, "Monte Carlo localization in dense multipath environments using UWB ranging," in *Proceedings of IEEE International Conference on Ultra-Wideband (ICU '05)*, pp. 314–319, September 2005.

[39] S. Venkatesh and R. M. Buehrer, "Multiple-access design for ad hoc UWB position-location networks," in *Proceedings of IEEE Wireless Communications and Networking Conference (WCNC '06)*, vol. 4, pp. 1866–1873, Las Vegas, Nev, USA, April 2006.

[40] D. Dardari, A. Conti, J. Lien, and M. Z. Win, "The effect of cooperation on UWB-based positioning systems using experimental data," *EURASIP Journal on Advances in Signal Processing*, vol. 2008, Article ID 513873, 2008.

[41] S. Venkatesh and R. M. Buehrer, "Power control in UWB position-location networks," in *Proceedings of IEEE International Conference on Communications (ICC '06)*, vol. 9, pp. 3953–3959, Istanbul, Turkey, June 2006.

[42] W. C. Headley, C.R.C.M. da Suva, and R. M. Buehrer, "Indoor location positioning of non-active objects using ultra-wideband radios," in *Proceedings of IEEE Radio and Wireless Symposium (RWS '07)*, pp. 105–108, Long Beach, Calif, USA, January 2007.

[43] D. B. Jourdan, D. Dardari, and M. Z. Win, "Position error bound for UWB localization in dense cluttered environments," *IEEE Transactions on Aerospace and Electronic Systems*, vol. 44, no. 2, pp. 613–628, 2008.

[44] Y. Shen and M. Z. Win, "Fundamental limits of wideband localization—part I: a general framework," *IEEE Transactions on Information Theory*, vol. 56, no. 10, pp. 4956–4980, 2010.

[45] Y. Shen, H. Wymeersch, and M. Z. Win, "Fundamental limits of wideband localization—part II: cooperative networks," *IEEE Transactions on Information Theory*, vol. 56, no. 10, pp. 4981–5000, 2010.

[46] Y. Shen and M. Z. Win, "On the accuracy of localization systems using wideband antenna arrays," *IEEE Transactions on Communications*, vol. 58, no. 1, pp. 270–280, 2010.

[47] S. Gezici, Z. Tian, G. B. Giannakis et al., "Localization via ultra-wideband radios: a look at positioning aspects for future sensor networks," *IEEE Signal Processing Magazine*, vol. 22, no. 4, pp. 70–84, 2005.

[48] I. Oppermann, M. Hämäläinen, and J. Iinatti, *UWB Theory and Applications*, John Wiley & Sons, 2004.

[49] H. Wymeersch, *Iterative Receiver Design*, Cambridge University Press, 2007.

[50] D. Fox, W. Burgard, H. Kruppa, and S. Thrun, "A Monte Carlo algorithm for multi-robot localization," Tech. Rep. CMU-CS-99-120, Computer Science Department, Carnegie Mellon University, Pittsburgh, Pa, USA, 1999.

[51] D. MacKay, *Information Theory, Inference and Learning Algorithms*, Cambridge University Press, 2003.

[52] A. Doucet, S. Godsill, and C. Andrieu, "On sequential Monte Carlo sampling methods for Bayesian filtering," *Statistics and Computing*, vol. 10, no. 3, pp. 197–208, 2000.

[53] Z. Botev, *Nonparametric Density Estimation Via Diffusion Mixing*, Postgraduate Series, The University of Queensland, 2007, http://espace.library.uq.edu.au/view/UQ:120006.

[54] J. Lien, *A framework for cooperative localization in ultra-wideband wireless networks [M.S. thesis]*, Department of Electrical Engineering and Computer Science, Massachusetts Institute of Technology, Cambridge, Mass, USA, 2007.

5

Indoor Localisation Using a Context-Aware Dynamic Position Tracking Model

Montserrat Ros,[1] Joshua Boom,[1] Gavin de Hosson,[1] and Matthew D'Souza[2]

[1] School of Electrical, Computer and Telecommunications Engineering, University of Wollongong, Wollongong, NSW 2522, Australia
[2] CSIRO ICT Centre, Brisbane, QLD 4069, Australia

Correspondence should be addressed to Montserrat Ros, montse@uow.edu.au

Academic Editor: Jinling Wang

Indoor wireless localisation is a widely sought feature for use in logistics, health, and social networking applications. Low-powered localisation will become important for the next generation of pervasive media applications that operate on mobile platforms. We present an inexpensive and robust context-aware tracking system that can track the position of users in an indoor environment, using a wireless smart meter network. Our context-aware tracking system combines wireless trilateration with a dynamic position tracking model and a probability density map to estimate indoor positions. The localisation network consisted of power meter nodes placed at known positions in a building. The power meter nodes are tracked by mobile nodes which are carried by users to localise their position. We conducted an extensive trial of the context-aware tracking system and performed a comparison analysis with existing localisation techniques. The context-aware tracking system was able to localise a person's indoor position with an average error of 1.21 m.

1. Introduction

The next generation of pervasive media applications, mobile social networking, and location-based services are increasingly reliant on accurate position localisation. Localisation for indoor environments has many applications for pervasive media. Low-powered or green efficient and inexpensive localisation will become important for the next generation of pervasive media applications that operate on battery-constrained mobile platforms.

Current localisation techniques depend on using sensing infrastructure already present in the environment such as visual markers, wireless LAN hotspots, cellular networks, or Global Position Systems' (GPS) satellite coverage. The popular use of GPS has led to a variety of mobile location-based services applications such as social networking, street map guide, or asset tracking. Recently, there has been great interest in localisation for indoor navigation applications. Indoor environments cause multipath interference to wireless communications because of the presence of physical obstacles such as metal beams or walls. Hence, this causes outdoor Radio-Frequency- (RF-) based localisation technologies such as GPS to function inaccurately indoors because of signal degradation. Other RF localisation methods such as Received Signal Strength or Time of Arrival also experience inaccuracies and reliability issues when operating indoors.

Wireless infrastructure that is currently used for both indoor and outdoor localisation, tends to be computationally intensive with high power consumption. Wireless sensor networks are an alternative form of wireless infrastructure that can be used for localisation but also operate at low power. Wireless sensor networks are used for a sensing and actuation applications including smart metering. As energy usage monitoring becomes an important lifestyle factor for workplaces and households, wireless smart metering networks will be more widely used. Wireless smart meters are being incorporated into new buildings for climate control and to improve power usage efficiency. Wireless smart metering infrastructure can potentially be used for low-powered indoor and outdoor localisation.

We designed and developed a wireless localisation tracking system that tracked people indoors. Our wireless localisation system used a low-powered wireless smart metering network infrastructure which consisted of power metering nodes placed at predetermined coordinates in a building level. The power metering nodes were used to determine the coordinates of the user within the region covered by the localisation network. Our wireless localisation tracking system consisted of users carrying a mobile node to triangulate their current position. A mobile phone interface was also developed to allow users to view their location.

One of the main drawbacks of wireless indoor localisation is reduced accuracy due to multipath fading and other Radio Frequency (RF) interferences. We found that using only received signal strength or other wireless channel propagation properties was not suitable for tracking users in real time, due to the lengthy time taken to calibrate for channel propagation parameters. To overcome this, we developed a context-aware tracking model for tracking people within a building. The context-aware tracking model incorporated "awareness" of the physical context of the surrounding environment (indoor building floor plans). Odometry information such as the estimated speed was also used to predict the next position of user. We extensively evaluated our system and investigated the following aspects:

(i) use of a low-powered, inexpensive smart metering network for indoor localisation and tracking,

(ii) analysis of the wireless indoor channel propagation on position accuracy,

(iii) development and testing of a context-aware tracking algorithm,

(iv) comparison of the accuracy of the context-aware tracking model with other wireless localisation technologies and protocols.

This paper is organised into 6 sections. Section 2 presents a review of related work. Sections 3 and 4 discuss the implementation of the context-aware tracking system. Section 5 presents the evaluation findings and analysis of the system. Conclusions and further areas of investigation are discussed in Section 6.

2. Related Work

Different types of wireless technologies, such as GPS, have been investigated for outdoor and indoor location systems. Unfortunately, GPS is not suitable for indoor use, and this has led to research into the use of other wireless technologies including UWB [1], ultrasonic and GSM [2] platforms. Regulations are not clear for the use of UWB, and ultrasonic location detection still requires RF transceivers. GSM uses existing cellular infrastructure; however, accurate position resolution indoors is difficult.

Received Signal Strength Indicators (RSSIs) are used for indoor and outdoor localisation, as outlined by Seco et al. [3]. The most common RSSI localisation techniques are RSSI Fingerprinting, RSSI triangulation and trilateration. RSSI Fingerprinting identifies specific positions with RSSI values, while RSSI triangulation, and trilateration associate RSSI with distance or angular trajectory between receiver and known transmitter positions in order to localise [3, 4]. Hightower et al. [5] describe the PlaceLab geophysical location

system in which users can determine their position in an urban environment. PlaceLab was an RSSI Fingerprinting technique that used wireless LAN hotspots and GSM broadcast towers to determine a user's position. The PlaceLab software used a database of known wireless LAN hotspots and GSM broadcast towers. The PlaceLab software can be used with a PDA or laptop with wireless LAN or GSM connectivity. Localisation accuracy is stated as being less then GPS, with 20–25 m using wireless LAN and 100 m to 150 m for GSM broadcast towers. A similar technique of using RSSI is employed by the power meter node network.

A classical case of using wireless beacons for navigation is presented by Want et al. [6]. The Active Badge project achieved a 5–10 m accuracy using infrared. The main drawback of this platform is that it required line of sight between beacons. An extension of the Active Badge Project was the ORL location system by Ward et al. [7] which developed a prototype network of ultrasonic beacons to perform realtime tracking of tagged mobile devices in an office environment. Other ultrasonic location systems such as the Cricket Mote [8] and the system by McCarthy et al. [9] describe how a network of ultrasonic beacons using time of flight analysis can determine distance position locations.

Klingbeil and Wark [10] developed a wireless sensor network for monitoring human motion and position in an indoor environment. Mobile nodes with inertial and heading sensors were worn by a person inside a building. A Monte-Carlo-based localisation algorithm that used a person's heading, indoor map information, and static node positions was developed and tested. One of the problems with this approach is the tedious sensor calibration required before use.

3. Indoor Localisation Network

Wireless smart meters are used primarily to monitor energy consumption and also for home automation applications. Smart metering protocols include the ZigBee protocol, Advanced Meter Infrastructure [15], Dash7 [16], and wireless M-Bus [17]. Wireless smart meter networks are designed to operate as short range networks like a home and or a large-scale "neighbourhood" area network. We used the ZigBee/802.15.4 wireless communications protocol to implement our smart meter network. ZigBee is a low data rate wireless communications protocol that can operate on devices with limited computing or power resources and cater for large networks of active devices [18].

Supported ZigBee features include Mesh Networking, 64 bit address, data rates: 20 kbps to 250 kbps and simple application profiles. ZigBee operates in the unlicensed ISM 2.4 GHz or 915 MHz frequency band [18]. Current ZigBee protocol radio transceivers have a large indoor range, up to 100 m. The use of RSSI allowed the ZigBee protocol to be adapted for use in the indoor localisation network. Each ZigBee transceiver has a 64 bit ID address which allows a ZigBee network to handle a large number of active nodes.

The localisation network as seen in Figure 1 consisted of three types of nodes: coordinator, power meter, and mobile. Mobile nodes were carried by users to determine their current location. The power meter nodes are used to determine

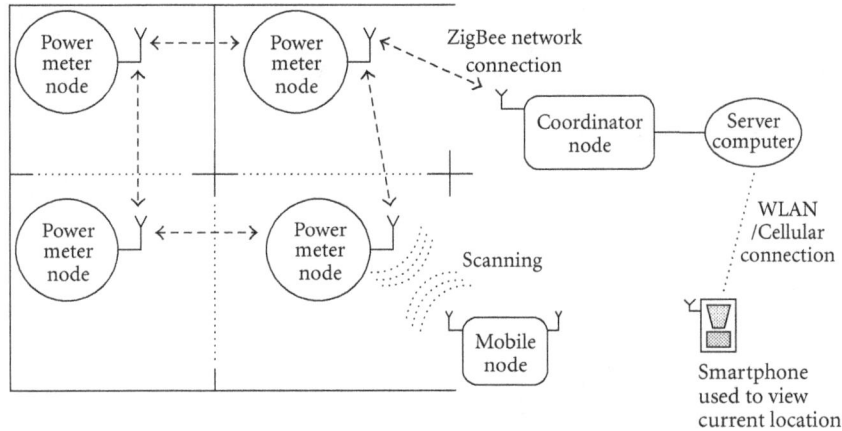

FIGURE 1: Overview of wireless smart metering network for context-aware tracking.

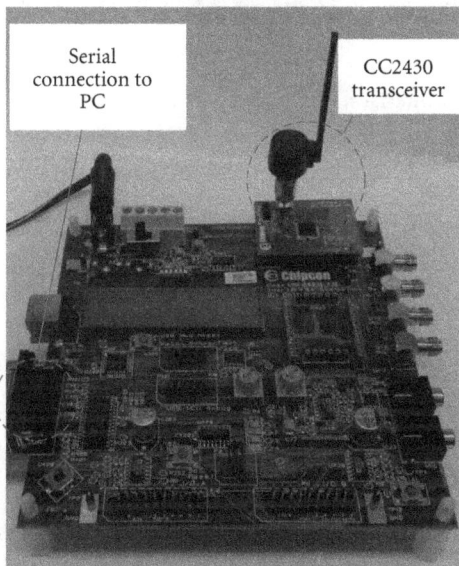

FIGURE 2: Coordinator node platform.

FIGURE 3: Power meter and mobile node platform.

a mobile nodes position via trilateration. The server connected to the coordinator node displays the current positions of the mobile nodes on a building floor plan.

3.1. Coordinator Node. The coordinator node, seen in Figure 2, is used to receive the location coordinates of each mobile node. The coordinator node communicates with the mobile node via the ZigBee mesh routing connection using the power meter nodes. The mobile nodes' positions received by the coordinator can be viewed using the Z-Location graphical user interface [19]. It displays the current locations of mobile nodes on a building floor plan. The coordinator node was implemented with a CC2430 Zig-Bee/802.15.4 module on a SmartRF development board [19]. The coordinator node is connected by a serial connection to a server computer. The server computer tracks the position of the mobile node using the context-aware tracking process,

as described in later sections. The coordinator node is also powered by standard mains electricity.

3.2. Power Meter Node. The power meter node, without power monitoring sensors, seen in Figure 3, communicates to the coordinator node via a ZigBee network connection. The position of each power meter node is known by the coordinator node. The power meter nodes are used by the mobile nodes for trilateration. The power meter node was implemented using the CC2430 ZigBee/802.15.4 wireless transceiver module from Texas Instruments [20]. Each CC2430 module has a unique 64 bit network address used as the power meter node's identifier.

3.3. Mobile Node. The function of a mobile node is to determine a user's position using received signal strength. The mobile node detects power meter nodes in near proximity. The mobile node uses the received signal strength from nearby power meter nodes to calculate its position. The predicted position is then transmitted to the coordinator nodes via

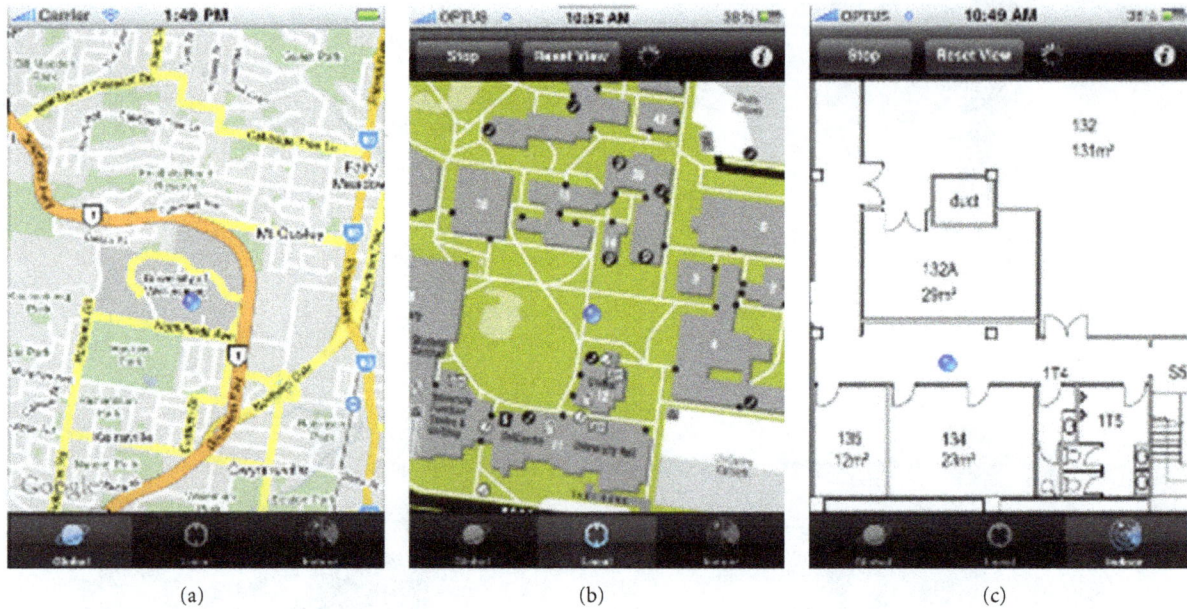

(a)   (b)   (c)

FIGURE 4: Mobile phone graphical user interface.

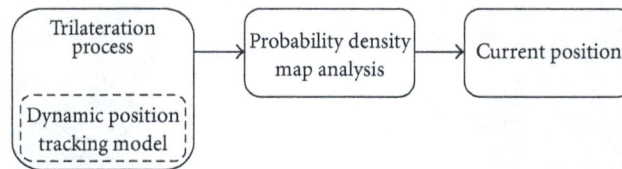

FIGURE 5: Overview of context-aware tracking process.

the power meter node network. The mobile node was implemented using the CC2431 ZigBee/802.15.4 Location Engine Module [20].

3.4. Mobile Phone User Interface. A mobile-phone-based graphical user interface was developed for the iPhone platform to allow users to monitor their indoor position, using the context-aware tracking system, via a cellular or wireless LAN link to the Server computer. A screenshot showing the map interface for indoor/outdoor areas can be seen in Figure 4.

4. Context-Aware Tracking

The context-aware tracking process calculated a person's position in real time using trilateration position estimation and a probability density map. Figure 5 shows an overview of the context-aware tracking process. First, the trilateration position estimation process is used to predict where a mobile node is located. The trilateration process used radio received signal-strength-based range-distance estimations and a Dynamic Position Tracking Model (DPTM) to compute an approximate position. A probability density map was used to determine if the position approximation was valid by using the context-aware information from the floor plan of the indoor environment.

4.1. Trilateration Position Estimation. The trilateration position estimation used the estimated range distances between

the mobile node and the surrounding power meter node network to calculate the mobile node's approximate position. The range distance estimator is first used to approximate the range distance between the mobile node and the power meter nodes. The estimated range distances are then processed by DPTM to predict a set of range distances based on human motion factors such as walking speed. The trilateration algorithm used the predicted range distances between the mobile and power meter nodes to estimate the coordinates of the mobile node.

4.1.1. Range Distance Estimator. The range distances are approximated using the RSSI and the coordinates of the power meter nodes of the indoor localisation network. Figure 6 shows how the mobile nodes interact with the power meter nodes. The mobile node periodically transmitted RSSI Measure messages to the nearest power meter nodes in range (Figure 6(a)). The power meter nodes use the RSSI Measure messages to calculate the received signal strength indicator value. Five messages are used to calculate an averaged RSSI value.

As seen in Figure 6(b), once a series of RSSI Measure messages has been transmitted, the mobile node will then transmit an RSSI and Position Request message to all power meter nodes in range. Each power meter node will then respond with its calculated RSSI value and position. A minimum of three power meter nodes must be in range of

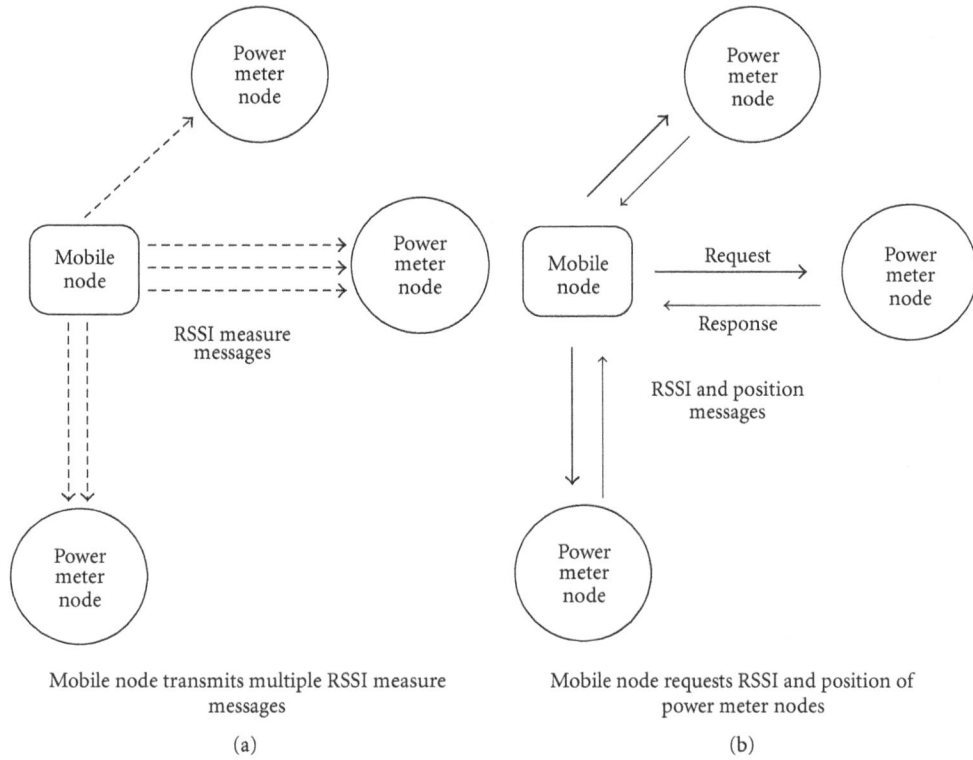

FIGURE 6: Mobile node received signal strength measuring process.

the mobile node in order to accurately approximate its position. If more than three power meter nodes are detected, the nodes with the strongest RSSI values are used. Power meter nodes with weak RSSI values were found to cause inaccuracies when calculating the range-distance estimate.

The RSSI value from the power meter node is used to calculate the range distance between the mobile and power meter nodes using the channel propagation equation as shown in (1). The pathloss channel coefficient is dependent on the surrounding environment. Experiments were conducted in different indoor environments to approximate the optimum value for pathloss channel coefficient:

$$r_i = 10^{(\text{RSSI}-A)/n}, \tag{1}$$

where RSSI is the received signal strength indicator of power meter node (dBm), r_i range distance between ith power meter node and mobile node, n the pathloss coefficient of the channel, and A absolute power received at a distance of 1 m from the transmitter (dBm).

4.1.2. Dynamic Position Tracking Model. In order to track a person's position in real time, a fast localisation process was required. One of the drawbacks of using the trilateration RSSI-based process to localise is that the received signal strength must be averaged over a period of time. Averaged RSSI values are needed to provide suitable accuracy. However, lengthy averaging periods can lead to localisation inaccuracies. For example, a moving mobile node will distort the average received signal strength values measured by the power meter nodes.

One way to improve the accuracy of real-time tracking is to use other odometry motion information, such as a person's directional heading and speed to predict the next position of user. Similar odometry motion-based tracking models have been used for localisation tracking by Klingbeil and Wark [10] and Lau and Chung [21]. The odometry information used in [10] was estimated using motion sensors, while in [21] the speed was estimated using the previous displacement distance of the user. We implemented a similar type of DPTM used in [21], to localise a person's position by using the previous displacement and the typical human walking speed to estimate the speed of the user. The DPTM used the speed and current position to predict the person's next position. Figure 7 shows an overview of the DPTM. The DPTM can be described using the following predictive odometry motion equations concerning distance and velocity:

$$\hat{R}_{\text{est}(i)} = \hat{R}_{\text{pred}(i)} + a\left(\hat{R}_{\text{prev}(i)} - \hat{R}_{\text{pred}(i)}\right), \tag{2}$$

$$\hat{V}_{\text{est}(i)} = \hat{V}_{\text{pred}(i)} + \frac{b}{T_S}\left(\hat{R}_{\text{prev}(i)} - \hat{R}_{\text{pred}(i)}\right), \tag{3}$$

$$\hat{R}_{\text{pred}(i+1)} = \hat{R}_{\text{est}(i)} + \hat{V}_{\text{est}(i)} \cdot T_S, \tag{4}$$

$$\hat{V}_{\text{pred}(i)} = \hat{V}_{\text{est}(i)}, \tag{5}$$

where $\hat{R}_{\text{est}(i)}$ is the estimated range, $\hat{R}_{\text{pred}(i)}$ predicted range, $\hat{R}_{\text{prev}(i)}$ measured range, $\hat{V}_{\text{est}(i)}$ estimated velocity, $\hat{V}_{\text{pred}(i)}$

(1) Estimate current position

$$\widehat{R}_{\text{est}(i)} = \widehat{R}_{\text{pred}(i)} + a\left(\widehat{R}_{\text{prev}(i)} - \widehat{R}_{\text{pred}(i)}\right)$$

a: gain constant

Use previous estimated
position and predicted position

(2) Estimate current velocity

$$\widehat{V}_{\text{est}(i)} = \widehat{V}_{\text{pred}(i)} + \frac{b}{T_s}\left(\widehat{R}_{\text{prev}(i)} - \widehat{R}_{\text{pred}(i)}\right)$$

b: gain constant

T_s: time segment

Use predicted velocity, previous
estimated position, and previous
predicted position

$$\widehat{R}_{\text{pred}(i+1)} = \widehat{R}_{\text{est}(i)} + \widehat{V}_{\text{est}(i)}T_s$$

Use estimated position and
velocity

$$\widehat{V}_{\text{pred}(i+1)} = \widehat{V}_{\text{est}(i)}$$

(4) Predict next position   (3) Predict next velocity

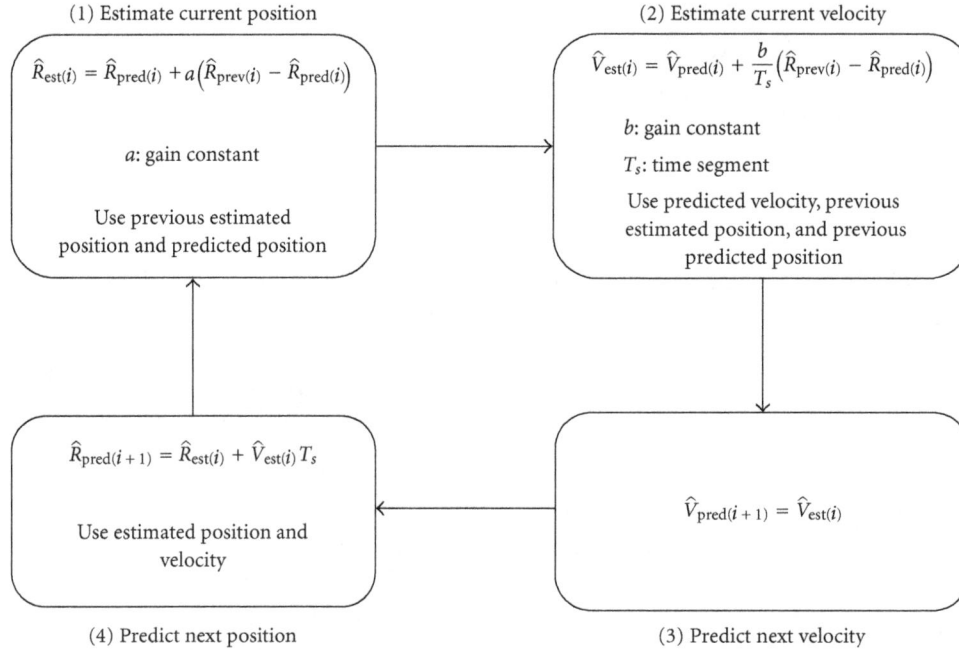

FIGURE 7: Overview of dynamic position tracking model.

predicted velocity, a, b gain constants, and T_S time update period.

The Predicted Velocity stated in (5) is an iterative process, which estimated the range using (3) to adjust the position calculated by the mobile node until the position error has been sufficiently reduced. The speed value was initially set to the average human walking speed of 1.3 m/s (Murray et al. [22]), for the predicted range in (4). A prototype of the DPTM was implemented on the Server Computer connected to the Coordinator node.

4.1.3. Trilateration Algorithm. Once the range distances between all detected power meter nodes and the mobile node has been estimated, a trilateration algorithm is used to calculate the position of the mobile node. The trilateration algorithm used each power meter nodes coordinates and predicted range distance, calculated from the DPTM. At least 3 power meter nodes are required to be detected. If less than 3 power meter nodes are detected within the time update period, then the position and range of the last detected power meter node is used. The detection time period can vary from 10 s to 30 s. The trilateration algorithm can be expressed as a typical linear system of

$$\begin{bmatrix} (x_1 - x)^2 + (y_1 - y)^2 \\ (x_2 - x)^2 + (y_2 - y)^2 \\ \vdots \\ \vdots \\ (x_n - x)^2 + (y_n - y)^2 \end{bmatrix} = \begin{bmatrix} r_1^2 \\ r_2^2 \\ \vdots \\ \vdots \\ r_n^2 \end{bmatrix}, \qquad (6)$$

x, y refer to the position coordinates and R is the vector of range distances between the mobile and power meter nodes. Equation (6) can be expressed as a Linear Least Squares System as seen in

$$\begin{bmatrix} x \\ y \end{bmatrix} = \left(A^T A\right)^{-1} A^T B, \qquad (7)$$

where A and B are as follows:

$$A = \begin{bmatrix} (x_1 - x_n) & (y_1 - y_n) \\ (x_2 - x_n) & (y_2 - y_n) \\ \vdots & \\ \vdots & \\ (x_{n-1} - x_n) & (y_{n-1} - y_n) \end{bmatrix},$$

$$\qquad (8)$$

$$B = \begin{bmatrix} r_1{}^2 - r_n{}^2 - x_1{}^2 + x_n{}^2 - y_1{}^2 + y_n{}^2 \\ r_2{}^2 - r_n{}^2 - x_2{}^2 + x_n{}^2 - y_2{}^2 + y_n{}^2 \\ \vdots \\ \vdots \\ r^2{}_{n-1} - r_n{}^2 - x^2{}_{n-1} + x_n{}^2 - y^2{}_{n-1} + y_n{}^2 \end{bmatrix}.$$

4.2. Context-Aware Tracking Algorithm. Indoor environments are characterised by unpredictable radio propagation channel parameters. These unpredictable channel parameters can cause distortion and multipath interference due to the presence of metallic structures within indoor environments. Since the trilateration algorithm uses RSSI for range

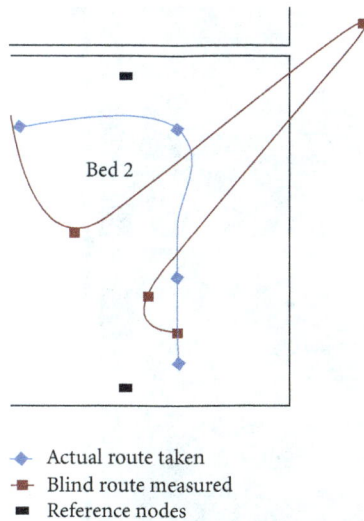

FIGURE 8: Example of an indoor navigation route predicted using RSSI-based trilateration only. Mobile node is predicted to be 0.75 m outside the floor plan.

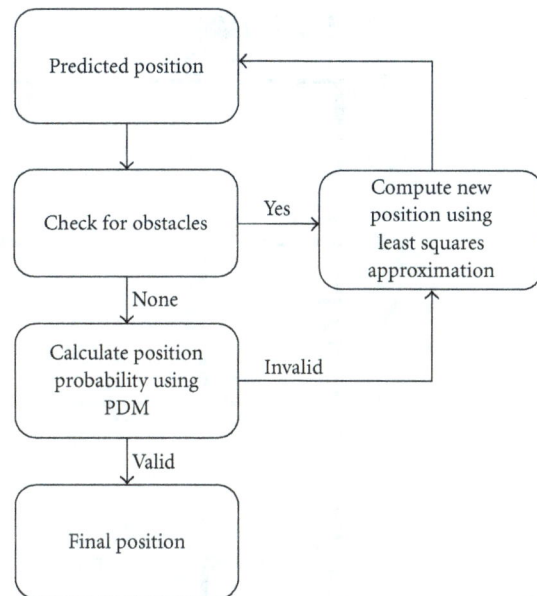

FIGURE 9: Overview of context-aware tracking algorithm.

estimations, wireless channel distortions will cause errors in the predicted position. An example of the influence of the RF interference can be seen in Figure 8 which compares the actual track of a mobile node with the track calculated using the trilateration position estimation. The actual path of the user was measured by walking a known surveyed path with visual observations of the current position being recorded on a regular interval by an observer. In order to overcome wireless-channel-interference-induced position errors, we developed a context-aware localisation process that used a probability density map.

The Probability Density Map (PDM) was used by the context-aware localisation process to determine the likelihood or probability of the mobile nodes estimated position. The PDM consisted of a floor plan with regions mapped with the probability of likely positions available. Position validity was approximated by detecting if the mobile node's track had to move through a wall or barrier, to its predicted position. Figure 9 shows an overview of the PDM process used to check the position of the mobile node. If the mobile node's predicted position is determined to be invalid, Least Squares approximation is used to calculate a new predicted position. The PDM process, as seen in Figure 9, is then repeated until the mobile node's position is found to be valid.

An example of the colour-coded floor plan can be seen in Figure 10. The colour red is used to highlight regions of high probability. The mobile node's position is considered to be valid if it lies in a region of high probability (red colour). The probabilities of each region are predetermined, but, in the future, this will be dynamically updated to reflect changes to the floor plan.

The wall or barrier collisions are also detected using the PDM. The barriers are considered as low probability regions (white colour) on the PDM. If the mobile node's predicted path crossed a barrier, then the DPTM would recursively reestimate the mobile node's predicted position until it becomes valid.

5. Evaluation

An initial trial of the localisation network used six power meter nodes with one mobile and coordinator node. The localisation network was deployed in an indoor area of 72 m^2. The aim of the trial was to evaluate the accuracy of the wireless trilateration mechanism and the context-aware tracking process. Specifically we investigated the following:

(i) the effect of wireless indoor channel propagation on position accuracy,

(ii) accuracy of the context-aware tracking model,

(iii) comparison of the context-aware tracking model accuracy with other existing wireless localisation techniques.

The localisation network of power meter nodes was deployed in a building as seen in Figure 11 on a 1 m spaced grid floor plan. Our tests consisted of a user walking two known paths whilst carrying a mobile node. Figure 11 shows both test paths—Route (A) and Route (B). Route (A) was 14 m long and was selected to have more physical obstacles such as walls, to test its effects on predicting the position of the mobile node. Route (B) was 12.5 m long and was selected to test how the localisation system would perform with a sparse deployment of the power meter nodes. Both routes were planned to be similar in length as to ensure that a similar distance could be walked on either routes. The ground truth or actual path of the user was measured by walking a known surveyed path with visual observations of the current position being recorded on a regular interval by an observer. We consider the test environment to be realistic for evaluating the indoor people tracker. The duration of the test path track was between 5 and 10 minutes (including time for ground truth measurements) and was repeated at least 10 times.

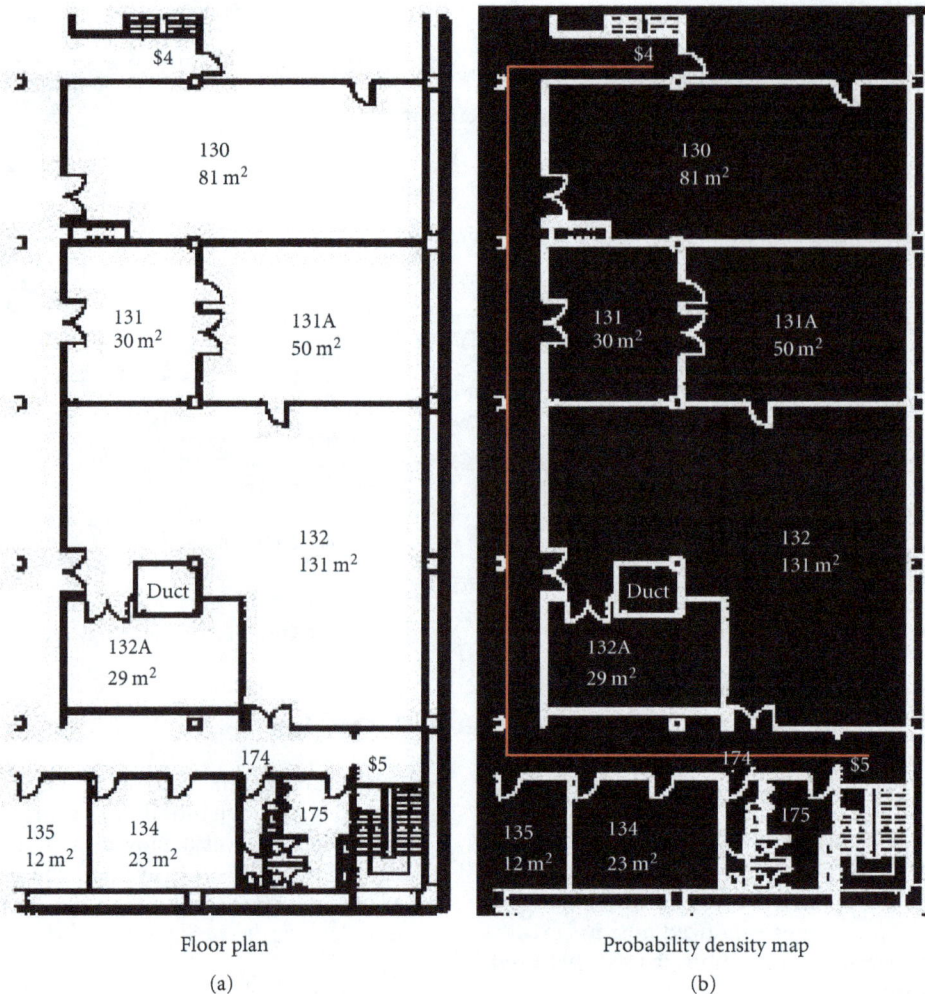

FIGURE 10: Example of floor plan and probability density map.

5.1. Wireless Trilateration. The wireless trilateration algorithm used the range distance estimator and dynamic position model. Figure 12 shows the predicted path tracks calculated using only wireless trilateration. Figure 13 shows the position error per distance travelled, using only the trilateration algorithm. Table 1 shows the average, minimum, and maximum position error only using the trilateration algorithm. The average position error was found to be higher for Route (B) than Route (A). This was because Route (B) had longer distances between the mobile and power meter nodes, which increased the occurrence of large range estimation errors.

5.2. Dynamic Position Tracking Model. We conducted a series of experiments to test the accuracy of the DPTM. The first experiment involved moving the mobile node in a 14 m straight line (not related to Route (A) or (B)) in the test deployment area at a constant walking speed. For initial testing purposes, walking in a straight line path was considered adequate to test the DPTM. The ground-truth positions of the mobile node were visually recorded by an observer. For further work, nondirect testing paths will

be considered. Figure 14 shows the actual, measured, and estimated travelled distances of the mobile node over time. In Figure 14, the distance was the displacement of the mobile node from a starting point. The measure distances are calculated directly from the mobile node's coordinates (using RSSI), and the estimated distance was calculated by using the DPTM. The maximum position error was 6 m using the mobile nodes coordinates. Using the DPTM, the maximum position error was reduced to 3 m.

The second experiment was similar to the first experiment except with a delay of 15 s introduced midway during the test. This tested how the DPTM responded to changes in movement. Figure 15 shows the actual, measured, and estimated distances with constant (continuous) velocity and changing velocity (from moving to stationary to moving again).

In order to calibrate the DPTM for good accuracy, the gain constants: a and b, in (2) and (3) had to be optimised. Figure 15 shows the large position inaccuracy (50%), if the gain constants a and b ((2), (3)) are not calibrated correctly. Using $a = 0.06$ and $b = 0.01$ was found to produce the best results.

TABLE 1: Trilateration and context-aware tracking accuracy.

Position error (m)	Route (A)			Route (B)		
	Mean	Max	Min	Mean	Max	Min
Trilateration algorithm only	0.91	2.59	0.22	2.36	7.49	0.25
Context-aware tracking algorithm	1.21	2.34	0.15	1.46	2.87	0.29

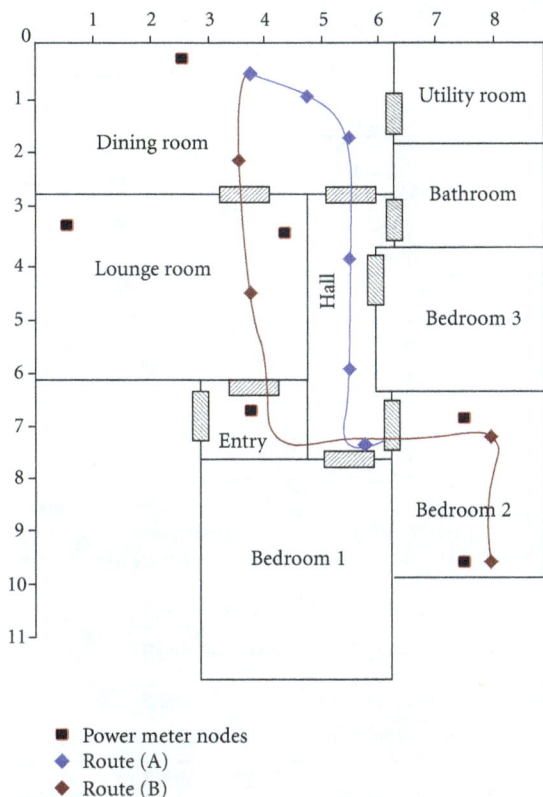

FIGURE 11: Test deployment shown on a 1 m spaced grid floor plan.

5.3. *Context-Aware Tracking.* We evaluated the context-aware tracking algorithm using the same path tracks as shown in Figure 11. Figure 12 shows the predicted path tracks calculated using the context-aware tracking algorithm. We measured the position error due to the distance travelled as seen in Figure 16. Table 1 shows the average, minimum, and maximum position error for the context-aware tracking algorithm.

As with only using the trilateration algorithm, the average position error was found to be higher for Route (B) than Route (A). Again, this was due to Route (B) having longer separation distances between the mobile and power meter nodes, which caused greater range estimation error. The context-aware algorithm compared to the trilateration estimated position had a reduced maximum position error by 9.6% for Route (A) and 67.7% for Route (B). The average position error for Route (A) was slightly higher than the trilateration estimate but was 38.56% lower for Route (B). The context-aware algorithm had a similar error for Route (A) but performed better for Route (B), when compared to the trilateration position estimates. Since the context-aware

algorithm used the probability density map, it was able to compensate for range estimation errors caused by long separation distances between the mobile and power meter nodes.

5.4. *Comparison with Existing Protocols.* There are many wireless protocols and technologies that can be used for indoor and outdoor localisation. We compare the different RF-based protocols to the context-aware tracking system. Table 2 displays a comparison of the different RF transceiver modules used by various localisation technologies.

Although assisted GPS is known to work primarily outdoors, it is able to operate indoors. However, this requires a significant amount of additional computation. We did observe a position error of 57 m indoors using the SigNav subATTO TM3 assisted-GPS module [11] which can operate indoors. Wireless LAN is being used by many localisation applications, as a supplement to assisted GPS. One of the drawbacks of using Wireless LAN is the large number of hotspots required. Biswas and Veloso [23] achieved an accuracy of 0.7 m but required the use of Bayesian filtering which needed a significant amount of computational processing and accurate calibration. Similarly, Raghavan et al. [24] also used Bayesian filtering with Bluetooth for localisation. The drawback of Bluetooth is the long scanning time (10 s–20 s). Localisation using GSM was found to be 20 m, by Otsason et al. [2]. One of the disadvantages of using GSM was its dependence on existing infrastructure.

In comparison, the context-aware tracking system consumes the least amount of power compared to the other protocols. As well, the context-aware system would require the least complex and low-powered infrastructure due to the use of wireless smart metering networks. Apart from GPS, these wireless communication protocols are designed to transfer data files rather than short messages and so require a large amount of processing power and transmit power. Hence, these protocols can be unsuitable for green pervasive computing applications.

5.5. *Discussion.* The context-aware tracking system was able to localise a user's position to an average error of 1.21 m and a maximum error of 2.34 m (Table 1). We have also shown that the context-aware tracking system can operate with widely and irregularly dispersed infrastructure. The advantage of the context-aware system was that it relied on wireless smart metering network infrastructure for localisation. The infrastructure for wireless smart metering is becoming more widely installed in built environments, due to the use of green energy efficiency monitoring. Compared to other systems such as Wireless LAN or GSM, using a smart

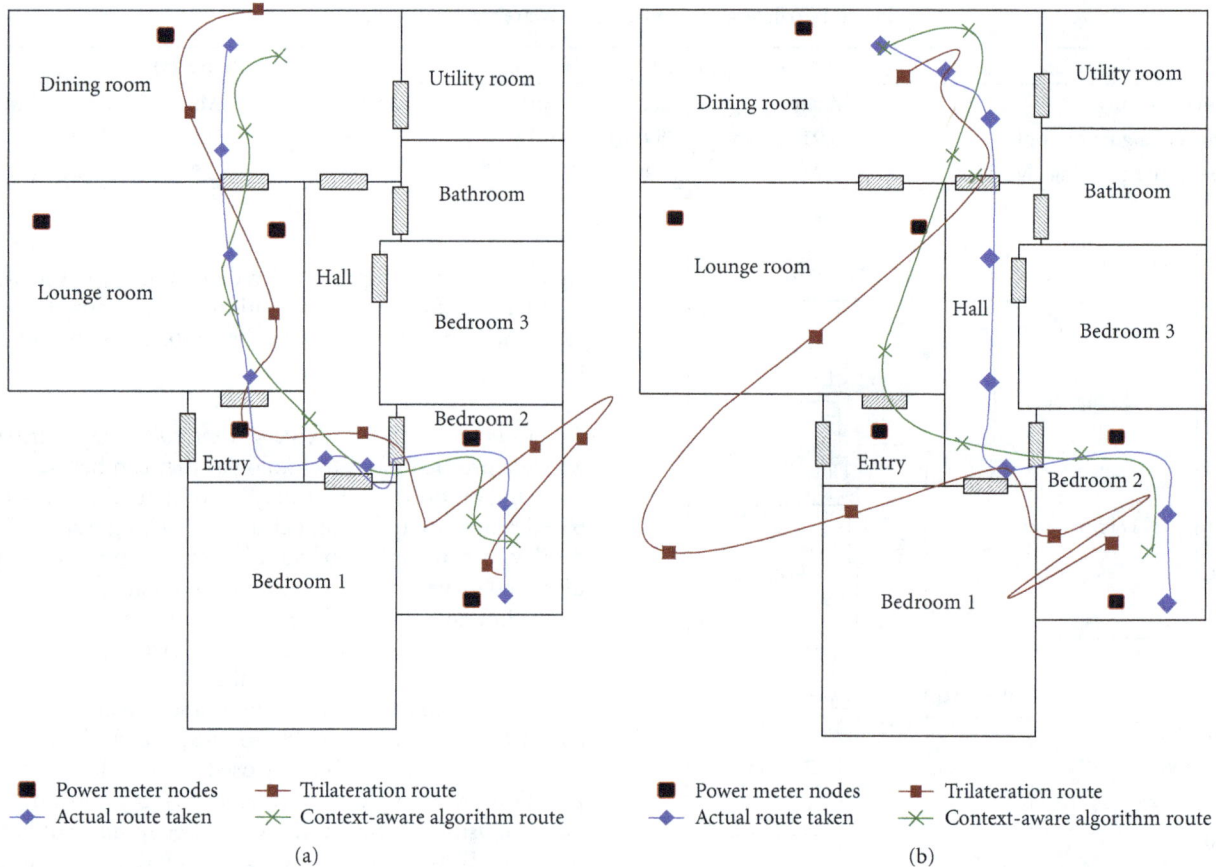

| Power meter nodes | Trilateration route |
| Actual route taken | Context-aware algorithm route |

(a)

| Power meter nodes | Trilateration route |
| Actual route taken | Context-aware algorithm route |

(b)

FIGURE 12: Predicted path track using the context-aware algorithm process and wireless trilateration.

TABLE 2: Comparison with assisted GPS—power and accuracy.

Method	Accuracy—position error (m)	Latency (s)	Power (mW)	TX power (mW)	RX sensitivity (dBm)	Indoor range
Assisted GPS[1]	57 m	30	300	—	−155 dBm	Global
GSM[2]	20 m	—	1162	3.16	−14 dBm	200 m
Wireless LAN[3]	3 m to 30 m	11 s	858	31.62	−74 dBm	100 m
Bluetooth[4]	3 m to 4 m	20 s	231	2.51	−88 dBm	30 m
Context-aware tracking algorithm	1.21 m (Table 1)	2 s	97.2	0 dBm	−92 dBm	30 m

References: [1][11], [2][12], [3][4, 13], [4][4, 14].

metering network provides a solution that is cost effective, has low energy usage, and is easy to install.

6. Conclusion and Further Work

We presented a context-aware tracking system that tracked users in an indoor environment. The context-aware system used a wireless smart metering network that consisted of power meter nodes placed throughout a building. A user carried a mobile node that tracked their current position. A smartphone could be used to view the mobile nodes current position, via a cellular or wireless LAN connection.

The context-aware tracking system localised a person's position by combining wireless trilateration, a dynamic posi-

tion tracking model, and a probability density map. The integral use of these three factors allowed the context-aware tracking system to achieve reasonable localisation accuracy with a sparsely and irregularly dispersed wireless network. This was advantageous compared to other RF-based localisation systems that rely on a dense wireless network to achieve good position accuracy. We found that only using wireless trilateration was not suitable for tracking users in real time due to the lengthy time required by the trilateration algorithm to overcome the effects of the wireless propagation channel. A dynamic position tracking model that estimated a position, based on a user's predicted velocity and heading, was found to improve the latency to localise a person's position. In order to further improve the position estimate,

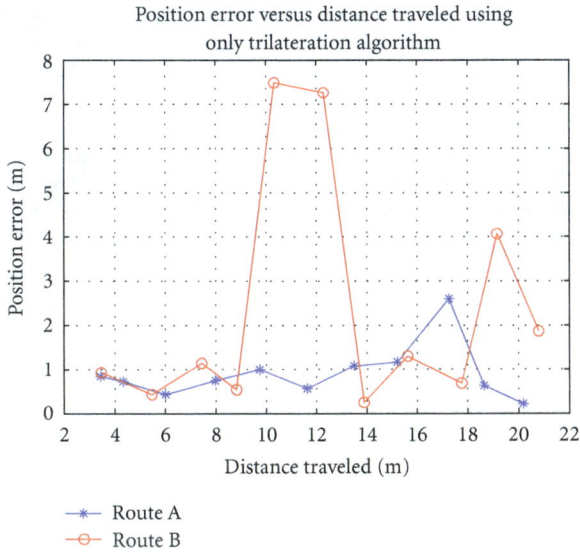

FIGURE 13: Position error versus distance traveled using only trilateration algorithm.

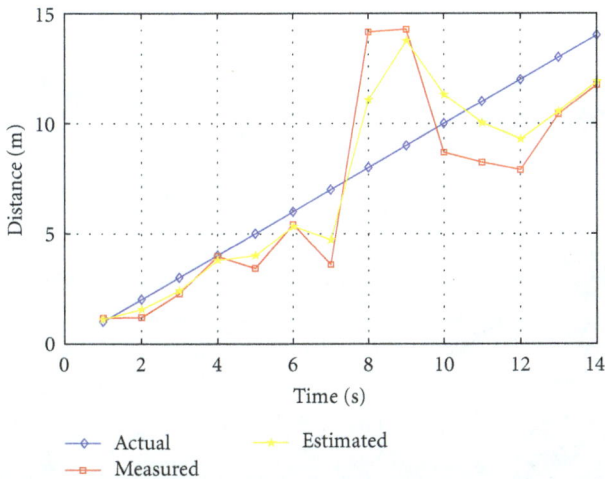

FIGURE 14: Distance versus time with a constant velocity.

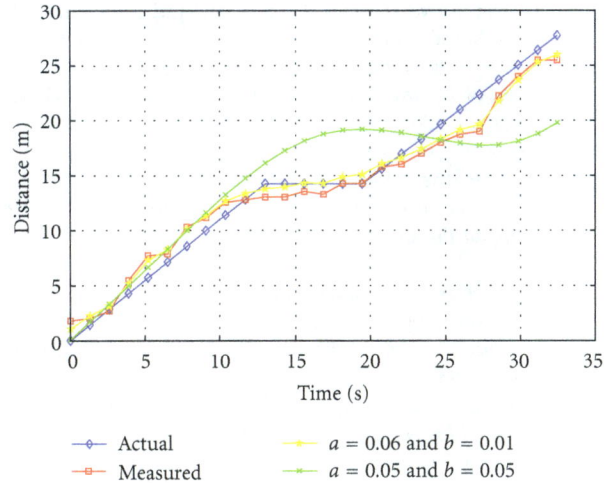

FIGURE 15: Distance versus time with a changing velocity.

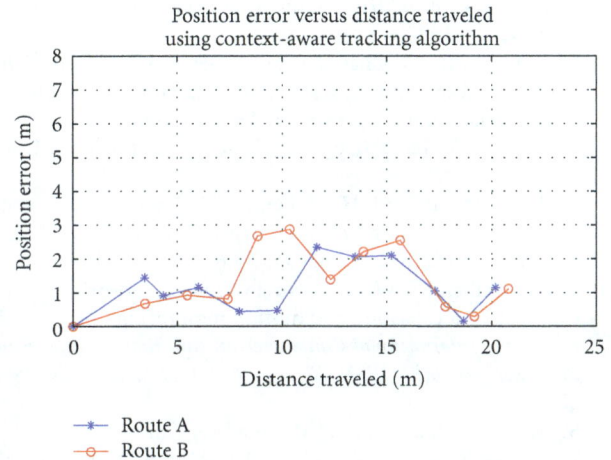

FIGURE 16: Position error versus distance traveled using context-aware tracking algorithm.

a probability density map, based on the floor plan of the indoor area, was used.

An initial trial of the localisation network was conducted using six power meter nodes, a mobile and coordinator nodes. We deployed the context-aware system in an indoor space of 72 m². We measured the channel propagation parameters and the dynamic position tracking model accuracy. We found that, by using the dynamic position tracking model, the error in position location was reduced from 6 m to 3 m. Combining the use of the probability density map allowed the context-aware system to localise a person's position to an average error of 1.21 m and a maximum error of 2.34 m.

The use of a wireless smart metering network for localisation is advantageous as this form of infrastructure will become more widely installed in buildings and other indoor or outdoor environments. This is due to the growing use of energy efficiency monitoring. Wireless smart metering infrastructure is also the least complex and low power consuming, when compared to other localisation network infrastructures. Compared to other localisation systems, the context-aware system is able to operate on little power while providing suitable accuracy for green pervasive applications.

Further work involves extensive testing with multiple operating mobile nodes and over a larger test region. Other areas of investigation involve looking at how 3-dimensional localisation can be achieved with more in-depth context awareness of the surrounding environment.

References

[1] V. Schwarz, A. Huber, and M. Tuchler, "Accuracy of a commercial UWB 3D location/tracking system and its impact on LT application scenarios," in *Proceedings of the IEEE International Conference on Ultra-Wideband (ICU '05)*, pp. 599–603, Windisch, Switzerland, 2005.

[2] V. Otsason, A. Varshavsky, A. LaMarca, and E. de Lara, "Accurate GSM indoor localization," in *Proceedings of the International Conference on Ubiquitous Computing (UBICOMP '05)*, vol. 3660 of *Lecture Notes in Computer Science*, pp. 141–158, Tokyo, Japan, 2005.

[3] F. Seco, A. R. Jimenez, C. Prieto, J. Roa, and K. Koutsou, "A survey of mathematical methods for indoor localization," in *Proceedings of the 6th IEEE International Symposium on Intelligent Signal Processing (WISP '09)*, pp. 9–14, Madrid, Spain, 2009.

[4] H. Liu, H. Darabi, P. Banerjee, and J. Liu, "Survey of wireless indoor positioning techniques and systems," *IEEE Transactions on Systems, Man and Cybernetics Part C*, vol. 37, no. 6, pp. 1067–1080, 2007.

[5] J. Hightower, A. LaMarca, and I. Smith, "Practical lessons from place lab," *IEEE Pervasive Computing*, vol. 5, no. 3, Article ID 1673364, pp. 32–39, 2006.

[6] R. Want, A. Hopper, V. Falcao, and J. Gibbons, "Active badge location system," *ACM Transactions on Information Systems*, vol. 10, no. 1, pp. 91–102, 1992.

[7] A. Ward, A. Jones, and A. Hopper, "A new location technique for the active office," *IEEE Personal Communications*, vol. 4, no. 5, pp. 42–47, 1997.

[8] N. B. Priyantha, A. Chakraborty, and H. Balakrishnan, "The Cricket location-support system," in *Proceedings of the 6th Annual International Conference on Mobile Computing and Networking (MobiCom '00)*, pp. 32–43, ACM, Boston, Mass, USA, 2000.

[9] M. McCarthy, P. Duff, H. Muller, and C. Randell, "Accessible ultrasonic positioning," *IEEE Pervasive Computing*, vol. 5, no. 4, Article ID 1717372, pp. 86–93, 2006.

[10] L. Klingbeil and T. Wark, "A wireless sensor network for real-time indoor localisation and motion monitoring," in *Proceedings of the International Conference on Information Processing in Sensor Networks (IPSN '08)*, pp. 39–50, St. Louis, Mo, USA, 2008.

[11] SigNav, TM3-02EH Timing Module User Guide, 2010.

[12] LinkSprite Technologies, GSM/GPRS Module User Manual, 2008.

[13] WI2WI, W2SW0001—802.11 b/g System-in-Package, 2008.

[14] Sena Technologies, OEM Bluetooth-Serial Module, Parani-ESD100, 2008.

[15] S. W. Luan, J. H. Teng, S. Y. Chan, and L. C. Hwang, "Development of a smart power meter for AMI based on ZigBee communication," in *Proceedings of the International Conference on Power Electronics and Drive Systems (PEDS '09)*, pp. 661–665, Taipei, Taiwan, 2009.

[16] Dash7 Alliance, Dash7, 2009, http://www.dash7.org/.

[17] M-Bus, "Wireless M Bus," 1999, http://www.m-bus.com/.

[18] ZigBee Alliance, ZigBee Specification, 2006, http://www.zigbee.org/.

[19] Texas Instruments, Texas instruments CC2431 zigbee development kit, 2007.

[20] Texas Instruments, CC2431 system-on-chip for 2.4 ghz zigbee/ieee 802.15.4 with location engine, 2007.

[21] E.-E.-L. Lau and W.-Y. Chung, "Enhanced rssi-based real-time user location tracking system for indoor and outdoor environments," in *Proceedings of the International Conference on Convergence Information Technology (ICCIT '07)*, pp. 1213–1218, IEEE Computer Society, 2007.

[22] M. Murray, A. Drought, and R. Kory, "Walking patterns of normal men," *The Journal of Bone and Joint Surgery*, vol. 46, no. 2, pp. 335–360, 1964.

[23] J. Biswas and M. Veloso, "WiFi localization and navigation for autonomous indoor mobile robots," in *Proceedings of the IEEE International Conference on Robotics and Automation (ICRA '10)*, pp. 4379–4384, Anchorage, Alaska, USA, 2010.

[24] A. Raghavan, H. Ananthapadmanaban, M. Sivamurugan, and B. Ravindran, "Accurate mobile robot localization in indoor environments using bluetooth," in *Proceedings of the IEEE International Conference on Robotics and Automation (ICRA '10)*, pp. 4391–4396, Anchorage, Alaska, USA, 2010.

6

GNSS Spoofing Detection Based on Signal Power Measurements: Statistical Analysis

V. Dehghanian,[1, 2] J. Nielsen,[1] and G. Lachapelle[1]

[1] *Position Location And Navigation (PLAN) Group, University of Calgary, Calgary, AB, Canada T2N 1N4*
[2] *Department of Math Physics and Engineering, Calgary, Mount Royal University, AB, Canada T3E 6K6*

Correspondence should be addressed to V. Dehghanian, vdehghanian@mtroyal.ca

Academic Editor: Jinling Wang

A threat to GNSS receivers is posed by a spoofing transmitter that emulates authentic signals but with randomized code phase and Doppler values over a small range. Such spoofing signals can result in large navigational solution errors that are passed onto the unsuspecting user with potentially dire consequences. An effective spoofing detection technique is developed in this paper, based on signal power measurements and that can be readily applied to present consumer grade GNSS receivers with minimal firmware changes. An extensive statistical analysis is carried out based on formulating a multihypothesis detection problem. Expressions are developed to devise a set of thresholds required for signal detection and identification. The detection processing methods developed are further manipulated to exploit incidental antenna motion arising from user interaction with a GNSS handheld receiver to further enhance the detection performance of the proposed algorithm. The statistical analysis supports the effectiveness of the proposed spoofing detection technique under various multipath conditions.

1. Introduction

The received GNSS signal power at the output of a 3 dB gain hemispherical linearly polarized antenna at ground level is approximately −130 dBm [1]. This makes GNSS receivers susceptible to nearby noise jammers and standoff spoofers (SS) that can easily transmit power levels well above −130 dBm. A high processing gain based on a long integration time is often the only option available to overcome a noise jammer. Nevertheless, if the GNSS receiver undergoes random motion, then the channel decorrelates quickly such that attaining such large processing gains to overcome jamming is neither feasible nor desirable from an operational perspective. Also a jammer is relatively easy to locate with radio direction finding and to potentially disable as its spectrum is significantly larger than the ambient noise [2, 3]. In addition, the noise jammer is at least detectable as the spectral power in the affected GNSS receiver band will be abnormally high. Hence the jammer can deny service but the user is aware of being jammed, limiting the damage potential of the jammer. A more insidious threat is the standoff spoofer that broadcasts a set of replicas of the authentic satellite vehicle (SV) signals visible to the mobile GNSS receiver [2]. Disruption of GNSS services is achieved by randomly modulating the code phase over a small region of the overall Code Delay Space (CDS) that is commensurate with a target area. The spoofing attack is assumed to happen during the acquisition stage. Therefore, it is not possible to identify the SS signal based on the code phase as corresponding to an outlier navigation solution. The SS is assumed to remain synchronized with currently visible GNSS signals and then transmit a set of signals that would correspond to the typical GNSS signals observable by a receiver in the target area. Note that an effective SS does not necessarily synthesize a specific counterfeit location for a specific GNSS receiver but rather aims to disrupt GNSS services over a general target area by matching the Doppler offset of the replicated SV signals and adjusting the code phase such that it is commensurate with the intended target region. Hence the GNSS receiver cannot easily detect the contribution of these counterfeit signals as obvious outliers. An unaware receiver computes the navigation solution based on the SS generated counterfeit signals which are passed on to the user as being reliable with potentially damaging consequences.

GNSS receivers tethered to a wireless data service provider will typically provide the user with an aided-GNSS (AGNSS) service, significantly reducing the CDS corresponding to a physical area of several square kilometres [4]. Hence there is a diminishing gain for the spoofer attempting to affect a larger target area than this. Hence the counterfeit SS navigation solutions will be construed as plausible. As such, receiver-autonomous integrity monitoring (RAIM) and fault detection and exclusion (FDE) are ineffective in discriminating signals sourced from the envisioned SS [5].

The typical handheld consumer GNSS receiver coherently integrates the signal for about 10 to 20 ms resulting in a correlation peak in the CDS that has a spread in Doppler of about 100 Hz, which is commensurate with the Doppler spread of typical urban traffic (<50 km/hr) [6]. Even if the GNSS receiver is equipped with other ancillary sensors such that the receiver velocity vector is independently known, this cannot be used to discriminate the SS signal as multipath Doppler spreading is approximately equivalent for both the SS and the authentic SV signals.

Note that the receiver processing gain used for suppressing a jamming signal is not effective in the case of the spoofer signal. Consequently, the spoofer transmit power can be orders of magnitude less than that of the noise jammer, which makes the spoofer source much more difficult to locate and disable through radio direction finding and beam forming.

The objective of this paper is to address a computationally efficient processing technique that can be added to relatively unsophisticated consumer grade GNSS receivers to discriminate the spoofer signals transmitted by an SS. The proposed processing is based on estimating and comparing the receiver signal power with a set of thresholds to verify the authenticity of the signal. The detection problem is formulated based on a Rayleigh fading multipath scenario. Nevertheless, it is shown that although suboptimal, the deduced expressions can be utilized for spoofing detection in a generalized Rician multipath channel with minimal performance degradation.

The proposed technique is further extended to include incidental motions of the handheld receiver, instigated through the user interaction with the handset device, in the form of spatial translation and polarization rotation. User interaction with the handheld creates variability in the antenna response, which can be transformed into a diversity gain that adds to the general processing gain of the receiver [7–10]. This processing gain enhances the estimation of the received signal power of the correlation peaks, that, is necessary information in spoofer discrimination. A case study based on GPS L1 C/A signals is developed to demonstrate the effectiveness of the proposed technique. Nevertheless, this technique can be directly extended for other GNSS signal formats such as GPS L2 C/A and GLONASS.

The rest of the paper is organized as follows. In Section 2 the system definition and the assumptions are given.

Section 3 formulates a multihypothesis detection problem and focuses on the statistical evaluation of the proposed technique, with the conclusions provided in Section 4.

2. System Definition

This paper considers the analysis of individual GNSS satellite signals, while realizing that simultaneous processing of the available GNSS signals provide extra diversity that can be used to further improve the performance of the proposed spoofer detection technique. The received complex GNSS baseband signal is denoted here by

$$g(t) = A(t)s_o(t) + w(t), \tag{1}$$

where the signal component of $g(t)$ is represented by $s(t) = A(t)s_o(t)$, where "t" is time, $A(t)$ is the channel response to the incident signal at the antenna, and $s_o(t)$ is the complex baseband component of the satellite signal, which can be written as

$$s_o(t; \tau, \Delta f) = d(t - \tau)c(t - \tau)e^{j(2\pi\Delta f t + \psi)}, \tag{2}$$

where $d(t)$ is the navigation data modulation, $c(t)$ is the Pseudo Random Noise (PRN) code, τ is the code phase, Δf represents carrier frequency offset (due to the Doppler of the GNSS signal as well as any frequency offset of the receivers local oscillator), and ψ is the initial phase offset. $s_o(t)$ is known to the receiver except for the navigation data, the code phase, the carrier frequency offset, and the initial phase offset ψ. The received signal, $g(t)$, is corrupted by additive noise (WGN) which has an equivalent complex baseband representation denoted by $w(t)$. It is assumed that $w(t)$ is a complex normal random process, independent of the signal, and has a Power Spectral Density (PSD) that is constant within the bandwidth of the received signal.

The GNSS receiver integrates a temporal snapshot of $g(t)$ over the interval of $t \in [0, T_I]$, where T_I is typically smaller than the duration of one navigation data bit (20 ms). The signal snapshot of $g(t)$ is collected by the receiver and then despread by a locally generated copy of $s_o(t)$ during the initial acquisition. The initial acquisition is typical of a multihypothesis detection in which the receiver searches the Code Doppler Space (CDS) for the frequency offset Δf and the code delay, τ [11, 12]. Note that the initial phase offset ψ is not known to the receiver during the initial acquisition and as such the output of the despreading matched filter is a random complex variable.

The despread baseband signal samples at a correlator output are represented by

$$
\begin{aligned}
x_{n;\tau,\Delta f} &= \frac{1}{T_I} \int_{(n-1)T_I}^{nT_I} g(t)s_o\left(t; \hat{\tau}, \Delta \hat{f}\right)^* dt \\
&= \frac{1}{T_I} \int_{(n-1)T_I}^{nT_I} A(t)dt \\
&\quad + \frac{1}{T_I} \int_{(n-1)T_I}^{nT_I} w(t)s_o\left(t; \hat{\tau}, \Delta \hat{f}\right)^* dt \\
&= s_n + w_n, \quad n = 1, \ldots, N,
\end{aligned}
\tag{3}
$$

where "$*$" is a complex conjugate, the subscript "n" denotes the nth signal sampling interval which extends over $t \in [(n-1)T_I, nT_I]$, and s_n, w_n are the postintegration signal and the WGN components, respectively. In addition, $\hat{\tau}$ and $\Delta\hat{f}$ represent the estimated code phase and Doppler based on the initial acquisition which consists of a maximum likelihood search over the CDS of a signal sample, $x_{n;\tau,\Delta f}$, such that

$$\left\{\hat{\tau}, \Delta\hat{f}\right\} = \max_{\arg\{\tau,\Delta f\}} \left(\left|x_{n;\tau,\Delta f}\right|^2\right), \tag{4}$$

where $\{\hat{\tau}, \Delta\hat{f}\}$ are the maximum likelihood (ML) estimates of the true code phase and Doppler frequency, respectively. The estimated code phase and Doppler are then passed on to the tracking loops to facilitate further receiver processing. Consequently, N signal samples, namely, $\mathbf{x} = [x_1, \ldots, x_N]$, can be collected and used for spoofer detection.

3. Theoretical Analysis of Spoofer Detection

A hypothetical scenario is considered based on an SS transmitting spoofing signals in an urban environment as shown in Figure 1. The authentic signal and the spoofer signal are affected by multipath fading and therefore, the received signal power is random in space and polarization. In other words, multipath fading results in signal power fluctuation when the receiver is spatially translated or undergoes polarization changes due to rotation. Unlike the authentic signal power which is insensitive to signal power variations arising from pathloss in the target area (this is due to the fact that the satellite-receiver separation is approximately unchanged over a period of several minutes), the spoofer signal power varies with variation in the spoofer-receiver separation. An empirical model of order n can be utilized to model the spoofer signal power variation due to pathloss as

$$\rho_d^{(sp)} = \rho_{R1}^{(sp)} - 10n\log_{10}\left(\frac{d}{R1}\right), \tag{5}$$

where $R1$ is a reference range, d is the spoofer-receiver range, n is the pathloss exponent, $\rho_d^{(sp)}$ is the average spoofer SNR at d, and $\rho_{R1}^{(sp)}$ is the average received spoofer SNR at $d = R1$ in dBs. Note that, for the spoofer to be effective, the average received spoofer signal power needs to be higher than that of the authentic signal in the target area. Therefore, the received signal power from a standoff spoofer varies significantly with range due to pathloss, meaning that the spoofer signal power is abnormally higher than that of the authentic signal when the receiver is in the proximity of the standoff spoofer. This characteristic of the spoofer signal can be exploited to limit the effectiveness of the SS in its target area based on comparing the measured signal power against a preset threshold.

As stated earlier, a receiver records N signal samples with each of these $n = 1, \ldots, N$ signal samples belonging to one of the three hypotheses, namely, the noise hypothesis H0, the authentic signal hypothesis H1, and the spoofer signal hypothesis H2 as

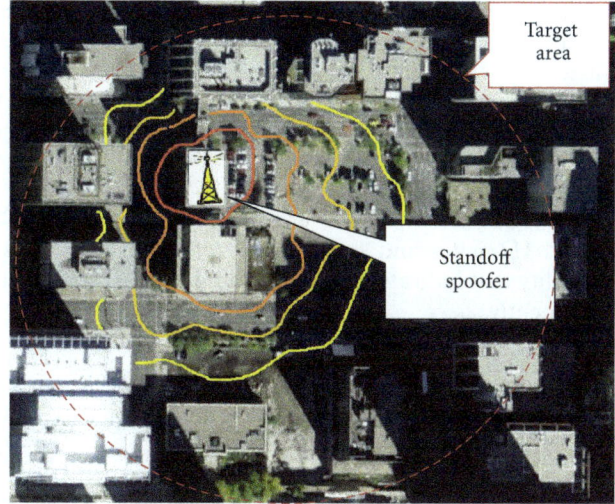

FIGURE 1: Hypothetical stand-off spoofer scenario in an urban canyon. The contours represent the random average spoofer signal power. The average authentic signal power is approximately constant over the entire area given that the receiver-satellite range is approximately unchanged and hence the pathloss.

$$\text{H0}: x_n = \frac{1}{T_I}\int_{(n-1)T_I}^{nT_I} w(t)s_o(t)^* dt \approx w_n$$

$$\text{H1}: x_n = \frac{1}{T_I}\int_{(n-1)T_I}^{nT_I} A^{(a)}(t)dt$$
$$+ \frac{1}{T_I}\int_{(n-1)T_I}^{nT_I} w(t)s_o(t)^* dt \approx s_n^{(a)} + w_n \tag{6}$$

$$\text{H2}: x_n = \frac{1}{T_I}\int_{(n-1)T_I}^{nT_I} A^{(sp)}(t)dt$$
$$+ \frac{1}{T_I}\int_{(n-1)T_I}^{nT_I} w(t)s_o(t)^* dt \approx s_n^{(sp)} + w_n,$$

where the normalization $\int_t^{t+T} |s_o(t)|^2 dt = 1$ is assumed and $\hat{\tau}$, $\Delta\hat{f}$ are suppressed for notational convenience. $A^{(a)}(t)$ and $A^{(sp)}(t)$ represent channel gains associated with the authentic and the spoofer signals, respectively. Consequently, a detection variable, $r = h(\mathbf{x})$, can be formulated to decide between the three hypotheses of (6) based on comparing "r" with a set of thresholds, ρ_1, ρ_2, as shown in Figure 2. Note that $h(\mathbf{x})$ is a function that maps the measured signal samples $\mathbf{x}$ to a single variable, r, which is a sufficient statistic with respect to H0, H1, and H2. As will be shown in Section 3.1, r can be found from the probability density functions (PDF) of $\mathbf{x}$ [13] or alternatively from the PDFs of "r" which are denoted here by

$$\text{H0}: f_{r|\text{H0}}(r)$$

$$\text{H1}: f_{r|\text{H1}}(r) \tag{7}$$

$$\text{H2}: f_{r|\text{H2}}(r),$$

where $f(\cdot)$ denotes a PDF.

One optimization criteria for determining the thresholds (ρ_1, ρ_2) is based on minimizing the probability of error, namely,

$$P_e = 1 - \left[\sum_{i=0}^{2} P(\mathrm{H}i \mid \mathrm{H}i) P(\mathrm{H}i) \right], \qquad (8)$$

where $P(\mathrm{H}i)$ for $i = 0, 1, 2$ are the probabilities of H0, H1, and H2 states and $P(\mathrm{H}i|\mathrm{H}i)$ denotes the conditional probability that indicates that of deciding $\mathrm{H}i$ if $\mathrm{H}i$ is correct. Consequently,

$$P_e = 1 - [P(\mathrm{H0})F_{r|\mathrm{H0}}(\rho_1) + P(\mathrm{H1})(F_{r|\mathrm{H1}}(\rho_2) - F_{r|\mathrm{H1}}(\rho_1))$$
$$+ P(\mathrm{H2})(1 - F_{r|\mathrm{H2}}(\rho_2))], \qquad (9)$$

where $F_{r|\mathrm{H}_i}(\cdot)$ denotes a cumulative distribution function (CDF) of the random variable "r" under H_i. As can be seen from (9), P_e is a function of the authentic signal, the spoofer, and the noise statistics. Therefore, any optimization based on minimizing the probability of error hinges on knowing the spoofer signal statistics, which is not available to an unsuspecting receiver given the capricious nature of a spoofer.

Alternatively, a second optimization can be made based on maximizing the probability of detection for a given probability of false alarm. Assuming $\rho_2 > \rho_1$, the threshold ρ_1 can be determined based on selecting a probability of false alarm P_{FA1} as

$$P_{\mathrm{FA1}} = \Pr\{r > \rho_1 \mid \mathrm{H0}\} \cup \Pr\{r > \rho_2 \mid \mathrm{H0}\}$$
$$= \Pr\{r > \rho_1 \mid \mathrm{H0}\}. \qquad (10)$$

Therefore,

$$\rho_1 = 1 - F_{r|\mathrm{H0}}(P_{\mathrm{FA1}}). \qquad (11)$$

As is evident from (11), ρ_1 depends on P_{FA1} and on the noise statistic, which is approximately known to the receiver.

As stated earlier, the average spoofer SNR is not known and varies with varying spoofer-receiver separation due to pathloss, spoofer transmit power variations, and so forth. However, the average authentic line of sight (LOS) SNR is approximately known, given that the average LOS CNR of GNSS signals at the ground level is typically within [40–50] (dB-Hz), which maps into a postprocessing SNR of approximately [10–20] dB based on 1 ms of coherent integration. This a priori information can be used to determine a second threshold, ρ_2, based on selecting a probability of false alarm associated with H2 as

$$P_{\mathrm{FA2}} = \int_{\rho_2}^{\infty} f_{r|\mathrm{H1}}(r) dr. \qquad (12)$$

Given that the satellite geometry is not known to an acquiring receiver, it is reasonable to assume that SVs are approximately uniformly distributed in the sky. Consequently, the PDF of the average post processing SNR of the authentic GNSS signals, $\rho^{(a)}$, can be approximated as

$$\rho^{(a)} \sim U(\rho_L, \rho_H), \qquad (13)$$

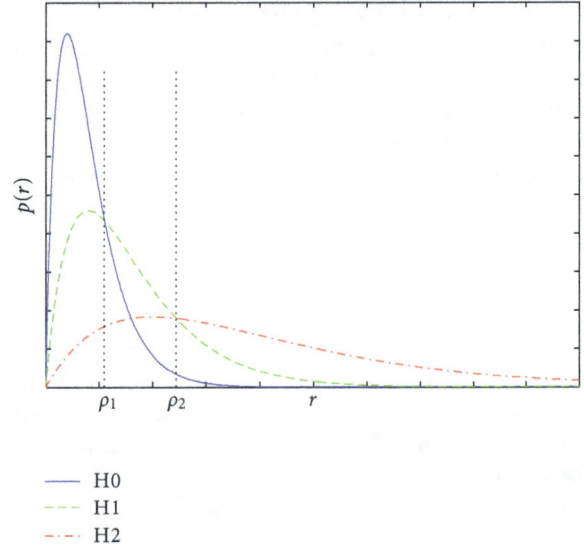

FIGURE 2: A diagram of the PDFs of the detection variable "r" under H0, H1, and H2 hypotheses.

where $U(\rho_L, \rho_H)$ denotes a uniform PDF and $\rho_L \approx 10, \rho_H \approx 20$ dB denotes the lower and the upper bounds of the uniform distribution. Consequently,

$$f_{r|\mathrm{H1}}(r) = \int f_{r|\mathrm{H1}}\left(r \mid \rho^{(a)}\right) f_{\rho^{(a)}}\left(\rho^{(a)}\right) d\rho^{(a)}$$
$$= \frac{1}{\rho_H - \rho_L} \int_{\rho_L}^{\rho_H} f_{r|\mathrm{H1}}\left(r \mid \rho^{(a)}\right) d\rho^{(a)}. \qquad (14)$$

ρ_2 can be numerically computed by inserting (14) into (12). Finally, the probability of detection associated with H2 can be computed as

$$P_{\mathrm{D2}}(\rho_2) = P_{22} = \int_{\rho_2}^{\infty} f_{r|\mathrm{H2}}(r) dr. \qquad (15)$$

3.1. Spoofer Detection Based on a Moving Antenna. As stated earlier, the typical usage mode of a handheld receiver includes incidental motion in the form of spatial translation, polarization rotation, and blocking of the receiver antenna. It is known that any temporal variation in the antenna response results in a temporal signal decorrelation in a multipath environment such that extra diversity branches can be made available for receiver processing [7–10].

To exploit the extra processing gain arising from antenna motion, the statistical properties of $\mathbf{x}$ need to be considered. Distribution of scatterers in many multipath environments such as indoors or urban areas approximately resembles a uniform sphere of scatterers [9, 14]. The correlation coefficient between signal samples, $\mathbf{s} = [s_1, \ldots, s_n]$, collected through spatially translating an antenna over an arbitrary trajectory in a Rayleigh fading environment that resembles a sphere of scatterers can be shown to be [7]

$$[\mathbf{C_s}]_{mn} = \eta \mathrm{sinc}(k_0 p_{mn}), \qquad (16)$$

where $k_0 = 2\pi/\lambda$ is the propagation constant, $p_{mn} = |\mathbf{p}_m - \mathbf{p}_n|$ is the spatial separation between the antenna positions at

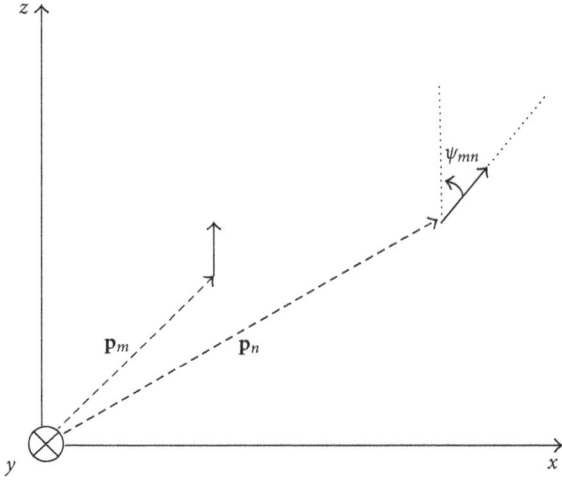

FIGURE 3: Spatial translation and polarization rotation of a GNSS handheld antenna.

which signal samples x_m and x_n are collected (see Figure 3), and η is the variance of $\mathbf{s}$. Consequently, x_m are statistically uncorrelated if the spatial separation between the antenna positions at which the samples are measured are greater than half a carrier wavelength (in GPS L1 frequency this maps into a spatial separation of 10 cm), resulting in the approximation of $\mathbf{C_s} \approx \eta \mathbf{I}_N$, where $\mathbf{I}_N$ is an $N \times N$ identity matrix.

Rotation is another form of user interaction with a handheld receiver that results in variation in antenna's polarization. Variation in antenna's polarization is known to result in signal decorrelation. It can be shown that the covariance of signal samples measured through polarization rotation of a handheld antenna follows from [15] as

$$[\mathbf{C_s}]_{mn} = \eta \cos \psi_{mn}, \tag{17}$$

where ψ_{mn} is the angular separation of the polarization vectors at which signal samples x_m and x_n are collected (see Figure 3). Note that only three degrees of freedom are realizable based on a polarization rotation of a linearly polarized antenna [16]. Therefore, $N \leq 3$ uncorrelated signal samples are realizable based on a polarization rotation.

A combination of polarization rotation and spatial translation can be utilized to further increase the number of diversity branches [9]. The cross-covariance arising from a combined spatial-polarization translation of a GNSS handheld antenna can be shown to be [9]

$$[\mathbf{C_s}]_{mn} = \eta \, \text{sinc}(k_0 p_{mn}) \cos \psi_{mn}. \tag{18}$$

As can be seen from (18), the receiver motion in the form of a combined translation in space and rotation of polarization decorrelates the received signal and therefore can be utilized to synthesize several diversity branches useful for receiver processing.

3.1.1. Uncorrelated Rayleigh Fading Channels. Assume that N uncorrelated signal samples are obtained based on a combined spatial and polarization translation of a GNSS handheld receiver in an uncorrelated Rayleigh fading channel such that $\mathbf{C_s} = \eta \mathbf{I}_N$. Consequently, $\mathbf{x} = \{x_1, \ldots x_N\}$ are jointly CN zero-mean RVs with $\mathbf{x} = \mathbf{s} + \mathbf{w} \sim \text{CN}(0, \mathbf{C_x})$. $\mathbf{C_x} = \mathbf{C_s} + \mathbf{C_w}$ denotes a covariance matrix of $\mathbf{x}$ with $\mathbf{C_w}$ as the noise covariance matrix. To simplify the expressions to follow and without any loss of generality, the noise covariance is normalized such that $\mathbf{C_w} = \mathbf{I}_N$. Therefore, the SNR can be written as

$$\rho = \eta. \tag{19}$$

Consequently, the signal samples collected by a moving antenna in an uncorrelated Rayleigh fading channel are distributed according to $\mathbf{x} \sim \text{CN}(0, (\eta+1)\mathbf{I})$. It can be shown that

$$r = \mathbf{x}^H \mathbf{x} \tag{20}$$

is a sufficient statistics with respect to the hypotheses H0, H1, and H2 and as such is the detection variable [13]. The thresholds (ρ_1, ρ_2) can be found by determining the PDF of r and substituting in (10)–(15) for any given P_{FA1}, P_{FA2}.

Note that r is a measure of the received signal power. Therefore, the detection problem is based on comparing the received signal power, r, with a set of thresholds, (ρ_1, ρ_2), to determine the authenticity of the received signal. For the spoofer to be effective, the spoofer signal power must be higher than that of the authentic signal in the target area such that the ML search in the CDS results in selecting the spoofer signal which has the largest correlation peaks. Therefore, r, which is a measure of the received signal power, can be utilized to discriminate the spoofer from the authentic signals.

3.1.2. Generalized Rician Channels. In a generalized Rician channel, the channel gain, $A(t)$, is a random variable distributed according to $\text{CN}(\mu, \eta/2)$ where $\mu = |\mu|\sqrt{2}\exp(j\alpha(t))$ is the complex mean with $\alpha(t)$ denoting the phase of the complex mean and $\eta/2$ is the variance of the in-phase and the quadrature-phase Gaussian components of the channel gain. Consequently, $\mathbf{x}$ are jointly CN RVs and are distributed according to $\mathbf{x} = \mathbf{s} + \mathbf{w} \sim \text{CN}(\overline{\mathbf{m}}, \mathbf{C_x})$, where $\overline{\mathbf{m}} = \mu[e^{j\alpha_1}, \ldots, e^{j\alpha_N}]^T$ is an $N \times 1$ vector with α_i denoting the phase, $\mathbf{C_x} = \mathbf{C_s} + \mathbf{C_w}$ is a covariance matrix of $\mathbf{x}$, and $\mathbf{C_w} = \mathbf{I}_N$ is the normalized noise covariance. In a Rician channel, the average SNR, ρ, can be defined as $\rho = 2|\mu|^2 + \eta$, and the magnitude of the mean, $|\mu|$, and the variance, η, are related through the Rician K-factor, κ, such that $\kappa = |\mu|^2/\eta$. Since the angle of arrival (AoA) of the dominant signal component is not known to the receiver, μ cannot be estimated and therefore $\overline{\mathbf{m}}$ and subsequently κ are unknown which makes it impossible to formulate a sufficient statistics based on a likelihood ratio test [13]. Nevertheless, as will be shown here, the performance of the spoofer detection is approximately insensitive to the variation in the K-factor, κ, and to the cross-correlation of signal samples $\mathbf{s}$ as long as the cross-correlation remains moderately low, for example, <0.7. This is reasonable since diversity gain arising from combining equal-power diversity branches remains mostly

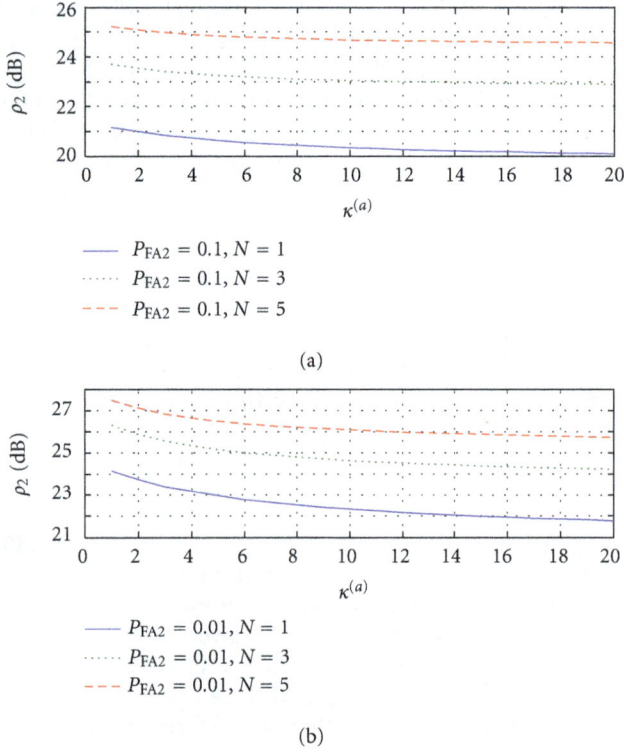

$P_{\mathrm{FA2}} = 0.1, N = 1$
$P_{\mathrm{FA2}} = 0.1, N = 3$
$P_{\mathrm{FA2}} = 0.1, N = 5$

(a)

$P_{\mathrm{FA2}} = 0.01, N = 1$
$P_{\mathrm{FA2}} = 0.01, N = 3$
$P_{\mathrm{FA2}} = 0.01, N = 5$

(b)

FIGURE 4: Threshold ρ_2 computed for various values of K-factor and different P_{FA2} and N at $\rho^{(a)} = 15$ (dB).

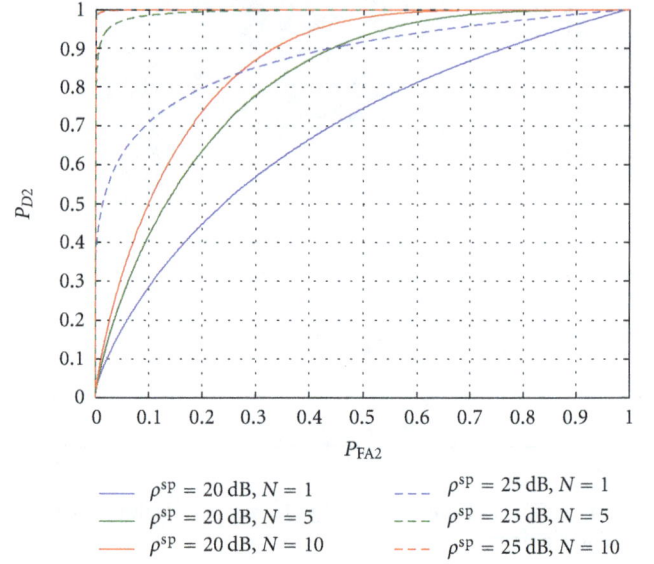

$\rho^{\mathrm{sp}} = 20\,\mathrm{dB}, N = 1$ $\rho^{\mathrm{sp}} = 25\,\mathrm{dB}, N = 1$
$\rho^{\mathrm{sp}} = 20\,\mathrm{dB}, N = 5$ $\rho^{\mathrm{sp}} = 25\,\mathrm{dB}, N = 5$
$\rho^{\mathrm{sp}} = 20\,\mathrm{dB}, N = 10$ $\rho^{\mathrm{sp}} = 25\,\mathrm{dB}, N = 10$

FIGURE 5: ROC curves based on $\rho^{(a)} = 15$ and $\kappa^{(a)} = \kappa^{(\mathrm{sp})} = 1$.

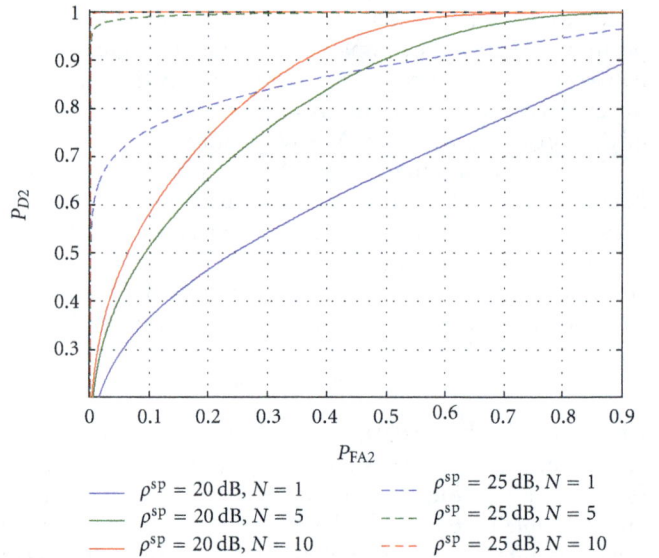

$\rho^{\mathrm{sp}} = 20\,\mathrm{dB}, N = 1$ $\rho^{\mathrm{sp}} = 25\,\mathrm{dB}, N = 1$
$\rho^{\mathrm{sp}} = 20\,\mathrm{dB}, N = 5$ $\rho^{\mathrm{sp}} = 25\,\mathrm{dB}, N = 5$
$\rho^{\mathrm{sp}} = 20\,\mathrm{dB}, N = 10$ $\rho^{\mathrm{sp}} = 25\,\mathrm{dB}, N = 10$

FIGURE 6: ROC curves based on $\rho^{(a)} = 15$ and $\kappa^{(a)} = 10$, $\kappa^{(\mathrm{sp})} = 1$.

unchanged for branch cross-correlations <0.7. Therefore, the suboptimal detection variable of (20) can be applied for spoofing detection in a generalized Rician channel with small performance degradation. Figure 4 shows ρ_2 being computed from (14) for various authentic and spoofer channel K-factors ($\kappa^{(a)}$ and $\kappa^{(\mathrm{sp})}$) and based on $\rho^{(a)} = 15$ (dB) and $\rho^{(\mathrm{sp})} = 20$ (dB), two typical $P_{\mathrm{FA2}} = 0.01, 0.1$, and for $N = 1,3,5$. As can be seen from Figure 4, smaller K-factors result in larger ρ_2 values. This is due to the increased uncertainty in the received signal power as the K-factor decreases. Nevertheless, the variation in ρ_2 is limited to a few dB and as such the K-factor does not play a major role in the optimization problem and may be ignored in the expense of slightly lower performance. Therefore, (20) can be applied to a generalized Rician channel as a suboptimal detector. In addition, as can be seen from this figure, a larger N results in a smaller ρ_2 for the same performance requirement of P_{FA2}. This is due to the diversity gain made available through the extra diversity branches for $N > 1$.

Figures 5 and 6 show the receiver operating characteristics (ROC) based on the detection variable of (20) and for $\rho^{(a)} = 15\,\mathrm{dB}$, various N and $\rho^{(\mathrm{sp})}$, and based on $\kappa^{(a)} = \kappa^{(\mathrm{sp})} = 1$ and $\kappa^{(a)} = 10$, $\kappa^{(\mathrm{sp})} = 1$. As can be seen in these figures, the detection performance improves with increasing the number of diversity branches, N. Also, larger $\rho^{(\mathrm{sp})}$ result in a better detection due to the further separation between the PDFs of the authentic and spoofing signals. When a stronger LOS signal component is present ($\kappa^{(a)} = 10$ in Figure 6), a better detection performance is realized due to

the reduced uncertainty in the authentic signal power. Note that setting $P_{\mathrm{FA2}} = 0$ in (12) results in $\rho_2 = \infty$ and therefore $P_{\mathrm{D2}} = 0$. This corresponds to a receiver not equipped with any spoofer detection.

To provide an alternative measure of performance improvement, the probability of error P_e of (9) can be used. Figure 7 shows P_e for various N, $\kappa^{(a)}$, and $\rho^{(a)} = 15$ and based on $\kappa^{(\mathrm{sp})} = 1$ and $\rho^{(\mathrm{sp})} = 25$ (dB). As can be seen from this figure, P_e is approximately independent of the exact value of K-factor, which emphasizes the previous observations of Figure 4 where it was shown that the threshold ρ_2 is not very sensitive to the variations of the K-factor. P_e decreases rapidly with increasing the number of diversity branches. The latter demonstrates the performance enhancement arising from

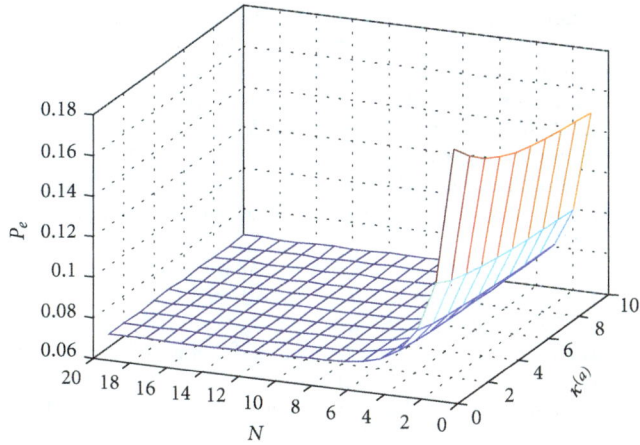

FIGURE 7: P_e for various N and $\kappa^{(a)}$, $\rho^{(a)} = 15$, and based on $P_{\text{FA1}} = P_{\text{FA2}} = 0.1$, $\kappa^{(\text{sp})} = 1$ and $\rho^{(\text{sp})} = 25$ (dB).

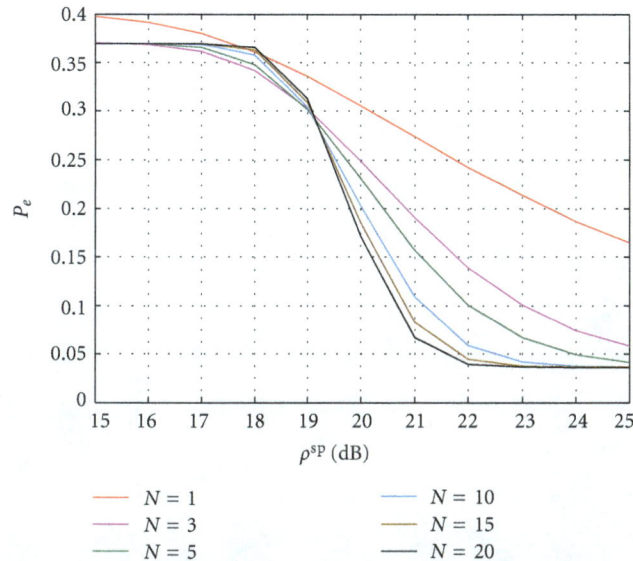

FIGURE 9: P_e for $N = 1$–20, $P_{\text{FA1}} = 0.01$ and $P_{\text{FA2}} = 0.1$, $\kappa^{(a)} = \kappa^{(\text{sp})} = 1$, $\rho^{(a)} = 15$, and $\rho^{(\text{sp})} = 10$–25 (dB).

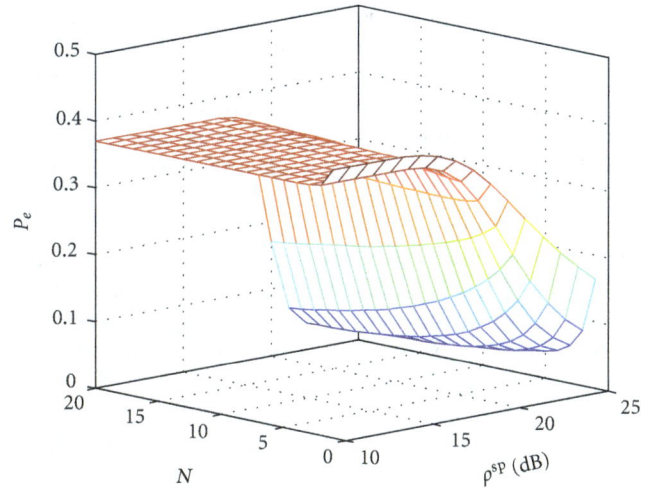

—— $N = 1$	—— $N = 10$
—— $N = 3$	—— $N = 15$
—— $N = 5$	—— $N = 20$

FIGURE 8: P_e versus $\rho^{(\text{sp})}$ and for $\rho^{(a)} = 15$ (dB) and $P_{\text{FA1}} = 0.01$ and $P_{\text{FA2}} = 0.1$, $\kappa^{(a)} = \kappa^{(\text{sp})} = 1$, and for $N = 1, 3, 5, 10, 15, 20$.

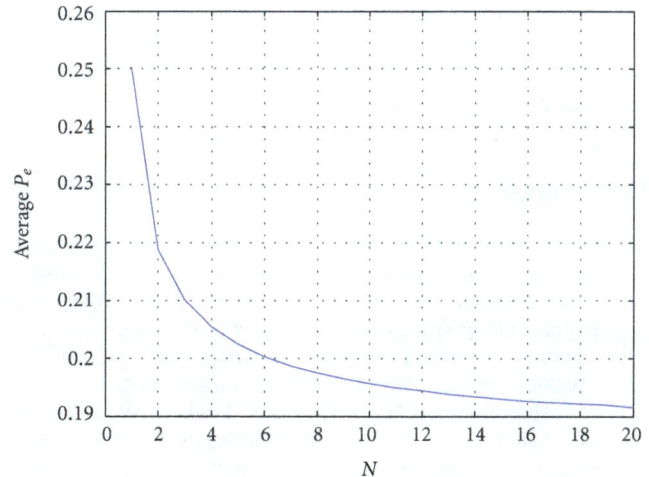

FIGURE 10: $\overline{P}_e$ for various N based on $P_{\text{FA1}} = 0.01$ and $P_{\text{FA2}} = 0.1$, $\kappa^{(a)} = \kappa^{(\text{sp})} = 1$ and $\rho^{(a)} = 15$ (dB), and for $\rho^{(\text{sp})}$ averages over $\rho_{\min} = 15$, $\rho_{\max} = 25$ (dB).

the extra diversity branches made available through utilising a moving antenna.

Figure 8 shows P_e for $\rho^{(a)} = 15$ (dB) and various $\rho^{(\text{sp})}$, $\kappa^{(a)} = \kappa^{(\text{sp})} = 1$, and for $N = 1, 3, 5, 10, 15, 20$. Similarly, larger N and larger $\rho^{(\text{sp})}$, which provide a better separation between the authentic and the spoofing signal PDFs, result in a smaller P_e. This is further demonstrated in Figure 9 where P_e is plotted for $N = 1$–20, $\kappa^{(a)} = \kappa^{(\text{sp})} = 1$, and $\rho^{(\text{sp})} = 10$–25 (dB). Note that, as the PDFs of the authentic and the spoofer signals become more alike, for example, $\rho^{(a)} = \rho^{(\text{sp})}$, P_e becomes larger.

As stated earlier, the spoofer signal power is affected by pathloss quantifiable by (5). As a result, the received spoofer SNR varies with the proximity to the spoofer transmitter. To provide an average measure of performance enhancement

arising from utilizing the proposed technique, the average probability of error, $\overline{P}_e$, can be defined as

$$\overline{P}_e = \int_{\rho_{\min}}^{\rho_{\max}} P_e\left(\rho^{(\text{sp})}\right) d\rho^{(\text{sp})}. \tag{21}$$

Figure 10 shows $\overline{P}_e$ for various N based on $\kappa^{(a)} = \kappa^{(\text{sp})} = 1$ and $\rho^{(a)} = 15$ (dB), and for $\rho_{\min} = 15$, $\rho_{\max} = 25$ (dB). The effect of diversity is further emphasized in this figure where the average probability of error decreases by increasing N such that $\overline{P}_e \simeq 0.19$ for $N = 20$, implying that the proposed technique is very effective in reducing the spoofer effectiveness in the target area.

4. Conclusions

A multi-hypothesis detection problem was formulated based on a likelihood ratio test applicable to GNSS spoofing detection. A straight forward spoofing detection technique based on signal power measurements was proposed and was shown to be effective for verifying the authenticity of the received GNSS signals in urban multipath environments, meaning that the spoofer signal power is abnormally higher than that of the authentic signal when the receiver is in the proximity of the standoff spoofer.

The proposed processing was further extended to exploit extra diversity branches made available based on a moving handheld receiver and was shown to further improve the spoofer detection performance. Unlike the previously proposed antispoofing techniques, the proposed technique does not require any hardware modification and can be readily applied to any handheld GNSS receiver with minimal firmware changes. It was shown that the proposed technique is largely insensitive to uncertainties in the statistical properties of the multipath channel as long as the collected signal samples are not strongly correlated. A suboptimal detector was proposed and effectively applied to a generalized Rician channel in which the channel parameters are not available to the receiver. An extensive statistical analysis was performed to assess the performance of the proposed technique. It was shown that the average probability of error can be reduced to less than 20% in a typical urban environment.

References

[1] E. D. Kaplan and C. J. Hegarty, *Understanding GPS: Principles and Applications*, Artech House, Norwood, Mass, USA, 2006.

[2] B. M. Ledvina, W. J. Bencze, B. Galusha, and I. Miller, "An in-line anti-spoofing device for legacy civil GPS receivers," in *Proceedings of the International Technical Meeting (ITM '10)*, pp. 868–882, San Diego, Calif, USA, January 2010.

[3] T. E. Humphreys, B. M. Ledvina, M. L. Psiaki, B. W. O'Hanlon, and P. M. Kintner, "Assessing the spoofing threat: development of a portable gps civilian spoofer," in *Proceedings of the 21st International Technical Meeting of the Satellite Division of the Institute of Navigation (ION GNSS '08)*, pp. 1198–1209, Savanna, Calif, USA, September 2008.

[4] F. S. T. V. Diggele, *A-GPS: Assisted GPS, GNSS, and SBAS*, Artech House, 2009.

[5] L. Scott, "Location assurance," *GPS World*, vol. 18, no. 7, pp. 14–18, 2007.

[6] W. C. Jakes, *Microwave Mobile Communications*, IEEE Press, New York, NY, USA, 1974.

[7] A. Broumandan, J. Nielsen, and G. Lachapelle, "Indoor GNSS signal acquisition performance using a synthetic antenna array," *IEEE Transactions on Aerospace and Electronic Systems*, vol. 47, no. 2, pp. 1337–1350, 2011.

[8] V. Dehghanian, *Generalized diversity gain of a mobile antenna [Ph.D. thesis]*, Electrical and Computer Engineering, University of Calgary, Calgary, Canada, 2011.

[9] V. Dehghanian, J. Nielsen, and G. Lachapelle, "Combined spatial-polarization correlation function for indoor multipath environments," *IEEE Antennas and Wireless Propagation Letters*, vol. 9, pp. 950–953, 2010.

[10] V. Dehghanian, J. Nielsen, and G. Lachapelle, "Diversity gain through antenna blocking," *International Journal of Antennas and Propagation*, vol. 2012, Article ID 735080, 6 pages, 2012.

[11] G. E. Corazza, C. Caini, A. Vanelli-Coralli, and A. Polydoros, "DS-CDMA code acquisition in the presence of correlated fading—part I: theoretical aspects," *IEEE Transactions on Communications*, vol. 52, no. 7, pp. 1160–1168, 2004.

[12] C. Caini, G. E. Corazza, and A. Vanelli-Coralli, "DS-CDMA code acquisition in the presence of correlated fading—part II: Application to cellular networks," *IEEE Transactions on Communications*, vol. 52, no. 8, pp. 1397–1407, 2004.

[13] S. Kay, *Fundamentals of Statistical Signal Processing: Detection Theory*, vol. 2, Prentice-Hall, Upper Saddle River, NJ, USA, 1998.

[14] H. L. VanTrees, *Detection Estimation and Modulation Theory: Part IV*, John Wiley & Sons, NewYork, NY, USA, 2002.

[15] V. Dehghanian, J. Nielsen, and G. Lachapelle, "Combined spatial-polarization correlation function for indoor multipath environments," in *Proceedings of the 7th International Symposium on Wireless Communication Systems (ISWCS '10)*, pp. 874–876, September 2010.

[16] A. S. Y. Poon and D. N. C. Tse, "Degree-of-freedom gain from using polarimetric antenna elements," *IEEE Transactions on Information Theory*, vol. 57, no. 9, pp. 5695–5709, 2011.

7

A Novel Quasi-Open Loop Architecture for GNSS Carrier Recovery Systems

Muhammad Tahir,[1] Letizia Lo Presti,[1] and Maurizio Fantino[2]

[1] *Dipartimento di Elettronica e Telecomunicazioni (DET), Politecnico di Torino, Corso Duca degli Abruzzi 24, 10129 Torino, Italy*
[2] *Navigation Laboratory, Istituto Superiore Mario Boella, Via P.C. Boggio 61, 10138 Torino, Italy*

Correspondence should be addressed to Muhammad Tahir, muhammad.tahir@polito.it

Academic Editor: Carles Fernández-Prades

The problem of designing robust systems to track global navigation satellite system (GNSS) signals in harsh environments has gained high attention. The classical closed loop architectures, such as phase locked loops, have been used for many years for tracking, but in challenging applications their design procedure becomes intricate. This paper proposes and demonstrates the use of a quasi-open loop architecture to estimate the time varying carrier frequency of GNSS signals. Simulation results show that this scheme provides an additional degree of freedom to the design of the whole architecture. In particular, this additional degree of freedom eases the design of the loop filter in harsh environments.

1. Introduction

In global navigation satellite systems (GNSSs) the relative motion of both GNSS satellites and the user causes a Doppler effect, which results in a large frequency shift in the carrier and in the code of the received signal [1]. Precise estimation of this frequency shift is one of the most demanding requirements for GNSS receivers, because only an accurate tracking of the carrier frequency and Doppler shift allows the receiver to work properly, enabling reliable estimates of position velocity and timing (PVT). In any GNSS receiver, the acquisition stage provides an initial coarse estimation of the frequency shift, which is subsequently refined by the tracking systems. They are generally implemented in the form of closed loops, that is, phase lock loops (PLLs) and frequency lock loops (FLL), which track respectively the phase and the frequency of the incoming carrier, [2].

The main building block of a closed loop architecture is the loop filter. The design of a loop filter has been extensively addressed in the literature regarding the continuous-time PLLs, and many results and methods exist for different scenarios. However, modern receivers work in the discrete-time domain, and so PLLs and FLLs are digital systems, whose loop filters are often designed starting from some equivalent analog prototypes, by adopting transformation techniques from the analog to the digital domain. These tracking loops are therefore de facto digital approximations of analog loops, whose quality breaks down as the integration time increases. A valid assumption for this approximation is that the product BT between the loop noise bandwidth B, and the integration time T remains close to zero. As this product increases, the loop becomes unstable, as discussed in [3]. However, in high dynamic and weak signal applications, it is necessary to work with large BT values. In these cases the design of the loop filters based on analog-to-digital transformations does not work properly. Therefore, other techniques have been proposed in literature, which are more robust when low update rates (long integration times) or large bandwidths are required. They are the controlled root method for the design of digital filters [4], the direct design of loop filters in digital domain by minimization criterion [5], the loop architectures based on Kalman filters [6], and the fuzzy loop architectures [7].

In this paper we propose an alternative solution, based on a novel quasi-open loop architecture, which relieves the stress on the loop filter in terms of stability. In the first step of our study we considered open loop techniques, since some advantages over the closed loop counterpart schemes can be achieved. For example, it is a known fact that PLLs are vulnerable to fading effects, typically associated with urban environments, and cycle slips [8]. Moreover, because of their closed loop structure, they need a long acquisition time before attaining the loop lock, and this may be a serious drawback when the Doppler significantly varies within this time interval. The use of open loop architectures can solve these problems. Several open loop architectures have been proposed in literature; most of them are based on the use of the Fast Fourier Transform (FFT) for the estimation of the frequency error between the incoming carrier and a local carrier replica. However, this block processing structure, typical of FFT-based methods, increases the system complexity. In our scheme we do not use FFT for frequency estimation; in our method we estimate the frequency by using conventional PLL/FLL discriminators, for example, the Kay's estimator [9], but with a different update of the frequency of the numerically controlled oscillator (NCO). In particular, instead of updating the NCO at each coherent integration time as in the closed loop architectures, the update is performed after NT integrations intervals, where N is an integer. Between two updating epochs the architecture works as an open loop. Using this approach we are actually working with a three-rate scheme, unlike the conventional two-rate closed loop systems. We have found that this additional degree of freedom can help us to ease the design of the loop filter in weak signal conditions, where it is necessary to extend the coherent integration time.

In a weak signal scenario the extension of the coherent integration time is the only possible option, since the increased processing gain allows the successful recovery of weak and extremely weak signals. However, the maximum coherent integration time in a GNSS receiver is limited by a variety of factors: the presence of navigation data modulation, the stability issues, and the demodulation losses due to a frequency mismatch. This effect is particularly detrimental when the loop update time increases (long integration time) and the incoming carrier has a high Doppler shift [3]. The first problem is usually solved by estimating the navigation data or using external assistance. The main contribution of this paper is related to the second issue, that is, the stability problem, which is solved by adopting an inherently stable architecture. We will show that the proposed architecture can work with extended integration times, which are not possible in the traditional closed loop schemes under the same conditions. In order to solve the third problem, methods of compensation of the user dynamics are necessary. For this purpose, PLLs are generally assisted by other systems, for example, an FLL or a tightly coupled inertial navigation system (INS).

The paper is organized as follows. The GNSS signal model is introduced in Section 2. Then, a general overview of the traditional GNSS carrier recovery systems is presented in Section 3, stressing the aspects which motivate the scheme proposed in this paper. Section 4 shows the novel quasi-open loop scheme for carrier recovery, proposed to overcome some limitations of the traditional schemes. Some simulation results are presented in Section 5, showing the feasibility of the proposed method in terms of tracking jitter performance,

in a dynamic scenario and with weak signals. On the basis of these results some possibilities to refine the proposed architecture are also discussed. Conclusions are drawn in Section 6.

2. Signal Model

The main purpose of this paper is to propose a novel architecture for tracking the frequency of the incoming GNSS signal which combines the good properties of both open loop and closed loop architectures. These systems recover the carrier from the received signal, which can be written, after downconversion to intermediate frequency (IF), sampling and quantization, as

$$x_w[n] = \sqrt{2P}a[n]\cos[2\pi(f_0 + f_d(nT_s))nT_s + \psi_s] + w[n],$$
$$\tag{1}$$

where P is the total received power of the useful signal, $a[n] = d[n]c[n]$ is the useful signal ($c[n]$ is the code and $d[n]$ the navigation message), f_0 is an intermediate frequency, $f_d(nT_s)$ is a frequency shift (which can be time variant), ψ_s is the initial phase (the phase for $n = 0$), and $w[n]$ is a noise component. This signal is obtained by sampling an analogical signal at a sampling frequency $f_s = 1/T_s$. The carrier signal is completely characterized by its *instantaneous* phase

$$\varphi_s(nT_s) = 2\pi(f_0 + f_d(nT_s))nT_s + \psi_s. \tag{2}$$

The noise term $w[n]$ is a realization of a Gaussian random process $W[n]$, with flat power spectral density $N_o/2$ over the receiver band B_r, and with power $\sigma_w^2 = N_oB_r$. $W[n]$ is not generally a white sequence with a flat power spectral density, due to front-end filtering. However, a white model is justified because the bandwidth of the front-end filter is usually close to the Nyquist sampling frequency, that is, $B_r = f_s/2$.

The purpose of the carrier recovery is to estimate $\varphi_s(nT_s)$ in order to construct a local oscillator (LO) of the type

$$x_{cr}[n] = \cos\hat{\varphi}_s(nT_s), \tag{3}$$

where $\hat{\varphi}_s(nT_s)$ is the estimate of $\varphi_s(nT_s)$, and the subscript cr denotes that this signal is recovered from the carrier. Therefore, the LO (producing $x_{cr}[n]$) can be considered the output of the carrier recovery system. We assume that both data and code can be wiped off and we focus on the operations performed by the carrier recovery system by considering the ideal signal

$$x_s[n] = x_s(nT_s) = \cos[2\pi(f_0 + f_d(nT_s))nT_s + \psi_s], \tag{4}$$

where $f_d(nT_s)$ is an unknown frequency, which varies with an unknown rule, ψ_s is also unknown, while f_0 is a known nominal frequency. In real cases, the signal $x_s(nT_s)$ is also affected by noise, and then the carrier has to be recovered from a noisy version of (4).

3. Traditional Tracking Architectures: Review and Limitations

The traditional GNSS carrier recovery systems can be classified into two categories: closed loop tracking systems and open loop phase and frequency estimators. In this section we briefly review the traditional closed loop systems and the open loop schemes, paying attention to the aspects which motivate the quasi-open loop scheme proposed in the next sections.

3.1. Closed Loop Architectures. Digital phase lock loops (DPLLs) and FLLs lie into the category of the classical closed loop architectures, whose basic scheme is shown in Figure 1. DPLLs are able to track both the instantaneous phase and frequency of the incoming carrier, while FLLs are only able to track the carrier frequency.

In the following we briefly describe the DPLL operations (the operations performed by FLLs are almost similar except for the discriminator). The estimation of the instantaneous phase is generally performed by a DPLL, which consists of a number of subsystems: (a) the phase estimator/discriminator, (b) the loop filters $F(z)$, (c) the local oscillator (LO) also called numerically controlled oscillator (NCO). The input signal $x_w[n]$ is first multiplied by the local carrier $x_L[n]$ generated by the NCO, and integrated by the integrate and dump block. During this integration process, L input samples are processed and used to produce the prompt correlator output $y_\phi[k]$. The corresponding integration time is $T = LT_s$. At this point the sampling rate of the system changes from $1/T_s$ (time domain n) to $1/(LT_s)$ (time domain k). The instantaneous phase is estimated from $y_\phi[k]$ by the phase discriminator. This estimate is filtered through the loop filter $F(z)$, with a loop noise bandwidth B, and the filter output is used to drive the LO for the carrier generation. This estimate is then progressively updated using the information provided by the new correlator output in a closed loop manner. Below we present two key points that motivate the use of a quasi-open loop scheme (for GNSS carrier recovery systems) rather than a closed loop one.

It is important to stress that, in the traditional closed loop carrier recovery systems, the LO role is twofold: in fact it is both the output of the system and an integral part of the phase estimator. In the quasi-open loop structure proposed in this paper these two roles will be decoupled, as explained in Section 4. This is the first key point of our method, as will be clear in the next sections.

The loop filter $F(z)$ is the most critical block, whose function is also twofold. Firstly, as the received signal and, thus, the discriminator output is corrupted by thermal noise, the filter is required to provide a degree of noise rejection. Secondly, it enables the processing of higher order dynamics. Transformation methods from the Laplace domain to the Z-domain are widely used to design loop filters [2]. These methods simply provide a discrete version of the loop filters that have been previously designed for the analog loops. However, this approach neglects both the inherent delay in the digital loop and the variation in the open loop gain due to the NCO update interval. In [3] it has been shown that the transformation methods properly work only if BT is close to zero, where B is the loop noise bandwidth and T is the integration time. As BT increases, the effective loop noise bandwidth and the closed loop pole locations deviate from the desired ones and eventually the loop becomes unstable,

as explained in [3]. The maximum achievable BT value depends on the type of the Laplace-Z transformation and on the characteristics of the original continuous-time filter. For most communication applications, this condition is satisfied because BT remains close to zero. Instead, for some new GNSS applications, such as for weak signal tracking and extremely high dynamic applications, larger BT values are required.

Simulation experiments have been carried out to analyze the behavior of an FLL for increasing values of BT. The incoming GNSS signal has been generated with a carrier to noise ratio $C/N_o = 40$ dB-Hz and with a ramp-type time varying frequency with a slope equal to 10 Hz/s. In order to track this frequency evolution, we have used a 2nd order FLL with a loop noise bandwidth $B = 10$ Hz, and we have changed BT (by modifying the integration time) to put progressively more stress on the loop. The frequency estimated by the FLL is shown in Figure 2 for three different values of BT, together with an estimate of C/N_o. The latter can be used as an indicator to check if the system is in lock state or not. These results show that the loop loses the lock and the FLL is no longer able to track the input frequency when $BT = 0.3$, as indicated by the estimated C/N_o. It is important to emphasize that the loop filter is not the only module in the closed loop scheme responsible for the loss of lock and the stability problems. The loop can also lose lock or can become unstable because of other factors, but here we concentrate on the effect of the loop filter design based on transformation methods on loop. To solve this problem, some techniques for designing the loop filters have been proposed in the literature, for example, the controlled root method, the direct design of loop filters in digital domain based on a minimization criterion, the loop architectures based on Kalman filters, and the fuzzy loop architectures. An alternative approach is to use the scheme proposed in Section 4, which simplifies the design of the loop filter. This is the second key point of our method.

3.2. Open Loop Architectures. In order to estimate the time delay and the carrier frequency from the incoming signal, the open loop schemes usually operate on batches of the incoming signal, as depicted in Figure 3. The open loop approach does not separate acquisition and tracking stages, as explained in [10]. An input signal batch is correlated with batches of a signal replica in order to obtain an entire 3-D image of the signal, whose dimensions are the code shift, the Doppler shift, and the signal energy. This batch-based correlation uses joint time-frequency domain techniques to allow some forms of parallel computing based on FFT. After the 3-D image has been obtained, batch estimators are applied to the 3-D function to compute the signal parameters. They operate by searching for the location of the maximum energy of the 3-D function and provide an estimate of the parameters $\Delta\psi$, $\Delta f[k]$, and the code delay corresponding to this location. Since these methods mainly rely on FFT-based correlators for the 3-D image computation, they usually exhibit a large computational complexity, which enormously increases as the batch size

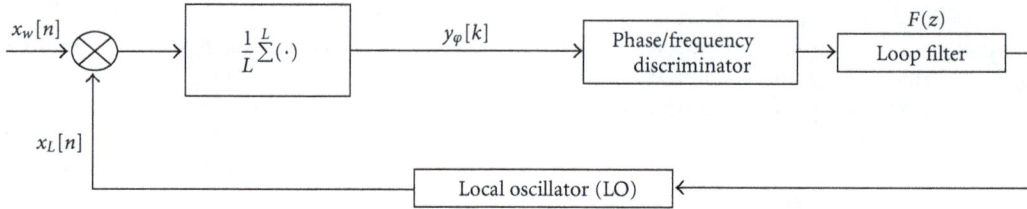

FIGURE 1: Basic classical closed loop carrier tracking architecture.

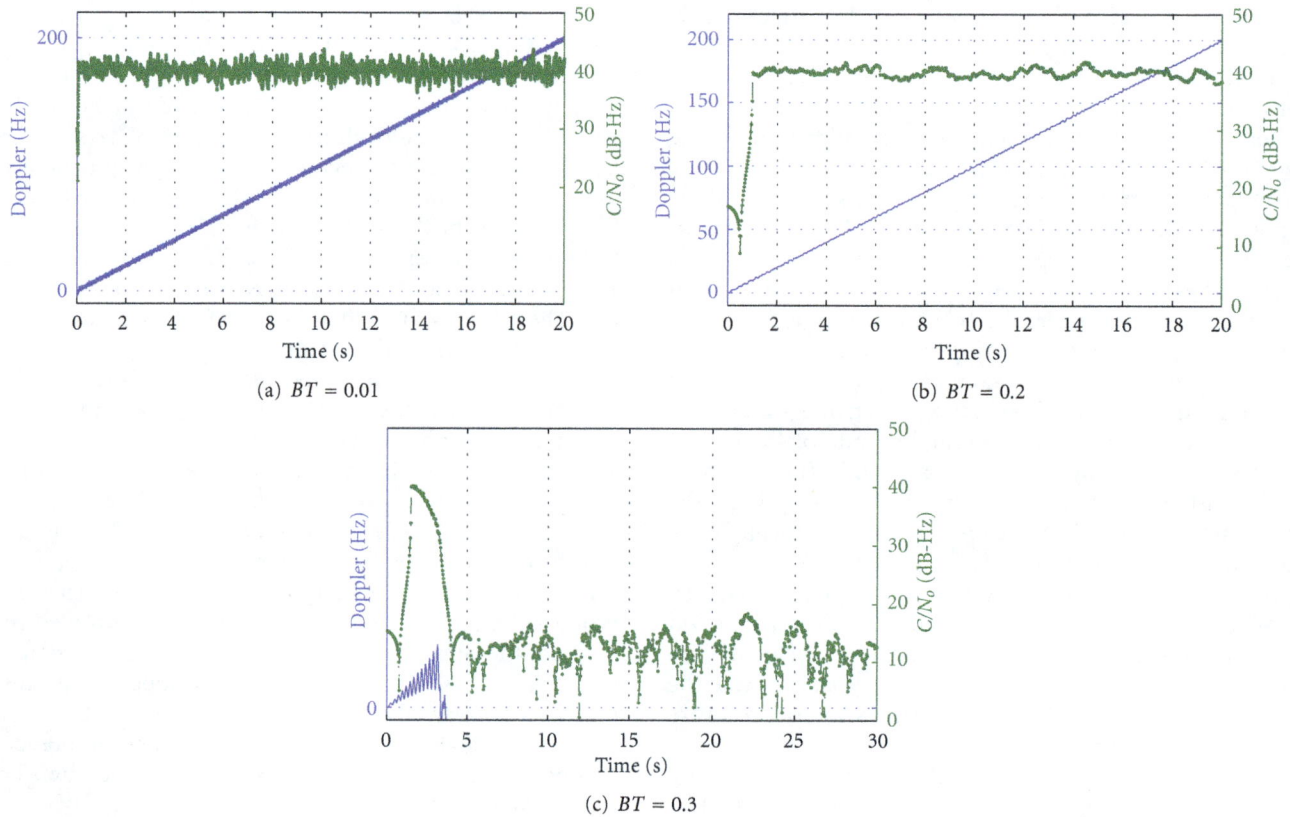

(a) $BT = 0.01$

(b) $BT = 0.2$

(c) $BT = 0.3$

FIGURE 2: 2nd order FLL tracking tesults for input ramp frequency of 10 Hz/s, $C/N_o = 40$ dB-Hz, $B = 10$ Hz and different integration Times.

increases. This is the main reason to avoid the use of these open loop processing techniques in real time receivers.

4. Proposed Quasi-Open Loop Architecture

In this section we propose a quasi-open loop scheme for carrier recovery in a GNSS receiver. We start our discussion from a basic open loop scheme, which utilizes a conventional discriminator. Then, after analyzing the behavior of the discriminator for the frequency estimation, we propose a quasi-open loop scheme for the continuous tracking of the incoming carrier frequency.

4.1. Basic Open Loop Scheme Utilizing Conventional Discriminators. The main idea of an open loop carrier recovery scheme that utilizes a conventional discriminator is depicted in Figure 4.

The main blocks of this scheme are the following.

(i) The local oscillator which generates the local signal $x_L[n]$. This complex LO is denoted by LO(E), to emphasize its role in the process of phase estimation.

(ii) The mixer which performs the multiplication between the local oscillator and the incoming signal.

(iii) The accumulator which accumulates L values of the signal $y[n]$; this block is often called *integrator* as it is equivalent to an integral in the continuous time domain. The integrator output is provided at the epochs $n = L, 2L, \ldots, kL, \ldots$.

(iv) The discriminator, which provides an estimate of the instantaneous phase/frequency of the incoming signal. It receives an input (the integrator output) at a rate which depends on the value of L.

(v) The open loop smoothing filter.

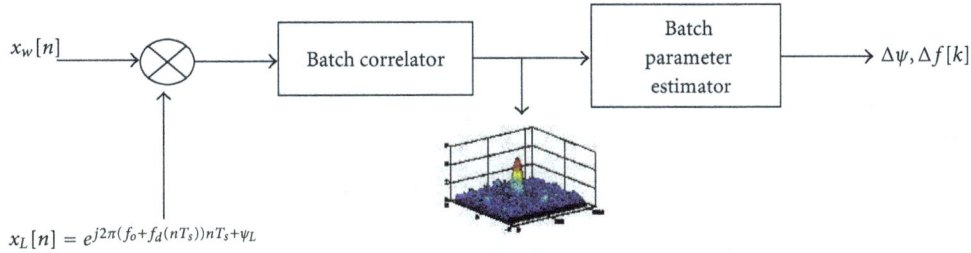

FIGURE 3: a generic open loop GNSS receiver.

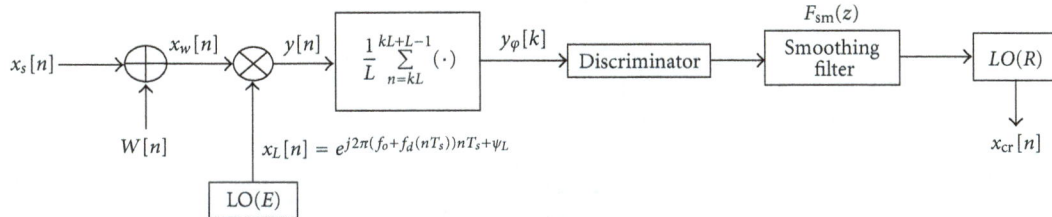

FIGURE 4: A generic open loop carrier recovery scheme utilizing conventional discriminators.

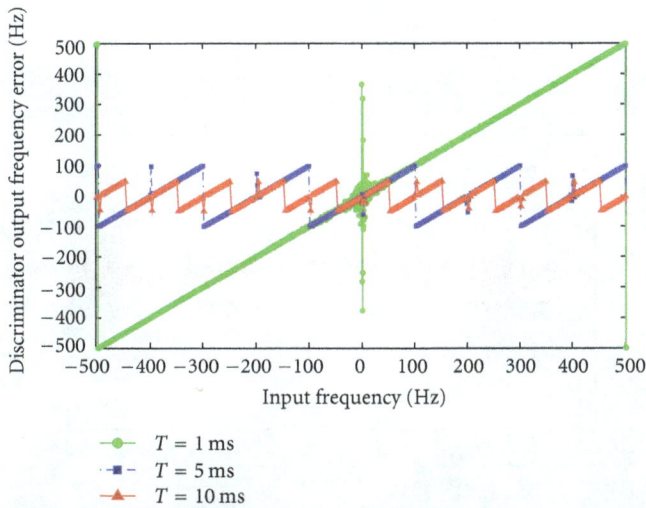

FIGURE 5: Frequency discriminator for different integration times.

(vi) The local oscillator LO(R), which provides a local carrier $x_{cr}[n]$, recovered from the incoming signal.

This scheme differs from its closed loop counterpart since we use each new phase/frequency estimate to update a separate local oscillator (the LO(R) block), instead of feeding back the local oscillator LO(E) that is embedded in the estimation module. This system works at two different rates, that is, in two different discrete-time domains: the mixer and the integrator works in the n domain with a rate r_n, while the discriminator works in the k domain, with a rate $r_k = r_n/L$. Once the discriminator has estimated the instantaneous phase/frequency of the incoming signal, the carrier can be recovered by building a sinusoidal signal, $x_{cr}[n]$. This is the recovered LO, denoted as LO(R), whose update rate is r_k. In a

DPLL, the LO(R) is also used to update the signal $x_L[n]$, then LO(E) and LO(R) coincide.

In our scheme we decouple LO(R) and LO(E), so as to design independently their denoising filters and update rates. This idea derives from the consideration that the update rate of LO(R) depends on the input dynamic, while the update rate of LO(E) depends on the frequency range the discriminator is able to process. It is evident that the two update rates are related, but the requirements can be different. Similar considerations can be done for the denoising filters.

Below we perform an analysis of the integrator and discriminator outputs which will help us to highlight the main factors that are needed to recover a continuous carrier. The analysis is performed in the ideal case of no noise. In this case, the integrator outputs of Figure 4 can be written in complex form as

$$y_\varphi[k] = \frac{1}{2L} \sum_{n=kL}^{(k+1)L-1} e^{j(2\pi\Delta f(nT_s)nT_s+\Delta\psi)}, \qquad (5)$$

where

$$\begin{aligned} \Delta f(nT_s) &= f_L(nT_s) - f_d(nT_s), \\ \Delta\psi &= \psi_L - \psi_s. \end{aligned} \qquad (6)$$

In theory (5), should contain also a double frequency component, which, however, can be neglected as it is filtered out by the integrator. If $\Delta f(nT_s)$ is a constant Δf, the summation in (5) can be written as a geometrical progression, from which a closed-form expression can be found. In fact by writing $i = n - kL$, (5) becomes

$$\begin{aligned} y_\varphi[k] &= \frac{1}{2L} \sum_{i=0}^{L-1} e^{j(2\pi(i+kL)\Delta f T_s+\Delta\psi)} \\ &= e^{j(2\pi\Delta f kLT_s+\Delta\psi)} e^{j\pi\Delta f T_s(L-1)} \frac{\sin\pi\Delta f LT_s}{2L\sin\pi\Delta f T_s} \end{aligned} \qquad (7)$$

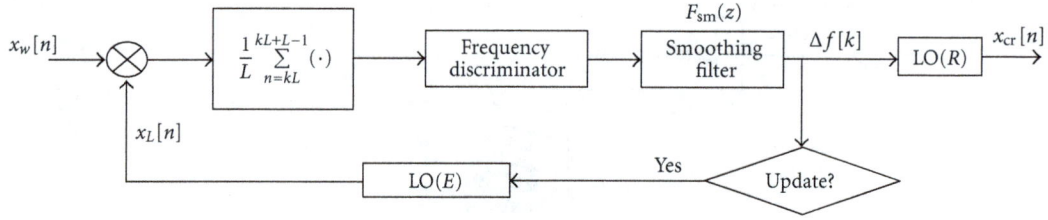

FIGURE 6: A quasi-open loop frequency estimator.

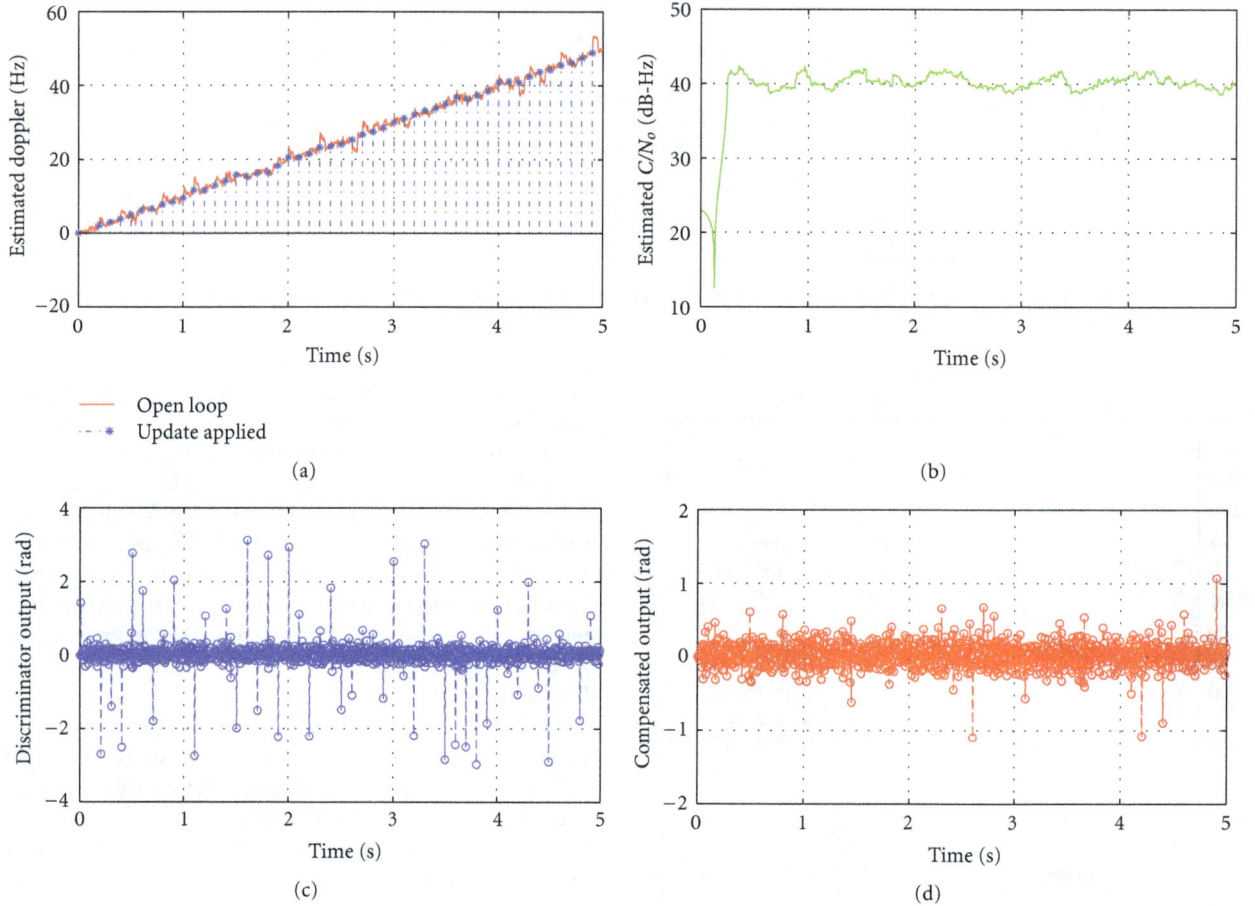

(a)

(b)

(c)

(d)

FIGURE 7: Proposed scheme tracking an input frequency ramp of 10 Hz/s, $C/N_o = 40$ dB-Hz, $T = 5$ ms.

whose instantaneous phase is

$$\varphi_{\text{inst}}[k] = 2\pi\Delta f\left(kL + \frac{L-1}{2}\right)T_s + \Delta\psi. \qquad (8)$$

If $\Delta f(nT_s)$ is approximately constant in the integration interval LT_s, that is $\Delta f(nT_s) \cong \Delta f[k]$ for $kL \leq n \leq (k+1)L - 1$, then it is possible to write

$$\varphi_{\text{inst}}[k] \cong 2\pi\Delta f[k]\left(kL + \frac{L-1}{2}\right)T_s + \Delta\psi. \qquad (9)$$

This phase contains an integer number L_k of cycles plus a fractional part, from which

$$\varphi_{\text{inst}}[k] = L_k 2\pi + \Phi(nT_s), \qquad (10)$$

where

$$\Phi(nT_s) = \text{mod}\left(\varphi_{\text{inst}}[k], 2\pi\right). \qquad (11)$$

In the classical DPLL schemes for GNSS applications, this phase is estimated at the time epochs $n = kL$, for $k = 0, 1, \ldots,$ as

$$\Phi(kLT_s) = \arctan\frac{Y_Q[k]}{Y_I[k]}, \qquad (12)$$

where $Y_I[k] = \mathcal{R}(y_\varphi[k])$ and $Y_Q[k] = \mathcal{I}(y_\varphi[k])$. This discriminator function can also be substituted by other operations which approximate the arctan function. At each

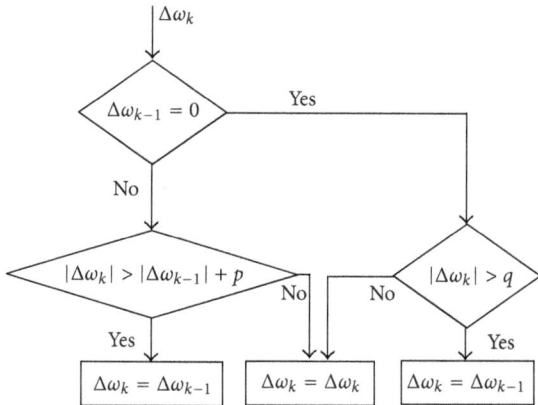

FIGURE 8: A simple algorithm to detect and remove discriminator output outliers.

epoch k, (12) provides a phase value available at the discrete-time instant $(k + 1)L$ and representative of the instantaneous residual phase at the time instant

$$l_k T_s = T_s \left[(k+1)L - \frac{(L+1)}{2} \right],$$
(13)

that is, the residual phase information is provided with a delay of $(L+1)/2$. Notice that the recovered carrier $x_{cr}[n]$ has to be generated at each time instants $t_n = nT_s$, as indicated in (3). This means that the missing values of the instantaneous phase have to be evaluated in some way. This problem can be easily solved if the unknown frequency $\Delta f(nT_s)$ can be estimated from the outputs of the discriminator function (12), used for the instantaneous phase estimation. This part is described below.

In order to estimate the instantaneous frequency, a quantity $y_\varphi[k] y_\varphi^*[k-1]$ is at first evaluated, and then its phase is extracted using an arctan phase discriminator. This is the usual operation performed by a 4-quadrant arctan discriminator of the type, [2],

$$\Delta f[k] = \frac{1}{2\pi T} \arctan \left(\frac{\mathcal{I}\left(y_\varphi[k] y_\varphi^*[k-1] \right)}{\mathcal{R}\left(y_\varphi[k] y_\varphi^*[k-1] \right)} \right).$$
(14)

Although this discriminator is termed as maximum likelihood estimator by [2], it is important to mention that this estimator only achieves the Cramer-Rao lower bound at a sufficiently high C/N_o, as we have demonstrated in [11]. Notice that (14) provides the frequency value without evaluating the instantaneous phase, just applying the integrator output to (14). Another possibility for the frequency evaluation is to operate on the output of the phase discriminator given in (12). However, since in a frequency discriminator only the phase difference is of interest, the discriminator given in (14) is generally more convenient, as it inherently reduces the phase wrapping problem [12]. Once the estimate of $\Delta f[k]$ is available, the carrier can be continuously recovered, at each instant, by the oscillator LO(R).

Before addressing the role of the filter in this scheme we want to highlight some limitations encountered when a carrier with a time varying Doppler frequency has to be tracked. The main problem with this type of scheme is that we cannot use it alone or without a closed loop updating because of the limited linear region of the discriminator function. In fact, as the difference $\Delta f[k]$ between the frequencies of the incoming and local carriers increases with time, the system tends to operate outside the linear region, where tracking is no longer possible. Moreover, this linear region reduces as the coherent integration time T increases. Therefore, it is not possible to track the frequency evolution without updating the frequency of the local oscillator LO(E). All the conventional discriminators exhibit the same behavior, as discussed in [2].

As an example, Figure 5 shows the output of the discriminator, given by (14), for different integration times. This behavior is observed when we try to use the scheme, given in Figure 4, for tracking a time varying input Doppler at $C/N_o = 50$ dB-Hz.

4.2. Modified Scheme: Quasi Open Loop Scheme. A possible method to overcome the problem of tracking a time varying frequency is to update the frequency of the local oscillator LO(E) after N epochs, instead of updating it at each epoch, as in the closed loop systems. This means that we are proposing a scheme working with three different rates, as opposed to the typical closed loop schemes working with two different rates. In the proposed scheme, shown in Figure 6, the additional update rate of the LO(E) frequency, equal to $r_u = r_k/N$, is introduced.

The real motivation behind this three rate schemes is to make the integration time T and the NCO update interval independent of each other, obtaining in turn an additional degree of freedom. The latter allows us to ease the design of the loop, as it will be shown below. The system can be considered *quasi-open*, as it works as an open loop between two updating epochs.

Notice that in our scheme only the frequency of the NCO is updated, while the phase is kept continuous at each epoch, regardless of whether the updating is or is not applied. The value of N depends on the specific application. For example, in case of high dynamics we need to update the NCO frequency more frequently, so a smaller value of N is necessary. Furthermore, we can also decide to have either regular update intervals, that is, to update after each N epoch, or irregular intervals which depend on the incoming Doppler evolution. In this study we have only considered regular update intervals.

4.3. Choice and Design of Loop Filter. The scheme of Figure 6 allows us to approach the design of the loop filter, taking into account that now its role is only to reduce the effect of noise on the frequency estimate. This is possible thanks to the available additional degree of freedom of the quasi-open scheme. The filter can be considered as a part of the discriminator, whose only task is to smooth the discriminator output. To this purpose, any reasonable choice

(a)

(b)

(c)

(d)

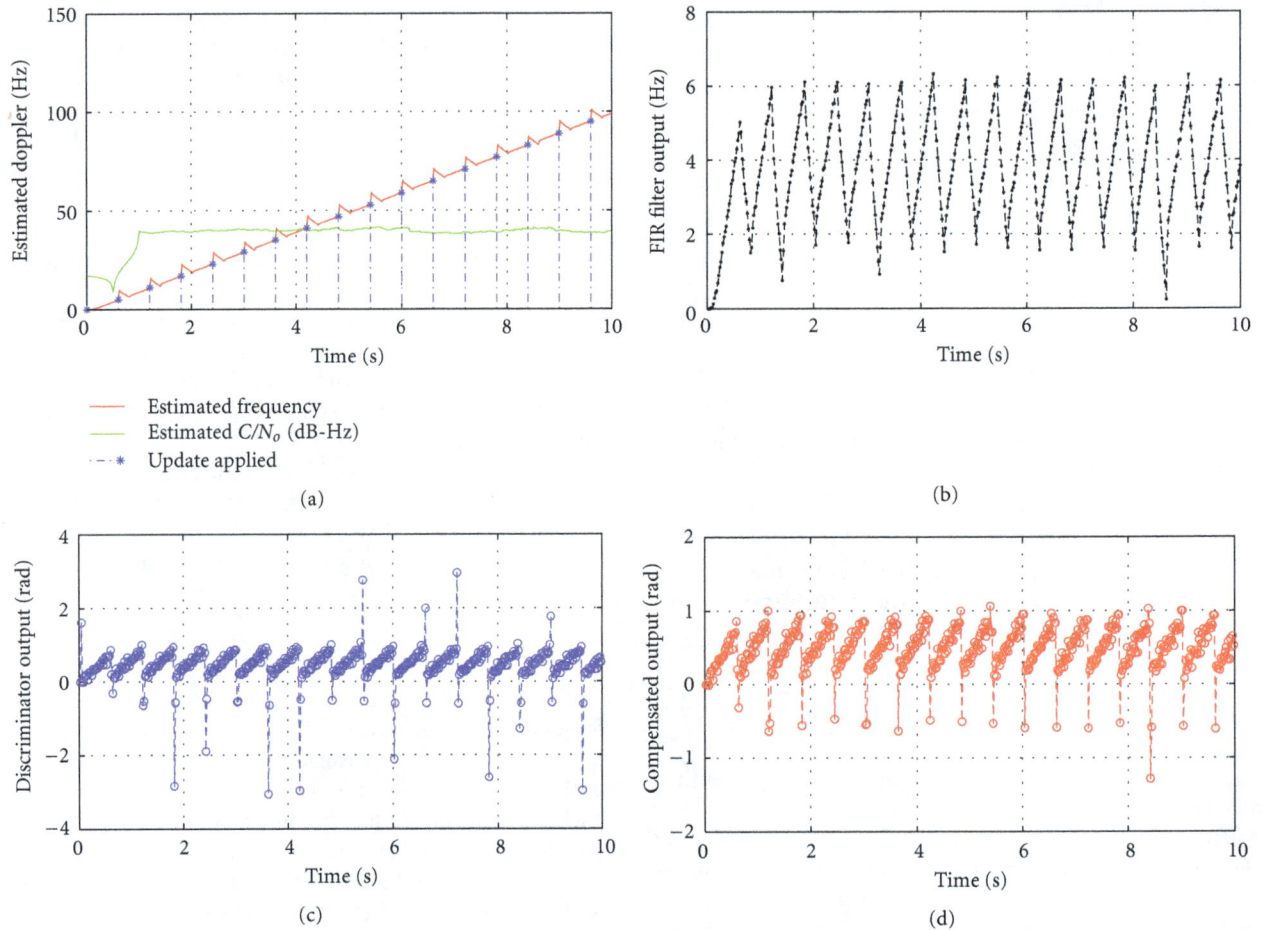

FIGURE 9: Proposed Scheme tracking an input frequency ramp of 10 Hz/s, $C/N_o = 40$ dB-Hz, $T = 20$ ms.

of the filter structure and parameters can be adopted, taking into account the specific application. Since finite impulse response (FIR) filters are simple to design, always stable, relatively insensitive to quantization and can have linear phase, we decided to explore the possibility to use them for denoising.

The transfer function $F_{sm}(z)$ of an FIR filter is

$$F_{sm}(z) = \sum_{i=0}^{L_{sm}-1} b_i z^{-i}, \qquad (15)$$

where L_{sm} denotes the number of taps and b_i are the filter coefficients, which become $b_i = 1/L_{sm}$ in the simplest case of a moving average (MA) filter. An MA-FIR filter is an excellent smoothing filter, but its frequency roll-off is slow and its stopband attenuation is ghastly, making it a scarcely effective low-pass filter. This is a typical result, as a digital filter can be generally optimized for time or frequency domain performance, but not for both. Since in our application the filter task is to mitigate the noise effect, the choice of an MA-FIR structure is justified. Notice that the purpose of this paper is to show the feasibility of a quasi-open architecture, then we did not concentrate on the optimization of the structure of the smoothing filter, which

could be also implemented with infinite impulse response (IIR) filters or with more optimized versions of FIR filters.

A possible consideration for choosing the type of filter could be to keep the computational complexity as low as possible. This complexity can be attributed to the number of operations (additions and multiplications) needed to compute the filter response. For digital filters, the computational complexity is more or less proportional to the number of filter coefficients. Usually, we need more than 5 taps to get good smoothing results in case of MA-FIR filters. With these values the computational complexity of a quasi-open loop scheme, which utilizes a smoothing MA-FIR filter, will be slightly higher than the one of an FLL utilizing 2nd order IIR filters. But, as mentioned above, we have many other advantages of using FIR filters as compared to IIR filters.

It is important to mention that the use of these filters changes the philosophy of the loop filter. As explained earlier, in a closed loop architecture the filter has a twofold task: one is to reduce the effect of the noise and the second one is to control the loop dynamics. In the schemes with a smoothing MA-FIR filter the loop dynamics is solely determined by the frequency discriminator. For example, it has been shown that the estimator proposed in (14) is unbiased if we have to track a constant frequency shift or a linear frequency ramp. But its

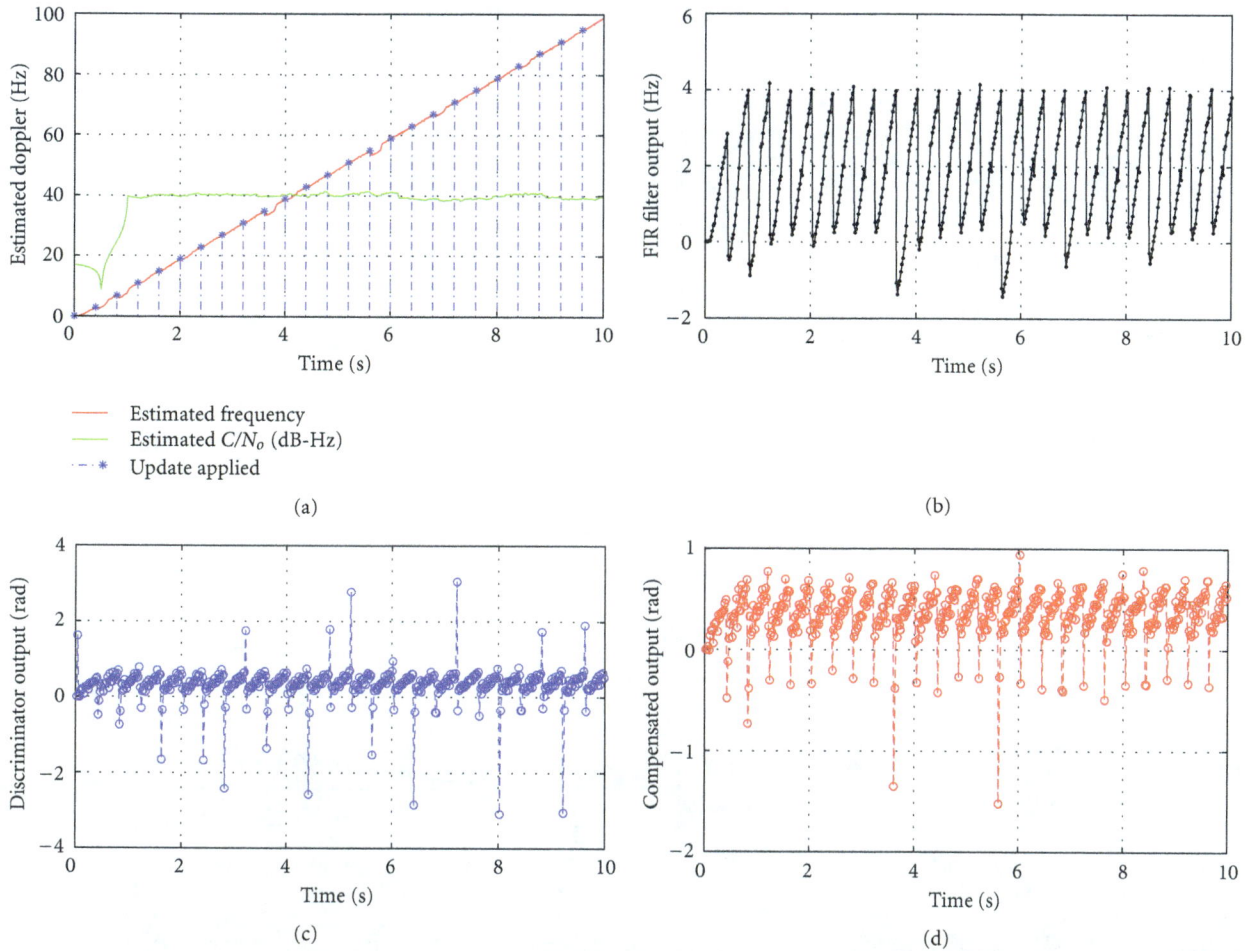

FIGURE 10: Proposed scheme tracking an input frequency ramp of 10 Hz/s, $C/N_o = 40$ dB-Hz, $T = 20$ ms.

estimate is biased when we have to track an input frequency with a quadratic component [8]. This is not a problem in our scheme as we can accommodate a higher order dynamics by taking additional measures. For example, we can either change the design of the discriminator, as discussed in [13], or we can insert a new block after the FIR filter to account for the higher dynamics. This problem has not been considered in this study, which is mainly devoted to the feasibility and performance of the quasi-open structure.

5. Results and Discussions

In this section we present some simulation results obtained by using this new type of quasi-open loop frequency estimator. We start from a basic scheme using an MA-FIR filter, we observe the results, and we introduce some modifications in the basic scheme to improve its performance. The system performance is described in terms of tracking jitter, also taking into account weak signal scenarios.

5.1. Some Implementation Aspects. In the first simulation example, a GNSS-like signal was generated with a ramp-type time-varying Doppler shift, with a slope of 10 Hz/s.

The signal was processed by a quasi-open loop frequency estimator, followed by a smoothing MA-FIR filter with L_{sm} = 10 taps. In the first stage (after switch-on) the system is completely open, and, after a time interval equal to the transient of the FIR filter, starts updating the loop. The updating is repeated at each integration interval NT.

The results with N = 20 are shown in Figure 7, which shows that the proposed scheme is able to successfully track the input Doppler frequency 7(a).

However, observing the results of Figure 7, we recognize that some problems are associated with this type of scheme. First of all, at each new update there is a jump in the discriminator output 7(c). These jumps give rise to undesired outliers which depend on the type of updating. These outliers can be compensated using the scheme shown in Figure 8, which adopts a simple strategy to eliminate any kind of anomaly, based on the comparison between the current and the previous discriminator outputs, indicated in Figure 8, respectively, as $\Delta\omega_k$ and $\Delta\omega_{k-1}$, and measured in rad units. The sudden jumps are eliminated by setting two thresholds p and q, whose values depend on the rate of the input Doppler frequency. We have set both of them equal to π in all the results presented in this paper. The

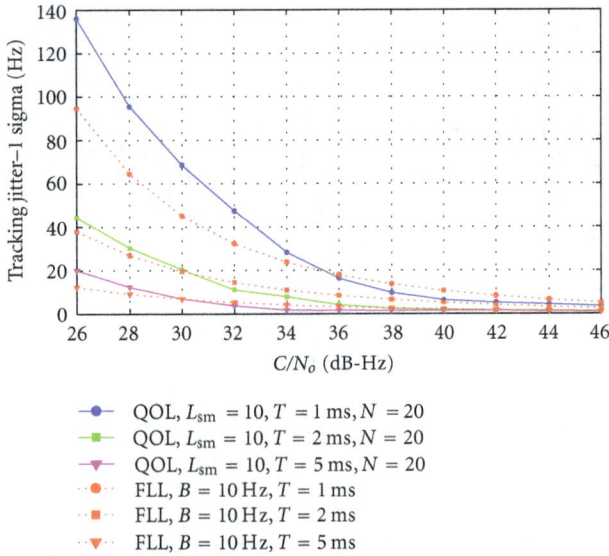

FIGURE 11: Tracking jitter performance.

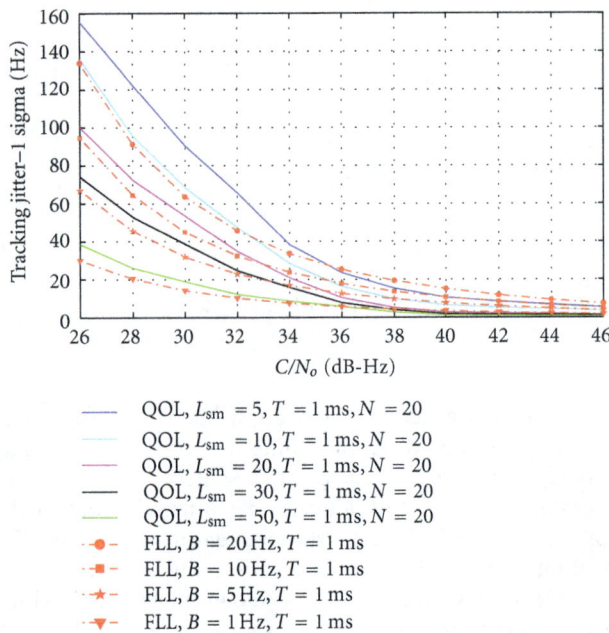

FIGURE 12: Tracking jitter performance with same integration time.

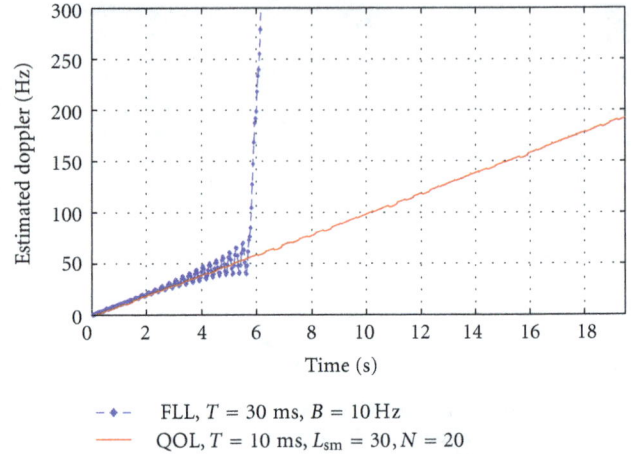

FIGURE 13: FLL versus QOL tracking results for incoming Doppler variation of 10 Hz/s, $C/N_o = 40$ dB-Hz.

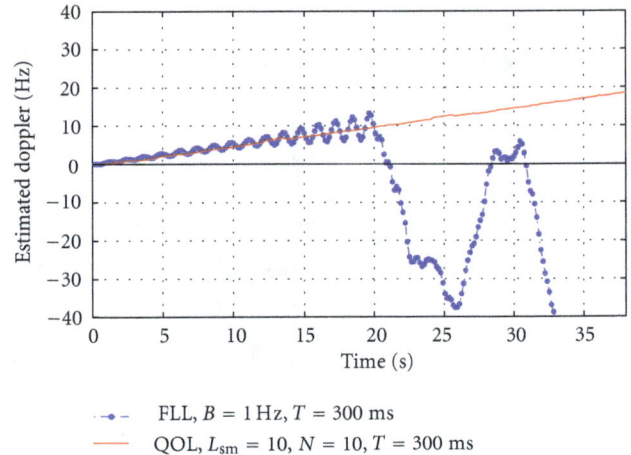

FIGURE 14: FLL versus QOL tracking results for incoming Doppler variation of 0.5 Hz/s, $C/N_o = 30$ dB-Hz.

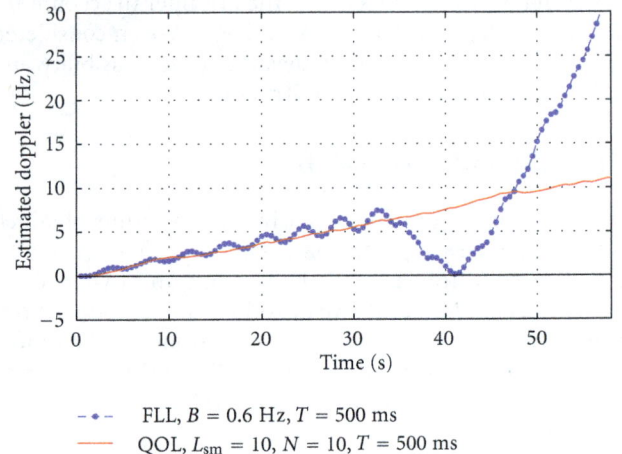

FIGURE 15: FLL versus QOL tracking results for incoming Doppler variation of 0.2 Hz/s, $C/N_o = 20$ dB-Hz.

compensated discriminator output shown in Figure 7(d) proves the effectiveness of the method.

The second problem associated to this type of scheme is related to the inherent time-transient of any FIR filter, which generates some regular slowly decreasing jumps at each new updating interval. This is due to the fact that the filter output experiences a transient, which vanishes only when the filter memory (equal to the filter length) is completely filled with input samples. This effect is clearly visible in Figure 9(a) where the integration time is $T = 20$ ms. A possible solution to mitigate this problem is to use a moving average FIR filter with a time-variant length. The number of the taps is set to one at each new updating, and then it is gradually increased to reach the specified maximum value L_{sm}. The results are presented in Figure 10, where we can see that we have no more jumps in the output.

At this point we have an architecture where it is possible to set the update rate of the NCO independently from the integration time. We have also demonstrated the feasibility of using an MA-FIR filter with a time-varying length as loop filter. The filter design is very easy and the resulting structure is always stable.

5.2. Performance of the Quasi-Open Scheme in User Dynamics.
The updating interval in the proposed scheme is strictly related to the dynamics under which the system has to work. In case of high dynamic applications the updating has to be applied more frequently, and this implies a smaller value of N, and vice versa. The dynamic range sustained by the quasi-open loop scheme does not only depend on N, but also on the linear region of the discriminator in (14), which, in turn, depends on the integration time T. We can relate the linear range of the discriminator, denoted by Δ_D (for a given T), and the rate of change of the input Doppler frequency, denoted by $\dot{f}_d$ and expressed in Hz/s as

$$\dot{f}_d T N < \frac{\Delta_D}{2} \qquad (16)$$

obtaining the condition for the quasi-open loop scheme to work properly. In this equation, the term $\dot{f}_d T N$ indicates the amount of Doppler accumulated between two updating intervals of the NCO. If this term lies inside the discriminator range, then the proposed scheme successfully tracks the Doppler variations.

As an example, for $T = 10$ ms, the linear range of the discriminator is about $\Delta_D = 100$ Hz (from -50 Hz to 50 Hz in Figure 5). Thus, for $T = 10$ ms and $N = 20$, the maximum Doppler variation that can be tracked successfully by the quasi-open loop scheme is given by

$$\dot{f}_d < \frac{\Delta_D}{2NT} = 250 \text{ Hz/s}. \qquad (17)$$

If the input Doppler variation exceeds this value, then the quasi-open loop is not able to track the incoming frequency. The problem can be solved by updating the NCO frequency more often. This can be achieved by decreasing the value of N.

5.3. Tracking Jitter Performance.
We carried out some simulation experiments to examine the tracking jitter performance of the proposed quasi-open loop scheme and we compared it with the theoretical jitter of a classical FLL. The results shown in Figure 11 were obtained for different integration times T and different C/N_o, by keeping the number L_{sm} of the MA-FIR filter taps constant, and by updating the NCO (local oscillator LO(E)) frequency after $N = 20$ epochs. In this way we were able to analyze the effect of increasing the integration time on the jitter performance.

To compare the results with those of a traditional closed loop scheme, we used the formula of the theoretical FLL tracking jitter, due to thermal noise, given in [2], that is

$$\sigma_{\text{FLL}} = \frac{1}{2\pi T} \sqrt{\frac{4FB}{C/N_o} \left[1 + \frac{1}{TC/N_o} \right]} \text{ [Hz]}. \qquad (18)$$

These values are also shown in Figure 11 for different integration times, and for a loop noise bandwidth $B = 10$ Hz. The choice of the loop noise bandwidth suitable for comparison is not an easy task and will be discussed in the next section.

It is clear from the obtained results that the tracking jitter performance of the quasi-open loop is better than that of an FLL for all the integration times at high C/N_o, but the performance degrades at low C/N_o. The main reason for this could be the poor quality of the frequency estimator (14) at a low C/N_o.

5.4. Loop Noise Bandwidth versus Filter Taps.
In this section, we discuss the equivalence between the loop noise bandwidth B of the closed loop schemes and the number L_{sm} of the filter taps in a quasi-open loop. In traditional closed loop schemes, like FLLs, the parameter B controls both the amount of noise rejected by the loop and the dynamic stress. If we increase B, the loop can sustain more dynamics, but more noise also affects the system, and vice versa.

A similar role is played by the number L_{sm} of filter taps in a quasi-open scheme. Increasing L_{sm} implies more smoothing, but imposes some constraints on the update interval N, which should be larger than L_{sm}. In fact, if $N < L_{sm}$ the filter never attains steady-state, thereby reducing the effectiveness of the smoothing, which will be incomplete. In other words, since the number of taps is variable and is set to one at each new update, the maximum number of taps will be N instead of L_{sm}. Another constraint on the value of N is the dynamic stress which can be sustained by the quasi-open loop scheme.

The equivalence between B and L_{sm} can be seen in Figure 12, which shows the tracking jitter curves for both schemes, with different values of B for the FLL and different values of L_{sm} for the quasi-open scheme. The curves were obtained by keeping the integration time constant. By observing these curves we can get an idea of the number of taps of the MA-FIR filter required to achieve the same jitter performance of an FLL with a given noise bandwidth. As shown in Figure 12, the jitter performance improves as L_{sm} increases, and B decreases.

It is very difficult to find the exact equivalence between these two parameters from these results. However, since we are usually interested in low C/N_o regions, we can somehow relate the two parameters. For example, to obtain the same performance of an FLL with $B = 10$ Hz, we need an MA-FIR filter with $L_{sm} = 20$ taps for $T = 1$ ms. On the other hand, for $B = 1$ Hz, the number of taps has to be $L_{sm} = 50$, which is quite complex in terms of computational cost. So, a reasonable choice could be to fix an approximate limit of $L_{sm} \leq 20$. In this way we will get almost the same performance of an FLL with $B = 10$ Hz.

In the following section we adopt this equivalence between B and L_{sm}, and, based on this, we show the advantages obtained with a quasi-open loop scheme in weak signal conditions.

5.5. Weak Signal Performance. In this section, we demonstrate that, by using a quasi-open scheme, we can work with extended coherent integration times which otherwise would not be possible with closed loop schemes under the same conditions, because of the resulting loop filter instability.

In the first simulation experiment, we again considered an incoming signal with a ramp-type Doppler frequency with a slope of 10 Hz/s, and with C/N_o = 40 dB-Hz. Although this is not a weak signal, we started with this value to better highlight the performance of both schemes, FLL and quasi-open loop, when they work under the same conditions. We set the update interval to N = 20, and the number of filter taps to L_{sm} = 10. For this choice of L_{sm}, a reasonable value of the FLL noise bandwidth could be around B = 10 Hz, as discussed in Section 5.4. The tracking results are shown in Figure 13 for T = 30 ms. Under these conditions BT = 0.3, which is high enough for the loop filter to become unstable. For the quasi-open loop scheme we can extend the integration time even more as long as the condition (16) remains valid.

In the next simulation experiment, we considered a weak signal with C/N_o = 30 dB-Hz, and with a Doppler varying at a rate of 0.5 Hz/s. Since C/N_o is very low, we set T = 300 ms, and the FLL bandwidth to B = 1 Hz, so as to have a suitable noise rejection performance for weak signals. For the quasi-open loop scheme, we chose L_{sm} = 10 and N = 10 to fulfill the condition (16). Here the value N = 20 used in the previous experiment cannot be kept because of condition (16). The results are shown in Figure 14. Again we observe the same situation observed in Figure 13: the FLL is no longer able to track the carrier, because of the loop filter instability, while the proposed scheme works well and successfully tracks the incoming Doppler variation.

Figure 15 shows the tracking results for an incoming signal with C/N_o = 20 dB-Hz, and a Doppler variation with a rate of 0.2 Hz/s. The integration time is set to T = 500 ms, since C/N_o is very low. For the quasi-open loop scheme, we chose N = 10, a value which does not violate the condition (16). The figure shows that the proposed scheme successfully tracks the frequency of the incoming signal even at this low C/N_o.

6. Conclusion

A novel quasi-open loop architecture has been proposed for tracking the frequency of received GNSS signals. The proposed architecture works with three different rates, unlike the classical closed loop schemes, PLLs and FLLs, which work with two different rates. The additional degree of freedom of the quasi-open scheme enables us to ease the design of the loop filter. Simulation results show that it is possible to design this filter with an FIR structure, by adopting some very simple rules for the design. Moreover,

the system results advantageous also in terms of stability when compared to a traditional closed loop architecture. It is also important to mention that if this type of scheme is used with classical DPLL schemes in an assisted manner, then a lower bandwidth and higher coherent integration times can be utilized in DPLL to track the incoming carrier in very weak signal conditions.

References

[1] B. Parkinson and J. Spilker, Eds., *Global Positioning System: Theory and Applications*, vol. 1, American Institute of Aeronautics and Astronautics, Washington, DC, USA, 1996.

[2] E. Kaplan and C. Hegarty, Eds., *Understanding GPS Principles and Applications*, Artech House, Norwood, Mass, USA, 2006.

[3] P. L. Kazemi, *Development of new filter and tracking schemes for weak gps signal tracking*, Ph.D. dissertation, Department of Geomatics Engineering, University of Calgary, 2010.

[4] S. A. Stephens and J. B. Thomas, "Controlled-root formulation for digital phase-locked loops," *IEEE Transactions on Aerospace and Electronic Systems*, vol. 31, no. 1, pp. 78–95, 1995.

[5] P. L. Kazemi, "Optimum digital filters for GNSS tracking loops," in *Proceedings of the 21st International Technical Meeting of the Satellite Division of the Institute of Navigation (ION GNSS '08)*, pp. 1188–1197, Center Savannah, Ga, USA, September 2008.

[6] M. L. Psiaki, "Smoother-based GPS signal tracking in a software receiver," in *Proceedings of the 14th International Technical Meeting of the Satellite Division of The Institute of Navigation (ION GPS '01)*, pp. 2900–2913, Salt Palace Convention Center, Salt Lake City, Utah, USA, 2001.

[7] A. M. M. Kamel, *Context aware high dynamics GNSS-INS for interference mitigation*, Ph.D. dissertation, Department of Geomatics Engineering, University of Calgary, 2011.

[8] J. Riba, J. Tom, and M. Lagunas, "Instantaneous open-loop frequency estimation methods for navigation receivers," in *Proceedings of the 2nd European Symposium on Global Navigation Satellite Systems (GNSS'98)*, Toulouse, France, 1998.

[9] S. Kay, "Fast and accurate single frequency estimator," *IEEE Transactions on Acoustics, Speech, and Signal Processing*, vol. 37, no. 12, pp. 1987–1990, 1989.

[10] F. van Graas, A. Soloviev, M. Uijt de Haag, and S. Gunawardena, "Closed-loop sequential signal processing and open-loop batch processing approaches for GNSS receiver design," *IEEE Journal on Selected Topics in Signal Processing*, vol. 3, no. 4, pp. 571–586, 2009.

[11] M. Tahir, M. Fantino, and L. L. Presti, "Characterizing different open loop fine frequency estimation methods for GNSS receivers," in *Proceedings of the International Technical Meeting of The Institute of Navigation (ION ITM '12)*, Newport Beach, Calif, USA, 2012.

[12] M. L. Fowler, "Phase-based frequency estimation: a review," *Digital Signal Processing*, vol. 12, no. 4, pp. 590–615, 2002.

[13] D. Simon and H. El-Sherief, "Fuzzy logic for digital phase-locked loop filter design," *IEEE Transactions on Fuzzy Systems*, vol. 3, no. 2, pp. 211–218, 1995.

8

Accurate GLONASS Time Transfer for the Generation of the Coordinated Universal Time

Z. Jiang and W. Lewandowski

Time Department, Bureau International des Poids et Mesures (BIPM), Pavillon de Breteuil, 92312 Sèvres Cedex, France

Correspondence should be addressed to Z. Jiang, zjiang@bipm.org

Academic Editor: Gonzalo Seco-Granados

The spatial techniques currently used in accurate time transfer are based on GPS, TWSTFT, and GLONASS. The International Bureau of Weights and Measures (BIPM) is mandated for the generation of Coordinated Universal Time (UTC) which is published monthly in the BIPM *Circular T*. In 2009, the international Consultative Committee for Time and Frequency (CCTF) recommended the use of multitechniques in time transfer to ensure precision, accuracy, and robustness in UTC. To complement the existing GPS and TWSTFT time links, in November 2009 the first two GLONASS time links were introduced into the UTC worldwide time link network. By November 2011, 6 GLONASS time links are used in the UTC computation. In the frame of the application in the UTC computation, we establish the technical features of GLONASS time transfer: the short- and long-term stabilities, the calibration process, and in particular the impact of the multiple GLONASS frequency biases. We then outline various considerations for future developments, including the uses of P-codes and carrier-phase information.

1. Introduction

GLONASS (from GLObal NAvigation Satellite System, GLN for short) is a radio-based satellite navigation system operated by the Russian Space Forces with the aim of providing real-time, all-weather, three-dimensional positioning, velocity measuring, and timing with a worldwide coverage. The completely deployed GLN constellation is composed of 24 satellites in three orbital planes of which the ascending nodes are 120° apart. Eight satellites are equally distributed in each plane. The first satellite was launched on 12 October 1982, and the constellation was completed in 1995, although until recent years it has not always been well maintained.

With respect to present and future techniques for accurate time transfers, GLN is comparable to other global navigation satellite systems (GNSSs): the United States' Global Positioning System (GPS), the upcoming Chinese Compass navigation system, and the Galileo positioning system of the European Union.

To guarantee the accuracy and robustness of UTC generation, a multitechnique strategy for UTC time transfer is indispensable. Over the last two decades much effort has been devoted to introducing GLN in UTC. However, earlier GLN studies [1–9] remained at an experimental stage because there were only a few operational GLN timing receivers, the GLN constellation was incomplete, and there were unsolved technical issues; among them the major difficulty was of the multiple GLN frequency biases.

The situation has greatly improved in recent years. As of 2008, there were 15 GLN timing receivers operating at UTC laboratories (see Table 9), and these were used to back up the regular GPS and TWSTFT links. Recent studies [10–12] have fixed the last remaining problems, and the first two GLN time links to be included in the generation of UTC were SU-PTB and UME-PTB, which were introduced in November 2009 (*Circular T* 263) [10–15]. Figure 1 shows the status of the time-transfer techniques used in UTC in November 2011 (*Circular T* 287). Here GLN&GPS stands for the combination of GLN and GPS code measurement data.

In this study, we investigate the receivers available at present in the UTC time transfer. The data in the numerical tests were collected mainly using the 3S Navigation and the AOS TTS GPS/GLN receivers. The conclusion obtained in this study is applicable in these two types of receivers. The

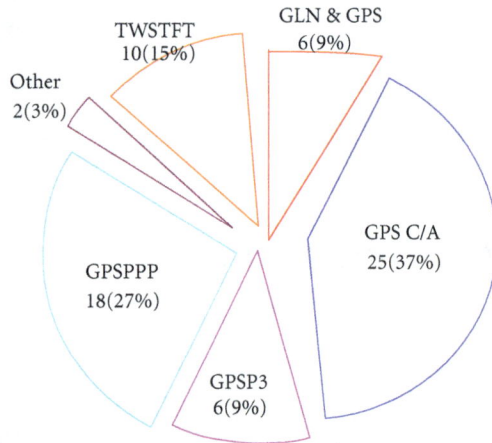

FIGURE 1: Status in November. 2011 (*Circular T* 287) of the 67 time transfer links used in UTC generation.

numerical analysis was carried out using the BIPM UTC/TAI software package Tsoft, with the usual monthly procedure. When the study was initialed, there were no TTS4 receiver data in the UTC databank. We have a couple of TTS4 receiver data recently and start to study them. As for Septentrio receivers, there is no software currently available to convert the receiver measurements to the CCTF CGGTS format used in UTC code time transfer. TTS-4 and Septentrio receivers are not investigated in this study.

In the following section we describe the technical features for the use of GLN in UTC, and then in Section 3 we present various ongoing studies at the BIPM and finally the conclusion.

In an earlier publication [16], we briefly reported the application of the GLN in UTC/TAI time and frequency transfers. In this paper, however, we present detailed considerations on this issue; in particular we discuss the impact of the GLN frequency biases in the UTC/TAI time links. For the readers who are not familiar with the concept of the accurate GNSS time transfer techniques, please refer to [16] that gives a simple explanation about GNSS CV and AV time transfers and to [17] where the AV is discussed in detail.

2. Use of GLN for UTC Time Transfer

GLN distributes three codes that can be used for time transfer: L1C, L1P, and L2P. The L1C code is authorized for civil applications in GLONASS ICD [18]. Although measurements are typically provided by the receivers, the P1 and P2 codes are primarily not intended for civil use [18]. A fourth code, L3P, free of ionospheric delays, is formed from a linear combination of L1P and L2P. The P-codes are of higher quality than the C-code, and logically one would thus expect them to have obvious advantages in time transfer. However, this was not observed in our previous investigations using the 3S Navigation receivers [6] nor in recent evaluations using the latest TTS-3 receivers [19]. We do not know the exact reason at present. Figure 2 illustrates a comparison of the

standard deviation of the smoothing residuals (σ) of the CV time links using GLN codes over different distances between 1200 and 9200 km. Here all the measurement data L1C, L1P, and L2P were corrected using the IGS precise orbit and ionosphere information. The CGGTTS data were collected in 2004 from the 3S receivers located at AOS (Poland), VSL (Netherlands), and CSIR (South Africa). It is seen that the mean values in the table obtained using the L1C and L1P data agree well with each other within the σ, implying that the same calibration applies across the same frequency band. The standard deviation obtained with the L1C code is statistically no bigger than that using the P-codes, and indeed for long distances, the L1C code results are slightly better than those of the P-codes. Similar results were obtained in more recent tests using TTS receivers [19].

The IGS analysis centres did not supply precise corrections for GLN satellite clocks (the IGS analysis center CODE recently announced the availability of the GLN clock product that we need to validate before using for UTC computation). hence the All in View (AV) technique [17] is not applicable for GLN at present. In GLN time transfer today: (1) Common View (CV) is still advantageous in cancelling the influence of the satellite clock and reducing the orbit and atmosphere delay uncertainties; (2) the state of the art of using the P-codes shows no obvious advantages over that of the L1C code, as unexpected biases and noises would degrade the quality of the P-code data. Further study is required.

The present study is therefore concentrated on L1C code CV time transfer and its application in UTC. In the following discussions, because the short-term measurement noise of the L1C time link is about 0.7 to 1.5 ns, as given in Tables 2–8 and Figures 3–7 in the following sections, the disturbing effects including that of the frequency biases with a magnitude well less than the measurement noise, saying 0.3 ns or less, will be considered negligible in the study.

Before GLN can be used in UTC, the following points need to be clarified:

(1) use of precise orbit and ionosphere corrections,

(2) biases due to the multiple GLN PRN and/or frequencies,

(3) short- and long-term stabilities,

(4) calibration and its long-term variation.

The first point has been fully discussed in earlier studies, such as [4–6]. Several analysis centres, including those of the IGS, ESA (European Space Agency), and IAC (Information Analysis Centre, Russian Federation), provide regular updates of the precise ephemerides of GLN satellites [14]. We currently use the IAC ephemeride products and the IGS ionosphere maps to compute the precise orbit and ionosphere corrections.

In the following sections, we discuss the three remaining points, based on test CGGTTS L1C data from UTC 1005 to UTC 1110 (May 2010 to October 2011), assuming that all the raw measurements have been corrected for precise orbits and atmosphere delays.

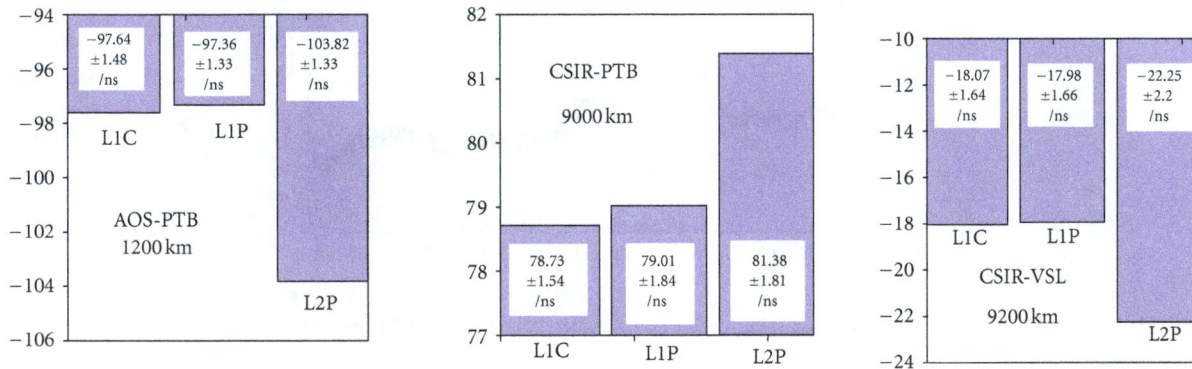

FIGURE 2: Histograms of the monthly mean value (Mean) of the CV clock differences and the standard deviations of the smoothing residuals ($\pm\sigma$) obtained using different GLN codes over different baselines. The values are given in form of Mean $\pm \sigma$ in ns. Data were collected from 3S receivers in 2004.

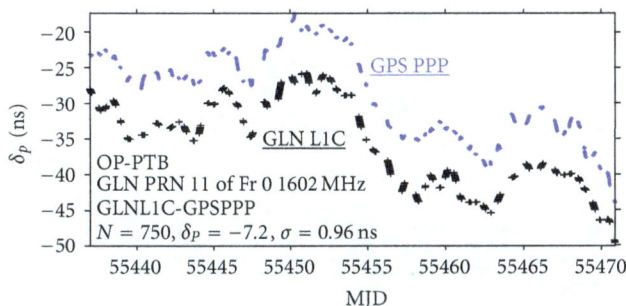

FIGURE 3: Bias of the GLN PRN 11 L1C link (+) relative to the GPS PPP link ($\bullet$) for the data set UTC 1009. Here $\sigma = 0.96$ ns is the standard deviation of the difference of GLNL1C-GPSPPP but that of the mean of the difference. The same definition is given to the σ throughout the paper.

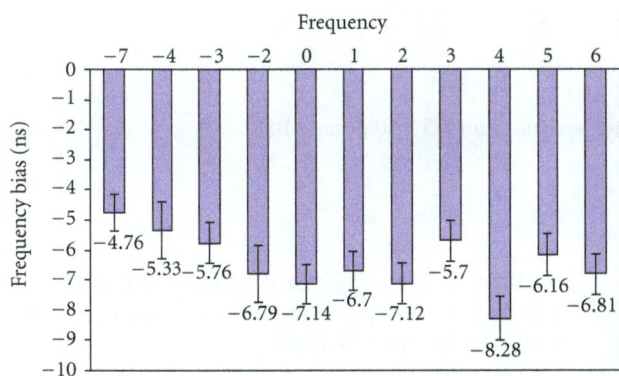

FIGURE 4: GLN Frequency L1C biases in order of the nominal frequencies corresponding to Table 2(b).

2.1. *The PRN and Frequency Biases in the CV Time Links.* Unlike GPS, the satellites of GLN is divided into groups according to the frequencies used. Early studies based on certain physical considerations and on first-generation GLN receivers, for example, 3S Navigation, addressed the question of so-called frequency biases disturbing the CV time transfer

[3, 4] and suggested a precorrection to each GLN frequency for the data to be used for UTC computation. A later study in 2005 based also on the 3S receivers [5, 6] however and a recent study in 2009 based on a new generation of GLN receivers, for example, TTS3 [10, 12], have found, though the detailed results have not been published, that the influence of the biases is negligible compared to the measurement noise (1 ns to 1.5 ns). This conclusion meant that in principle the GLN CV time transfer technique could be used directly as GPS without the need of the frequency bias corrections for the computation of UTC; that is, comparing the gain and the complexity of the computation, it is not worth to make the frequency bias corrections in the monthly UTC computation. Further investigations corresponding to the P-codes of both GLN and GPS can be found in [11].

We estimate PRN and/or frequency biases, based on the most acceptable hypothesis, for example, [3], that different frequencies emitted by different satellites through different channels of a receiver causes different biases, which perturb the GLN CV time transfer. It is important to establish whether the biases are well below the measurement noise and are therefore negligible, or alternatively if a calibration or correction is needed for each frequency in a GLN CV time link.

2.1.1. PRNs and Frequencies of GLN. As of mid-2010, the GLN system comprises 20 satellites operating the L1C code. Table 1 lists the operational GLN PRNs observed using receivers TTS-2 and TTS-3. A total of 20 operational PRNs are recorded using TTS-3 receivers and only 11 PRNs using TTS-2 receivers. A further PRN, 09, is listed in the official catalogue [18] as operating only in L1C but is not observed by the TTS receivers. Table 1 listed the satellites in order of the frequency codes. In total 11 coded frequencies are emitted by 1 or 2 satellites each. Excluding PRN 03, there are on average about 900 L1C observations in a typical UTC monthly data file using a TTS-3 receiver.

2.1.2. Biases of PRNs and Frequencies. Our main interest is the influence of the so-called frequency biases on the CV time links. According to previous studies, we assume first that

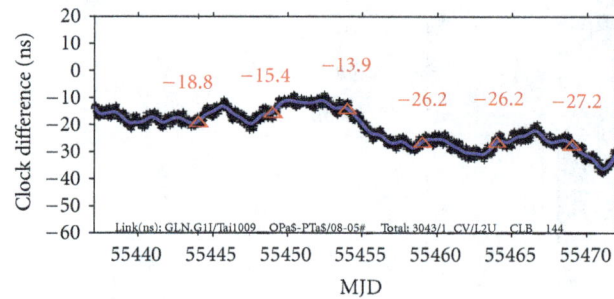

(a) GLN L1C time link OP-PTB 1009 where all the frequency biases have been corrected

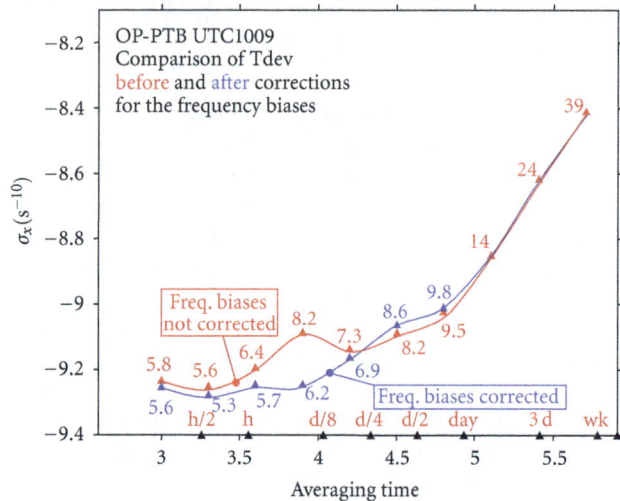

(b) Comparison of the time deviations σ_x of the same data as in Figure 5(a) before and after correction for all the frequency biases. Here h stands for hour, d for day, $d/2$ for half day, 3d for three days, and wk for a week. The x and y axis are labeled with log numbers and those in the graph are the real numbers. The same notations are used in all the TDev plots below

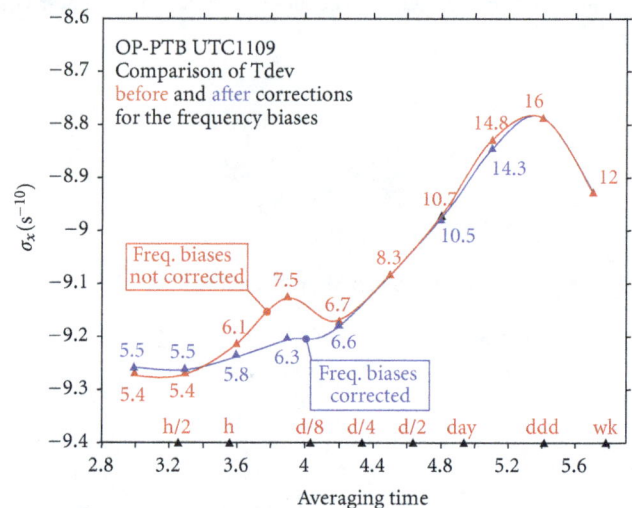

(c) Comparison of the Time deviations σ_x of the same baseline before and after correction for all the frequency biases (here 1109 is one year after 1009 as shown in Figure 5(b))

FIGURE 5: GLN L1C time links OP-PTB and Time deviations for UTC 1009 and 1109.

the frequency biases exist and are physically caused by the GLN frequencies, significantly receiver dependent, and are constant. The frequency biases should therefore be universal and could be corrected for in the UTC time transfer. We focus our analysis on the SU-PTB and OP-PTB baselines because both are UTC links, and for the latter we also have GPS PPP and TW links, which are more precise and provide good references for the evaluation of the GLN links. All three laboratories are equipped with TTS-3 receivers. To study the physical cause(s) of the frequency biases, we proceeded as follows:

(i) we first split the raw data file containing all the PRNs into subfiles for each PRN and then compute the one-PRN links;

(ii) we then compare the one-PRN links to the GPS PPP link to compute the frequency biases and use them to calibrate the raw link data;

(iii) we study if there are gains by comparing the time deviations and the differences versus GPS PPP and TW;

(iv) finally, we apply the "frequency biases" obtained from a month of a baseline to "calibrate" the raw data from other months and other baselines to see if the biases are "universal" (independent of receivers, months, and locations).

Figure 3 illustrates the bias of GLN PRN 11 L1C computed by comparing the OP-PTB CV link to that of GPS PPP for the data set UTC 1009. The bias of the PRN 11 is −7.20 ns

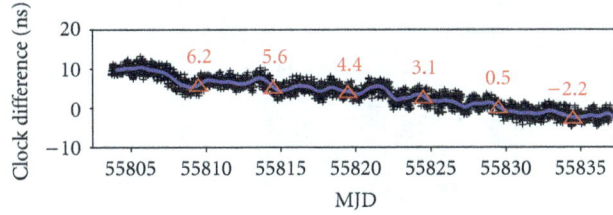

(a) The GLN L1C UTC time link SU-PTB 1109 after correction for the frequency biases

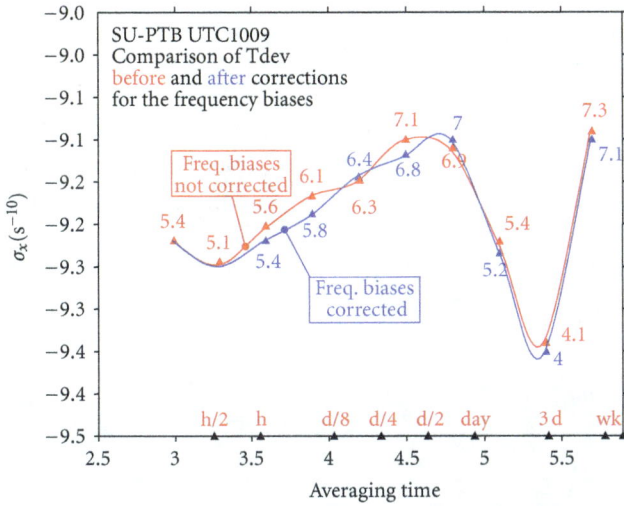

(b) Comparison of time deviations of the time link SU-PTB 1009 with and without correction for frequency biases

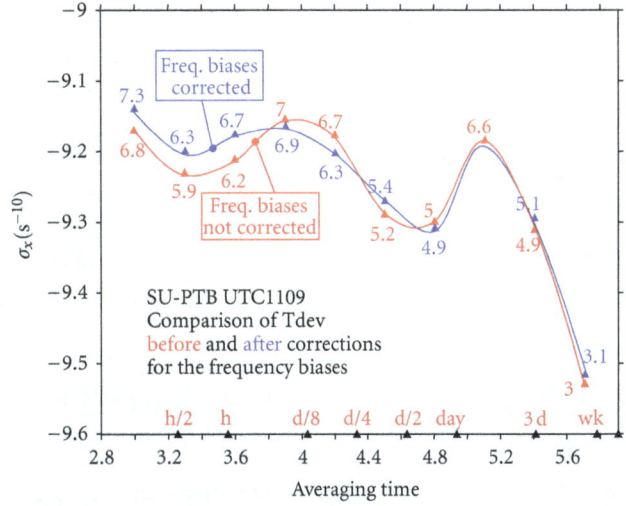

(c) Comparison of time deviations of the time link SU-PTB 1109 with and without correction for frequency biases (here 1109 is one year after 1009 in Figure 6(b))

FIGURE 6: GLN L1C time links SU-PTB and Time deviations for UTC 1009 and 1109.

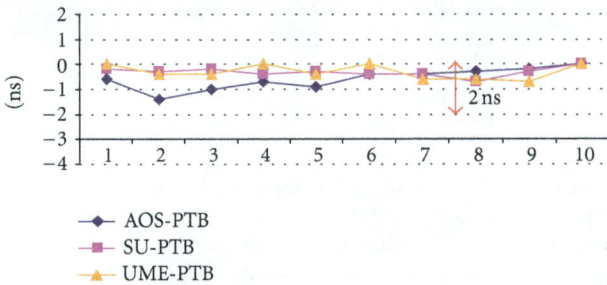

FIGURE 7: Consistency of UTC links between GPS C/A and GLN L1C (10-month comparison corresponding to Table 8).

±0.96 ns, including the calibration difference between GPS PPP and GLN L1C. What is important is not the size of the bias but whether or not it depends significantly on the GLN frequency, the receiver, and time.

Table 2(a) lists the PRN biases in the CV links with respect to the GPS PPP for the baseline OP-PTB (data set 1009). Observing the relation between the biases and the frequencies, Table 2(b) and the corresponding Figure 4 show the values in the increasing order of the nominal frequencies. The number of common points of the comparison (N) is typically about 750, with the exception of 220 for PRN 03. The standard deviation, (σ_P), of the bias determined for each PRN is slightly smaller than the measurement noise in the GLN L1C code (typically 1 ns to 1.5 ns as mentioned previously).

In Tables 2(a) and 2(b) and Figure 4, it is seen that the standard deviation σ_F of the frequency bias δ_F (cf. the caption of the Figure 3) is about 0.7 ns, while the maximum difference between the δ_F is 3.5 ns, bigger than the measurement noise. On the other side, the differences between the PRNs using the same frequency are mostly less than 0.3 ns, much smaller than the measurement noise. This would indicate that the biases vary with the frequency codes but the satellites.

TABLE 1: Operational GLN PRNs recorded using TTS-2 and TTS-3 time receivers (N is the number of observations by TTS3).

PRN	Fr. code	N	Receiver
GLN 11	0	904	TTS-3
GLN 15	0	915	TTS-3
GLN 01	1	900	TTS-3/TTS-2
GLN 05	1	906	TTS-3/TTS-2
GLN 20	2	901	TTS-3/TTS-2
GLN 24	2	912	TTS-3/TTS-2
GLN 13	−2	944	TTS-3
GLN 19	3	921	TTS-3/TTS-2
GLN 23	3	903	TTS-3
GLN 18	−3	920	TTS-3
GLN 22	−3	892	TTS-3
GLN 17	4	895	TTS-3/TTS-2
GLN 21	4	905	TTS-3/TTS-2
GLN 02	−4	893	TTS-3
GLN 03	5	251	TTS-3/TTS-2
GLN 07	5	885	TTS-3/TTS-2
GLN 04	6	851	TTS-3/TTS-2
GLN 08	6	902	TTS-3/TTS2
GLN 10	−7	915	TTS-3
GLN 14	−7	912	TTS-3

We also estimated the so-called frequency biases using other references such as P3 and TW, and the results are almost the same as those listed in Table 2; that is, the standard deviation is mainly due to the noise in L1C. There seems to be no obvious correlation between the amplitudes of the biases and the nominal frequencies.

2.1.3. Corrections for Frequency Biases in GLN CV Time Transfer.

It is expected that application of the frequency bias corrections to the raw GLN measurements should lead to a significant reduction in noise level and improvement in the short-term stability of the link. In Figure 5, Figure 5(a) shows the GLN L1C link of OP-PTB 1009 where all the frequency biases have been corrected; Figure 5(b) shows a comparison of the time deviations of the time links before and after correction for the frequency biases calculated for the UTC months of 1009.

Figure 5(c) is the comparison of the time deviations of the data 1109 (one year after 1009) with and without the bias corrections. Similar to Figure 5(b), an improvement in the time transfer stability is observed for the averaging time of 2 to 3 hours. The time deviation is an indicator of the time stability in a link. Comparing the time deviations estimated before and after the bias corrections, it is seen in Figures 5(b) and 5(c) that after correction the little knolls at about 2-3 hour averaging time in the uncorrected plot disappear. Assuming the trajectory of the GLN satellite is on average symmetric around the observers, 2-3 hours correspond to the half-time of the observable passage of the satellite. The results show a gain in time transfer quality for an averaging time of 2-3 hours. In consequence, the time deviation of the

TABLE 2: (a) GLN PRN/Fr L1C biases relative to GPS PPP for the link OP-PTB 1009. (b) GLN Frequency L1C biases in increasing order of the nominal frequencies.

(a)

PRN	Fr	f/MHz	N	δ_P/ns	σ_P/ns
11	0	1602.0	750	−7.204	0.963
15	0		744	−7.083	0.920
01	1	1602.5625	727	−6.882	0.995
05	1		733	−6.518	0.896
20	2	1603.125	740	−7.071	0.968
24	2		745	−7.175	1.016
13	−2	1600.8750	757	−6.793	0.975
19	3	1603.6875	754	−5.559	0.928
23	3		730	−5.851	1.077
18	−3	1600.3125	759	−5.724	0.906
22	−3		732	−5.800	1.034
17	4	1604.25	745	−8.287	1.031
21	4		736	−8.277	1.070
02	−4	1599.7500	751	−5.328	0.983
03	5	1604.8125	220	−6.206	1.081
07	5		722	−6.120	1.017
04	6	1605.375	710	−6.727	0.931
08	6		720	−6.907	1.069
10	−7	1598.0625	753	−4.706	0.888
14	−7		748	−4.823	0.966

(b)

Fr	Fr′/MHz	N	δ_F/ns	σ_F/ns
−7	1598.0625	753	−4.76	0.65
−4	1599.7500	751	−5.33	0.98
−3	1600.3125	759	−5.76	0.69
−2	1600.8750	757	−6.79	0.98
0	1602.0	750	−7.14	0.66
1	1602.5625	727	−6.70	0.67
2	1603.125	740	−7.12	0.70
3	1603.6875	754	−5.70	0.70
4	1604.25	745	−8.28	0.74
5	1604.8125	220	−6.16	0.74
6	1605.375	710	−6.81	0.71

TABLE 3: Gains in the standard deviation of the smoothing residuals for the GLN L1C baseline OP-PTB after correction for the frequency biases calculated for the period 1009 (comparison over 18 months).

Period yymm	σ/ns raw link	$\underline{\sigma}$/ns bias calibrated	Gain
1005	1.199	1.073	11%
1008	1.199	1.110	7%
1009	**1.260**	**1.150**	**9%**
1109	1.180	1.134	4%

one month data set is slightly improved for averaging times within one day.

TABLE 4: Gains in standard deviation of the smoothing residuals after correction for the frequency biases calculated for the OP-PTB UTC 1009.

Baseline	Distance/km	σ/ns raw link	σ/ns bias calibrated	Gain
AOS-PTB	500	1.721	1.381	20%
NIS-PTB	3000	1.338	1.276	5%
OP-PTB	**700**	**1.260**	**1.150**	**9%**
SG-PTB	6300	2.500	2.557	−2%
UME-PTB	1900	1.398	1.403	0

TABLE 5: GLN PRN/Fr L1C biases computed with SU-PTB 1009 versus GPS C/A (a constant of −200 ns is subtracted from the δ).

PRN	Fr.	N	δ_P/ns	σ_P/ns
11	0	700	−8.699	0.639
15	0	697	−8.331	0.649
01	1	683	−8.809	0.743
05	1	693	−8.507	0.684
20	2	691	−8.673	0.691
24	2	696	−8.578	0.760
13	−2	711	−8.857	0.711
19	3	701	−8.833	0.627
23	3	685	−8.700	0.751
18	−3	703	−9.988	0.690
22	−3	684	−9.730	0.698
17	4	703	−10.495	0.756
21	4	678	−10.365	0.701
02	−4	692	−8.927	0.674
03	5	200	−8.871	0.613
07	5	663	−8.811	0.686
04	6	663	−9.895	0.673
08	6	689	−9.880	0.735
10	−7	708	−9.329	0.744
14	−7	713	−9.078	0.763

TABLE 6: Gains in standard deviation of the smoothing residuals before and after corrections for the frequency biases calculated for the SU-PTB link for the period 1009.

Period yymm	σ/ns raw link	σ/ns bias calibrated	Gain
1005	1.252	1.254	0%
1008	1.068	1.043	2%
1009	**1.066**	**1.022**	**4%**
1109	1.150	1.177	−2%

The standard deviation of the smoothing residuals is also an index of the gains. If the frequency biases are constant for that baseline, they should be applicable to the raw data of other periods. We used the frequency bias corrections listed in Table 2(a), based on the 1009 data, to correct the raw data of 1005, 1008, and 1109 for the same baseline, OP-PTB. The result is given in Table 3. A considerable gain of 7% to 11%

is seen within 4 months from 1005 to 1009. The gain seems reduced with time if we compare the σ of 9% in 1009 and 4% in 1109, one year after 1009. This 4% is probably the physical gain due to the hardware delay between different frequencies, which impact the CV time links. Given the σ of 1.2 ns, 4% of the σ is 48 ps. Obviously 48 ps is numerically negligible for the GNSS code time transfer.

Because the same type of the receiver TTS3 is used (hence the hardware delay for same frequency is similar if not equal) we may further assume that the frequency corrections obtained from OP-PTB can be used for other receivers at AOS, NIS, SU, UME, and SG. We may expect a global gain of about 9%. Table 4 lists the results obtained for the five baselines of different distances. Two of the links show no improvement after correction: SG-PTB (−2%) and UME-PTB (0%), while three of the links (AOS-PTB, OP-PTB, and NIS-PTB) show a marked decrease in the standard deviations of the smoothing residuals, of 11% on average. We may have two explanations for this conflicted result. (1) A set of bias corrections is applicable only for a particular pair of receivers, that is, baseline dependent. The 11% gain is accident. (2) The frequency biases are not receiver only dependent but affected by some unknown factor which is common for AOS, OP, and NIS but not for SG and UME. The ionosphere influence is location, direction, and frequency dependent. The residual influence of the IGS ionosphere correction used in this study might be one of such factors. However further investigation is required.

2.1.4. Case of the UTC Link SU-PTB. We can use the same method to study the GLN UTC link SU-PTB. Because neither GPS PPP nor TW data exist for this baseline, we have to use GPS C/A as the reference to compute the so-called frequency biases.

Table 5 lists the frequency biases computed for GLN SU-PTB 1009 referenced to GPS C/A. As we assume the frequency biases are receiver dependent hence constant with time, we can apply these values obtained from the SU-PTB GLN data for the period 1009 to correct the corresponding data for 1008 and 1005 as well 1109.

Figure 6(a) shows the time link SU-PTB 1109, and Figure 6(b) illustrates the time deviations before and after correction for the frequency biases on the same baseline on 1009. Figure 6(c) shows that of 1109, one year after 1009. Not as seen for the baseline OP-PTB, Figures 6(b) and 6(c) show no obvious improvement in the time deviation for averaging time of 2-3 hours.

The standard deviations of the smoothing residuals for the months 1005, 1008, 1009, and 1109 are listed in Table 6. There is a slight, statistically not meaningful, variation in the standard deviations, 1% on average. Taking the value $\sigma = 1.2$ ns, 1% means 12 ps. The 1005 and 1009 data are separated by 4 months and 1009 and 1109 by 12 months. The gains of the application of the biases to the 1005 and 1109 data are 0% and −2%, that is, no gain in applying the so-called frequency-bias corrections. The frequency-bias corrections obtained from 1009 might not be really or completely caused by the frequency-bias but, at least partially, by some other frequency dependent biases. For this

TABLE 7: Comparing GLN PRN biases computed using OP-PTB and SU-PTB 1009.

PRN	Fr	$d\delta_{P1}$	σ_1	$d\delta_{P2}$	σ_2	Mean$_1$	Mean$_2$
11	0	0	0.963	0	0.639	0.0	0.0
15	0	0	0.920	0	0.649		
01	1	0.322	0.995	−0.110	0.743	0.4	−0.1
05	1	0.565	0.896	−0.176	0.684		
20	2	0.133	0.968	0.026	0.691	0.0	−0.1
24	2	−0.092	1.016	−0.247	0.760		
13	−2	0.411	0.975	−0.158	0.711	0.4	−0.2
19	3	1.524	0.928	−0.502	0.627	1.4	−0.3
23	3	1.353	1.077	−0.001	0.751		
18	−3	1.359	0.906	−1.657	0.690	1.4	−1.3
22	−3	1.404	1.034	−1.031	0.698		
17	4	−1.204	1.031	−2.164	0.756	−1.1	−1.9
21	4	−1.073	1.070	−1.666	0.701		
02	−4	1.755	0.983	−0.596	0.674	1.8	−0.6
03	5	0.998	1.081	−0.172	0.613	1.0	−0.3
07	5	0.963	1.017	−0.480	0.686		
04	6	0.477	0.931	−1.196	0.673	0.3	−1.4
08	6	0.176	1.069	−1.549	0.735		
10	−7	2.498	0.888	−0.630	0.744	2.4	−0.7
14	−7	2.260	0.966	−0.747	0.763		

baseline, it seems the frequency biases are statistically not baseline dependent.

According to Tables 3 and 6, the gains on average are about 0–4% or 0–50 ps for OP-PTB and for SU-PTB correspondingly. Even if we apply them to correct the frequency biases, such small values will be masked by the measurement noise and other frequency dependant biases.

2.1.5. Discussion. The previous results do not fully support the previous studies summarized in the beginning of Section 2.1 that the frequency biases should be precorrected for UTC time transfer within the L1C uncertainty. Would there exist other frequency dependent (or independent) factors, in addition to the receiver only dependent ones, that affect the frequency biases? Let us by the way point out that receiver dependent must lead to baseline dependent because the baseline is composed of a pair of receivers.

Let us use the exclusion method to examine a seemed impossible possibility.

If the biases are physically caused by the GLN signal frequencies alone, they should be constant with time, isotropically equivalent, and independent of receivers and baselines. As we now have two sets of frequency biases, obtained from the baselines OP-PTB and SU-PTB (Tables 2 and 5), both computed using the same data set UTC 1009, we can examine this hypothesis. In Table 7, $d\delta_P$ is obtained by subtracting the bias of the frequency "0" (PRN 11 and 15): $d\delta_{P1}$ is from the baseline OP-PTB (Table 2(a)) and $d\delta_{P2}$ from SU-PTB (Table 5). The Mean is the mean value of the $d\delta_P$ of the different PRNs using the same frequency. To hold the assumption, the values of Mean$_1$ and Mean$_2$ should agree

with each other within measurement noise (1 ns). As seen in Table 7, for more than half of the frequencies coded (Fr 3, −3, 4, −4, 5, 6, and −7) the same values are not found for the two baselines. For example, for the Frequency (−7), the difference of Mean$_1$ and Mean$_2$ is 3.1 ns which is much bigger than measurement noise.

This numerical evaluation based on two CV links does not prove the existence of the impact of the biases which are bigger than the measurement noise and depend on the GLN frequencies. Again, we cannot exclude the effects of other frequency-dependent factor(s) including the impact of the temperature variations. Considering the gain in applying the frequency bias corrections is not significant and the complexity of the computation is, it has been decided [12] not to use these corrections in the computation of UTC.

2.2. Calibration and Long-Term Stability of the GLN Time Links. A time link technique can be used in UTC only when it is calibrated, and its short- and long-term stabilities are proven. In the following study we use GPS as reference.

Table 8 and Figure 7 present the results of a ten-month comparison and list the differences between the GPS AV C/A links and GLN CV L1C links on the three UTC baselines AOS-PTB, SU-PTB, and UME-PTB between May 2009 and February 2010. All the data were collected using the same type of receivers (TTS-3). The GLN and the GPS raw data were corrected using the IGS/ESA precise ephemeride and ionosphere maps. The GLN links were calibrated and aligned to GPS in May 2009 [10, 12]. The calibrations of GPS and GLN links are stable and perfectly consistent. The mean values of the differences are −0.3 ns and −0.6 ns with

TABLE 8: Calibration consistency of GPS C/A versus GLN L1C links over 10 months (values given in the table are the mean of the differences between GPS and GLN links and its standard deviation).

YYMM	AOS-PTB/ns	SU-PTB/ns	UME-PTB/ns
1002	-0.6 ± 1.6	-0.2 ± 1.4	0.0 ± 1.4
1001	-1.4 ± 1.6	-0.3 ± 1.6	-0.4 ± 1.4
0912	-1.0 ± 1.5	-0.2 ± 1.6	-0.4 ± 1.4
0911	-0.7 ± 1.6	-0.4 ± 1.6	
0910	-0.9 ± 1.4	-0.3 ± 1.6	-0.4 ± 1.3
0909	-0.4 ± 1.6	-0.4 ± 1.6	-0.0 ± 1.4
0908	-0.4 ± 1.6	-0.4 ± 1.6	-0.6 ± 1.4
0907	-0.3 ± 1.6	-0.7 ± 1.6	-0.6 ± 1.4
0906	-0.2 ± 1.6	-0.3 ± 1.6	-0.7 ± 1.4
0905	-0.0 ± 1.6	-0.0 ± 1.6	-0.0 ± 1.4
Mean	-0.6	-0.3	-0.3
σ	0.4	0.2	0.3

TABLE 9: UTC laboratories operating two or three time and frequency transfer facilities as of 2008 [20].

Lab	GPS	GLN	TW
AOS	√	√	√
AUS	√		√
CH	√		√
IT	√	√	√
KRIS	√	√	√
LDS	√	√	
MIKE	√	√	
NICT	√		√
NIM	√		√
NIS	√	√	
NIST	√	√	√
NMIJ	√		√
NPL	√		√
NPLI	√	√	
NTSC	√		√
OP	√		√
PTB	√	√	√
ROA	√		√
SG	√	√	√
SP	√		√
SU	√	√	√
TL	√		√
UME	√	√	
USNO	√	√	√
VSL	√	√	√
ZA	√	√	

standard deviations between 0.2 ns and 0.4 ns and the RMS 0.4 and 0.7 ns. The GPS and GLN data are well consistent within their measurement uncertainties.

As the short- and long-term stabilities of GPS are well proven and GPS and GLN are completely independent

systems, this close consistency between the data sets demonstrates that the GLN time transfer technique is as stable as GPS in both the short and long terms. The same conclusion holds for the long-term variations in their calibrations (cf. [10, 12]).

2.3. Combination of GLN and GPS for UTC Time Transfer. Since January 2011, a combination of the GLN L1C and GPS C/A code time links has been used for SU-PTB and UME-PTB in UTC time transfer [21]. This is the first time that data from different GNSS have been combined for a UTC time link. By the end of 2011, 6 combined links are used for UTC computation. The discussion in the following focuses on introducing the weighted combination.

The UTC time transfer strategy until the end of 2010 was the so-called *primary* UTC *time transfer technique*, meaning that only the "best" techniques are used for UTC generation and others are kept as backup. Thus TWSTFT links are used in preference to GNSS links, and GPS links in preference to GLN links, and so forth. The coexisting multitechniques strategy has led to a rapid increase in the level of redundancy in the UTC data bank, with new techniques being added all the time. The tendency to use multitechniques for UTC time transfer is unavoidable. As of 2008 there were 26 UTC laboratories operating multifacilities of time transfer [20]; among them 15 were equipped with both GPS and GLN receivers. Table 9 summarizes the availability of the GNSS and TW facilities at some of the national laboratories contributing to UTC, where at least two time and frequency transfer techniques are equipped.

As discussed previously, (cf. Table 8 and Figure 7), the calibrations of GPS and GLN links agree well with each other and are stable with time. We can therefore take the mean values of sets of GLN L1C code CV and GPS C/A code AV data as (GPS C/A+GLN L1C)/2 or depending on the measurement quality of GPS and GLN, take the weighted combination as $[n \times (\text{GPS C/A}) + m \times (\text{GLN L1C})]/(n + m)$, namely, GLN&GPS standing for the time link combination using GLN and GPS data. Here n and m are the weights of the GPS and GLN.

In the numerical tests, we use the more precise TW and GPS PPP links as references to estimate the gains. Both are available for the baseline OP-PTB. In March 2010, the measurement uncertainty u_A of the GPS PPP and TW links for this baseline are, respectively, 0.3 ns and 0.7 ns (TW degraded somewhat since the beginning of 2010 from its previous conventional value, 0.5 ns). It should be pointed out that taking GPS PPP as reference may somewhat disfavour the GLN L1C CV links because GPS and GLN are independent systems while the GPS C/A and GPS PPP are not completely independent. We use the data sets of UTC 1002, 1005, and 1009 as well as a-15 month long-term data set 1007–1109. We test also the UTC baselines SU-PTB and INPL-PTB using the arbitrarily selected data sets of UTC 1102 and 1110.

Table 10 shows the standard deviations of the GPS-only, the GLN-only, and the combination GLN&GPS links against TW and GPS PPP. Here σ is the standard deviation of a single technique and $\underline{\sigma}$ is that of the GLN&GPS. The

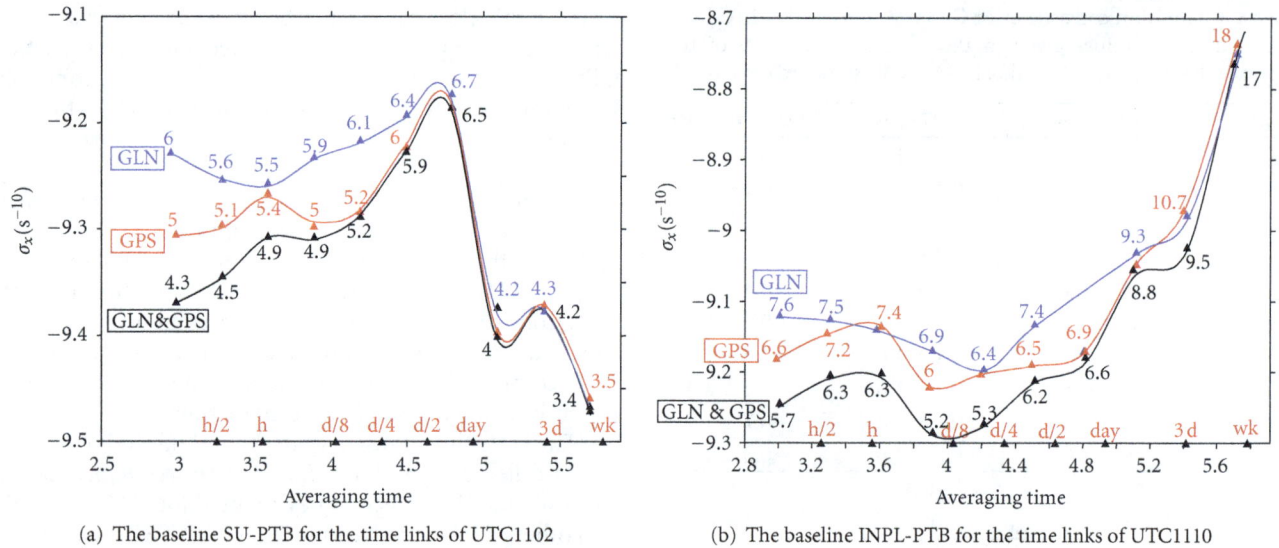

(a) The baseline SU-PTB for the time links of UTC1102   (b) The baseline INPL-PTB for the time links of UTC1110

FIGURE 8: Comparison of the time deviations between the GPS-only, GLN-only, and GLN&GPS links for the baselines SU-PTB (Figure 8(a)) and INPL-PTB (Figure 8(b)). Both are the UTC time links.

TABLE 10: Comparison of the standard deviations of the clock differences for the GPS-only, GLN-only, and GLN&GPS links for the baseline OP-PTB 1005 (MJD 55313 to 55346).

Compared to	GPS-only σ/ns	GLN-only σ/ns	(GLN&GPS) $\underline{\sigma}$/ns	Gains $(\sigma - \underline{\sigma})/\sigma$
TW	1.240	1.369	1.215	6.5%
PPP	1.182	1.285	1.149	7%

gain is computed by the equation $(\sigma - \underline{\sigma})/\sigma$. The standard deviations of the differences of the GPS-only, GLN-only, and GLN&GPS time links relative to TW are 1.240 ns, 1.369 ns, and 1.215 ns, respectively. The averaged gain in GLN&GPS versus GLN-only and GPS-only with respect to TW is 6.5%. Similarly, taking PPP as reference, the standard deviations are 1.182 ns, 1.285 ns, and 1.149 ns, respectively. The gain with respect to PPP is 7%. The combination thus confers an average gain of 7%. Knowing the measurement uncertainties of TW and of GPS PPP and the simplicity of the combination computation, the gain here is hence conservative and the operation is worthy.

Figures 8(a) and 8(b) illustrate the time deviation of the time links of GPS-only, GLN-only, and the combination GLN&GPS for the baseline SU-PTB 1102 and INPL-PTB 1110. The short-term stability of the GPS-only link is slightly better than that of the GLN-only, probably as a result of the advantage of the AV technique against the CV. The stability of the combined solution GLN&GPS is better in the short term than that of the GPS-only and the GLN-only. For averaging times of beyond 20 hours, the three time deviation curves converge.

Figure 9(a) shows the (GLN&GPS) data for the UTC baseline OP-PTB for the period UTC 1009 (corresponding to MJD 55437 to 55472). Figure 9(b) compares the time

deviations between the corresponding GPS-only, GLN-only, and GLN&GPS links. The comparison shows that for averaging times of up to half day the combined (GLN&GPS) link is much more stable than the data from either of the single techniques: less noisy and less biased.

To compare the long-term stabilities, we look at the GPS-only, GLN-only, and GLN&GPS data over a 15-month period (1007–1109: MJD 55378–55834) for the UTC baseline OP-PTB. Figure 10 shows the comparisons of the corresponding time deviations. After the better averaging based on the increased number of data points, we see here more clearly that the stability of the combined link GLN&GPS is better than the single techniques, at least for averaging times of up to 1 day.

The combination thus leads to an improvement in the short-term stability for averaging times of up to 1 day. Since January 2011 combined solutions have therefore been applied in UTC generation. We gave some examples of the links based on a combination of two fully independent techniques to be used in UTC time transfer [21].

3. Future Development in GLN Time Transfer

The possible use of P3-code clearly merits further investigation. Other open issues are the use of the carrier phase, the calibrations, and the raw data recording. We briefly outline our considerations for the coming future studies at the BIPM.

3.1. Use of the GLN Carrier Phase. Given the success of GPS PPP [22], GLN PPP is certainly worthy of study. At present few authors work on this topic [23] and as yet there is not a good enough solution to be able to use in UTC time transfer. In GNSS PPP, the P-codes and the carrier-phase (CP) data are dealt with together. In contrast to PPP, we are investigating a different approach, namely, the postcombination. We first

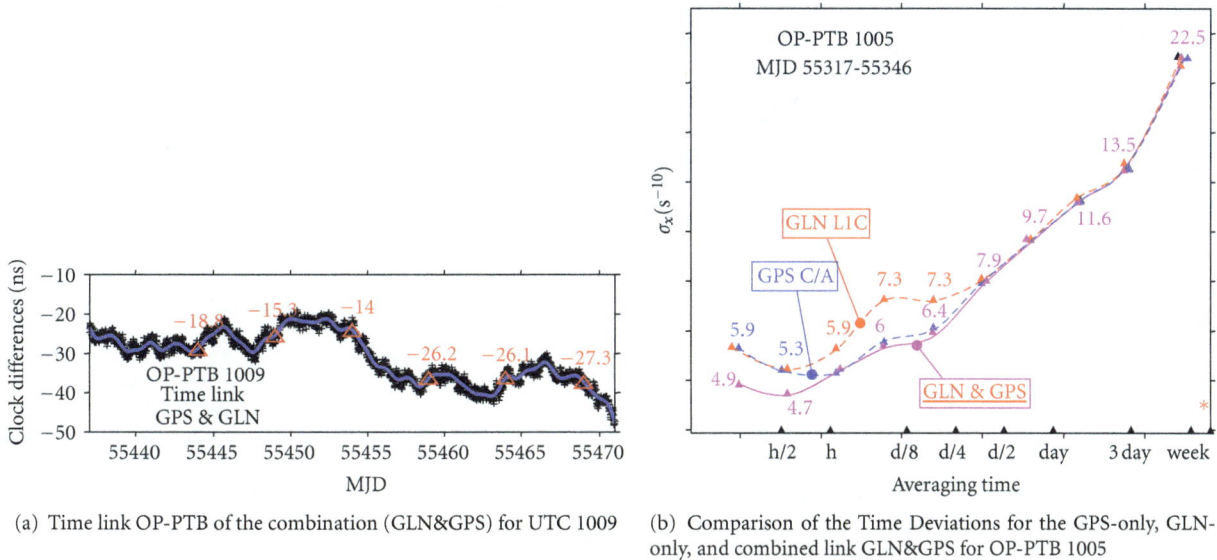

(a) Time link OP-PTB of the combination (GLN&GPS) for UTC 1009

(b) Comparison of the Time Deviations for the GPS-only, GLN-only, and combined link GLN&GPS for OP-PTB 1005

FIGURE 9: The GLN&GPS time transfer baseline OP-PTB.

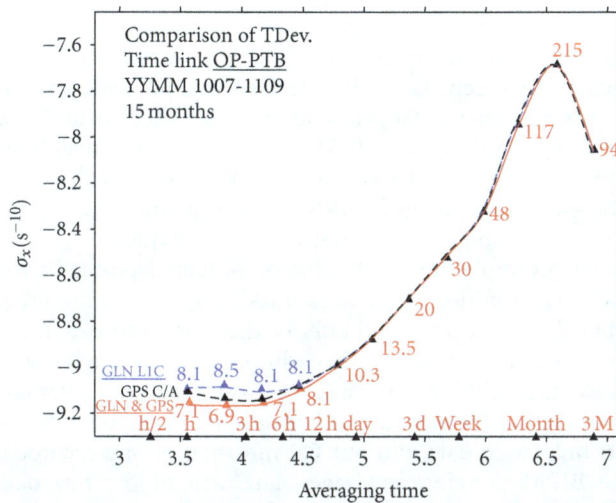

FIGURE 10: 15-month long-term comparison of the TDev of GLN/L1C, GPS C/A, and GLN&GPS over the baseline OP-PTB between 1007 and 1109. The GLN&GPS is the most stable one and is used as the official UTC link. The TDev of the three links converges up to 1 day.

compute the code and CP separately and then to combine the code and CP solutions.

One difficulty with the PPP is the ambiguity of the carrier-phase information. In addition, PPP relies on the Earth geocentric reference and related quantities, such as the geocentric coordinates of the satellites in space and of the antenna centres of the receivers on the ground, and the processing is complex.

The result of a time link is the *clock difference* (CD) between the two master clocks on the two ends of a baseline. In a clock comparison, the CD is given by the code data. If we can generate a carrier-phase solution that gives the

rates of the CD(RCD), we can use these rates to smooth the code solution CD. The advantage of this approach is that the carrier phase is two orders more precise than the code which generates the clock difference. This method of smoothing is not only precise but also easy. Further, the ambiguity in the simple difference of the CP solution, that is, in the RCD, is cancelled, and the absolutely determined geocentric terms required in the PPP/CP solution are simplified. Mathematically, the problem is to smooth a series of measurements using its derivatives. As the method (namely, combined smoothing) and its application in time transfer have been fully discussed in [24], we will not repeat them here.

Study of the GLN RCD option is an ongoing activity at the BIPM. One way to generate the difference in rates between two clocks is to differentiate the PPP data [25]. Our interest hereafter is not in combining GLN code and GPS CP data but lies in the method of the combination of the GLN L1C code and the GLN CP information (or RCD exactly) which is not available. As a simulation test, we use the GPS CP to replace the GLN CP. In the following discussion we examine the method of smoothing GLN code with the RCD and estimate the potential gains and the achievable uncertainty, assuming that the GLN CP is as precise as GPS CP. We then present the result of the combined smoothing of the GLN L1C and the RCD, namely, GLN RCD, which has the advantage of maintaining the calibration defined by the GLN L1C and the short-term stability assigned by the CP. It should be pointed out that the instabilities of the P-codes and the coarse code L1C are of the same order of magnitude, while the CP is two orders of magnitude more precisely. Earlier studies using GPS data proved that using the RCD to smooth either the coarse codes or the precise P-codes gives the same result in terms of stability. The following numerical test shows the same for GLN data. More details can be found in [25].

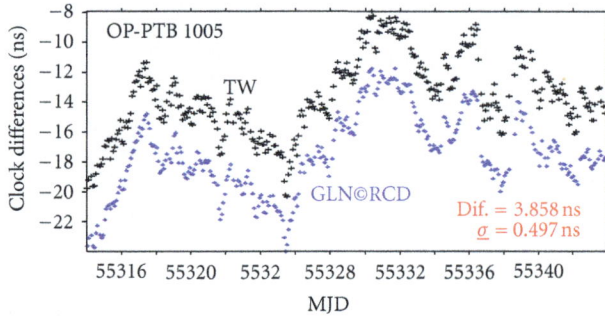

Figure 11: Comparison between the TW (+) and GLN©RCD (•) time links for the baseline OP-PTB.

Table 11: Comparison of the TW, GLN-only, GPS PPP, and GLN©RCD time links for OP-PTB 1005.

Link differences	N	Mean/ns	σ/ns
TW-(GLN-only)	324	3.873	1.346
TW-GLN RCD	324	3.858	0.497
GPSPPP-GLN RCD	2870	−0.921	0.100

Table 11 compares the time links using TW, GLN-only, GPS PPP, and the combined smoothing GLN©RCD for the baseline OP-PTB using the data set UTC1005 (MJD 55313–55345). The mean values obtained for (TW-(GLN-only)) and (TW-GLN©RCD) are 3.873 ns and 3.858 ns, respectively. The difference between these two results, 0.015 ns, is well below the measurement noise in GLN L1C, confirming that the GLN RCD method keeps the calibration of the GLN L1C. The respective standard deviations are 1.346 ns and 0.497 ns; that is, the measurement noise is well reduced. This is also supported by the standard deviation of the difference (GPSPPP-GLN©RCD) which is only 0.1 ns. Figure 11 shows a comparison between the TW and GLN©RCD time links for the baseline OP-PTB.

Figure 12 shows the corresponding time deviations for the links based on GLN-only, GPS PPP, and GLN©RCD for the same baseline and period. The stability of the GLN©RCD and GPS PPP links is almost identical. In general, the characteristics of the combined smoothing data are dominated in the long term by that of the code used, and the CP dominates the short terms.

3.2. Improvement of the Calibration Uncertainty. The total uncertainty in (UTC-UTC(k)) is dominated by the uncertainty of the time-transfer calibration. Currently, the best calibration uncertainties in GNSS time transfer are 5 ns [15, 21]. Hence a key factor in the reduction of the uncertainty in UTC products is to improve the GNSS calibrations. A BIPM calibration scheme has been proposed, aiming to achieve a calibration uncertainty of less than 2 ns [26]. A pilot project improving the Asian links organized by the BIPM is ongoing, and a significant improvement in GLN calibrations is expected.

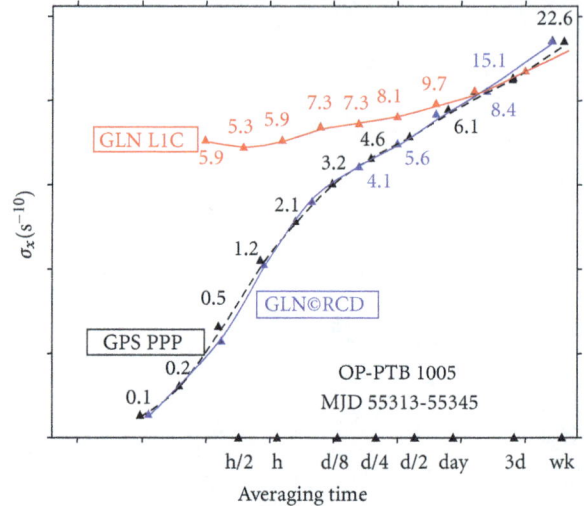

Figure 12: Time deviations of the GLN-only, GPS PPP, and the combined GLN©RCD links for OP-PTB 1005.

3.3. Raw Data Recording in the CCTF GGTTS Format. The CCTF GGTTS data format was designed in the early 1980s when GPS was introduced into time transfer using the receivers available at the time. The format has since been updated to accept GLN data as well but its basic specifications remain unchanged, and it is still used to facilitate the computation of UTC/TAI. However, some conventions defined in the GGTTS are now outdated due to the ever-progressing technology in GNSS receiver manufacturing and the introduction of new time-transfer techniques.

For example, one of the major outdated points in the GGTTS convention is that for a tracking arc of 16 minutes of data collection only 13 minutes of them are recorded and 3 minutes of data are wasted. In addition, the time tagging with a fixed interval of 780s and a lag of about 4 minutes every day is impractical for most users. The data are round off at 0.1 ns and only code data without CP information are recorded. The BIPM therefore envisages a reform of the raw data collection conventions and an update of the GGTTS format [27] to take account of current and future improvements in GLN time and frequency transfer.

4. Conclusion

To guarantee the precision, the accuracy, and the robustness of UTC generation, the multitechnique strategy for UTC time transfer is indispensable. Efforts towards introducing GLN to complement GPS and TW in the generation of UTC began in the early 1990s, and in November 2009 the first two GLN time links were introduced into the UTC worldwide time link network.

In this paper we present the technical features of GLN time transfer as important for UTC production: a study of the so-called frequency biases, the short- and long-term stabilities, the calibration process, and the advantages of combining GLN and GPS. We also describe various ongoing projects at the BIPM, particularly concerning the use of carrier-phase data.

The present study is focused on the application of GLN L1C code in the generation of UTC, which yields a short-term stability of 1 ns to 1.5 ns. The calibration uncertainty is 5 ns, and the long-term stability is about the same as for GPS. The combination of the GLN L1C and GPS C/A codes makes sense in reducing the short-term stability and particularly in increasing the accuracy and the robustness in the UTC links.

The cause of the so-called frequency biases remains unclear for the authors. Although correction for estimated frequency biases leads to some slight gains for certain baselines, these gains are not seen ubiquitously, and, pending further research, it has been decided not to apply such corrections for GLN links used in the computation of UTC.

Notation

UTC:	Coordinated Universal Time
BIPM:	International Bureau of Weights and Measures
GLN:	GLONASS (GLObal Navigation Satellite System) [18]
GPS:	Global Positioning System
GNSS:	Global Navigation Satellite Systems
IGS:	International GNSS Service
TW:	TWSTFT (Two-Way Satellite Time and Frequency Transfers)
PRN:	PseudoRandom Noise code signal. Each GPS satellite transmits a unique code sequence (Code Division Multiple Access) and may be identified according to its PRN number. All GLN satellites transmit the same PRN signals using different frequencies (Frequency Division Multiple Access). In the UTC/TAI data format (CGGTTS), PRN is the nominal number of a GLN satellite
Fr:	Frequency or frequency code
δ_P:	Bias in time delay of a GLN PRN
δ_F:	Bias in time delay of a GLN frequency
CV:	Code-based common view time transfer
AV:	Code-based and/or carrier phase all in view time transfer [17]
P3:	Time transfer (CV and/or AV) using the linear combination of L1 and L2 measurements to achieve ionosphere-free code measurements
PPP:	Time transfer using carrier-phase precise point positioning technique [22]
GLN&GPS:	Time transfer combining GPS C/A and GLN L1C codes
Gain:	In percentage to indicate the improvement in time transfer quality. The gain in σ versus σ is computed by the equation $(\sigma - \underline{\sigma})/\sigma$
CP:	Carrier phase
CD:	Clock difference
RCD:	Rate of CD
yymm:	Year and month (an UTC computation month), for example, 0910 for 2009 October and 1005 for 2010 May.

Acronyms Used for the National UTC Laboratories

AOS:	Astrogeodynamical Observatory, Borowiec (Poland)
CSIR:	National Metrology Institute of South Africa (NMISA, South Africa)
INPL:	National Physical Laboratory, Jerusalem (Israel)
NIS:	National Institute for Standards, Cairo (Egypt)
OP:	Observatoire de Paris (France)
PTB:	Physikalisch-Technische Bundesanstalt, Braunschweig and Berlin (Germany)
SG:	Agency for Science Technology and Research (A*STAR) (Singapore)
SU:	Institute for Physical-Technical and Radiotechnical Measurements, Rostekhregulirovaniye of Russia (VNIIFTRI), Moscow, (Russian Federation)
UME:	Ulusal Metroloji Enstitüsü/National Metrology Institute, Gebze-Kocaeli (Turkey)
VSL:	Dutch Metrology Institute, Delft (Netherlands).

Acknowledgment

The authors are grateful to the UTC contributing laboratories for the data used in this study and the reviewers for their constructive scientific suggestions.

References

[1] P. Daly, N. B. Koshelyaevsky, W. Lewandowski, G. Petit, and C. Thomas, "Comparison of GLONASS and GPS time transfers," *Metrologia*, vol. 30, no. 2, pp. 89–94, 1993.

[2] W. Lewandowski, J. Azoubib, and A. G. Gevorkyan, "First results from GLONASS common-view time comparisons realized according to the BIPM international schedule," in *Proceedings of the 28th PTTI*, pp. 357–366, 1996.

[3] J. Azoubib and W. Lewandowski, "Test of GLONASS precise-code time transfer," *Metrologia*, vol. 37, no. 1, pp. 55–59, 2000.

[4] W. Lewandowski, J. Nawrocki, and J. Azoubib, "First use of IGEX precise ephemerides for intercontinental GLONASS P-code time transfer," *Journal of Geodesy*, vol. 75, no. 11, pp. 620–625, 2001.

[5] Z. Jiang and G. Petit, "Evaluation of the effects of the IGEX/IGS corrections on the GLN time and GLN time transfer," *BIPM Technical Memorandum TM135*, 2005, ftp://tai.bipm.org/TimeLink/LkC/VAR/Doc/GLN/.

[6] Z. Jiang and W. Lewandowski, "Recent study on GLONASS time transfer—application of Tsoft for the GLN calculations," *BIPM Technical Memorandum TM136*, 2005, ftp://tai.bipm.org/TimeLink/LkC/VAR/Doc/GLN/.

[7] A. Foks, W. Lewandowski, and J. Nawrocki, "Frequency biases calibration of GLONASS P-code time receivers," in *Proceedings of the 19th EFTF*, Besancon, France, 2005.

[8] W. Lewandowski, A. Foks, Z. Jiang, J. Nawrocki, and P. Nogaś, "Recent progress in GLONASS time transfer," in *Proceedings of the Joint IEEE International Frequency Control Symposium*

(FCS) and Precise Time and Time Interval (PTTI) Systems and Applications Meeting, pp. 728–734, August 2005.

 [9] J. Nawrocki, W. Lewandowski, P. Nogaś, A. Foks, and D. Lemański, "an experiment of GPS+GLONASS common-view time transfer using new multi-system receivers," in *Proceedings of the 20th EFTF*, Braunschweig, Germany, 2006.

[10] Z. Jiang and W. Lewandowski, "New evaluation of Glonass time transfer," *BIPM Technical Memorandum TM170*, 2009, ftp://tai.bipm.org/TimeLink/LkC/VAR/Doc/GLN/.

[11] Z. Jiang and W. Lewandowski, "On the PRN/Frequency Offsets for GLN Time Transfer in UTC Computation," *BIPM Technical Memorandum TM184*, 2010, ftp://tai.bipm.org/TimeLink/LkC/VAR/Doc/GLN/.

[12] W. Lewandowski and Z. Jiang, "Use of Glonass at the BIPM," in *Proceedings of the PTTI*, pp. 5–13, 2009.

[13] Z. Jiang, L. Tisserand, A. Harmegnies, W. Lewandowski, and G. Petit, "Use of the IAC GLN products in UTC time link computation," *BIPM Technical Memorandum TM183*, 2010, ftp://tai.bipm.org/TimeLink/LkC/VAR/Doc/GLN/.

[14] Z. Jiang, G. Petit, A. Harmegnies, and W. Lewandowski, "Comparison of the GLONASS orbit products for UTC time transfer," in *Proceedings of the EFTF*, 2011.

[15] BIPM Circular T 263, 2009, ftp://ftp2.bipm.org/pub/tai/publication/cirt.263.

[16] Z. Jiang and W. Lewandowski, "Use of Glonass for UTC time transfer," *Metrologia*, vol. 49, pp. 57–61.

[17] Z. Jiang and G. Petit, "Time transfer with GPS all in view," in *Proceedings of the Asia-Pacific Workshop on Time and Frequency*, pp. 236–243, 2005.

[18] GLONASS Interface Control Document Navigational radiosignal in bands L1, L2 (Edition 5.1) Moscow, Russia, 2008, http://rniikp.ru/en/pages/about/publ/ICD_GLONASS_eng.pdf.

[19] Z. Jiang, "A remark on the TTS3 GLN L3P codes," *BIPM Technical Memorandum TM175*, 2010, ftp://tai.bipm.org/TimeLink/LkC/VAR/Doc/GLN/.

[20] BIPM Annual Report on Time Activities, 2008.

[21] BIPM Circular T 287, 2011, ftp://ftp2.bipm.org/pub/tai/publication/cirt.287.

[22] G. Petit and Z. Jiang, "Precise point positioning for TAI computation," *International Journal of Navigation and Observation*, vol. 2008, Article ID 562878, 8 pages, 2008.

[23] P. Defraigne, Q. Baire, and N. Guyennon, "GLONASS and GPS PPP for time and frequency transfer," in *Proceedings of the IEEE International Frequency Control Symposium Joint with the 21st European Frequency and Time Forum*, pp. 909–913, June 2007.

[24] Z. Jiang and G. Petit, "Combination of TWSTFT and GNSS for accurate UTC time transfer," *Metrologia*, vol. 46, no. 3, pp. 305–314, 2009.

[25] Z. Jiang, "Combination of GPS and GLN," *BIPM Technical Memorandum TM179*, 2010, ftp://tai.bipm.org/TimeLink/LkC/VAR/Doc/GLN/.

[26] Z. Jiang, G. Petit, F. Arias, and W. Lewandowski, "BIPM calibration scheme for UTC time links," in *Proceedings of the EFTF*, pp. 1064–1069, 2011.

[27] Z. Jiang and W. Lewandowski, "Some remarks on the CCTF CGGTTS format," in *Proceedings of the EFTF*, pp. 317–322, 2011.

9

Nonlinear Bayesian Tracking Loops for Multipath Mitigation

Pau Closas,[1] Carles Fernández-Prades,[1] José Diez,[2] and David de Castro[2]

[1] *Centre Tecnològic de Telecomunicacions de Catalunya (CTTC), Parc Mediterrani de la Tecnologia, Avenida Carl Friedrich Gauss 7, Barcelona, 08860 Castelldefels, Spain*
[2] *DEIMOS Space S.L.U, Ronda de Poniente 19, Tres Cantos 28760, Madrid, Spain*

Correspondence should be addressed to Pau Closas, pclosas@cttc.es

Academic Editor: Maarten Uijt de Haag

This paper studies Bayesian filtering techniques applied to the design of advanced delay tracking loops in GNSS receivers with multipath mitigation capabilities. The analysis includes tradeoff among realistic propagation channel models and the use of a realistic simulation framework. After establishing the mathematical framework for the design and analysis of tracking loops in the context of GNSS receivers, we propose a filtering technique that implements Rao-Blackwellization of linear states and a particle filter for the nonlinear partition and compare it to traditional delay lock loop/phase lock loop-based schemes.

1. Introduction

Global Navigation Satellite Systems (GNSS) are the general concept used to identify those systems that allow user positioning based on a constellation of satellites. Specific GNSS are the well-known American GPS, the Russian GLONASS, or the forthcoming European Galileo. All those systems rely on the same principle: the user computes its position by means of measured distances between the receiver and the set of in-view satellites. These distances are calculated estimating the propagation time that synchronously transmitted signals take from each satellite to the receiver. Therefore, GNSS receivers are only interested in estimating the delays of signals which are received directly from the satellites, referred to as line-of-sight signal (LOSS), since they are the ones that carry information of direct propagation time. Hence, reflections distort the received signal in a way that may cause a bias in delay and carrier-phase estimations. Multipath is probably the dominant source of error in high-precision applications, especially in urban scenarios, since it can introduce a bias up to a hundred of meters when employing a 1-chip wide (standard) delay lock loop (DLL) to track the delay of the LOSS, which is a common synchronization method used in spread-spectrum receivers. This error might be unacceptable in many applications.

Sophisticated synchronization techniques estimate not only LOSS parameters but those of multipath echoes.

This results in enhanced, virtually bias-free pseudorange measurements. In this paper, we investigate multipath estimating tracking loops in realistic scenarios, where this effect is known to be severe. The analysis is driven in two directions. Firstly, a review of statistical characterization of the channel model in such situations is performed and a commercial signal simulator. Secondly, a novel multipath estimating tracking loop is discussed, providing details on the implementation, as well as comparisons to state-of-the-art techniques when different channel characteristics are considered. This tracking loop resorts to the Bayesian nonlinear filtering framework, sequentially estimating the unknown states of the system (i.e., parameters of the LOSS and echoes) and providing robust pseudorange estimates, subsequently used in the positioning solution. The so-called multipath estimating particle filter (MEPF) considers Rao-Blackwellization of signal amplitudes and the use of a suitable nonlinear filter for the rest of nonlinear states, for example, time-delays and their rate. More precisely, Rao-Blackwellization involves marginalization of linear states and the use of a standard Kalman filter to track signal amplitudes with the goal of reducing the estimation variance, since (i) the dimensionality of the problem that nonlinear filters solve is reduced and (ii) linear states are optimally tackled. For the nonlinear part of the state space we consider sequential Monte-Carlo methods (specifically, the standard particle filtering) as one of the most promising alternatives in advanced

GNSS receiver designs. Realistic computer simulation results are presented using the GRANADA FCM signal simulator and the performance of the MEPF is evaluated.

The remainder of the paper is organized as follows. Section 2.1 provides a brief overview of the fundamentals of GNSS, their signal structure, available channel models, and receivers' architecture and describes a realistic simulation platform. Section 3 sketches the basics of particle filters, and Section 4 is devoted to their application to GNSS signal synchronization in the presence of multipath. Section 5 presents computer simulations, and finally Section 6 concludes the paper. For the sake of completeness, the paper shows in the Appendix the equivalence between precorrelation and postcorrelation processing of GNSS signals. Notice that in this paper, the MEPF method operates after correlation is performed in order to operate at a lower data rate.

2. Fundamentals of Global Navigation Satellite Systems

GNSS space vehicles broadcast a low-rate navigation message that modulates continuous repetitions of pseudorandom spreading codes, that in turn are modulating a carrier signal allocated in the L band. The navigation message, after proper demodulation, contains among other information the so-called ephemeris, a set of parameters that allow the computation of the satellite position at any time. These positions, along with the corresponding distance estimations, allow the receiver to compute its own position and time, as we will see hereafter. Basically, a GNSS receiver performs trilateration, a method for determining the intersections of three or more sphere surfaces given the centers and radii of the spheres. In this case, the centers of the spheres are the satellites, whose position can be computed from the navigation message, and the radii of the spheres are the distances between the satellites and the receiver, estimated from the time of flight.

The distance between the receiver and a given satellite can be computed by

$$\rho_i = c\left(t_i^{\text{Rx}} - t_i^{\text{Tx}}\right), \qquad (1)$$

where $c = 299792458 \text{ m/s}$ is the speed of light, t_i^{Rx} is the receiving time in the receiver's clock, and t_i^{Tx} the time of transmission for a given satellite i. Receiver clocks are inexpensive and not perfectly in sync with the satellite clock, and thus this time deviation is another variable to be estimated. The clocks on all of the satellites belonging to the same system s, where $s = \{\text{GPS}, \text{Galileo}, \text{GLONASS}, \ldots\}$, are in sync with each other, so the receiver's clock will be out of sync with all satellites belonging to the same constellation by the same amount $\Delta t^{(s)}$. In GNSS, the term *pseudorange* is used to identify a range affected by a bias, directly related to the bias between the receiver and satellite clocks. There are other factors of error: since propagation at speed c is only possible in the vacuum, atmospheric status affects the propagation speed of electromagnetic waves modifying the propagation time and thus the distance estimation. For instance, the ionosphere, that is the part of the atmosphere above 60 km until 2000 km of the Earth surface, is a plasmatic medium that causes a slowdown in the group velocity and a speed up of the phase velocity, having an impact in code and phase delays and, thus, impeding precise navigation when its effects are not mitigated. Actually, errors can be on the order of tens of meters in geomagnetic storm episodes [1].

For each in-view satellite i of system s, we can write

$$\rho_i = \sqrt{\left(x_i^{\text{Tx}} - x\right)^2 + \left(y_i^{\text{Tx}} - y\right)^2 + \left(z_i^{\text{Tx}} - z\right)^2} + c\Delta t^{(s)} + \sigma_e, \qquad (2)$$

where $(x_i^{\text{Tx}}, y_i^{\text{Tx}}, z_i^{\text{Tx}})$ is the satellite's position (known from the navigation message), (x, y, z) the receiver's position, and σ_e gathers other sources of error. Since the receiver needs to estimate its own 3D position (three spatial unknowns) and its clock deviation with respect to the satellites' time basis, at least $3 + N_s$ satellites must be seen by the receiver at the same time, where N_s is the number of different navigation systems available (in-view) at a given time. Each received satellite signal, once synchronized and demodulated at the receiver, defines one equation such as the one defined in (2), forming a set of nonlinear equations that can be solved algebraically by means of the Bancroft algorithm [2] or numerically, resorting to multidimensional Newton-Raphson and weighted least square methods [3]. When *a priori* information is added we resort to Bayesian estimation, a problem that can be solved recursively by a Kalman filter or any of its variants. The problem can be further expanded by adding other unknowns (for instance, parameters of ionospheric and tropospheric models), sources of information from other systems, mapping information, and even motion models of the receiver. In the design of multi-constellation GNSS receivers, the vector of unknowns can also include the receiver clock offset with respect to each system in order to take advantage of a higher number of in-view satellites and using them jointly in the navigation solution, therefore increasing accuracy.

2.1. Signal Model. A general signal model for most navigation systems consists of a direct-sequence spread-spectrum (DS-SS) signal [4], synchronously transmitted by all the satellites in the constellation. This type of signals enables code division multiple access (CDMA) transmissions, that is, satellite signals are distinguished by orthogonal (or quasi-orthogonal) codes. At a glance, these signals consists of two main components: a ranging code (the PRN spreading sequence) and a low rate data link (broadcasting necessary information for positioning such as satellites orbital parameters and corrections). The complex baseband model of the signal transmitted by a GNSS space vehicle reads as

$$s_T(t) = \sqrt{P_T}\left(\gamma \sum_{u=-\infty}^{\infty} d_I(u) p_I(t - uT_{b_I}) \right.$$

$$\left. + j\sqrt{1 - \gamma^2} \sum_{l=-\infty}^{\infty} d_Q(l) p_Q\left(t - lT_{b_Q}\right) \right), \qquad (3)$$

where

$$p_I(t) = \sum_{n=0}^{N_{c_I}-1} q_I(t - nT_{\text{PRN}_I}),$$

(4)

$$q_I(t) = \sum_{k=0}^{L_{c_I}-1} c_I(k)g_{T,I}(t - kT_{c_I}),$$

being P_T the transmitting power, γ a parameter controlling the power balance, $d_I(m) \in \{-1, 1\}$ the data symbols, T_{b_I} the bit period, N_{c_I} the number of repetitions of a full codeword that spans a bit period, $T_{\text{PRN}_I} = T_{b_I}/N_{c_I}$ the codeword period, $c_I(k) \in \{-1, 1\}$ a chip of a spreading codeword of length L_{c_I} chips, $g_{T,I}(t)$ the transmitting chip pulse shape, which is considered energy normalized for notation clarity, and $T_{c_I} = T_{b_I}/N_{c_I}L_{c_I}$ is the chip period. Figure 1 aims at clarifying the relation between those bits/chips parameters. Subindex I refers to the in-phase component, and all parameters are equivalently defined for the quadrature component, referred to with the subindex Q. This signal model describes all GNSS's signals-in-space, for instance GPS L1, GPS L5, Galileo E1, and Galileo E5. Refer to [5] for the details.

2.2. Propagation Channel Model. A key aspect in the definition of the propagation channel model between satellites' antenna and the user's receiver antenna is whether it can be considered narrowband or wideband, which depends on the bandwidth of the propagation channel in which a given signal is transmitted, being assessed with respect to the channel *coherence bandwidth*. The coherence bandwidth is defined as the frequency band within which all frequency components are equally affected by fading due to multipath. In narrowband systems, all the components of the signal are equally influenced by multipath, while in wideband systems the various frequency components of the signal are differently affected by fading. Narrowband systems, therefore, are affected by nonselective fading, whereas wideband systems are affected by selective fading. The coherence bandwidth depends on the environment and is given by

$$B_c = \frac{1}{2\pi T},$$

(5)

where T is the *delay spread*, which is the time span between the arrival of the first and the last multipath signals that can be sensed by the receiver. In a fading environment, a propagated signal arrives at the receiver through multiple paths. For a typical GNSS multipath propagation channel in which $T < 0.5\,\mu s$ (the limit can be greater in nonurban areas, but in general it is not lower), we obtain that the system is wideband if transmitted signals are wider than 320 kHz, which is the case for GNSS waveforms (in the order of MHz). Hence, we conclude that we need to define propagation channel models considering wideband systems. Another important definition within this context concerns *coherence time*. The coherence time, T_{coh}, is defined as the time interval

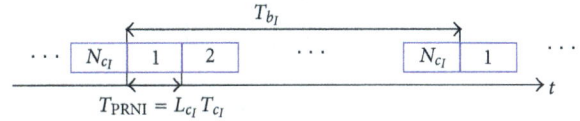

FIGURE 1: Relation among the parameters defining bits and spreading sequences in a generic navigation signal (in-phase component).

during which the characteristics of the propagation channel remain approximately constant, and it is given as

$$T_{\text{coh}} = \frac{1}{2f_m},$$

(6)

where f_m is the maximum Doppler shift. The Doppler shift is given as v/λ, where v is the radial speed of the mobile terminal with respect to the satellite and λ is the signal wavelength. A channel is considered WSSUS (wide-sense stationary with uncorrelated scatterers) during the coherence time.

In the following, we describe four of the most relevant satellite channel models found in the literature.

2.2.1. Jahn's Channel Characterization. Jahn et al. provided a wideband channel model for land mobile satellite services [6]. The model was derived from a channel measurements campaign performed in the L band at 1820 MHz. An aircraft transmitted a spread spectrum signal of 30 MHz, being received by a mobile receiver (handheld or car terminal). From those measurements, authors characterized the channel assuming WSSUS and modeling it as a filter structure with delay taps. Then, they provided statistical models for LOS (Rician probability density function for the amplitude of the direct path), shadowing (ray amplitude following a Raileigh distribution with a lognormal distributed mean power), near echoes (the number of the near echoes follows a Poisson distribution, with delays being exponentially distributed and amplitudes following a Rayleigh distribution), and far echoes (same distributions than near echoes but with other parameters). Table 1 summarizes the main features of Jahn's statistical channel model.

2.2.2. Loo's Channel Characterization. The Loo's land mobile satellite channel model [7] is a statistical model that assumes that the LOS component under foliage attenuation (shadowing) is lognormally distributed and that the multipath effect is Rayleigh distributed. This model provides complete statistical descriptions for different shadowing and multipath conditions based on an extensive measurement campaign for different frequency bands. For the L band, the "Inmarsat's Marecs A" satellite was used as transmitter, while a mobile laboratory was considered for signal reception, resulting in a fixed 19° elevation. Many more investigations on L-band measurements are also referred to in [8], obtaining results for other elevation angles. Table 1 summarizes the main features of Loo's statistical channel model.

2.2.3. Pérez-Fontán's Channel Characterization. The model presented by Fontán et al. in [9] addressed the statistical

TABLE 1: Comparison between channel models.

	Jahn	Pérez-Fontán	Loo	Steingass/Lehner
Frequency	1820 MHz (L-band)	L, S, Ka band	UHF, L, S, Ka band	GPS L1, but others can be selected
Bandwidth	30 MHz (wideband)	Narrowband, wideband	Depends on meas. campaign	Wideband
User	(i) Handheld at 4 km/h. (ii) Vehicle at 40 km/h	(i) Handheld. (ii) Vehicle	Depends on meas. campaign	(i) Vehicle (ii) Pedestrian
Dynamics	Constant velocity	Not provided for L band, wideband	Not provided	Variable velocity
Types	(i) Direct path (ii) Near echoes (iii) Far echoes	(i) Direct signal (ii) Multipath echoes. (iii) Diffuse components	(i) Direct signal (ii) Multipath echoes	(i) Direct Path, up to 3 diffracted LOS (ii) Reflected echoes
Number	Scenario dependant	50 rays	Not provided	Up to 50
Attenuation	10 to 30 dB	9 to 16 dB	<30 dB	Rice fading
Doppler	Not straightforward	Included in generated time series	Not straightforward	Deterministic included in the generated time series
Shadowing	(i) LOS (ii) Shadowing (iii) NLOS	(i) LOS (ii) Moderate shadowing (iii) Deep shadowing conditions	(i) LOS (ii) Light shadowing (iii) Heavy shadowing	(i) LOS (ii) House front diffraction (iii) Trees (iv) Lamp posts
Elevation dependant	$15°, 25°, 35°, 45°, 55°$	$15°, 30°, 45°, 60°, 70° 80°$	$19°$ (other based on different references)	$[5°, 90°]$, based on 9 levels of measurement data: $5°, 10°, 20°, 30°, 40°, 50°, 60°, 70°, 80°$
Azimuth dependant	None.	Uniform azimuth distribution.	None.	Distribution extracted from measurements and depends on elevation
Environments	(i) Open (ii) Rural (iii) Suburban (iv) Urban (v) Highway	(i) Open (ii) Suburban (iii) Urban (iv) Tree shadowed	(i) Suburban (ii) Tree shadowed (iii) Other based on different references	Urban and suburban for car and pedestrian
Pros.	(i) Different scenarios available (ii) Complete statistical model (iii) Good shadowing and NLOS approach (iv) Available parameters for GNSS L-wideband	(i) Time series are obtained prior to statistical model generation (ii) Complete statistical model for both LOS and multipath rays	(i) Complete statistical model (ii) To be used as baseline for other statistical models	(i) Downloadable model available (ii) Urban, suburban and rural (iii) Vehicle and pedestrian (iv) Realistic model (v) Includes different elevations (vi) Includes houses, trees and lamp posts (vii) Low computational cost. (viii) Takes into account generated scenario (ix) Generates MP profile (x) Developed for GNSS

TABLE 1: Continued.

	Jahn	Pérez-Fontán	Loo	Steingass/Lehner
Cons.	(i) Measurements-based model parameters (ii) Not possible to obtain parameters for user-defined scenarios (iii) Constant velocity with fixed values (iv) No elevations over 55° (v) Rays Doppler not fully defined	(i) Measurements-based model parameters (ii) Not possible to obtain parameters for user-defined scenarios (iii) Not enough measured data to extract significant model parameters (iv) Limited data for applicable GNSS conditions: L band, wideband	(i) Measurements-based model parameters (ii) Not straightforward multipath parameters (iii) Not enough available data to extract parameters for different scenarios (iv) No detailed user dynamics description (v) Limited data for applicable GNSS conditions: L band, wideband	(i) The license does not allow the modification of the code. Although own propagation data can be used to modify the model (ii) Scenario needs to be reexecuted for each considered satellite

modeling of shadowing and multipath effects in land mobile satellite applications for a wide range of environments with different clutter densities (from open to dense urban areas) and elevation angles (from 5° to 90°) at L, S, or Ka Bands, using a comprehensive experimental database to extract the model parameters for the different bands, environments, and elevations. One of its main contributions consists of producing time series of any channel parameter whose study is required, instead of just cumulative distribution functions. These ones may be computed later from the generated series. The model uses a first-order Markov chain to describe the slow variations of the direct signal, basically due to shadowing/blockage effects. The overall signal variations due to shadowing and multipath effects within each individual Markov state are assumed to follow a Loo distribution with different parameters for each shadowing condition (Markov state). Up to this point the model is of the narrow-band type since it does not account for time dispersion effects. These effects are introduced by using an exponential distribution to represent the excess delays of the different echoes. Table 1 summarizes the main features of Pérez-Fontán's channel model.

2.2.4. Steingass/Lehner's Channel Characterization.
The Steingass/Lehner land mobile channel model presented in [10] was developed using data recorded in a high-resolution measurement campaign carried out in Munich in 2002. Different types of environments (urban, suburban, and rural) were measured for car and pedestrian applications. It has been approved as standard by the ITU [11]. For the measurements, a 100 MHz signal near the GPS L1 band was used. This signal provided a time resolution of about 10 ns. The received signal was processed using a super-resolution algorithm to extract the single reflections. With this information, the probability density distribution of the parameters of the reflected rays, such as Doppler shift, power of echoes, duration of a reflector, and number of echoes, were extracted. In urban environments, three major obstacles influence the propagation of the LOS signal: house fronts, trees, and lamp posts. The model is comprised of a deterministic part with a generated scenery, which computes geometrically the LOS signal shadowing and knife-edge diffraction for house fronts, lamp posts, and trees. The other observables like the number of coexisting echoes, life span of reflectors, and the mean power of the echoes are generated stochastically, using the probability density distribution extracted from the measurements. The output of the model is a complex time-variant channel impulse response recalculated each time step. Table 1 summarizes the main features of Steingass/Lehner's channel model.

2.3. A Realistic Signal/Channel Simulator.
When transmitted, satellite's signals travel through a propagation channel which modifies its amplitude, phase, and delay. Indeed, many replicas of the same transmitted signal can reach the receiver's antenna due to multipath propagation. In general, these replicas are caused by reflections of the direct signal in surrounding obstacles (e.g., buildings, trees, and

ground etc.). As shown above, such propagation channel is generically modeled by a linear time-varying impulse response with M_i propagation paths:

$$h_i(t; \xi) = \sum_{m=0}^{M_i-1} \alpha_{i,m}(t) e^{j\phi_{i,m}(t)} \delta(\xi - \tau_{i,m}(t)), \qquad (7)$$

where $\alpha_{i,m}$, $\phi_{i,m}(t)$ and $\tau_{i,m}(t)$ are the amplitude, phase, and delay of the mth propagation path for the ith satellite, ξ is the multipath delay axis and the index $m = 0$ stands for the line-of-sight signal. These channel parameters can be seen as realizations of random processes with underlying probability density functions $f_{\alpha_p}(\alpha)$, $f_{\phi_p}(\phi)$, and $f_{\tau_p}(\tau)$, respectively, whose shape and parameters are approximated by the models outlined above.

Therefore, considering M_s visible satellites, the signal $r(t)$ received at the receiver's antenna is the superposition of the transmitted signals, as propagated through the corresponding channel, and corrupted by additive noise, $w(t)$. This reads as

$$r(t) = \sum_{i=0}^{M_s-1} s_{T,i}(t) * h_i(t; \xi) + w(t)$$

$$= \sum_{i=0}^{M_s-1} \int_{-\infty}^{+\infty} s_{T,i}(t - \xi) h_i(t; \xi) d\xi + w(t) \qquad (8)$$

$$= \sum_{i=0}^{M_s-1} \sum_{m=0}^{M_i-1} \alpha_{i,m}(t) e^{j\phi_{i,m}(t)} s_{T,i}(t - \tau_{i,m}(t)) + w(t),$$

where $s_{T,i}(t)$ is the transmitted signal $s_T(t)$ corresponding to the i-th satellite.

As shown in [12], the term $\phi_{i,m}(t)$ can be approximated by its first-order Taylor expansion as $\phi_{i,m}(t) \approx 2\pi f_{d_{i,m}}(t)t + \phi_{i,m,0}$. Hence, the general baseband equivalent model that will be used along this paper is

$$r(t) = \sum_{i=0}^{M_s-1} \sum_{m=0}^{M_i-1} \alpha_{i,m}(t) e^{j2\pi f_{d_{i,m}}(t)t}$$
$$\cdot e^{j\phi_{i,m,0}} s_{T,i}(t - \tau_{i,m}(t)) + w(t). \qquad (9)$$

The first element in the receiver RF chain is a right hand circularly polarized (RHCP) antenna, usually with nearly hemispherical gain coverage, with the mission to receive the radionavigation signals of all the satellites in view. The RF signals collected by the antenna are immediately amplified by a low noise amplifier (LNA), a key element which is the most contributing block to the noise figure of the receiver. The LNA also acts as a filter, minimizing out-of-band RF interferences and setting the sharpness of the received code. After the LNA, the amplified and filtered RF signals are then downconverted to an intermediate frequency (IF) using signal mixing frequencies from local oscillators (LOs). These LOs are derived from a receiver reference oscillator, often an oven-stabilized clock with typical accuracies of 10^{-8}. There is a need for one LO per down-conversion stage. Two or three down-conversion stages are commonly devoted to reject mirror frequencies or large out of band jamming signals, in particular the 900 MHz used by the GSM mobile communication system. However, depending on the subsequent analog-to-digital converter (ADC) characteristics, a one-stage downconversion or even a direct L-band sampling is also possible [13]. The lower sideband generated by the mixer process is selected, while the upper sideband is filtered by a postmixer bandpass filter. It is important to point out that signal Doppler's and PRN codes are preserved after the mixing stage, only the carrier frequency is lowered.

In the sequel, we focus on the contribution of a single satellite and thus omit the dependence with i of the signal model. Considering a generic data sequence d, chip code c, chip-shaping pulse $g_T(t)$, chip period T_c, N_c full codes in a whole bit, and data period T_b, the baseband equivalent received signal for a channel model as in (7) but particularized to $M_i = 1$ (i.e., only one line of sight signal) can be put in the form

$$
\begin{aligned}
\tilde{r}_0(t) &= a_0(t) \sum_{u=-\infty}^{\infty} d(u) \sum_{n=0}^{N_c-1} \sum_{k=0}^{L_c-1} c(k) \\
&\quad \cdot \tilde{g}_R(t - \tau_0(t) - kT_c - nT_{\mathrm{PRN}} - uT_b) + \tilde{w}(t) \\
&= a_0(t) \sum_{u=-\infty}^{\infty} d(u)\tilde{p}(t - \tau_0(t) - uT_b) + \tilde{w}(t) \\
&= |a_0(t)| e^{j(2\pi f_d(t)t + \phi_0)} \sum_{u=-\infty}^{\infty} d(u)\tilde{p}(t - \tau_0(t) - uT_b) \\
&\quad + \tilde{w}(t),
\end{aligned}
\tag{10}
$$

where $\tilde{g}_R(t)$ is the pulse received at the antenna and then filtered by a precorrelation filter (usually the LNA), $\tilde{p}(t)$ is the filtered version of $p(t) = p_I(t) + j p_Q(t)$, and the term $\tilde{w}(t)$ stands for the filtered thermal noise and other unmodeled terms. The objective of a synchronization method is to estimate the time delay $\tau_0(t)$, Doppler shift $f_d(t)$ and the carrier phase information ϕ_0 embedded into the phase of the complex amplitude $a_0(t) = |a_0(t)| e^{j(2\pi f_d(t)t + \phi_0)}$.

The analog-to-digital conversion and the automatic gain control (AGC) processes take place at IF or baseband, where all the signals from GNSS satellites in view are buried in thermal noise. Once the received signal is digitized, it is ready to feed each of the N digital receiver channels. Every receiver channel is intended to acquire and track the signal of a single GNSS satellite; typical receivers are equipped with $N = 12$ channels. The multiplication of the IF digitized signal by a local replica of its carrier frequency allows to produce the in-phase (I) and quadrature-phase (Q) components of the digitized signal.

Assuming $\tilde{w}(t)$ as additive white Gaussian noise (AWGN), at least in the band of interest, it is well known that the optimum receiver is the code matched filter, expressed as

$$
\begin{aligned}
h_{\mathrm{MF}}\left(t; \hat{\tau}_0, \hat{f}_{d_0}, \hat{\phi}_0\right) &= \sum_{k=0}^{L_c-1} c^*(k) g_R^*(-t - kT_c + \hat{\tau}_0 + L_c T_c) \\
&\quad \cdot e^{-j\hat{\phi}_0} e^{-j2\pi \hat{f}_{d_0}(t)t} \\
&= q_R^*(-t + \hat{\tau}_0 + L_c T_c) e^{-j\hat{\phi}_0} e^{-j2\pi \hat{f}_{d_0}(t)t},
\end{aligned}
\tag{11}
$$

where $\hat{\tau}_0, \hat{f}_{d_0}, \hat{\phi}_0$ are local estimates of the time delay, Doppler shift, and carrier phase of the received signal, and $(dot)^*$ stands for the complex conjugate operator. Theoretically $g_R(t) = g_T(t)$, but actual implementations make use of approximated versions: while $g_T(t)$ is a rectangular pulse filtered at the satellite, $g_R(t)$ is digitally generated at the receiver and therefore not filtered. In addition, $g_T(t)$ is usually filtered again by a precorrelation filter before the matched filter, as expressed in (10) with $\tilde{g}_R(t)$. The code matched filter output can be written in the form

$$
y\left(t; \check{\tau}_0, \check{f}_{d_0}, \check{\phi}_0\right) = \tilde{r}_0(t; \tau_0, f_{d_0}, \phi_0) * h_{\mathrm{MF}}\left(t; \check{\tau}_0, \check{f}_{d_0}, \check{\phi}_0\right).
\tag{12}
$$

Notice that, in the matched filter, we have substituted the estimates $\hat{\tau}_0$, $\hat{f}_{d_0}$, and $\hat{\phi}_0$ for trial values obtained from previous (in time) estimates of these parameters which we have defined as $\check{\tau}_0$, $\check{f}_{d_0}$, and $\check{\phi}_0$, respectively. This is the usual procedure in GNSS receivers, since the estimates are not really available, but to be estimated after correlation.

In DS-SS terminology, the matched filter is often referred to as *correlator*, while the processing it performs is called *despreading*. Since the correlators perform accumulation of the sampled signal during a period T_{int} and then release an output, we can write the discrete version of the signal as

$$
y_n = \sum_{s=0}^{\lfloor T_{\mathrm{int}}/T_s - 1 \rfloor} y\left(nT_{\mathrm{int}} - sT_s; \check{\tau}_0, \check{f}_{d_0}, \check{\phi}_0\right),
\tag{13}
$$

where T_s is the sampling period, T_{int} is the integration time (usually, $T_{\mathrm{int}} = T_c L_c$) and $\lfloor \cdot \rfloor$ stands for the nearest integer towards zero.

Equation (13) can be expressed more conveniently by solving the convolution in (12), which yields [14]

$$
\begin{aligned}
y_{n,I} &= \frac{|a_0|}{2} K \frac{\sin\left(\pi \overline{\Delta f} T_{\mathrm{int}}\right)}{\pi \overline{\Delta f} T_{\mathrm{int}}} d\left([n]_{T_b/T_{\mathrm{int}}}\right) R_{\tilde{p}q}\left(\overline{\Delta \tau_0}\right) \\
&\quad \cdot \cos\left(\pi \overline{\Delta f} T_{\mathrm{int}} + \overline{\Delta \phi}\right) + v_I(n),
\end{aligned}
\tag{14}
$$

where we defined $\overline{\Delta f} = f_{d_0} - \check{f}_{d_0}$, $\overline{\Delta \phi} = \phi_0 - \check{\phi}_0$ and $\overline{\Delta \tau_0} = \tau_0 - \check{\tau}_0$ (i.e., the estimation errors), $\lfloor \cdot \rfloor$ stands for the nearest integer toward zero, and $[n]_{T_b/T_{\mathrm{int}}}$ means the integer part of nT_{int}/T_b, being T_b the navigation bit period, and

$$
R_{\tilde{p}q}(\xi) = \frac{1}{T_{\mathrm{PRN}}} \int_{\xi}^{T_{\mathrm{PRN}}+\xi} \tilde{p}(t) q^*(t - \xi) dt
\tag{15}
$$

is the correlation function. An equivalent derivation for the Q arm leads to

$$
\begin{aligned}
y_{n,Q} &= \frac{|a_0|}{2} K \frac{\sin\left(\pi \overline{\Delta f} T_{\mathrm{int}}\right)}{\pi \overline{\Delta f} T_{\mathrm{int}}} d\left([n]_{T_b/T_{\mathrm{int}}}\right) R_{\tilde{p}q}\left(\overline{\Delta \tau_0}\right) \\
&\quad \cdot \sin\left(\pi \overline{\Delta f} T_{\mathrm{int}} + \overline{\Delta \phi}\right) + v_Q(n).
\end{aligned}
\tag{16}
$$

Terms $\overline{\Delta f}$, $\overline{\Delta \phi}$, and $\overline{\Delta \tau_0}$ should be regarded as the average local phase error over the integration interval, that is, $\overline{\Delta \phi} = \Delta \phi + 2\pi \Delta f(T_{\mathrm{int}}/2)$, assuming a frequency rate error $\Delta \dot{f}$

(i.e., a phase acceleration error) equal to zero. In case of inclusion of such effect in the model, the average phase error can be expanded as

$$\overline{\Delta\phi} = \Delta\phi + 2\pi\Delta f \frac{T_{\text{int}}}{2} + 2\pi\frac{T^2}{6}\Delta\dot{f}. \qquad (17)$$

In this expression, the terms $\Delta\phi$, Δf, and $\Delta\dot{f}$ are referred to the error values at the beginning of the integration interval.

In the following, we will consider $K = (T_{\text{int}}/T_s)$ as the integer number of samples collected in an accumulation. This number will not be integer in receiver configurations having a sample rate incommensurable with the chip rate, and thus some integration blocks will have $K + 1$ samples instead of K. This effect can be considered negligible for the analysis presented in this paper.

In the case of $M_i > 1$ (i.e., in the presence of multipath), (12) becomes a sum of all the replicas convoluted with a filter matched to the line of sight signal, whose estimated parameters are possibly biased by the presence of multipath. Since the convolution is a linear operator, the correlator output will be a linear combination of the contributions made by each signal path.

Note that an arbitrary number of correlators (very early, early, prompt, late, very late, etc.) can be used in the filter update, just adding or subtracting the correlator offset to the argument of $R_{\tilde{p}q}$ (i.e., $R_{\tilde{p}q}(\Delta\tau_{0,n} + \delta)$, $R_{\tilde{p}q}(\Delta\tau_{0,n} - \delta)$, etc.). The correlators' output can be stacked in a vector $\mathbf{y}_n$, which will be the measurements used in next section.

In the context of this work, we used the GRANADA (Galileo Receiver ANAlysis and Design Application) simulation platform to simulate realistic channel and receiver scenarios. The GRANADA Factored Correlator Model (FCM) blockset (see Figure 2) is a MATLAB/Simulink (MATLAB and Simulink are registered trademarks of The MathWorks, Inc.) library that provides a swift, flexible, and realistic way of simulating different signal processing architectures, either of standalone GNSS receivers or multisystem solutions. The FCM was included in a Simulink blockset, which, since 2007, has been commercially available as part of the GRANADA product family, whose remaining products were developed by DEIMOS Space in the frame of the Galileo Receiver Development activities (GARDA), funded by the Galileo Joint Undertaking (now European GNSS Agency, GSA) under the 6th Framework Program of the European Union.

The FCM separates the effects of carrier and code Doppler and misalignment on a GNSS receiver's correlator outputs into several multiplicative factors and allows the inclusion (or not) of each factor independently. Since it is an analytical model, the computation rate can be as low as the tracking loop rate, dramatically increasing simulation speed: the FCM provides directly the correlators' output, precluding the need of simulating the lower-level signal processing stages, significantly reducing the computational load and hence decreasing processing and memory requirements, while still accounting for various effects (as filtering, carrier phase and frequency errors, code delay error, code Doppler, noise, and multipath), thus keeping a high level of realism [15]. Since, statistically speaking, it is equivalent to work with samples before or after the correlation process (proof in the

Appendix), we take advantage of working at the correlator output since it considerably reduces the computational load.

Once configured (type of signal, propagation channel, user dynamics, sampling frequency before correlation, number of correlators and their spacing, integration period, environment, etc., see Figure 3), FCM provides the measurements $\mathbf{y}_n$ used in the simulations presented in Section 5.

3. Particle Filtering

Bayesian filtering involves the recursive estimation of states $\mathbf{x}_n \in \mathbb{R}^{n_x}$ given measurements $\mathbf{y}_n \in \mathbb{R}^{n_y}$ at time n based on all available measurements, $\mathbf{y}_{1:n} = \{\mathbf{y}_1, \ldots, \mathbf{y}_n\}$. To that aim, we are interested in the filtering distribution $p(\mathbf{x}_n|\mathbf{y}_{1:n})$, which can be recursively expressed as

$$p(\mathbf{x}_n \mid \mathbf{y}_{1:n}) = \frac{p(\mathbf{y}_n \mid \mathbf{x}_n)p(\mathbf{x}_n \mid \mathbf{x}_{n-1})}{p(\mathbf{y}_n \mid \mathbf{y}_{1:n-1})}p(\mathbf{x}_{n-1} \mid \mathbf{y}_{1:n-1}), \qquad (18)$$

with $p(\mathbf{y}_n|\mathbf{x}_n)$ and $p(\mathbf{x}_n|\mathbf{x}_{n-1})$ referred to as the likelihood and the prior distributions, respectively. Unfortunately, (18) can only be obtained in closed-form in some special cases. For instance, when the model is linear and Gaussian, the Kalman Filter (KF) [16] provides the optimal solution. In more general setups—nonlinear and/or non-Gaussian—we should resort to more sophisticated methods [17]. In this paper we consider particle filters (PFs) [18, 19].

PFs approximate the filtering distribution by a set of N weighted random samples, forming the so-called set of particles $\{\mathbf{x}_n^{(i)}, w_n^{(i)}\}_{i=1}^N$. These random samples are drawn from the importance density distribution, $\pi(\cdot)$,

$$\mathbf{x}_n^{(i)} \sim \pi\left(\mathbf{x}_n \mid \mathbf{x}_{0:n-1}^{(i)}, \mathbf{y}_{1:n}\right), \qquad (19)$$

and weighted according to the general formulation

$$w_n^{(i)} \propto w_{n-1}^{(i)} \frac{p\left(\mathbf{y}_n \mid \mathbf{x}_{0:n}^{(i)}, \mathbf{y}_{1:n-1}\right)p\left(\mathbf{x}_n^{(i)} \mid \mathbf{x}_{n-1}^{(i)}\right)}{\pi\left(\mathbf{x}_n^{(i)} \mid \mathbf{x}_{0:n-1}^{(i)}, \mathbf{y}_{1:n}\right)}. \qquad (20)$$

Algorithm 1 outlines the operation of the Standard PF (SPF) when a new measurement $\mathbf{y}_n$ becomes available. After particle generation, weighting, and normalization, a minimum mean square error (MMSE) estimate can be obtained by a weighted sum of particles. A typical problem of PFs is the degeneracy of particles, where all but one weight tend to zero. This situation causes the particle to collapse to a single state point. To avoid the degeneracy problem, we apply resampling, consisting in eliminating particles with low importance weights and replicating those in high-probability regions [20, 21]. In this work, we consider a multinomial sampling scheme for the resampling step.

3.1. Rao-Blackwellized Particle Filter. In this paper, we analyze a way to alleviate the dimensionality problem based on the marginalization of linear states. The basic idea is that a KF can optimally deal with these states, while reducing the dimension of the state space that the nonlinear filter

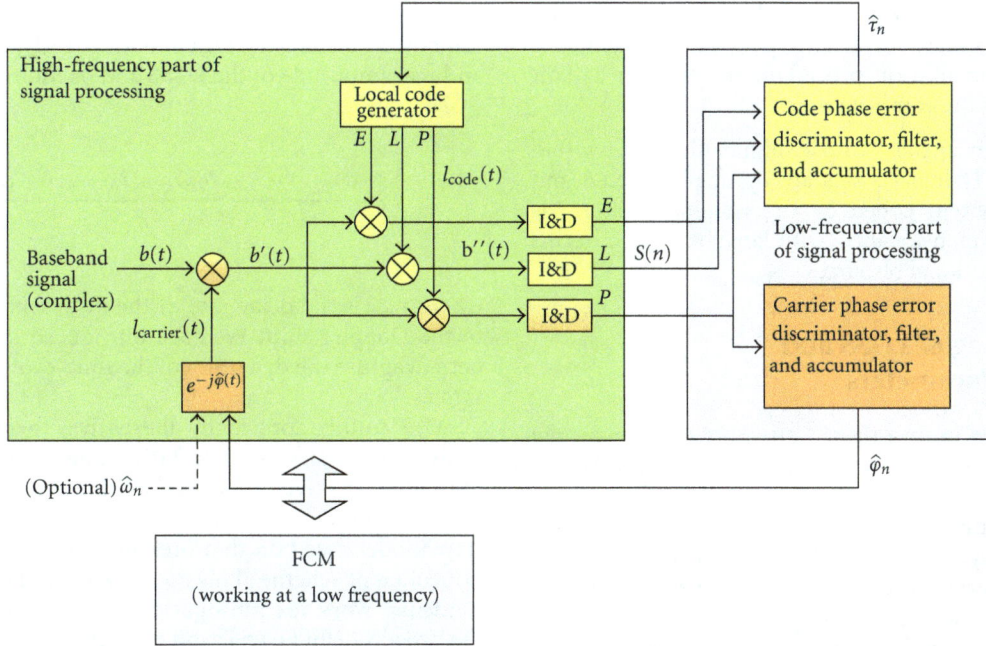

FIGURE 2: Schematic of the tracking stage of a GNSS receiver and substitution of the high frequency stage of the receiver (correlation and carrier wipe-off) with the FCM blockset [22].

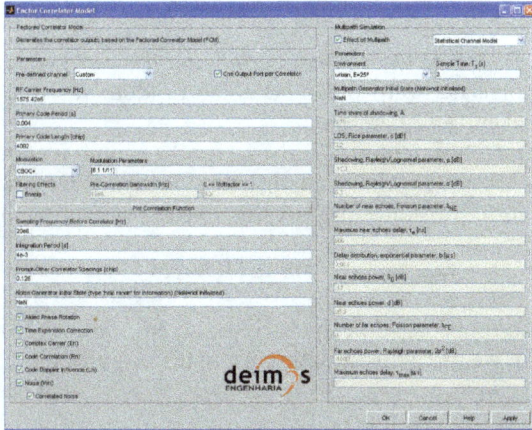

FIGURE 3: Configuration screen of the FCM blockset.

Require: $\{\mathbf{x}_{n-1}^{(i)}, w_{n-1}^{(i)}\}_{i=1}^N$ and $\mathbf{y}_n$

Ensure: $\{\mathbf{x}_n^{(i)}, w_n^{(i)}\}_{i=1}^N$ and $\widehat{\mathbf{x}}_n$

1: **for** $i = 1$ to N **do**

2:      Generate $\mathbf{x}_n^{(i)} \sim \pi(\mathbf{x}_n | \mathbf{x}_{0:n-1}^{(i)}, \mathbf{y}_{1:n})$

3:      Calculate $\widetilde{w}_n^{(i)} = w_{n-1}^{(i)} \dfrac{p(\mathbf{y}_n | \mathbf{x}_{0:n}^{(i)}, \mathbf{y}_{1:n-1}) \, p(\mathbf{x}_n^{(i)} | \mathbf{x}_{n-1}^{(i)})}{\pi(\mathbf{x}_n^{(i)} | \mathbf{x}_{0:n-1}^{(i)}, \mathbf{y}_{1:n})}$

4: **end for**

5: **for** $i = 1$ to N **do**

6:      Normalize weights: $w_n^{(i)} = \dfrac{\widetilde{w}_n^{(i)}}{\sum_{j=1}^N \widetilde{w}_n^{(j)}}$

7: **end for**

8: MMSE state estimation: $\widehat{\mathbf{x}}_n = \sum_{i=1}^N w_n^{(i)} \mathbf{x}_n^{(i)}$

9: $\{\mathbf{x}_n^{(i)}, 1/N\}_{i=1}^N = \text{Resample}(\{\mathbf{x}_n^{(i)}, w_n^{(i)}\}_{i=1}^N)$

ALGORITHM 1: Standard particle filtering (SPF).

has to explore. The procedure was proposed in [23, 24] for the case of dealing with the nonlinear states with a PF. The algorithm was termed Marginalized particle filter (MPF), although the same concept is also referred to as Rao-Blackwellized PF (RBPF) in other works [25, 26]. The latter nomenclature is because marginalization resorts to a general result due to [27, 28] referred to as the Rao-Blackwell theorem, which shows that the performance of an estimator can be improved by using information about conditional probabilities. The Rao-Blackwell theorem states let $\widehat{\theta} = g(\mathbf{x})$ represent any unbiased estimator for θ and $T(\mathbf{x})$ be a sufficient statistic for θ under $p(\mathbf{x}, \theta)$. Then the conditional expectation $\widehat{\theta}^{\text{RB}} = \mathbb{E}\{g(\mathbf{x}) | T(\mathbf{x})\}$ is independent of θ, and it is the uniformly minimum variance unbiased estimator (cf. [29, 30] for the details) The result of a corollary points

out that the use of a Rao-Blackwellized estimator effectively reduces the variance of the estimation error. Therefore, when possible, it is desirable to apply marginalization procedures. Corollary: let $\widehat{\theta}$ be an unbiased estimator and let $\widehat{\theta}^{\text{RB}}$ be the Rao-Blackwell estimator, then

$$\mathbb{E}\left\{ \left(\theta - \widehat{\theta}\right)^2 \right\} \geq \mathbb{E}\left\{ \left(\theta - \widehat{\theta}^{\text{RB}}\right)^2 \right\}. \tag{21}$$

Final remarks on Rao-Blackwellization are worth mentioning.

(i) Rao-Blackwellization is a procedure suitable when linear substructures are present in the dynamical model.

(ii) It is a variance reduction technique, in the sense that the estimation variance of a filter considering this marginalization procedure is less than a filter estimating the complete state space.

(iii) Filtering linear states with a Kalman filter has twofold benefits: (1) linear states are optimally filtered and (2) the system coped by the nonlinear filter has reduced dimensionality (with large benefits in terms of computational resources).

4. Joint Filtering of LOSS and Multipath Parameters

The technique herein investigated attempts to estimate the synchronization parameters of both the LOSS and $M - 1$ multipath components. We refer to the algorithm as the multipath estimating particle filtering, or MEPF for short. Here the term Bayesian means that the algorithm is using some sort of a *priori* information regarding these parameters (such as interdependencies and time evolution models). This approach was first introduced in [31] and further refined in [32], although other papers might be found following the same scheme [33] with more complex time-evolving models. The application of Bayesian filtering techniques becomes straightforward when one describes the problem at hand in terms of a measurement equation and a process equation (i.e., how unknowns evolve randomly over time).

4.1. Observations. A receiver implementing such Bayesian tracking loops typically processes each satellite independently, and most of the work in the literature discusses architectures using IF signal. Here we are interested in operating at the output of the bank of correlators.

Observations for the i-th satellite are gathered into a random vector $\mathbf{y}_n$, where we omitted the subindex i for the sake of clarity. The ℓth element in $\mathbf{y}_n$ corresponds to the sample of the ℓ-th correlator, and it is expressed as

$$y_{n,\ell}(\boldsymbol{\alpha}_n, \boldsymbol{\phi}_n, \boldsymbol{\tau}_n) = \sum_{m=0}^{M-1} \alpha_{m,n} e^{j\Delta\phi_{m,n}} R_{n,\ell}(\Delta\tau_{\ell,m,n}) + v_{n,\ell}, \quad (22)$$

accounting that $\Delta\tau_{\ell,m,n} = \tau_{m,n} - \hat{\tau}_{0,n-1} + \delta_\ell$ corresponds to the point where the ℓ-th early/late sample is evaluated. As usual, $m = 0$ denotes LOSS. Here we consider a noncoherent tracking architecture that operates with the squared outputs. This scheme avoids the estimation of carrier phases, and thus it reduces the state-space dimension. In our implementation, a conventional PLL/FLL network is used in parallel to the MEPF. Therefore, the observations are the parallel outputs of the correlation bank, which we denote as

$$\mathbf{y}_n = \left(\left| y_{n,1}(\boldsymbol{\alpha}_n, \boldsymbol{\tau}_n) \right|^2, \ldots, \left| y_{n,L}(\boldsymbol{\alpha}_n, \boldsymbol{\tau}_n) \right|^2 \right)^T, \quad (23)$$

where L is the total number of correlators used at the receiver. We made apparent the dependence of measures on unknown states: real amplitude ($\boldsymbol{\alpha}_n$) and time delay ($\boldsymbol{\tau}_n$) of each replica m of the signal.

4.2. Process Dynamics. The state space is composed of the unknown parameters of the model, namely, delay, delay rate, and real amplitude of the LOSS and its multipath replica:

$$\mathbf{x}_n = \Big(\underbrace{\tau_{0,n}, \ldots, \tau_{M-1,n}}_{\boldsymbol{\tau}_n}, \underbrace{\dot{\tau}_{0,n}, \ldots, \dot{\tau}_{M-1,n}}_{\dot{\boldsymbol{\tau}}_n}, \underbrace{\alpha_{0,n}, \ldots, \alpha_{M-1,n}}_{\boldsymbol{\alpha}_n} \Big)^T, \quad (24)$$

where $\dot{\tau}_{m,n}$ is the delay rate of the m-th component, related to the Doppler shift. We have introduced this delay rate to better capture the dynamics of the time-evolving delay of the signals.

One could adopt many alternatives to specify the time-evolving processes for each state, ranging from the simplistic (although effective in some situations) autoregressive model to more sophisticated models. Here, we adopt a channel state model based on that presented in [34], adapted to the noncoherent scheme. This model was motivated by channel modeling work for multipath prone environments such as the urban satellite navigation channel [35].

The dynamics of time delay and delay rate for the LOSS (i.e., $m = 0$) are described by

$$\begin{pmatrix} \tau_{0,n} \\ \dot{\tau}_{0,n} \end{pmatrix} = \begin{pmatrix} 1 & T_{\text{int}} \\ 0 & 1 \end{pmatrix} \begin{pmatrix} \tau_{0,n-1} \\ \dot{\tau}_{0,n-1} \end{pmatrix} + \mathbf{u}_{0,n}^{\tau}, \quad (25)$$

where T_{int} is the integration period and the process noise is an uncorrelated zero-mean Gaussian random variable with diagonal entries $\sigma_{0,\tau}^2$ and $\sigma_{0,t}^2$.

The evolution of $\tau_{m,n}$ and $\dot{\tau}_{m,n}$ for the echoes is modeled with a truncated Gaussian distribution as in [31], which allows us to introduce the fact that due to physical reasons

$$\tau_{m,n} > \tau_{0,n} \quad \forall m \in \{1, \ldots, M-1\}, \quad (26)$$

in outdoor propagation channels [6, 11, 36]. Taking (26) into account, we force this situation using the evolution

$$\begin{aligned} \tau_{m,n} &= \tau_{0,n} + \left| \tau_{m,n-1} + u_{m,n}^{\tau} \right|, \\ \dot{\tau}_{m,n} &= \dot{\tau}_{0,n} + u_{m,n}^{t}, \end{aligned} \quad (27)$$

with $u_{m,n}^{\tau}$ and $u_{m,n}^{t}$ being zero-mean Gaussian random variables with variances $\sigma_{m,\tau}^2$ and $\sigma_{m,t}^2$, respectively. For the evolution of each $\alpha_{m,n}$ we consider independent autoregressive models with variance $\sigma_{m,\alpha}^2$. The overall covariance matrix of the process is denoted as $\boldsymbol{\Sigma}_x$ and is constructed with $\sigma_{0,\tau}^2$, $\sigma_{m,\tau}^2$, $\sigma_{0,t}^2$, $\sigma_{m,t}^2$, $\sigma_{0,\alpha}^2$, and $\sigma_{m,\alpha}^2$ in its diagonal.

4.3. Algorithm Implementation. From the previous modeling, we realize that the state space can be partitioned into linear and nonlinear subspaces. Clearly, these can be identified as

$$\begin{aligned} \mathbf{x}_n^{\text{l}} &= \boldsymbol{\alpha}_n, \\ \mathbf{x}_n^{\text{nl}} &= \left(\boldsymbol{\tau}_n^T, \dot{\boldsymbol{\tau}}_n^T \right)^T. \end{aligned} \quad (28)$$

By the chain rule of probability, linear states can be analytically marginalized out from $p(\mathbf{x}_n \mid \mathbf{y}_{1:n})$:

$$p\left(\mathbf{x}_n^l, \mathbf{x}_{0:n}^{nl} \mid \mathbf{y}_{1:n}\right) = p\left(\mathbf{x}_n^l \mid \mathbf{x}_{0:n}^{nl}, \mathbf{y}_{1:n}\right) p\left(\mathbf{x}_{0:n}^{nl} \mid \mathbf{y}_{1:n}\right) \quad (29)$$

and, taking into consideration that $\mathbf{x}_n^l$ generates a linear Gaussian state-space, $p(\mathbf{x}_n^l \mid \mathbf{x}_{0:n}^{nl}, \mathbf{y}_{1:n})$ can be updated analytically via a KF conditional on $\mathbf{x}_{0:n}^{nl}$ and only the non-linear part of $\mathbf{x}_n$ needs to be estimated with a nonlinear filter. In the proposed scheme, an SPF is run to characterize $p(\mathbf{x}_{0:n}^{nl} \mid \mathbf{y}_{1:n})$ and a KF is executed to obtain $p(\mathbf{x}_n^l \mid \mathbf{x}_{0:n}^{nl}, \mathbf{y}_{1:n})$.

Notice that both linear and nonlinear states are interdependent, thus the algorithm has to be aware of this coupling. The details might be consulted in [23] for the general algorithm and in [12] for the specific GNSS setup considered here. At a glance, each particle in the PF has an associated KF that tracks amplitudes. Then, before particle generation, KF prediction is run and the results are used in the particle filter. Similarly, once particles are weighted this information is used in the update step of the KF.

5. Results in Realistic Scenarios

We used the GRANADA FCM blockset of Simulink to simulate the GPS L1 C/A signal, the propagation channel, and the inaccuracies of the receiver front end. An initial set of controlled scenarios is simulated to analyze the method. Then, from the set of reviewed channel models, we have selected Jahn's to show simulation results in a realistic environment. The GPS signal is spread spectrum with a code length of 1023 chips and a chip rate of 1.023 Mchips/s (notice that a chip of the signal corresponds to approximately 300 meters in length and the duration of an entire codeword is one millisecond). The carrier frequency of the transmitted signal was 1575.42 MHz and the receivers precorrelation bandwidth was 2 MHz. Estimates of time delay were performed at a rate of 50 Hz, which corresponds to an integration time of 20 milliseconds, assuming bit synchronization. The carrier-to-noise density ratio (C/N$_0$) of the simulated satellite was 38 dB-Hz. The dynamics of the scenario were due to the relative motion of the satellite-receiver, which is completely simulated by the GRANADA FCM blockset, and the receiver performed a pedestrian-like trajectory at 1 m/s. Simulation time was 50 seconds.

We compared the performance of the MEPF with the results of a narrow 0.125-chip spacing DLL (state-of-the-art in GNSS receivers) with an equivalent noise bandwidth of 1 Hz. This architecture uses 3 correlators. Also, the benchmark receiver implements a coherent phase lock loop (PLL) carrier phase discriminator using a second-order filter and an error accumulator with equivalent noise bandwidth 10 Hz. The initial time-delay ambiguity at which the filter was initialized was drawn from $\mathcal{N}(\tau_{0,0}, T_{\text{chip}}/2)$, with T_{chip} the chip period.

It has been reported in [37] that the number of correlators (L) used in the PF plays an important role. For instance, in AWGN on the order of $L = 11$ correlators are required to obtain stable results. Also, the algorithm improves its

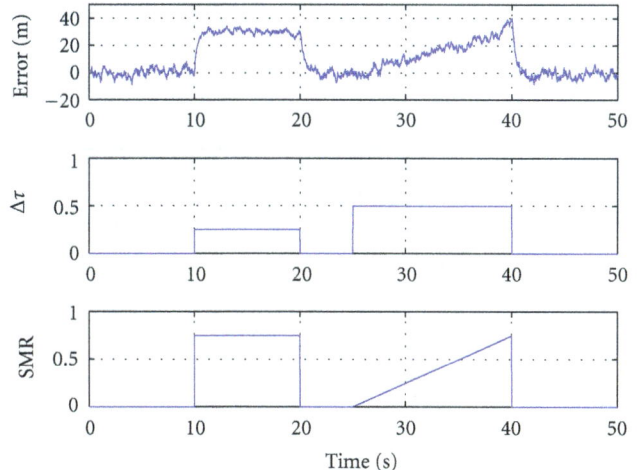

FIGURE 4: Evolution of pseudorange error in DLL/PLL scheme under severe multipath propagation.

performance with the number of particles although this improvement saturates at 300 particles.

Figures 4–7 show the behavior of the classical DLL/PLL scheme and the proposed MEPF, respectively, in a multipath scenario. In this experiments, we used $L = 21$ correlators for the MEPF in order to span correlators along regions of interest in terms of multipath estimation and mitigation. The results are organized as follows. Top figure represents the obtained pseudorange error. Central figure is the relative delay between the LOSS and the multipath replica, in the first representative interval ($t \in [10, 20]$) it has been set to 0.25 chips and in the second interval ($t \in [25, 40]$) to 0.5 chips. Bottom figure plots the signal-to-multipath ratio (SMR) in linear scale of the simulated scenario. During the first interval, the SMR was abruptly kept constant to 0.75 and during the second interval it grew linearly from 0 to 0.75. Since the MEPF is very sensitive to the tuning of process covariance matrix—as many Bayesian filtering solutions,—we have investigated three different setups with $N = 1000$ particles. Namely, (i) in Figure 5 we used standard deviations $\sigma_{0,\tau} = .03/c$, $\sigma_{0,\dot{t}} = 0.03/c$, $\sigma_{m,\tau} = 100/c$, $\sigma_{m,\dot{t}} = 0.03/c$, $\sigma_{0,\alpha} = 0.0001$, and $\sigma_{m,\alpha} = 0.01$; (ii) in Figure 6 we used $\sigma_{0,\tau} = 30/c$, $\sigma_{0,\dot{t}} = 0.3/c$, $\sigma_{m,\tau} = 30/c$, $\sigma_{m,\dot{t}} = 0.3/c$, $\sigma_{0,\alpha} = 0.0001$, and $\sigma_{m,\alpha} = 0.0001$; and finally (iii) in Figure 7 we used $\sigma_{0,\tau} = 3/c$, $\sigma_{0,\dot{t}} = 0.3/c$, $\sigma_{m,\tau} = 30/c$, $\sigma_{m,\dot{t}} = 0.3/c$, $\sigma_{0,\alpha} = 0.0001$, and $\sigma_{m,\alpha} = 0.01$. At the light of the results, the latter configuration provided a good performance as it allowed for sufficient delay excursions to explore the state space and fast variations in multipath amplitude were coped. A summary of results in terms of bias, variance, and RMSE over the entire simulation can be consulted in Table 2. We can observe that, compared to DLL schemes, a remarkable performance improvement can be obtained after properly adjusting the covariances.

Finally, we tested the algorithm in a more realistic scenario. We selected the Jahn's channel model with the same receiver parameters as before. Particularly, the considered channel was that of a satellite at an elevation angle of 55° in an urban scenario with an average C/N$_0$ of 38 dB-Hz.

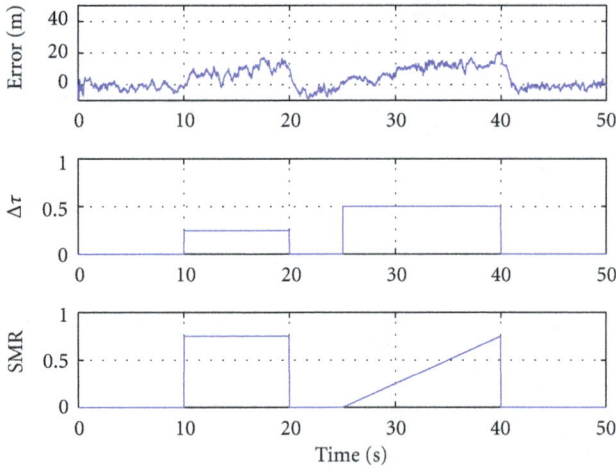

FIGURE 5: Evolution of pseudorange error in MEPF scheme under severe multipath propagation, setup number 1.

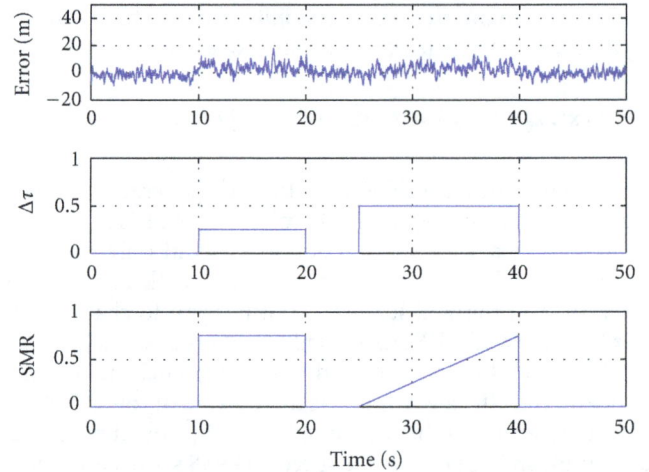

FIGURE 6: Evolution of pseudorange error in MEPF scheme under severe multipath propagation, setup number 2.

TABLE 2: Comparison of pseudorange error metrics in meters.

	DLL/PLL	MEPF (no. 1)	MEPF (no. 2)	MEPF (no. 3)
Bias (m)	10.66	3.97	5.48	1.24
std dev (m)	13.23	6.42	8.75	4.05
RMSE (m)	17.00	7.55	10.33	4.24

The results can be consulted in Figure 8, where it can be observed that MEPF requires an initial convergence time (depending on the covariance matrix set) larger than DLL schemes. Conversely, it appears more robust to channel impairments. Numerically, the RMSE in the overall simulation is of 8.48 m and 4.82 m for DLL and MEPF, respectively. For the MEPF we used $M = 2$ paths, $N = 1000$ particles, and $\sigma_{0,\tau} = .03/c$, $\sigma_{0,\dot{t}} = 0.03/c$, $\sigma_{m,\tau} = 3/c$, $\sigma_{m,\dot{t}} = 0.03/c$, $\sigma_{0,\alpha} = 0.0001$, and $\sigma_{m,\alpha} = 0.001$.

6. Conclusions

In this paper we have analyzed an advanced tracking loop for time-delay and carrier-phase estimation in a GNSS

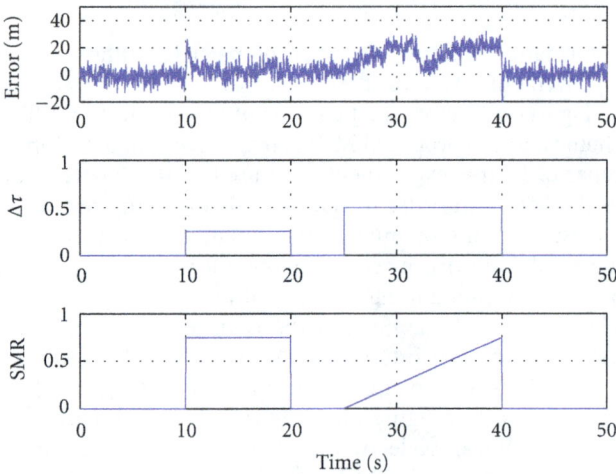

FIGURE 7: Evolution of pseudorange error in MEPF scheme under severe multipath propagation, setup number 3.

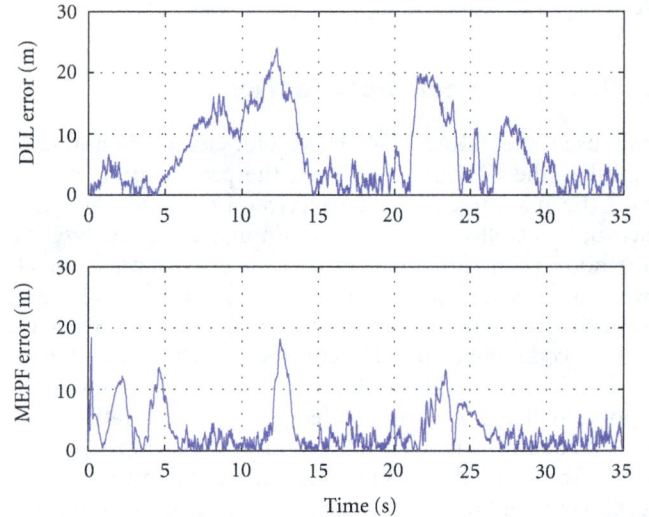

FIGURE 8: Pseudorange error in DLL and MEPF schemes in an urban scenario at an elevation of 55°, as modeled by Jahn's model.

receiver based on sequential Monte-Carlo methods. The algorithm builds upon previous work by the authors on Rao-Blackwellized particle filtering while introducing more realistic process dynamics and the usage of postcorrelation observations, that reduce the computational burden at the receiver. The paper presents the general signal model, GNSS concept, and trade-offs the most common propagation channel models. A realistic scenario simulator based on the FCM blockset of Simulink was used Section 5. Results point out the need for properly setting not only the number of particles but the number of correlation outputs used as observations. Also, degradation of conventional DLL/PLL schemes in multipath-rich scenarios became clear. Nevertheless, the correct selection of a process covariance matrix was seen to affect significantly the performance of the MEPF and future work should be devoted in self-adjustment of such matrix.

Appendix

Equivalence of Pre/Postcorrelation Receiver Architectures

In this appendix we establish a basic result showing the equivalence between processing pre- and postcorrelation signals. That is to say, from a statistical point of view, an estimator of a given parameter (e.g., time delay) computed using a bunch of snapshots taken at the IF signal level $(r_0(t; \boldsymbol{v}))$ is the same as that which is derived using the output of the correlators $(y(t; \boldsymbol{v}))$. It is a well-known result in statistical signal processing that both signals are sufficient statistics, and thus one is able to derive an estimator of $\boldsymbol{v} = (\tau_0, f_{d_0}, \phi_0)^T$ using either. However, we will see that this equivalence becomes evident when one examines the likelihood distribution (the density where the information from measurements is gathered) for each approach.

If we analyze first the case of using the IF signal we should be aware of the following.

(i) This approach does not force an implementation based on early, prompt, and late samples; as observations are directly the baseband signal at the sampling frequency.

(ii) It is necessary to use a sufficiently large set of IF data to be able to infer any parameter from it. That is, one has to integrate over a certain integration time, T_{int}, since the signal-to-noise ratio of GNSS signals is typically well below the noise level.

The term $\mathbf{r}_n \in \mathbb{C}^K$ stands for the vector of snapshots of the IF signal, as gathered for the nth integration interval, defined as

$$\mathbf{r}_n = \begin{pmatrix} r_0(nT_{\text{int}}) \\ \vdots \\ r_0(nT_{\text{int}} + (K-1)T_s) \end{pmatrix}, \quad (A.1)$$

using the same notation conventions used along the document. Then, the likelihood can be decomposed as the independent contribution of each snapshot

$$\log(p(\mathbf{r}_n \mid \boldsymbol{v}_n)) = \prod_{k=1}^K p\left(\underbrace{r_0(nT_{\text{int}} + (k-1)T_s)}_{r_{0,n}(k)} \mid \boldsymbol{v}_n \right), \quad (A.2)$$

and assuming Gaussianity for the noise term, we could identify that

$$\log(p(\mathbf{r}_n \mid \boldsymbol{v}_n)) \propto -\sum_{k=1}^K \left\| r_{0,n}(k) - h_n^{(1)}(k; \boldsymbol{v}_n) \right\|^2, \quad (A.3)$$

where $h_n^{(1)}$ stands for the precorrelation signal model, which was defined earlier as

$$h_n^{(1)}(t; \boldsymbol{v}_n) = |a_0(t)| e^{j(2\pi f_{d_0} t + \phi_0)} d_n \tilde{p}(t - \tau_0). \quad (A.4)$$

Further manipulation of the loglikelihood yields to

$$\log(p(\mathbf{r}_n \mid \boldsymbol{v}_n)) \propto \sum_{k=1}^K |r_{0,n}(k) \tilde{p}^*(t - \boldsymbol{v}_n)|^2 - |r_{0,n}(k)|^2$$

$$\propto |y_n(\boldsymbol{v} - \boldsymbol{v}_n)|^2, \quad (A.5)$$

with the latter step being clear if one accounts for the definition of y_n as the output of a correlator. Recall that $\boldsymbol{v}$ is the true unknown parameter of the signal. An ML estimator of $\boldsymbol{v}$ could be obtained after maximizing the latter equation.

Drawbacks of this approach are twofold.

(i) It might be computationally expensive as large data sets need to be processed to increase the signal-to-noise ratio, and thus K might be large depending on T_s.

(ii) There is a requirement for performing signal processing operations at a high rate, since it operates at the sampling frequency.

If we turn our attention to the conventional approach in which one uses samples at the output of a bank of correlators, we should see the following.

(i) This approach forces an implementation based on early, prompt, and late samples; this means that samples are taken assuming a previous estimation (prompt) of the parameters, denoted as $\check{\boldsymbol{v}} = (\check{\tau}_0, \check{f}_{d_0}, \check{\phi}_0)^T$.

(ii) Few samples are sufficient to infer estimates of $\boldsymbol{v}$. After correlation an integration over a certain interval is already done, T_{int}, and therefore the signal-to-noise ratio is relatively high.

In this case, measurements can be expressed as $y_n(\boldsymbol{v} - \check{\boldsymbol{v}})$ at the output of the n-th integration interval. In this measurement we explicitly expressed that samples are taken with respect to the error between true and prompt parameters, $\Delta\boldsymbol{v} = \boldsymbol{v} - \check{\boldsymbol{v}}$. Notice that we considered that only the prompt is used for the sake of clarity. It is easy to obtain a similar result, as the one shown here, when one accounts for several early and late samples.

Then, the log-likelihood under the Gaussian assumption is

$$\log(p(y_n \mid \boldsymbol{v}_n)) \propto \left\| y_n(\Delta\boldsymbol{v}) - h_n^{(2)}(\boldsymbol{v}_n - \check{\boldsymbol{v}}) \right\|^2$$

$$\propto \Re\left\{ y_n(\Delta\boldsymbol{v}) \left(h_n^{(2)}(\boldsymbol{v}_n - \check{\boldsymbol{v}}) \right)^* \right\}, \quad (A.6)$$

with $h_n^{(2)}$ being the postcorrelation signal model

$$h_n^{(2)}(\Delta\boldsymbol{v}) = \frac{|a_0|}{2} K \frac{\sin(\pi \Delta f T_{\text{int}})}{\pi \Delta f T_{\text{int}}} d\left([n]_{T_b/T_{\text{int}}}\right)$$

$$\cdot R_{\tilde{p}q}(\Delta\tau_0) \cos(\pi \Delta f T_{\text{int}} + \Delta\phi), \quad (A.7)$$

and $\boldsymbol{v}_n$ the unknown parameter we want to estimate at n.

If we set $\boldsymbol{v}_n = \breve{\boldsymbol{v}}$, we can identify that

$$
\log(p(y_n \mid \boldsymbol{v}_n)) \propto \Re\left\{ y_n(\boldsymbol{v} - \boldsymbol{v}_n)\left(h_n^{(2)}(0)\right)^* \right\}
$$
$$
\propto \left| y_n(\boldsymbol{v} - \boldsymbol{v}_n) \right|^2 . \tag{A.8}
$$

From the latter mathematical derivations, we can conclude an important result:

$$
\boxed{\log(p(\mathbf{r}_n \mid \boldsymbol{v}_n)) \propto \log(p(y_n \mid \boldsymbol{v}_n))} \tag{A.9}
$$

for a given integration interval T_{int} considering KT_s snapshots. As said, similar results apply for larger integration and more early/late samples.

As a consequence, we can state the following: the ML estimator of $\boldsymbol{v}$ computed from the data sets $\mathbf{r}_n$ and y_n is equivalent.

To sum up, from a statistical point of view, both approaches are equivalent and the choice should be made considering implementation aspects. For instance, it is clear that using precorrelation measurements $\mathbf{r}_n$ involves larger computational burden than using post-correlation samples. Another important conclusions is that since in the pre-correlation approach we also need to integrate in order to increase the signal-to-noise ratio, effects happening faster than T_{int} will not be captured by the estimation algorithm. The same happens in the post-correlation case. Therefore, the limitation of which phenomena could be tracked is inherent to the GNSS signal, instead of the way it is processed (i.e., pre-or postcorrelated samples).

Acknowledgment

P. Closas and C. Fernández-Prades were supported by the European Commission under COST Action IC0803 (RFCSET).

References

[1] M. Hernández-Pajares, J. M. J. Zornoza, J. S. Subirana, R. Farnworth, and S. Soley, "EGNOS test bed ionospheric corrections under the October and November 2003 storms," *IEEE Transactions on Geoscience and Remote Sensing*, vol. 43, no. 10, pp. 2283–2293, 2005.

[2] S. Bancroft, "An algebraic solution of the GPS equations," *IEEE Transactions on Aerospace and Electronic Systems*, vol. 21, no. 1, pp. 56–59, 1985.

[3] G. Strang and K. Borre, *Linear Algebra, Geodesy, and GPS*, Wellesley Cambridge Press, 1997.

[4] B. Hofmann-Wellenhof, H. Lichtenegger, and E. Wasle, *GNSS -Global Navigation Satellite Systems: GPS, GLONASS, Galileo & More*, Springer-Verlag, Wien, Austria, 2008.

[5] C. Fernández-Prades, L. L. Presti, and E. Falletti, "Satellite radiolocalization from GPS to GNSS and beyond: novel technologies and applications for civil mass market," *Proceedings of the IEEE*, vol. 99, no. 11, pp. 1882–1904, 2011.

[6] A. Jahn, H. Bischl, and G. Heiss, "Channel characterization for spread spectrum satellite communications," in *Proceedings of the 4th International Symposium on Spread Spectrum Techniques & Applications (ISSSTA '96)*, pp. 1221–1226, September 1996.

[7] C. Loo and J. S. Butterworth, "Land mobile satellite channel measurements and modeling," *Proceedings of the IEEE*, vol. 86, no. 7, pp. 1442–1462, 1998.

[8] M. A. V. Castro, F. P. Fontan, A. A. Villamarín, S. Buonomo, P. Baptista, and B. Arbesser, "L-band Land Mobile Satellite (LMS) amplitude and multipath phase modeling in urban areas," *IEEE Communications Letters*, vol. 3, no. 1, pp. 12–14, 1999.

[9] F. P. Fontán, M. Vázquez-Castro, C. E. Cabado, J. P. García, and E. Kubista, "Statistical modeling of the LMS channel," *IEEE Transactions on Vehicular Technology*, vol. 50, no. 6, pp. 1549–1567, 2001.

[10] A. Steingass and A. Lehner, "A channel model for land mobile satellite navigation," in *Proceedings of the the European Navigation Conference*, pp. 2132–2138, German Institute of Navigation (DGON), July 2005.

[11] Recommendation ITU-R P.681-7, "Propagation data required for the design of Earth-space land mobile telecommunication systems," 2009, http://www.itu.int/rec/R-REC-P.681-7-200910-I/en/.

[12] P. Closas, *Bayesian signal processing techniques for GNSS receivers: from multipath mitigation to positioning [Ph.D. dissertation]*, Universitat Politècnica de Catalunya (UPC), Department of Signal Theory and Communications, Barcelona, Spain, 2009.

[13] D. M. Akos, M. Stockmaster, J. B. Y. Tsui, and J. Caschera, "Direct bandpass sampling of multiple distinct RF signals," *IEEE Transactions on Communications*, vol. 47, no. 7, pp. 983–988, 1999.

[14] B. Parkinson and J. Spilker, Eds., *Global Positioning System: Theory and Applications*, vol. 1 of *Progress in Astronautics and Aeronautics*, American Institute of Aeronautics, Washington, DC, USA, 1996.

[15] J. S. Silva, P. F. Silva, A. Fernández, J. Diez, and J. F. M. Lorga, "Factored correlator model: a solution for fast, flexible, and realistic GNSS receiver simulations," in *Proceedings of the 20th International Technical Meeting of the Satellite Division of The Institute of Navigation (ION GNSS '07)*, pp. 2676–2686, Fort Worth, TX, USA, September 2007.

[16] R. E. Kalman, "A new approach to linear filtering and prediction problems," *Transactions of the ASME-Journal of Basic Engineering*, vol. 82, pp. 35–45, 1960.

[17] Z. Chen, "Bayesian filtering: from Kalman filters to particle filters, and beyond," Tech. Rep., Adaptive Systems Laboratory, McMaster University, Ontario, Canada, 2003.

[18] M. S. Arulampalam, S. Maskell, N. Gordon, and T. Clapp, "A tutorial on particle filters for online nonlinear/non-Gaussian Bayesian tracking," *IEEE Transactions on Signal Processing*, vol. 50, no. 2, pp. 174–188, 2002.

[19] P. M. Djurić, J. H. Kotecha, J. Zhang et al., "Particle filtering," *IEEE Signal Processing Magazine*, vol. 20, no. 5, pp. 19–38, 2003.

[20] M. Bolić, P. M. Djuric, and S. Hong, "Resampling algorithms for particle filters: a computational complexity perspective," *Eurasip Journal on Applied Signal Processing*, vol. 2004, no. 15, pp. 2267–2277, 2004.

[21] R. Douc, O. Cappé, and E. Moulines, "Comparison of resampling schemes for particle filtering," in *Proceedings of the 4th International Symposium on Image and Signal Processing and Analysis (ISPA '05)*, pp. 64–69, Zagreb, Croatia, September 2005.

[22] GRANADA Galileo Receiver ANalysis And Design Application. The Reference Galileo Simulation Toolkit for GNSS Receiver Research And Development. Factored Correlator

Model Blockset v2.0 User Manual, Deimos Engenharia, S.A., 2009.

[23] T. Schön, F. Gustafsson, and P. J. Nordlund, "Marginalized particle filters for mixed linear/nonlinear state-space models," *IEEE Transactions on Signal Processing*, vol. 53, no. 7, pp. 2279–2289, 2005.

[24] R. Karlsson, *Particle filtering for positioning and tracking applications [Ph.D. dissertation]*, Linköping University, Linköping, Sweden, 2005.

[25] R. Chen and J. S. Liu, "Mixture Kalman filters," *Journal of the Royal Statistical Society B*, vol. 62, no. 3, pp. 493–508, 2000.

[26] A. Doucet, N. de Freitas, and N. Gordon, Eds., *Sequential Monte Carlo Methods in Practice*, Springer, 2001.

[27] C. Rao, "Information and the accuracy attainable in the estimation of statistical parameters," *Bulletin of Calcutta Mathematical Society*, vol. 37, pp. 81–91, 1945.

[28] D. Blackwell, "Conditional expectation and unbiased sequential estimation," *The Annals of Mathematical Statistics*, vol. 18, no. 1, pp. 105–110, 1947.

[29] E. Lehmann, *Theory of Point Estimation. Probability and Mathematical Statistics*, John Wiley & Sons, 1983.

[30] A. Papoulis and S. U. Pillai, *Probability, Random Variables and Stochastic Processes*, McGraw-Hill, New Delhi, India, 4th edition, 2001.

[31] P. Closas, C. Fernández-Prades, and J. A. Fernández-Rubio, "Bayesian DLL for multipath mitigation in navigation systems using particle filters," in *Proceedings of the IEEE (ICASSP '06)*, Toulouse, France, May 2006.

[32] P. Closas, C. Fernández-Prades, and J. A. Fernández-Rubio, "A Bayesian approach to multipath mitigation in GNSS receivers," *IEEE Journal on Selected Topics in Signal Processing*, vol. 3, no. 4, pp. 695–706, 2009.

[33] M. Lentmaier, B. Krach, and P. Robertson, "Bayesian time delay estimation of GNSS signals in dynamic multipath environments," *International Journal of Navigation and Observation*, vol. 2008, Article ID 372651, 11 pages, 2008.

[34] B. Krach, P. Robertson, and R. Weigel, "An efficient two-fold marginalized Bayesian filter for multipath estimation in satellite navigation receivers," *Eurasip Journal on Advances in Signal Processing*, vol. 2010, Article ID 287215, 2010.

[35] A. Steingass and A. Lehner, "Measuring the navigation multipath channel—a statistical analysis," in *Proceedings of the 17th International Technical Meeting of the Satellite Division of the Institute of Navigation (ION GNSS '04)*, pp. 1157–1164, Long Beach, Calif, USA, September 2004.

[36] M. Irsigler, J. A. Ávila-Rodríguez, and G. W. Hein, "Criteria for GNSS multipath performance assessment," in *Proceedings of the International Technical Meeting of the Institute of Navigation(ION GPS/GNSS '05)*, Long Beach, Calif, USA, September 2005.

[37] P. Closas, C. Fernández-Prades, J. Diez, and D. de Castro, "Multipath estimating tracking loops in advanced GNSS receivers with particle filtering," in *Proceedings of the IEEE Aerospace Conference*, Big Sky, Mont, USA, March 2012.

10

Effectiveness of GNSS Spoofing Countermeasure Based on Receiver CNR Measurements

J. Nielsen, V. Dehghanian, and G. Lachapelle

Position Location and Navigation Group, University of Calgary, Calgary, AB, Canada T2N 1N4

Correspondence should be addressed to V. Dehghanian, vdehghan@ucalgary.ca

Academic Editor: Dennis M. Akos

A perceived emerging threat to GNSS receivers is posed by a spoofing transmitter that emulates authentic signals but with randomized code phase and Doppler over a small range. Such spoofing signals can result in large navigational solution errors that are passed onto the unsuspecting user with potentially dire consequences. In this paper, a simple and readily implementable processing rule based on CNR estimates of the correlation peaks of the despread GNSS signals is developed expressly for reducing the effectiveness of such a spoofer threat. Consequently, a comprehensive statistical analysis is given to evaluate the effectiveness of the proposed technique in various LOS and NLOS environments. It is demonstrated that the proposed receiver processing is highly effective in both line-of-sight and multipath propagation conditions.

1. Introduction

GNSS satellites are approximately 20,000 km away and transmit several watts of signal power such that at the ground level, the power output of a 3-dB gain linearly polarized antenna is nominally −130 dBm [1]. As such, a modest jammer can easily disrupt GNSS signals by increasing the noise floor, making the acquisition of GNSS signals rather difficult. A high processing gain based on a long integration time is one of the possible countermeasures to overcome a noise jammer. Nevertheless, if the GNSS receiver undergoes random motion and is subjected to multipath fading as in a typical urban environment, then the channel decorrelates quickly such that attaining such large processing gains to overcome the jamming is not feasible. However, the noise jammer is at least detectable as the spectral power in the affected GNSS receiver band will be abnormally high. Hence, the jammer can deny service but the user is aware of being jammed, limiting the damage potential of the jammer. Also the jammer is relatively easy to locate with radio direction finding and to potentially disable as its spectrum is significantly larger than the ambient noise [2, 3].

A more insidious threat is the standoff spoofer which broadcasts a set of replicas of the authentic SV signals currently visible to the mobile GNSS receiver [2]. The unaware receiver computes the navigation solution based on these counterfeit signals which are passed on to the user as being reliable with potentially damaging consequences. GNSS-based location estimates that are inaccurate but assumed to be accurate are potentially more damaging to the user than in the jamming case where at least the user knows that the service is temporarily unavailable. As the receiver processing gain used for suppressing the jammer is not applicable in the case of the spoofer signal, the spoofer transmit power can be orders of magnitude less than that of the noise jammer. This makes the spoofer signal much more difficult to locate and disable.

There are essentially two categories of spoofer threats envisioned. The first is the self-intentional spoofer that provides the user a means of compromising its GNSS position. An example is a fishing vessel wishing to enter prohibited areas undetected by a GNSS-based monitoring system. A collocated spoofer could provide counterfeit signals to fabricate navigation solution that falls outside the prohibited area [4, 5]. Another example is that of an offender required to wear a mandatory GNSS tracker to ensure compliance with travel restrictions [2].

The second type of spoofers is the standoff spoofer (SS) that could be used in urban areas for malicious purposes ranging from sporadic disruptive hacking to sophisticated organized terrorist activities. The SS is illustrated in Figure 1 which covers a target area as a sector of an annulus ring. Multiple SS devices could potentially be used to collectively cover a given area such as an urban downtown core. Based on this, the perceived spoofer threat is a network of terrestrial SSs that can cause widespread disruption of GNSS-based location services in dense urban areas.

The SS is of interest in this paper specifically for the scenario of a terrestrial transmitter source that broadcasts replicas of the GNSS signals that are visible in the target area illustrated in Figure 1. Disruption of GNSS services in the target area is achieved by randomly modulating the code phase over a small region of the overall Code-Delay Space (CDS) that is commensurate with the target area. Therefore, at least two correlation peaks will be observed in the CDS. An unsuspecting receiver detects the larger of the correlation peaks which can belong to the spoofer signal. The code phase and the Doppler associated with the spoofer signal are then passed onto the tracking segment and consequently a false navigation solution is generated. Note that, while the target area depicted in Figure 1 has hard boundaries, such boundaries are generally blurry and not well defined. The effectiveness of the SS is considered to drop off outside the depicted annulus sector region with vague boundaries between radii R_1 and R_2. In a typical scenario, R_1 and R_2 are envisioned to be of the order of about 500 m and 2 km such that each SS covers an area of several square kilometres. A modest network of SS devices can then adequately cover a downtown core area. However, for sake of simplicity, only a single isolated SS will be considered in this paper.

The SS is assumed to remain synchronized with currently visible GNSS signals and then synthesize a set of GNSS signals corresponding to the target area. The objective of the SS is not to synthesize a specific counterfeit location for a specific GNSS receiver within the target area. This is not possible as the location of the GNSS receiver is not known to the SS. Furthermore, the objective of the SS is disruption over the general target area rather than affecting specific receivers. As such, the SS transmission signal synthesis does not have to be overly sophisticated. It matches the Doppler offset of the replicated SV signals and adjusts the code phase such that it is commensurate with the intended target region. Note that an urban area is a primarily non-line-of-sight (NLOS) multipath channel. Therefore, the Doppler spectrum as perceived by the GNSS receiver will be spread by an amount commensurating with the magnitude of the receiver velocity but will not be sensitive to direction. Hence, other than the deterministic Doppler offset of the SV to stationary ground-based receiver, no further modulation of the Doppler is required by the SS to ensure a plausible counterfeit signal. The typical handheld consumer GNSS receiver coherently integrates the signal for about 10 to 20 ms. Based on this, the correlation peak in the CDS will have a spread in Doppler of about 100 Hz which is commensurate with the Doppler spread of typical urban traffic (<50 km/hr) [6]. Even if the GNSS receiver is equipped with other inertial means such that the receiver velocity vector is known, this cannot be used to discriminate the SS signal as multipath Doppler spreading occurs for both the SS and the authentic signals.

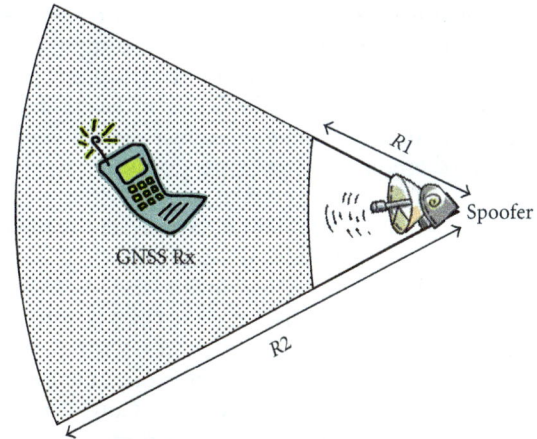

FIGURE 1: Standoff Spoofer (SS) illuminating a target area which is a sector of an annulus extending from R_1 to R_2.

The code phase of the SS transmissions matches the nominal code phase of the authentic GNSS signals in the target area. Note that the target area is limited to one or two kilometres and hence, the code phase only differs by several chips from one extreme of the target area to the other. For example, in a 90-degree sector with $R_1 = 500$ m and $R_2 = 1500$ m, the average spread is only about four chips. The SS generated code phase will correspond to a random location within the target area generated by slowly and randomly modulating the code phase over a small domain commensurating with the dimensions of the target area. Note that a sophisticated GNSS receiver can potentially discriminate against the SS signal based on the code phase corresponding to an outlier navigation solution. However, as the target region is not very large, the counterfeit SS navigation solutions will be plausible and cannot be easily dismissed as outliers. Furthermore, the typical consumer grade GNSS unit does not possess processing to track multiple candidate navigation solutions let alone discriminate plausible outliers. Also, receiver autonomous integrity monitoring (RAIM) and fault detection and exclusion (FDE) are not effective in detecting such navigationally consistent spoofing signals [4]. Finally, it should also be mentioned that typically GNSS receivers tethered to a wireless data service provider will typically provide the user with an aided GNSS (AGNSS) service, significantly reducing the CDS corresponding to a physical area of several square kilometres [7]. Hence, there is a diminishing gain for the spoofer attempting to affect an area larger than this.

As stated earlier, current consumer-grade receivers are equipped with RAIM and FDE which are not effective in mitigating the navigationally consistent spoofing attacks. A more sophisticated countermeasure to the SS with a random code delay modulation is to carefully tracking all combinations of possible navigation solutions and then dismissing solutions that are less likely based on tracking records spanning several tens of seconds up to the current time. This solution likelihood can be augmented with the use of ancillary sensors and other prior knowledge or belief maps [8]. However, the consumer-grade GNSS receivers considered herein are assumed not to possess this level of sophistication. Rather, the objective is to address a computationally efficient processing method that can be added to relatively unsophisticated consumer grade GNSS receivers and that will be effective in discriminating against the SS. Such processing is based on the received carrier-to-noise ratio (CNR) measurements of the received GNSS signals. CNR measurement is an integrated part of all GNSS receivers as the navigation algorithm heavily relies on determining the weight of the observables based on measuring the instantaneous CNR. A simple discriminant is that if the CNR is implausibly high then an SS is suspected. Such processing is easily implemented with essentially minor firmware changes to the receiver or an in-line filter component [2]. However, there is the question of how to optimally set the threshold used for CNR comparison. The optimum threshold is easily determined and justified for LOS propagation with a known antenna gain and orientation. However, for a handheld unit operating in an urban canyon with a compromised multi-band antenna that is randomly oriented and potentially shadowed, setting the optimum threshold is no longer deterministic nor trivial. Optimization is necessarily based on a statistical analysis, which is the focus of this paper.

The rest of the paper is organized as follows. In Section 2, the system definition and simplifying assumptions are given. A difficulty encountered with the statistical assessment of the SS effectiveness is the plethora of disparate parameters and plausible scenarios encountered. For this paper, a constrained set of idealized parameters and assumptions is necessary to obtain fundamental insights. In Section 3, the effectiveness of the SS and the receiver countermeasures is considered for a variety of LOS and NLOS scenarios. Section 3.5 relates these findings to the plausible physical coverage range of the SS. Finally, Section 4 states the major conclusions.

2. System Description and Assumptions

The performance of spoofer detection based on a threshold applied to the CNR in conjunction with a simple decision rule is analyzed for various propagation conditions. To do this in a comprehensive manner that is not obscured by details, it is necessary to use simplifying assumptions and constraints. While these may erode generality, the benefit is a set of insights gained that are applicable to less idealized and more realistic scenarios.

It is assumed that the GNSS receiver performs a reduced search over the CDS based on traditional despreading correlation processing for each candidate GNSS signal that is potentially visible to the receiver. Assuming that both the authentic and SS signals are present at the receiver for a given despread GNSS signal, the outcome is a set of two correlation peaks corresponding to the spoofer and the authentic signal. The complex amplitude of the authentic and spoofer correlation peaks is represented as

$$x^a = \sqrt{\rho_{a_0}} h_a + w_a,$$
$$x^s = \sqrt{\rho_{s_0}} h_s + w_s,$$
(1)

where ρ_{a_0} and ρ_{s_0} are the average CNRs of the authentic and SS signals, respectively. The complex channel gains are denoted by h_a and h_s with $E[|h_a|^2] = E[|h_s|^2] = 1$ where E denotes the expected value operation. Also w_a and w_s represent the normalized white Gaussian noise samples distributed according to $CN(0, 1)$ with $CN(\mu, \sigma^2)$ denoting a circularly normal multivariate distribution with a mean of μ and a variance of σ^2. Note that the noise variance is normalized to simplify the expressions to follow.

It is assumed that there are nominally two correlation peaks in the CDS hypothesis space that correspond to the spoofer and the authentic signal for a specific GNSS signal with sample-based CNRs denoted as ρ_s and ρ_a, respectively, namely,

$$\rho_a \equiv |x^a|^2 - 1,$$
$$\rho_s \equiv |x^s|^2 - 1.$$
(2)

There are many variations as to how the receiver implements the correlation search over the CDS; however, this assumption of the correlator structure simplifies the system description and subsequent analysis. Furthermore, the possibility of the authentic signal resulting in two distinct correlation peaks due to resolvable multipath or poor receiver design is not considered. The GNSS receiver cannot determine which correlation peak corresponds to the desired authentic signal. However, recognizing that there are two possible choices from which it suspects spoofer activity, it can impose the following simple heuristic rule for selecting the authentic signal:

> Choose the larger of the two peaks as the authentic peak if $(\rho_s < \rho_T) \cap (\rho_a < \rho_T)$, otherwise choose the smaller peak.

Here ρ_T is a threshold CNR that ρ_s and ρ_a will be compared to, which is the subject of some adaptive optimization process. Based on this formulation, the probability of a selection error can be evaluated. An error occurs every time the spoofer correlation peak is selected instead of the authentic peak with the Doppler and code delay coordinates passed on to the navigation solution processor. As such there are two types of errors described as

$$\text{type I error: } \{(\rho_s < \rho_T) \cap (\rho_a < \rho_T)\} \cap (\rho_a < \rho_s),$$
$$\text{type II error: } \{(\rho_s > \rho_T) \cup (\rho_a > \rho_T)\} \cap (\rho_s < \rho_a).$$
(3)

A graphical aid is introduced in Figure 2 which provides a method of calculating the probability of receiver error as the sum of the probabilities of the two types of errors. This probability will be denoted as P_e and is a measure of the effectiveness of the spoofer; that is, the higher P_e is over a given target area of the spoofer, the more effective it is, and is therefore a suitable metric for quantifying the effectiveness of the SS. P_e depends on the probability density function (PDF) of the CNRs of the authentic and spoofing correlation peaks.

To proceed further, the following definitions are made:

$f_a(\rho_a; \rho_{a_0})$: PDF of ρ_a with the parameter ρ_{a_0};

$f_s(\rho_s; \rho_{s_0})$: PDF of ρ_s with the parameter ρ_{s_0};

$F_a(\rho_a; \rho_{a_0}) = \int_0^{\rho_a} f_a(\lambda; \rho_{a_0}) d\lambda$: cumulative distribution of the authentic signal;

$F_s(\rho_s; \rho_{s_0}) = \int_0^{\rho_s} f_s(\lambda; \rho_{s_0}) d\lambda$: cumulative distribution of the authentic signal.

Assuming that the authentic and the spoofer CNR samples, $\{\rho_a, \rho_s\}$, are statistically independent random variables, then the joint PDF can be expressed as the product of

$$f_{a,s}(\rho_a, \rho_s; \rho_{a_0}, \rho_{s_0}) \approx f_a(\rho_a; \rho_{a_0}) f_s(\rho_s; \rho_{s_0}). \tag{4}$$

This assumption is based on the authentic SV original signal and the terrestrial source SS signal coming from different bearings and hence, in a dense urban area, the fast fading and nominal path-loss is independent. As the bearings are sufficiently different, the longer-term fading or shadowing is not correlated [6]. Hence, the assumption of independence implied by (4) is made herein. However, there are instances where shadowing does become correlated especially if the bearings of the authentic and SS signals are similar. Based on the graphic shown in Figure 2, P_e is given by

$$
\begin{aligned}
P_e &= \int_0^{\rho_T} f_s(\rho_s) \left(\int_0^{\rho_s} f_a(\rho_a) d\rho_a \right) d\rho_s \\
&\quad + \int_{\rho_T}^{\infty} f_a(\rho_a) \left(\int_0^{\rho_a} f_s(\rho_s) d\rho_s \right) d\rho_a \\
&= \int_0^{\rho_T} f_s(\rho_s) (F_a(\rho_s) - F_a(0)) d\rho_s \\
&\quad + \int_{\rho_T}^{\infty} f_a(\rho_a) (F_s(\rho_a) - F_s(0)) d\rho_a,
\end{aligned}
\tag{5}
$$

where the simplified notation omits the parameters ρ_{s_0} and ρ_{a_0} which are initially assumed to be known parameters. Using $F_a(0) = F_s(0) = 0$, (5) becomes

$$P_e = \int_0^{\rho_T} f_s(\rho_s) F_a(\rho_s) d\rho_s + \int_{\rho_T}^{\infty} f_a(\rho_a) F_s(\rho_a) d\rho_a. \tag{6}$$

The minimum value of P_e can be determined by setting $(\partial / \partial \rho_T) P_e = 0$ such that the condition

$$\frac{\partial}{\partial \rho_T} \int_0^{\rho_T} f_s(\rho_s) F_a(\rho_s) d\rho_s + \frac{\partial}{\partial \rho_T} \int_{\rho_T}^{\infty} f_a(\rho_a) F_s(\rho_a) d\rho_a = 0 \tag{7}$$

emerges and reduces to

$$\frac{f_s(\rho_T)}{F_s(\rho_T)} = \frac{f_a(\rho_T)}{F_a(\rho_T)} \tag{8}$$

which is then solved for the optimum value of ρ_T. Equation (8) is mathematically equivalent to

$$f_s(\rho_T) F_a(\rho_T) - f_a(\rho_T) F_s(\rho_T) \equiv \frac{\partial}{\partial \rho_T} \left(\frac{F_s(\rho_T)}{F_a(\rho_T)} \right) = 0. \tag{9}$$

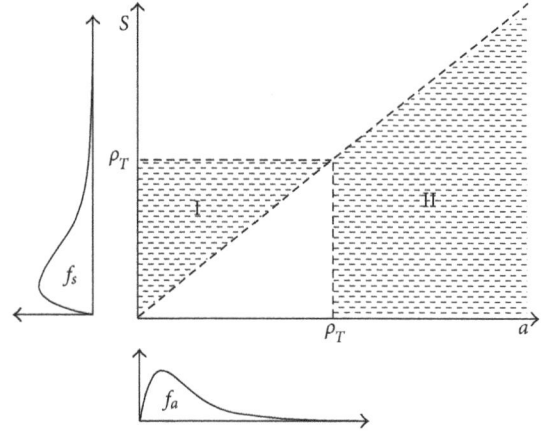

FIGURE 2: Graphical integration regions for the two error types.

A useful observation is that if the PDFs of the authentic and the spoofer signals are scaled versions of each other, that is, $F_a(\rho_T) = F_s(\rho_T / c)$; then (9) holds only if $\rho_T = 0$ and $\rho_T = \infty$, since a cumulative distribution function (CDF) is a monotonically increasing function. This means that a finite threshold other than $\rho_T = 0$ and $\rho_T = \infty$ does not exist. In other words, for the common case when $f_a(\rho_a)$ is a monomodal function then it is easily shown that $f_a(\rho_a)/F_a(\rho_a)$ is a monotonically decreasing function. Hence, if $f_a(\rho)$ is approximately a translation of the function $f_s(\rho)$, then the intersection points of $f_s(\rho_T)/F_s(\rho_T)$ and $f_a(\rho_T)/F_a(\rho_T)$ can only be at $\rho_T = 0$ and $\rho_T = \infty$. This observation will be used in the next section. Note that a threshold of $\rho_T = \infty$ is equivalent to having no threshold rather than applying a nonrealistically large threshold.

3. Performance of Antispoofing for LOS and NLOS Conditions

In this section, P_e is determined for LOS and NLOS scenarios. This is generally done by first solving for the optimum threshold ρ_T and then determining P_e.

3.1. LOS with Additive Noise. As defined in (1), the in-phase and quadrature components of the demodulated signal are normalized such that the additive noise is of unit variance for the in-phase and quadrature Gaussian components. With this, the LOS signal from the authentic signal will have a mean square magnitude of $2\rho_{a_0}$. Likewise the LOS from the SS will have a mean square magnitude of $2\rho_{s_0}$. Hence, the PDF of the square magnitudes of the correlation peaks corresponding to the authentic and spoofer signals will then be given as

$$
\begin{aligned}
f_a(\rho_a; \rho_{a_0}) &= \chi_2'^2(\rho_a; 2\rho_{a_0}, 1), \\
f_s(\rho_s; \rho_{s_0}) &= \chi_2'^2(\rho_s; 2\rho_{s_0}, 1),
\end{aligned}
\tag{10}
$$

where $\chi_N'^2(x; \mu, \sigma^2)$ is the noncentral chi-square PDF of variable x with N degrees of freedom (DOF), the noncentrality parameter μ, and the corresponding variance of the Gaussian

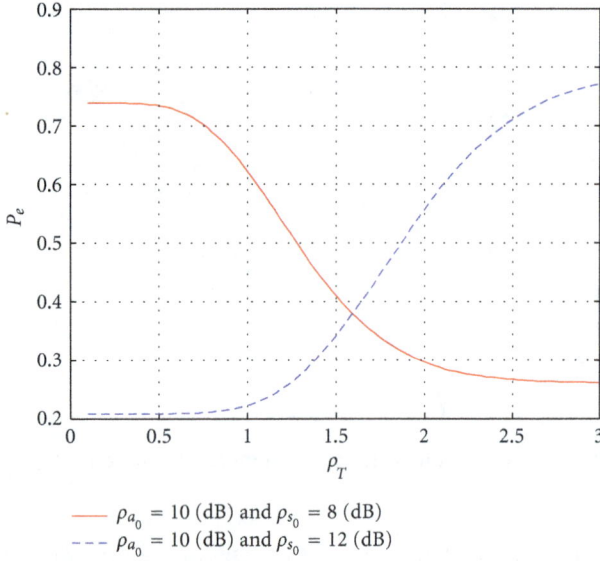

FIGURE 3: P_e as a function of ρ_T.

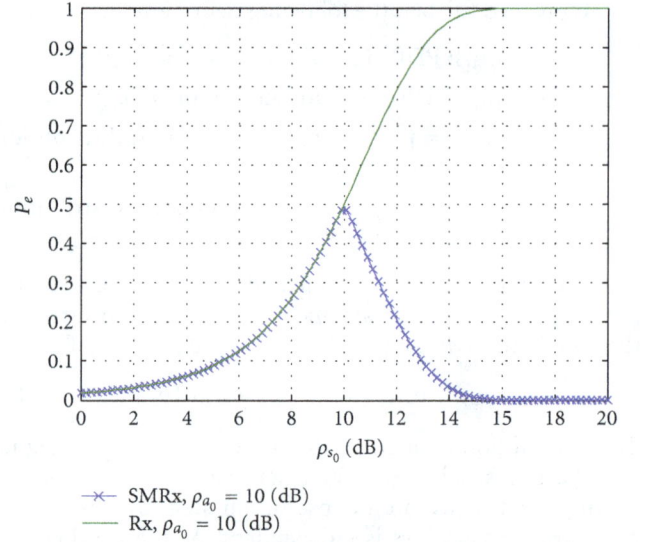

FIGURE 4: P_e as a function of ρ_{s_0} for a conventional receiver (Rx) and a spoofer mitigated receiver (SMRx).

parameter σ^2 [9]. P_e is plotted in Figure 3 as a function of ρ_T for specific cases where $\rho_{a_0} > \rho_{s_0}$ and $\rho_{a_0} < \rho_{s_0}$. As stated earlier, when $\rho_{a_0} > \rho_{s_0}$ the optimum threshold is $\rho_T = \infty$, while for $\rho_{a_0} < \rho_{s_0}$ the optimum threshold is $\rho_T = 0$. This is tantamount to selecting the larger of the two peaks if the average power of the authentic signal is larger than the average power of the spoofer. Otherwise, choose the smaller of the two peaks if the average power of the spoofer is larger than the average power of the authentic signal. This trivial conclusion is a manifestation of the assumption that ρ_{a_0} and ρ_{s_0} are known, which is not generally the case.

Note that as $f_a(\rho)$ is approximately a translation of the function $f_s(\rho)$ then the intersection points of $f_s(\rho_T)/F_s(\rho_T)$ and $f_a(\rho_T)/F_a(\rho_T)$ can only be at $\rho_T = 0$ and $\rho_T = \infty$ as observed before.

Figure 4 shows a plot of P_e for a receiver with no spoofer mitigation, herein denoted by Rx, compared to the P_e for a receiver with spoofer mitigation, herein denoted by SMRx, with $\rho_T = \infty$ for $\rho_{a_0} > \rho_{s_0}$ and $\rho_T = 0$ for $\rho_{a_0} < \rho_{s_0}$. The GNSS receiver with no spoofer mitigation is equivalent to setting $\rho_T = \infty$. As such there is no difference in the performance of the GNSS receivers with and without spoofer mitigation when $\rho_{a_0} > \rho_{s_0}$. However, for the case of $\rho_{a_0} < \rho_{s_0}$, the effectiveness of the spoofer mitigation is clearly evident in the reduction of P_e.

3.2. NLOS with Additive Noise.

In this section, it is assumed that ρ_{a_0} and ρ_{s_0} are again deterministic and known to the receiver. The PDFs of the magnitude of the correlation peaks corresponding to the authentic and spoofer signals are then be given as

$$f_a(\rho_a; \rho_{a_0}) = \chi_2^2(\rho_a; \rho_{a_0}),$$
$$f_s(\rho_s; \rho_{s_0}) = \chi_2^2(\rho_s; \rho_{s_0}), \tag{11}$$

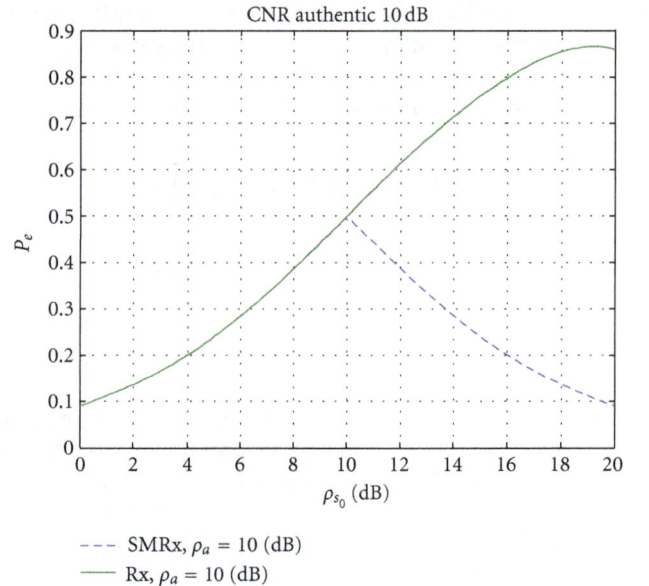

FIGURE 5: Comparison of the conventional and the spoofer mitigation receiver based on 2 DOF in a NLOS Rayleigh fading channel.

where $\chi_2^2(x; \sigma^2)$ is the central chi-square PDF of variable x with 2 DOF, with a variance of each DOF of $\rho_{a_0} + 1$ for the authentic signal and $\rho_{s_0} + 1$ for the spoofing signal.

Figure 5 shows a plot of P_e for a receiver with no spoofer mitigation (Rx) compared to the P_e for a receiver with spoofer mitigation (SMRx) with $\rho_T = \infty$ for $\rho_{a_0} > \rho_{s_0}$ and $\rho_T = 0$ for $\rho_{a_0} < \rho_{s_0}$. Comparing Figure 5 with Figure 4, it is evident that the spoofer mitigation is more effective when a LOS rather than a NLOS scenario is encountered. Hence, when the spoofer and authentic signals are more random as in the NLOS case, distinguishing them based on the sample CNR is more difficult and hence, subject to higher P_e.

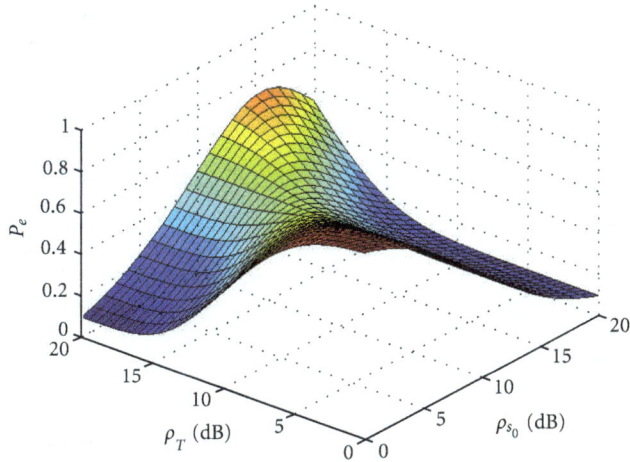

FIGURE 6: P_e as a function of ρ_T and ρ_{s_0}, for $\rho_{a_0} = 10$ and NLOS Rayleigh conditions based on 2DOF.

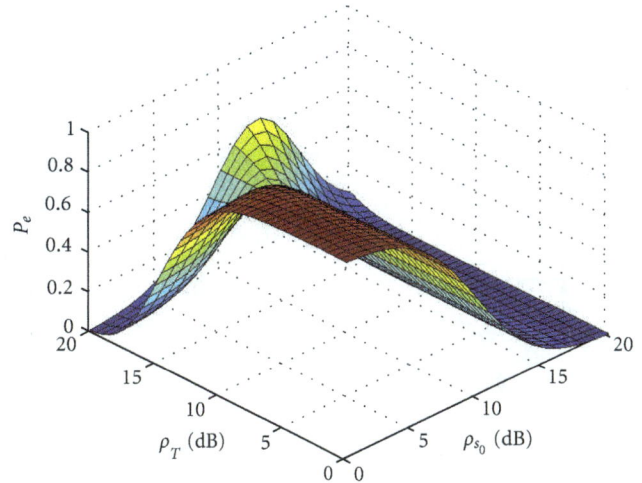

FIGURE 7: P_e as a function of ρ_T and ρ_{s_0}, for $\rho_{a_0} = 10$ and NLOS conditions based on $M = 3$ (6DOF).

Figure 6 shows P_e as a function of ρ_T and for various ρ_{s_0}. The effectiveness of the spoofer countermeasure is again evident in the region where $\rho_{a_0} < \rho_{s_0}$. The same behavior as before occurs, namely, that the optimum ρ_T for spoofer power less than authentic power is $\rho_T = \infty$ while for spoofer power greater than authentic power is $\rho_T = 0$, which is again a manifestation of the assumed known average powers.

3.3. Diversity NLOS with Additive Noise.

Assuming a ring or a sphere of scatterers to model a typical urban environment, the signals arriving at antennas with an approximate separation of half a carrier wavelength, are statistically uncorrelated. Consequently, M statistically independent samples of the receiver correlator output can be made available through accumulating M successive samples of the correlator outputs as the receiver is moving. The CNR of each correlation sample is ρ_{a_0} and ρ_{s_0} for the authentic and spoofing signals, respectively, which are again assumed to be deterministic and known to the receiver.

A plot of P_e based on $M = 3$ independent samples is shown in Figure 7. Similar to the no diversity case with $M = 1$, the optimum ρ_T for spoofer power less than the authentic power is $\rho_T = \infty$, while for spoofer power greater than the authentic power, the maximum is $\rho_T = 0$. Again, this is reasonable as the spoofer and authentic signal is identically distributed except for the deterministic and known average powers. Clearly, if it is known that $\rho_{a_0} > \rho_{s_0}$ then the larger peak would correspond to the authentic signal more often than the lower peak.

3.4. Measurement Uncertainty and Unknown Spoofer Average Power.

In the previous sections, the outcome was a trivial optimization of ρ_T as $\rho_T = 0$ if $\rho_{a_0} < \rho_{s_0}$ and $\rho_T = \infty$ if $\rho_{a_0} > \rho_{s_0}$, which resulted from the assumption that $\{\rho_{s_0}, \rho_{a_0}\}$ was known to the receiver. In this section, the more realistic multipath propagation case is considered where the average spoofer CNR is completely unknown. This is reasonable as the spoofer could be of arbitrary transmit power and range

from the receiver. However, it will be assumed that ρ_{a_0} is known approximately to the receiver. This is reasonable as the average power of a GNSS SV signal is approximately known in a multipath environment with the exception of factors such as shadowing and building penetration losses. Antenna orientation is typically not a factor as the multipath is distributed across a large angular sector.

As ρ_{s_0} is unknown, it is reasonable to assume a uniform PDF for ρ_s such that $f_s(\rho_s) = c_s$ where c_s is a constant. Consequently, P_e can be found from (6) as

$$P_e = c_s \int_0^{\rho_T} F_a(\rho_a)\,d\rho_a + c_s \int_{\rho_T}^{1/c_s} a f_a(\rho_a)\,d\rho_a. \qquad (12)$$

Now the optimum ρ_T can be found from $\partial P_e(\rho_T)/\partial \rho_T = 0$ which simplifies to

$$F_a(\rho_T) - \rho_T f_a(\rho_T) = 0. \qquad (13)$$

Equation (13) can be solved to find the optimum ρ_T. Figure 8 shows $F_a(\rho_T) - \rho_T f_a(\rho_T)$ for $M = 1, \ldots, 4$ based on a Rayleigh fading channel and $\rho_{a_0} = 10$ (dB). As can be seen from this figure, $\rho_T = \infty$ is optimum for $M = 1$. This means that a finite threshold does not exist for $M = 1$ and as such the proposed spoofing countermeasure does not reduce the spoofer effectiveness as $\rho_T = \infty$ is equivalent to a receiver with no spoofing countermeasure. However, as the diversity order increases, an optimum ρ_T other than 0 or ∞ can be found from (13). As will be shown in the next section, the optimum value of ρ_T reduces P_e and as such reduces the spoofer effective range.

3.5. Relating Observations of Spoofer Effectiveness to Physical Range.

Having evaluated P_e for various scenarios, it is of interest to determine the spoofer effectiveness as a function of the physical range. The potential target area of the spoofer as illustrated in Figure 1 is conceptually the physical region in which P_e is large enough to impact the navigation solution. In this section, an approximation of the physical range of

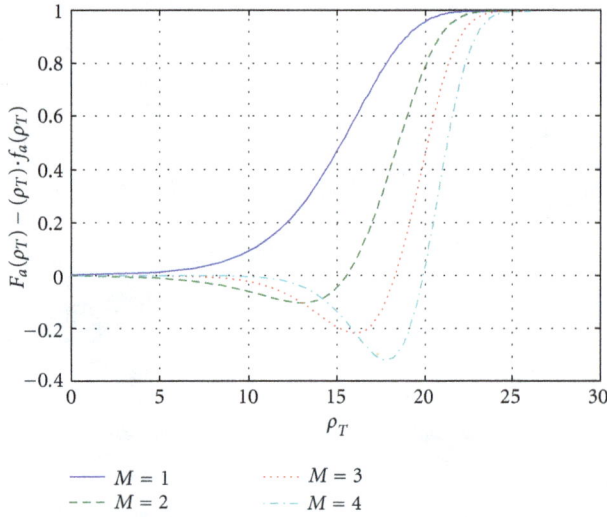

FIGURE 8: $F_a(\rho_T) - \rho_T f_a(\rho_T)$ as a function of ρ_T for various number of diversity branches based on a NLOS Rayleigh fading channel and $\rho_{a_0} = 10$ (dB).

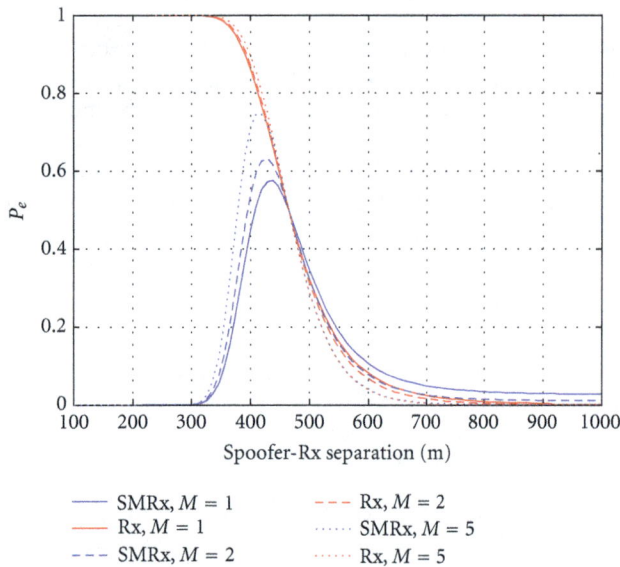

FIGURE 9: P_e as a function of spoofer-Rx separation in a LOS channel with measurement errors and based on $\rho_{a_0} = 10$ (dB), $\rho_{s_0}(R_1) = 30$ dB, and a path-loss exponent of $n = 3$.

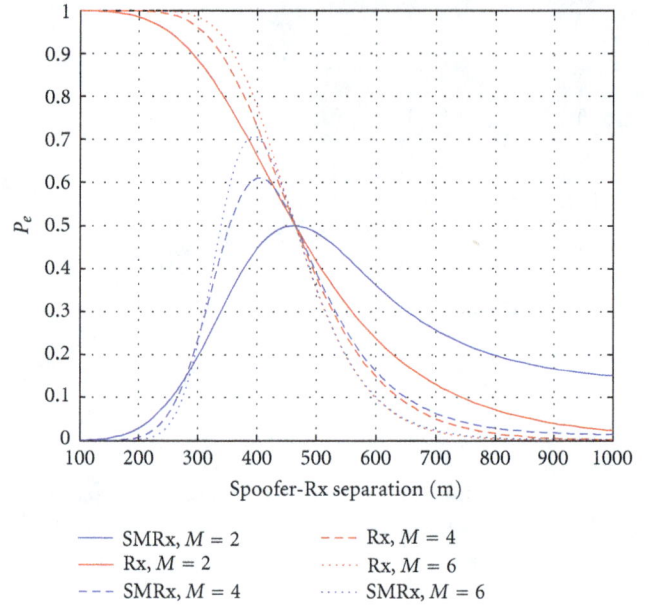

FIGURE 10: P_e as a function of spoofer-Rx separation in a Rayleigh channel and based on $\rho_{a_0} = 10$ (dB), $\rho_{s_0}(R_1) = 30$ dB, and a path-loss exponent of $n = 3$.

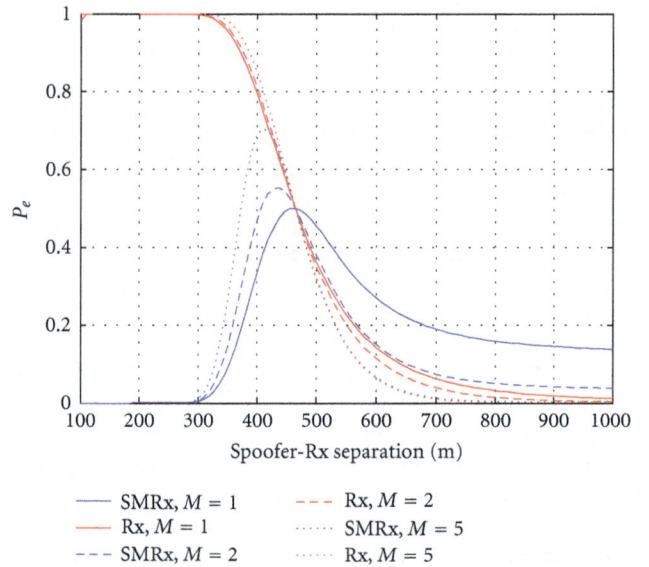

FIGURE 11: P_e as a function of spoofer-Rx separation in a Rician channel with $K_a = K_s = 1$ and based on $\rho_{a_0} = 10$ (dB), $\rho_{s_0}(R_1) = 30$ dB, and a path-loss exponent of $n = 3$.

spoofer effectiveness is determined based on the empirical path-loss model of order n as

$$\rho_{s_0} = \rho_{s_0}^{(R_1)} - 10n \log_{10}\left(\frac{d}{R_1}\right), \qquad (14)$$

where R_1 is a reference range, d is the spoofer-receiver range, n is the path-loss exponent, and $\rho_{s_0}^{(R_1)}$ is the average received spoofer CNR at $d = R_1$.

For a LOS scenario with measurement errors, the PDFs of the SS CNR and SV CNR estimates are noncentral chi-square with $2M$ DOF with M denoting the number independent

diversity branches used to estimate the CNR. P_e can therefore be found by computing ρ_T using (13) and substituting it in (6). Figure 9 shows P_e for the spoofer mitigated receiver (SMRx) as well as a conventional Rx for various spoofer-receiver separations and M. As can be seen from this figure, aSMRx significantly reduces the effectiveness of the spoofer through reducing P_e. Also observed is that the higher the diversity order M is, the more effective the spoofer mitigation

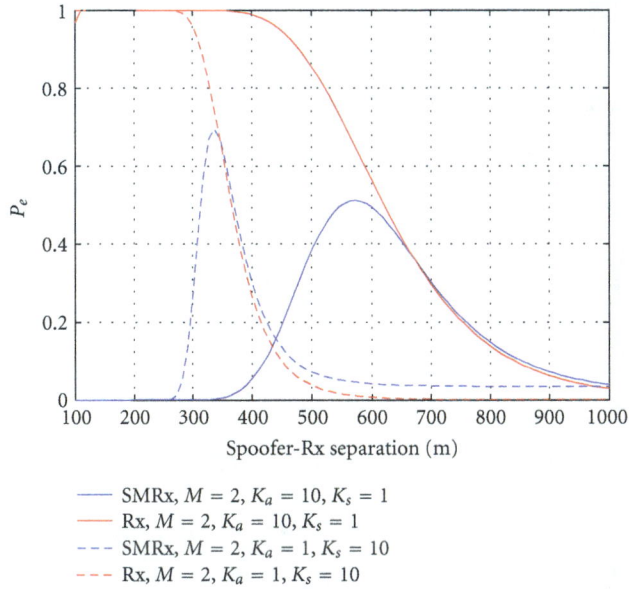

FIGURE 12: P_e as a function of spoofer-Rx separation in a Rician channel based on $\rho_{a_0} = 10\,\text{(dB)}$, $\rho_{s_0}(R_1) = 30\,\text{dB}$, and a path-loss exponent of $n = 3$.

TABLE 1: Spoofer range reduction factor (SRRF) computed for various channel scenarios based on $\rho_{a_0} = 10\,\text{dB}$, $\rho_{s_0}(R_1) = 30\,\text{dB}$.

	No. of diversity branches	Path-loss exponent (n)	K_a	K_s	SRRF (%)
Rician Ch.	$M = 1$	3	1	1	60
	$M = 2$	3	1	1	70
	$M = 5$	3	1	10	75
	$M = 5$	3	10	1	70
Rayleigh Ch.	$M = 2$	3	0	0	45
	$M = 5$	3	0	0	60
LOS	$M = 1$	3	NA	NA	74
	$M = 5$	3	NA	NA	75

is. For a Rayleigh fading channel, the PDFs of the spoofer and the authentic CNRs are central chi-square with $2M$ DOF. P_e can be found by numerically computing ρ_T from (13) and setting it in (6). Figure 10 shows P_e for an SMRx as well as a conventional Rx with no spoofing countermeasures. Note that the performance of the SMRx is significantly better than that of a conventional Rx with higher diversity branches resulting in better performance. In addition, Figures 11 and 12 compare the P_e of SMRx and Rx under a generalized Rician channel with various K-factors such that [10]

$$f_a(\rho_a; \rho_{a_0}) = \chi_{2M}'^2\left(\frac{K_a}{K_a + 1}\rho_{a_0}^2, \frac{1}{K_a + 1}\rho_{a_0} + 1\right),$$

$$f_s(\rho_s; \rho_{s_0}) = \chi_{2M}'^2\left(\frac{K_s}{K_s + 1}\rho_{s_0}^2, \frac{1}{K_s + 1}\rho_{s_0} + 1\right),$$

(15)

where K_a and K_s are the Rician K-factors associated with the SV and the SS channels, respectively. Similar to the LOS and the Rayleigh channels, a noticable improvement spoofer mitigation is realizable. In order to quantify the reduction in spoofer effective range, a heuristic metric is introduced here as

$$\text{SRRF} = \left(\frac{\int_{R_1}^{R_2} P_e^{\text{Rx}}dR - \int_{R_1}^{R_2} P_e^{\text{SMRx}}dR}{\int_{R_1}^{R_2} P_e^{\text{Rx}}dR}\right) \times 100,$$

(16)

where SRRF denotes the spoofer range reduction factor. The SRRF is computed for various channel scenarios and diversity branches and the results are summarized in Table 1.

4. Conclusions

It was shown that a relatively unsophisticated standoff spoofer can effectively disrupt a large physical area. However, processing based on estimating the CNR of the spoofer and the authentic received signals and applying a straightforward threshold rule can significantly reduce the effectiveness of the standoff spoofer. This was shown for LOS, NLOS, and Ricean multipath conditions. If the average spoofer and authentic signal power is known then the setting of ρ_T is trivial. However, if ρ_s is completely unknown then it has a finite optimum, that is, a function of ρ_a and the type of propagation environment detected by the receiver. An expression for computing the optimum ρ_T was deduced and applied to various channels. The results demonstrated the effectiveness of the proposed spoofer mitigation technique. A heuristic metric of spoofer effectiveness (SRRF) was proposed. It was shown that SRRF is reduced by up to 75% for LOS, 45% for NLOS Rayleigh $M = 2$, and 60% for NLOS Rayleigh $M = 5$ and 70% based on a Rician channel with $K_a = K_s = 1$ for $M = 2$, hence aptly demonstrating the effectiveness of the proposed countermeasure approach.

References

[1] E. D. Kaplan and C. J. Hegarty, *Understanding GPS: Principles and Applications*, Artech House, Norwood, Mass, USA, 2006.

[2] B. M. Ledvina, W. J. Bencze, B. Galusha, and I. Miller, "An in-line anti-spoofing device for legacy civil GPS receivers," in *Institute of Navigation—International Technical Meeting (ITM '10)*, pp. 868–882, San Diego, Calif, USA, January 2010.

[3] T. E. Humphreys, B. M. Ledvina, M. L. Psiaki, B. W. O'Hanlon, and P. M. Kintner, "Assessing the spoofing threat: development of a portable gps civilian spoofer," in *Proceedings of the 21st International Technical Meeting of the Satellite Division of the Institute of Navigation (ION GNSS '08)*, pp. 1198–1209, Savanna, Calif, USA, September 2008.

[4] L. Scott, "Location assurance," *GPS World*, vol. 18, no. 7, pp. 14–18, 2007.

[5] L. Scott, "Anti-spoofing and authenticated signal architetures for civil navigation systems," in *Proceedings of the ION GPS/ GNSS*, Portland, Ore, USA, September 2003.

[6] W. C. Jakes, *Microwave Mobile Communications*, IEEE Press, New York, NY, USA, 1974.

[7] F. S. T. V. Diggele, *A-GPS: Assisted GPS, GNSS, and SBAS*, Artech House, 2009.

Legend for Figure 12:
- SMRx, $M = 2$, $K_a = 10$, $K_s = 1$
- Rx, $M = 2$, $K_a = 10$, $K_s = 1$
- SMRx, $M = 2$, $K_a = 1$, $K_s = 10$
- Rx, $M = 2$, $K_a = 1$, $K_s = 10$

[8] S. Thurun, W. Burgard, and D. Fox, *Probabilistic Robotics*, MIT Press, 2006.

[9] S. Kay, *Fundamentals of Statistical Signal Processing: Detection Theory*, vol. 2, Printice-Hall, Upper Saddle River, NJ, USA, 1998.

[10] J. G. Proakis, *Digital Communications*, McGraw-Hill, New York, NY, USA, 2001.

11

Augmented Kalman Filter and Map Matching for 3D RISS/GPS Integration for Land Vehicles

Matthew Cossaboom,[1] Jacques Georgy,[2] Tashfeen Karamat,[3] and Aboelmagd Noureldin[1,3]

[1] Navigation and Instrumentation Research Group (NavINST), Electrical and Computer Engineering Department,
  Royal Military College of Canada, Kingston, ON, Canada K7K 7B4
[2] Trusted Positioning Inc., Calgary, AB, Canada T2L 2K7
[3] Navigation and Instrumentation Research Group (NavINST), Electrical and Computer Engineering Department,
  Queen's University, Kingston, ON, Canada K7L 3N6

Correspondence should be addressed to Jacques Georgy, jgeorgy@trustedpositioning.com

Academic Editor: Olivier Julien

Owing to their complimentary characteristics, global positioning system (GPS) and inertial navigation system (INS) are integrated, traditionally through Kalman filter (KF), to obtain improved navigational solution. To reduce the overall cost of the system, microelectromechanical system- (MEMS-) based INS is utilized. One of the approaches is to reduce the number of low-cost inertial sensors, decreasing their error contribution which leads to a reduced inertial sensor system (RISS). This paper uses KF to integrate GPS and 3D RISS in a loosely coupled fashion to enhance navigational solution while further improvement is achieved by augmenting it with map matching (MM). The 3D RISS consists of only one gyroscope and two accelerometers along with the vehicle's built-in odometer. MM limits the error growth during GPS outages by restricting the predicted positions to the road networks. The performance of proposed method is compared with KF-only 3D RISS/GPS integration to demonstrate the efficacy of the proposed technique.

1. Introduction

Low-cost navigation applications are highly dependent on satellite navigation systems, primarily global positioning system (GPS). It is composed of a constellation of 24 (with room to spare for some additional) satellites covering the globe in a manner that ensures continuous worldwide coverage. To obtain accurate positioning data, one must be in direct line of sight with at least four satellites. The main advantage of the GPS is that it can determine one's location, accurate to within a range of 30 m when using a single point positioning technique, and to a few centimeters when using a differential GPS technique [1–4]. However, the satellite signal can be blocked in GPS-denied environments such as urban canyons and tunnels. This is a major problem because there will be an interruption in the real-time positioning information. To overcome this navigational data gap, GPS is usually integrated with an inertial navigation system (INS) because it does not rely on any external sources [1–3]. The INS is a self-contained system consisting of three accelerometers and three gyroscopes which is mounted on the moving platform to monitor linear accelerations and angular velocities. Given the initial values of navigation parameters, the measurements from INS can be processed to determine current position, velocity, and attitude of the moving platform with respect to a certain frame of reference [4, 5]. Since higher-end INS are very expensive therefore not suitable for low-cost applications, contemporary research is focused on micro-electromechanical system- (MEMS-) based INS [6–8]. They are the key to the navigation applications where size, weight, and cost are the main concern, such as land vehicle and pedestrian navigation. However, the MEMS-based INS sensors suffer from noise, bias, and drift errors which are much more serious than the higher-grade sensors [9, 10]. Therefore, when MEMS-based INS works unaided, the performance will degrade very quickly compared to

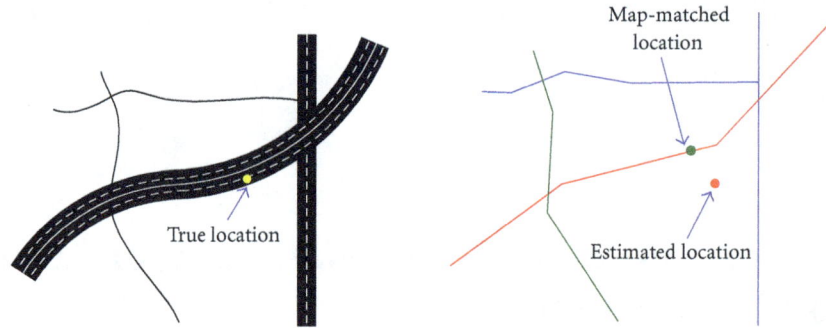

FIGURE 1: The map matching problem (adapted from [31]).

higher-grade INS [11]. Since a bias in accelerometer contributes to an error in position which is proportional to t^2 and a bias in gyroscope causes an error in position which is proportional to t^3, this research utilizes one gyroscope and two accelerometers along with vehicle's built-in odometer to get a full 3D navigational solution [12].

Integrating INS with GPS has several advantages because they possess complementary error characteristics. GPS bounds the INS drift in the long run whereas INS fills the GPS data gaps during GPS signal interruption. The traditional method of INS/GPS integration is Kalman filtering (KF) which can be implemented in a loosely coupled or a tightly coupled manner. The loosely coupled scheme of integration requires at least four satellites for GPS measurement update whereas tightly coupled integration can benefit from GPS even when only one satellite is available. However, tightly coupled approach is much more complex to implement and hardly any superior when more than three satellites are visible. Both the aforementioned approaches can be implemented in open- and closed-loop fashion. Open-loop filters do not use feedback; the input data does not use corrections whereas closed-loop filters use the previous corrections to minimize the approximation errors [13, 14].

KF uses a linearized system and measurement models. KF techniques suffer from divergence during outages due to approximations during the linearization process, especially when utilizing MEMS-based inertial measurement units (IMUs). This problem can be avoided by using particle filter (PF) which enhances the performance of the MEMS-based INS by including the nonlinearities in the system and measurement models [15, 16]. Particle filtering is a nonlinear filtering technique that does not require the system model to be linearized. It can accommodate for arbitrary sensor characteristics, motion dynamics, and noise distribution because of its ability to deal with nonlinear non-Gaussian models [17, 18]. Other methods of integration have been investigated based on artificial intelligence (AI), also known as neural network (NN) [6, 19–24]. The major challenge of PF or the AI method is the fact that they are computationally expensive and may not be useful in some applications.

Map matching (MM) is the process of utilizing a digital road network map database to improve the predicted position errors during integration [25–29]. Motivated by the simplicity and drawbacks of KF, this research will focus on

reducing the KF integration errors by utilizing MM. The goal of MM is to match the estimated location with the road network map [30]. Figure 1 gives a good representation of the MM approach [31]. The left diagram displays the person's actual location on the actual streets whereas the right diagram displays the set of estimated arcs (digital road networks) with the estimated location and the MM location. This example uses a piecewise linear solution to estimate the arcs in the roads.

2. RISS/GPS Integration Using KF

KF uses a linearized system model and has several limitations. It requires a stochastic model of the inertial sensor errors and a priori information about the data covariance provided by both inertial system and GPS [6, 32]. KF techniques suffer from divergence during outages due to approximations during the linearization process, especially when utilizing MEMS-based IMUs. As a result, the inertial-based position and velocity errors could grow quite significantly. The details of traditional KF and derivation of its equations can be found in some excellent texts such as [33–36]. However, a brief overview of KF equations is presented.

KF operates in two distinctive stages: (1) prediction stage and (2) update stage. In the prediction stage, a new prediction of the error states (1) and error covariance (2) are determined for the next time step. The equations for the prediction stage are as follows [35, 36].

Prediction of error states:

$$\hat{x}_k^- = \Phi_{k,k-1}\hat{x}_{k-1}^+, \tag{1}$$

where $\hat{x}_k$ is the error state vector, $(-)$ indicates the *a priori* estimate while *posteriori* estimate is indicated with $(+)$. The Φ is the state transition matrix. The predicted error covariance is expressed as follows.

Prediction of the error covariance:

$$P_k^- = \Phi_{k,k-1}P_{k-1}^+\Phi_{k,k-1}^T + G_{k-1}Q_{k-1}G_{k-1}^T, \tag{2}$$

where G is the noise distribution matrix and Q is the covariance of the system noise.

In the update stage, the KF makes corrections to the predicted state estimate based on new information from the GPS measurements. These corrections are appropriately weighed though Kalman gain (3) which determines if the prediction or the measurement should be trusted more. Then the Kalman gain is used to update the state estimate (4) and error covariance matrix (5) as the posteriori estimate for the next prediction stage.

Kalman gain:

$$K_k = \left(P_k^-\right)H_k^T\left(H_k P_k^- H_k^T + R_k\right)^{-1}, \tag{3}$$

where R is the covariance of the measurement noise and H is the measurement design matrix.

Updating of error states:

$$\hat{x}_k^+ = \hat{x}_k^- + K_k\left(Z_k - H_k\hat{x}_k^-\right), \tag{4}$$

where Z is the difference between the INS and GPS position and velocity components.

Updating of error covariance matrix:

$$P_k^+ = (I - K_k H_k)P_k^-, \tag{5}$$

where I is the identity matrix.

This research used a loosely coupled 3D RISS/GPS integration approach. Loosely coupled integration helps assessing the map matching better because if we use the tightly coupled integration then during the partial GPS outage there will be two or three satellites which will help the solution as well as the map matching. Therefore to be able to better assess the map matching, we focus on loosely coupled integration because in loosely coupled integration, there is no satellites at all during the outage and enhancement contribution will come from map matching algorithm. As mentioned earlier, due to the complex error characteristics of MEMS-based sensors, this paper uses a different configuration of inertial sensors where only one gyroscope and two accelerometers along with the vehicle's built-in odometer are used to obtain a three-dimensional navigation solution [12, 37]. This is termed as *reduced inertial sensor system* (RISS) as opposed to *full IMU* which uses three gyroscopes and three accelerometers. The 2D RISS was first introduced in [38] where a KF was used for 2D RISS/GPS integration, a PF for 2D RISS/GPS integration was proposed in [39]. The 3D RISS was first introduced in [12], together with its full derivation, and its detailed advantages over a full-IMU-based solution and over 2D solutions. A tightly coupled KF 3D RISS/GPS integration solution was proposed in [37]. As explained in the aforementioned literature, there are only three sensors contributing to the errors versus six. A gyroscope is mounted so that its axis is aligned with the vertical axis of the vehicle to obtain the azimuth, and the vehicle odometer provides the forward speed [37]. Two accelerometers, instead of gyroscopes, are used to compute the pitch and roll angles. They are aligned with the forward and transversal axes of the vehicle body frame. The pitch and azimuth angles are used to calculate the velocities and then the position components can be calculated.

The azimuth angle is calculated by integrating the gyroscope measurement ω_z, as shown in (6). This measurement includes the component of the Earth rotation and rotation of the local level frame on the Earth's curvature, these quantities are removed from the measurement before integration [40],

$$\dot{A} = -\left[\omega_z - \omega^e \sin\varphi - \frac{v_e\tan\varphi}{R_N + h}\right], \tag{6}$$

where ω^e is the Earth's rotation rate, φ is latitude, v_e is the east velocity, R_N is the normal radius of the earth ellipsoid, and h is altitude.

When the vehicle is moving, the forward accelerometer measures the forward vehicle acceleration as well as the component due to gravity. Therefore, the following relationship is used to calculate the pitch angle:

$$p = \sin^{-1}\left(\frac{f_y - a_{od}}{g}\right), \tag{7}$$

where f_y is the forward accelerometer measurement, a_{od} is the odometer-derived acceleration, and g is the Earth's gravity.

The transversal accelerometer measures the normal component of the vehicle acceleration and the component due to gravity. Therefore roll angle is computed as follows:

$$r = -\sin^{-1}\left(\frac{f_x + v_{od}\omega_z}{g\cos p}\right), \tag{8}$$

where f_x is the transversal accelerometer measurement and v_{od} is the odometer measurements.

The three velocities (east v_e, north v_n, and up v_u) are calculated using A, p, and v_{od} through the following relationship:

$$\mathbf{v} = \begin{bmatrix} v_e \\ v_n \\ v_u \end{bmatrix} = \begin{bmatrix} v_{od}\sin A\cos p \\ v_{od}\cos A\cos p \\ v_{od}\sin p \end{bmatrix}. \tag{9}$$

Then the time rate of change of the position components can be obtained as follows:

$$\dot{\mathbf{r}}^l = \begin{pmatrix} \dot{\varphi} \\ \dot{\lambda} \\ \dot{h} \end{pmatrix} = \begin{pmatrix} 0 & \frac{1}{R_M + h} & 0 \\ \frac{1}{(R_N + h)\cos\varphi} & 0 & 0 \\ 0 & 0 & 1 \end{pmatrix} \begin{pmatrix} v^e \\ v^n \\ v^u \end{pmatrix} = D^{-1}\mathbf{v}^l, \tag{10}$$

where λ is the longitude and R_M is the meridian radius of the earth ellipsoid.

When RISS is integrated with GPS using a KF to create a 3D position solution, the error state vector has nine error states. They are latitude, longitude, and altitude errors ($\delta\varphi$, $\delta\lambda$, δh), the east, north and up velocity errors (δv_e, δv_n, δv_u), the azimuth error δA, the gyroscope error $\delta\omega_z$, and the error from the odometer acceleration δa_{od}. The stochastic errors associated with the gyroscope and the odometer-derived acceleration are modeled by Gauss-Markov model where γ_{od} is the inverse of the autocorrelation time for the odometer-derived acceleration noise, σ_{od}^2 is the variance of odometer-derived acceleration noise, β_z is the inverse of the autocorrelation time for the gyroscope noise, and σ_z^2 is the variance of the gyroscope noise. The complete error state

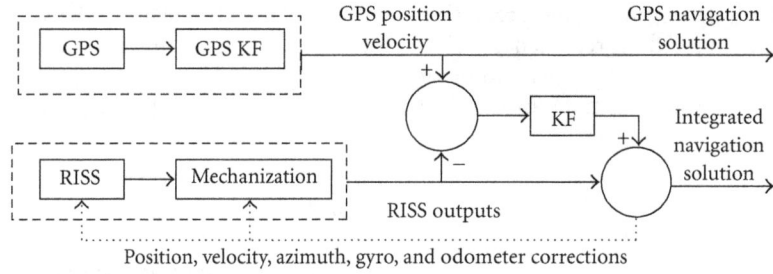

FIGURE 2: Loosely coupled RISS/GPS KF integration diagram.

system model is expressed as follows with complete detail shown in (12):

$$\delta\dot{\mathbf{x}}_{\text{RISS}} = F_{\text{RISS}}\delta\mathbf{x}_{\text{RISS}} + G_{\text{RISS}}w_{\text{RISS}}, \qquad (11)$$

$$\delta\dot{\mathbf{x}}_{\text{RISS}} = \begin{bmatrix} \delta\dot{\varphi} \\ \delta\dot{\lambda} \\ \delta\dot{h} \\ \delta\dot{v}_e \\ \delta\dot{v}_n \\ \delta\dot{v}_u \\ \delta\dot{A} \\ \delta\dot{a}_{\text{od}} \\ \delta\dot{\omega}_z \end{bmatrix} = \begin{bmatrix} 0 & 0 & 0 & 0 & \dfrac{1}{(R_M+h)} & 0 & 0 & 0 & 0 \\ 0 & 0 & 0 & \dfrac{1}{(R_N+h)\cos\varphi} & 0 & 0 & 0 & 0 & 0 \\ 0 & 0 & 0 & 0 & 0 & 1 & 0 & 0 & 0 \\ 0 & 0 & 0 & 0 & 0 & 0 & a_{\text{od}}\cos A\cos p & \sin A\cos p & 0 \\ 0 & 0 & 0 & 0 & 0 & 0 & -a_{\text{od}}\sin A\cos p & \cos A\cos p & 0 \\ 0 & 0 & 0 & 0 & 0 & 0 & 0 & \sin p & 0 \\ 0 & 0 & 0 & 0 & 0 & 0 & 0 & 0 & 1 \\ 0 & 0 & 0 & 0 & 0 & 0 & 0 & -\gamma_{\text{od}} & 0 \\ 0 & 0 & 0 & 0 & 0 & 0 & 0 & 0 & -\beta_z \end{bmatrix} \begin{bmatrix} \delta\varphi \\ \delta\lambda \\ \delta h \\ \delta v_e \\ \delta v_n \\ \delta v_u \\ \delta A \\ \delta a_{\text{od}} \\ \delta\omega_z \end{bmatrix} + \begin{bmatrix} 0 \\ 0 \\ 0 \\ 0 \\ 0 \\ 0 \\ 0 \\ \sqrt{2\gamma_{\text{od}}\sigma_{\text{od}}^2} \\ \sqrt{2\beta_z\sigma_z^2} \end{bmatrix} w_R. \qquad (12)$$

where $\delta\dot{\mathbf{x}}_{\text{RISS}}$ is the state vector, F_{RISS} is the 9×9 dynamic coefficient matrix, G_{RISS} is the 9×1 noise coupling vector, and w_{RISS} is the unit variance white Gaussian noise.

In order to provide optimal estimation of the above error state vector $\delta\dot{\mathbf{x}}_{\text{RISS}}$, observations for the above system can be provided in the following form:

$$\delta\mathbf{z} = H\delta\mathbf{x}_{\text{RISS}} + \gamma, \qquad (13)$$

where $\delta\mathbf{z}$ is the observations vector giving the difference between the RISS and GPS positions and velocities, H is the design matrix giving the ideal noiseless relationship between the observations vector and the state vector, and γ is the vector of observations random noise, which is assumed to be white sequence not correlated with the RISS system noise. For the RISS proposed in this study, the parameters of the measurement model are given as follows:

$$\delta\mathbf{z} = \begin{bmatrix} \varphi_{\text{RISS}} - \varphi_{\text{GPS}} \\ \lambda_{\text{RISS}} - \lambda_{\text{GPS}} \\ h_{\text{RISS}} - h_{\text{GPS}} \\ v_{e,\text{RISS}} - v_{e,\text{GPS}} \\ v_{n,\text{RISS}} - v_{n,\text{GPS}} \\ v_{u,\text{RISS}} - v_{u,\text{GPS}} \end{bmatrix}. \qquad (14)$$

The measurement design matrix H would be 6×9 for the position and velocity error states, and can be written as follows:

$$H = \begin{bmatrix} 1 & 0 & 0 & 0 & 0 & 0 & 0 & 0 & 0 \\ 0 & 1 & 0 & 0 & 0 & 0 & 0 & 0 & 0 \\ 0 & 0 & 1 & 0 & 0 & 0 & 0 & 0 & 0 \\ 0 & 0 & 0 & 1 & 0 & 0 & 0 & 0 & 0 \\ 0 & 0 & 0 & 0 & 1 & 0 & 0 & 0 & 0 \\ 0 & 0 & 0 & 0 & 0 & 1 & 0 & 0 & 0 \end{bmatrix}. \qquad (15)$$

The covariance of the measurement noise matrix R would be a 6×6 matrix consisting of the position and velocity measurement error covariance. Figure 2 shows the loosely coupled RISS/GPS KF integration scheme.

3. Map Matching

Map matching (MM) is the process of utilizing a digital road network map database to improve the predicted position errors during integration. Motivated by the simplicity and drawbacks of KF, this research will focus on reducing the KF integration errors by utilizing MM. The goal of MM is to match the estimated location with the road network map. There have been many different approaches and algorithms to the MM problem that have been researched [26, 27]. This paper focuses on three main algorithms from [31]. These are point-to-point matching, point-to-curve matching, and curve-to-curve matching.

The point-to-point matching algorithm is basically like a search problem [30]. The algorithm matches the estimated location, P, to the closest node or point in the network. This could take a lot of time to calculate the distance from P to every node in the network. Therefore, the user must identify those nodes that are within a certain distance of P, and only calculate those distances. The distance is dependent on the type of data being use and it is up to the user to determine it. In this research, a distance of 1000 meters from the current prediction solution was used, which will be discussed later in the next section. Point-to-point matching is very simple to implement and fast, but it does have some problems during execution. The algorithm is very sensitive to how the network is digitized.

The point-to-curve matching algorithm tries to identify the curve (arc) that is closest to P, rather than the point. The same problem arises with the amount of time to calculate the distance from P to every arc in the network. Therefore, the user must identify those arcs that are within a certain distance of P, and only calculate those distances. The network uses a piecewise linear solution to estimate the arcs in the roads, hence the algorithm must find the minimum distance from P to each of the line segments and select the smallest. The method used in the research is a combination of point-to-point matching and point-to-curve matching because of the format of the map data used.

The final approach is curve-to-curve matching, which considers the estimated location as a curve, P, consisting of points $P^0, P^1, P^2, \ldots, P^m$. Then it matches it to the closest arc, which requires some measure of distance between curves. There are different ways to measure the distance between two curves. One way is determining the minimum distance and matching it to that curve. Another technique is measuring the average distance between the curves.

As described above, there are many different techniques for MM. The algorithm is heavily dependent on how the data or network is structured. This was a very important challenge to overcome during this research. The method used in this paper is a combination of point-to-point matching and point-to-curve matching, which is dictated by the format of the map data used.

4. Development of the Augmented KF/MM for RISS/GPS Integration for Land Vehicles

The map data used in this research is integrated with inertial sensor measurements through KF for reliable positioning during GPS outages. The map data was provided by the Queen's university, Kingston ON, which was the 2009 street data as a part of the Arc Geographic Information System (ArcGIS) software produced by Environmental Systems Research Institute Incorporated (ESRI). The data was in shape files that consisted of latitude and longitude coordinates and included all types of road ways: highways, rural roads, and urban roads. The coordinates were the start points

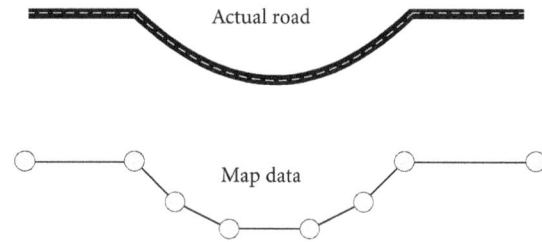

FIGURE 3: Representation of the map data during a turn.

and end points of line segments of every road. The data used a piecewise linear solution to estimate the arcs in the roads. Whenever there was a turn in the road a new line segment was started and completed. Therefore the length of the line segments varied depending on how straight or curved the road was. Figure 3 gives a representation of the map data. The size of the data was limited to the area of the trajectory that was being experimented, which was the Kingston area. This was very important because it reduced the actual size of the data, which would affect the process time of the MM algorithm. The map data was a large database of street line segments.

The initial setup of the map data was a very crucial step in this research. The data was already in latitude and longitude coordinates which was a very good start. It was in a shape file format, which was easily loaded into MATLAB 2009 using the Mapping Toolbox. A shape file is a geospatial vector data format for geographic information systems software. The data was then converted into x and y coordinates in metres. The x and y coordinates are the distances being travelled along the East and North directions. This conversion had to take into account a certain reference point, which was chosen as the start point of the trajectory. The equations below were used for the Easting and Northing calculations into metres:

$$y = \text{North}_{\text{metres}} = (\varphi - \varphi_{\text{initial}})(R_M + h), \quad (16)$$

$$x = \text{East}_{\text{metres}} = (\lambda - \lambda_{\text{initial}})(R_N + h)\cos\varphi, \quad (17)$$

where φ_{initial} and, λ_{initial} are the latitude and longitude of the point chosen to be the origin of the Cartesian coordinates and h is the altitude.

Then the slope (m) and the y-intercept (b) for each line segment were calculated. The slope was calculated using the following equation:

$$m = \frac{(y_2 - y_1)}{(x_2 - x_1)}. \quad (18)$$

The y-intercept was calculated using the equation of a straight line, $y = mx + b$, which was rearranged to solve for b as shown below:

$$b = y - (mx). \quad (19)$$

All of these calculations were completed in a simple algorithm with MATLAB, and once completed the results were stored and saved in a database. A representation of how the

TABLE 1: Map data setup.

Line segment start point		Line segment end ooint		Slope (m)	Y-intercept (b)
X_1	Y_1	X_2	Y_2	m of the line segment	b of the line segment
X_2	Y_2	X_3	Y_3	m of the line segment	b of the line segment
X_3	Y_3	X_4	Y_4	m of the line segment	b of the line segment
X_4	Y_4	X_5	Y_5	m of the line segment	b of the line segment
$\vdots$	$\vdots$	$\vdots$	$\vdots$	$\vdots$	$\vdots$
New road begins					
X_1	Y_1	X_2	Y_2	m of the line segment	b of the line segment

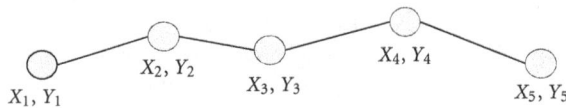

FIGURE 4: Road using four line segments.

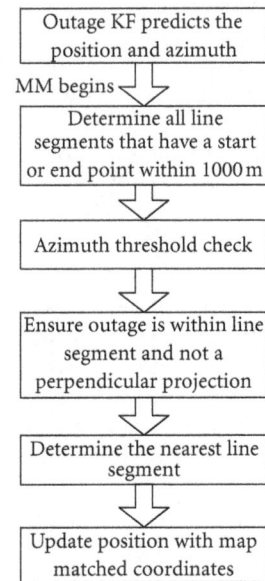

FIGURE 5: Map matching algorithm flowchart.

data was stored is shown in Table 1. The first four rows in Table 1 represent a road or street that contains four line segments and five sets of coordinates as shown in Figure 4. The database was quite large and would be used as look-up table as part of the MM algorithm.

The MM algorithm is greatly dependent on the accuracy of the map data. The data acquired was 2009 street data which seems fairly new but there are more things to consider. Highways and roads always have maintenance and construction frequently going on. An example is Highway 401 near Kingston, ON, which is being expanded to accommodate more lanes of traffic. Changes like this will greatly affect the accuracy of the results of MM. However, regularly updating the MM data would mitigate this effect.

Moreover, the size of the data is another limitation of using map data. The size of the data for Kingston, ON, and the surrounding area including Gananoque, ON, is approximately 2.1 megabytes. This does not seem very large but when it is being used as a look-up database, processing time will be increased, especially when including larger areas to cover.

4.1. Map Matching Algorithm. The map matching algorithm developed during this research was the main contribution. During GPS outages, the KF solution still had an error drift due to the inertial sensors errors (including bias drift and scale factor instability). The purpose for the development of the MM algorithm was to improve this position error drift during GPS outages. The method used in the paper is a combination of point-to-point matching and point-to-curve matching because of the setup of the map data used. The results will compare the standalone KF results to the KF/MM results.

Figure 5 is a flowchart of the MM algorithm that was developed. The algorithm consists of five steps. Initially when there is a GPS outage, the KF will go into the prediction stage, and it will predict the position errors, velocity errors, and the azimuth errors based on the dynamic error model.

The position, velocity, and azimuth components are then obtained after removing these errors. The MM algorithm will be then called as shown in Figure 5.

The first step is to determine all the line segments that have a start or end point within 1000 metres from the GPS outage, and store these line segments. The second step is the azimuth threshold check, which stores all the remaining line segments that pass this check. The third step is to ensure that the GPS outage is within the line segment and does not perpendicularly intersect the line outside of the line segment. The fourth step is determining the nearest line segment from the GPS outage. This step could contain many line segments or only a few, depending on how many segments made it through the first three steps. Final step is the map matching step where position, latitude, and longitude are updated or matched with the coordinates on the nearest line segment. These five steps will be discussed in detail in the next five sub-sections.

4.1.1. Finding All Line Segments within a Certain Distance. The first step in the MM algorithm is to determine all the line segments that have a start point or end point within 1000

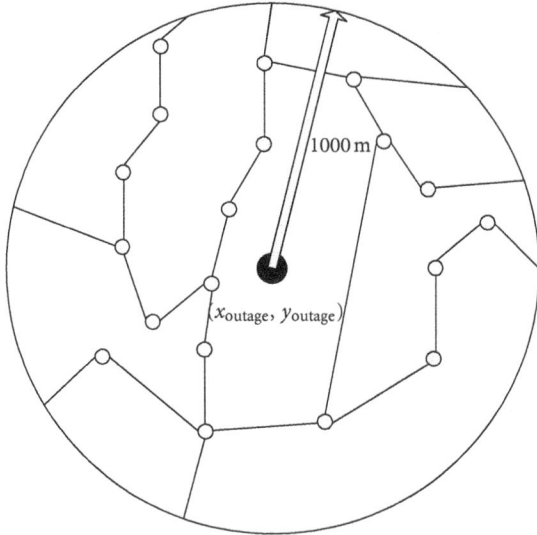

FIGURE 6: Map matching representation after step one.

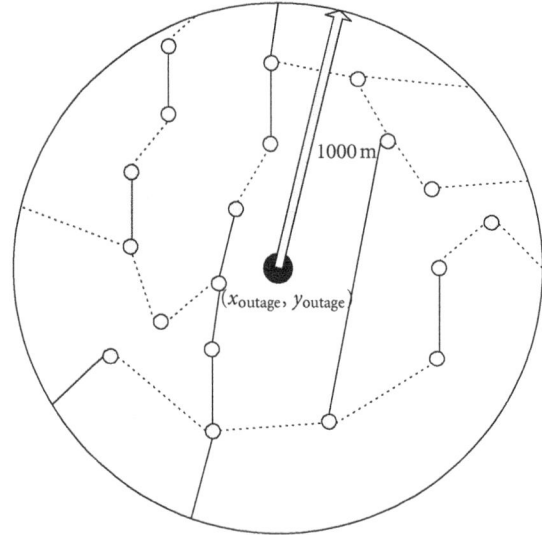

FIGURE 7: Map matching representation after step two.

metres from the GPS outage. When a line segment meets these criteria, all the line segment information, start and end point coordinates, slope, and y-intercept are stored in a new database. The distance equations used are given as follows:

$$\text{Distance}_{\text{start}} = \sqrt{\left(x_{\text{outage}} - x_1\right)^2 + \left(y_{\text{outage}} - y_1\right)^2},$$

$$\text{Distance}_{\text{end}} = \sqrt{\left(x_{\text{outage}} - x_2\right)^2 + \left(y_{\text{outage}} - y_2\right)^2}, \tag{20}$$

where the GPS outage latitude and longitude coordinates are converted to x_{outage} and y_{outage} using (16) and (17), respectively.

Figure 6 is the start of a map matching example that will be used to demonstrate the five steps of the MM algorithm. After step one, the line segments are reduced to the segments that have a start or end point within 1000 metres from the GPS outage location.

As displayed in Figure 6, there is a possibility to have many line segments that meet the criteria in step one of the algorithm which are stored and carried over to step two.

4.1.2. Azimuth Threshold Check. The second step of the MM algorithm is the azimuth angle threshold check. Azimuth or heading is defined as the horizontal angle measured clockwise from any fixed reference plane.

During this step the azimuth angle of each line segment, carried over from step one, is calculated. The azimuth angle is calculated as follows:

$$dx = x_2 - x_1,$$
$$dy = y_2 - y_1,$$
$$Az_{\text{line segment}} = \tan^{-1}\left(\frac{dx}{dy}\right). \tag{21}$$

Both directions of the line segment are compared to predicted KF azimuth angle. These two directions have heading

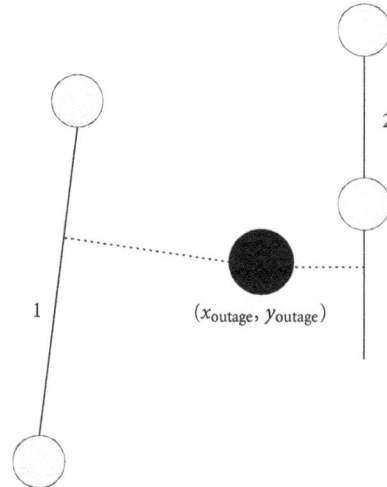

FIGURE 8: Perpendicular line verification.

of $Az_{\text{line segment}}$ and $Az_{\text{line segment}} + 180°$. The threshold of azimuth verification could be changed but typically a threshold of $40°$ was used. Only the line segments below this threshold are kept and the rest are rejected. Figure 7 indicates the line segments with dashes that are removed by the azimuth threshold check.

4.1.3. GPS Outage Position Check. The third step of the algorithm is to find between which line segments the GPS outage actually lies. Figure 8 gives an excellent example of the perpendicular line verification. The outage is closer to line segment 2 but it is not in between the two points of the segment. This step will remove the line segments that the GPS outage does not fall within. However, this is done with a tolerance so that line segments which are much closer could also be considered.

This verification is completed by calculating the coordinates where a perpendicular line from the GPS outage would

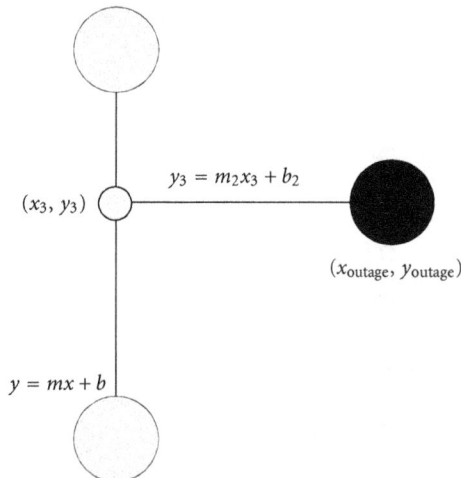

FIGURE 9: Representation of perpendicular intersection.

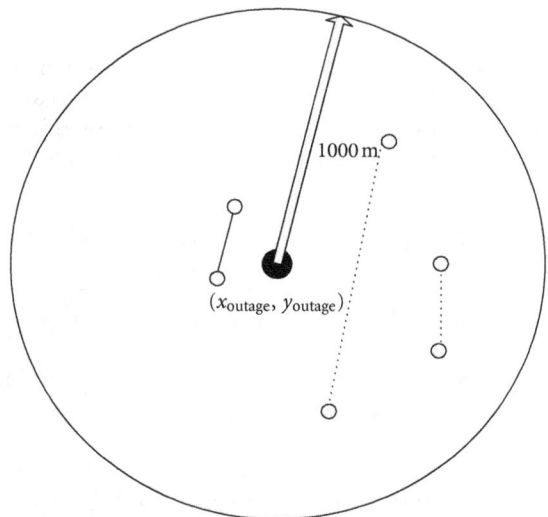

FIGURE 10: Map matching representation after step three.

intersect the line segment. If these coordinates fall within the line segment, that line segment is stored and carried over to the fourth step. The following equations are used to complete this step. We first start with calculating the normal slope of the perpendicular line as follows:

$$m_2 = -\left(\frac{1}{m_{\text{line segment}}}\right). \tag{22}$$

This is then followed by calculating the y-intercept of the perpendicular line with respect to the GPS outage,

$$b_2 = y_{\text{outage}} - \left(m_2 x_{\text{outage}}\right). \tag{23}$$

Consequently, if the two straight lines are made equal, x_3 can be solved for,

$$mx_3 + b = m_2 x_3 + b_2,$$
$$x_3 = \frac{(b_2 - b)}{(m - m_2)}. \tag{24}$$

To solve for y_3, just use the equation of a straight line,

$$y_3 = mx_3 + b. \tag{25}$$

In the above equation, (x_3, y_3) are the coordinates where a perpendicular line from the outage would intersect the line segment. Figure 9 gives an illustration of the above procedure.

Then to verify if (x_3, y_3) are on the line segment, the length of the line segment is compared to the distance of the start point to (x_3, y_3). The same comparison is done with the end point of the line segment. If the length of the line segment is greater than both, then the line segment is stored and carried over to step four.

Figure 10 displays the line segments that are removed (dashed line segments) by the outage position check.

4.1.4. Determine the Nearest Line Segment.
The fourth step of the algorithm is just a basic calculation to determine the

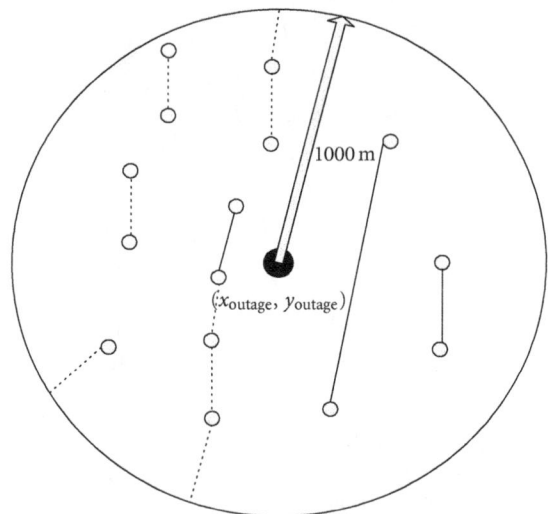

FIGURE 11: Map matching representation after step four.

closest line segment to the GPS outage. The perpendicular distance from the outage to the line segment is calculated as shown:

$$d_{\text{perpendicular}} = \left|\frac{-mx_{\text{outage}} + y_{\text{outage}} - b}{\sqrt{m^2 + 1}}\right|. \tag{26}$$

This perpendicular distance is calculated for all the remaining line segments and then the line segment with the smallest perpendicular distance is selected as the match. Figure 11 displays the line segments that are removed (dashed line segments). The solid line segment is selected for the map matching.

4.1.5. Update the Position.
The fifth step is the final step of the proposed algorithm and provides the actual map matching. Here the position, latitude, and longitude are updated or matched with the perpendicular coordinates, (x_3, y_3), on the nearest line segment. This is shown in Figure 12.

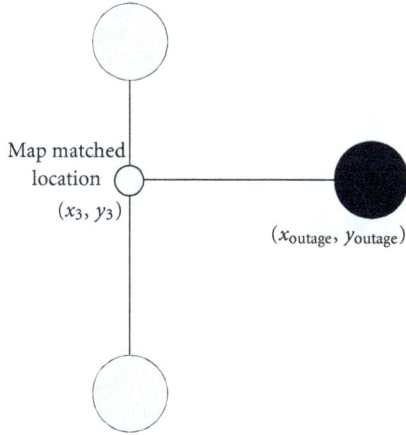

FIGURE 12: Map matching step.

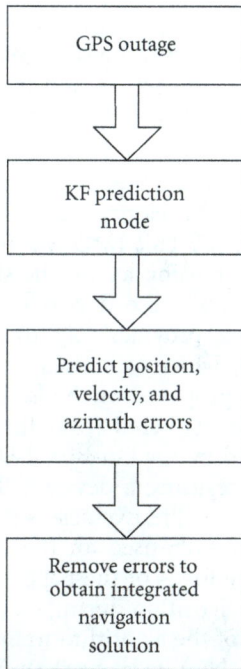

FIGURE 13: KF prediction stage block diagram.

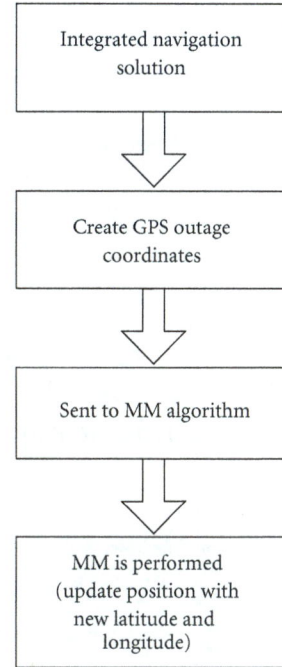

FIGURE 14: Integration of MM algorithm with KF-based solution.

FIGURE 15: Data collection equipment mounted inside the road test vehicle.

The (x_3, y_3) coordinates must be converted to latitude and longitude. For the latitude, this is done by rearranging (16) as follows:

$$\varphi = \left(\frac{\text{North}_{\text{metres}}}{R_M + h} \right) + \varphi_{\text{initial}}. \tag{27}$$

For the longitude, the conversion is done by rearranging (17) as follows:

$$\lambda = \left(\frac{\text{East}_{\text{metres}}}{(R_N + h) \cos \varphi} \right) + \lambda_{\text{initial}}, \tag{28}$$

where $\text{North}_{\text{metres}}$ is y_3 and $\text{East}_{\text{metres}}$ is x_3.

4.2. Integration of MM Algorithm with KF. The above MM algorithm is integrated with the KF-based method of RISS/GPS integration. While at least 4 GPS satellites are visible, GPS will provide update to the KF. Initially when there is a GPS outage, the KF will go into the prediction stage and it will predict the position errors, velocity errors, and the azimuth errors based on the dynamic error model. The integrated navigated solution (position, velocity, and azimuth components) is then obtained after removing these errors. Thle above procedure is shown by the block diagram in Figure 13.

The latitude and longitude from the integrated navigation solution are sent to the MM algorithm as the GPS outage coordinates. Then the MM algorithm commences and the integrated navigation solution is updated by the map matched position. This is shown by the block diagram in Figure 14.

The map matching algorithm discussed above is very intuitive which was implemented with the KF algorithm that was already developed by our research group. As will be

Figure 16: Kingston downtown trajectory, ten outages are shown by circles.

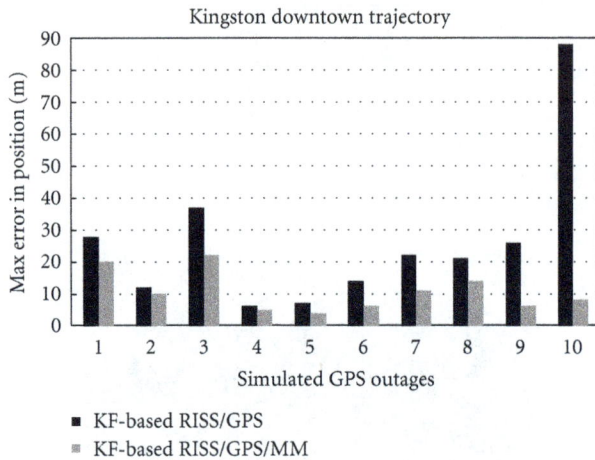

Figure 17: Maximum error in position (m) for Kingston downtown trajectory.

shown in the next section, the KF results could not compensate all the errors caused by the inertial sensors whereas MM algorithm mitigated most of the errors and improved the navigational solution to a great extent which was limited mostly by the accuracy of the map data used.

5. Experimental Results

This section introduces the equipment used and describes the road tests performed to assess the efficacy of the MM algorithm. The results will be shown with the different trajectories that were examined. The focus will be on Kingston area trajectories, including downtown, rural, and highway driving. The results of the proposed method, augmented KF/MM integration, will be discussed in detail and compared to the results of the traditional method of KF-based RISS/GPS integration for land vehicles. The developed method was examined through real road test trajectories by introducing GPS outage at various places encompassing the scenarios of a typical road trajectory.

Crossbow IMU300CC MEMS-based inertial sensors were used for the experiments [41]. The IMU is a six-degree-of-

freedom inertial system that uses solid-state devices to measure the angular rate and linear acceleration. This IMU was utilized in RISS architecture, and the performance was examined on real road data collected over various trajectories.

The reference solution used to evaluate the proposed method is based on the Honeywell HG1700 AG11 high-end tactical-grade IMU. This IMU was integrated with the NovAtel GPS receiver using an off-the-shelf assembly, the G2 Pro-Pack SPAN unit, also developed by NovAtel [42]. This integrated system provides a tightly coupled RISS/GPS navigation solution, which was used as the reference for comparisons of the proposed methods. The forward speed (odometer data) was gathered from the vehicle's built in sensors and collected by the On-Board Diagnostics version II (OBD II) interface using a device called CarChip [43]. The setup inside the road test vehicle is shown in Figure 15. It may be noted that GPS used for the system is of higher quality; however, the focus of the paper was not to see the performance of the algorithm during inaccurate readings of GPS but the ability of the algorithm to bridge the complete GPS outages. Since the outages were simulated, the quality of GPS is not a main factor to consider here, especially when the outages simulate total blockage of the GPS signals.

It may be noted that the trajectory figures were created using GPS Visualizer [44] which uses Google maps and suffers from small errors due to which even the reference trajectories sometimes seems off road, especially when zoomed in. However, for calculation purposes, reference trajectory is considered as the best solution. Another point worth noting is that the map data does not match perfectly with the reference solution. One of the most obvious reasons for this is that the map data uses a piecewise linear solution to estimate the arcs in the roads, whereas the reference solution was taken from the integrated RISS/GPS solution provided by the NovAtel SPAN unit which is mounted inside the vehicle and produces data at 100 Hz producing which is virtually continuous. Also, the map data used one single line segment (down the centre line) for urban and rural streets. It did not have a line segment for both directions. Therefore, there is

FIGURE 18: GPS outage three during the Kingston downtown trajectory.

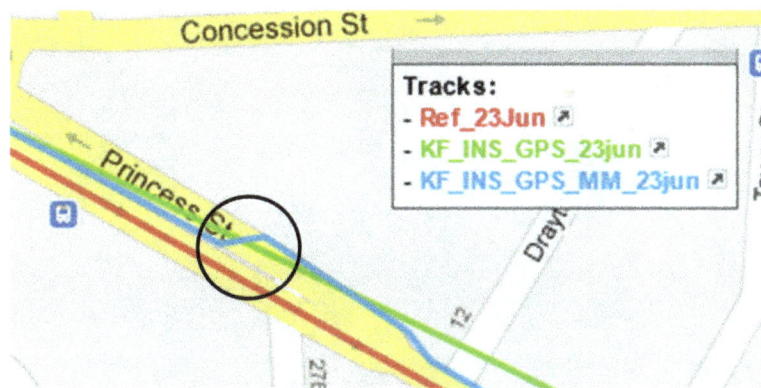

FIGURE 19: Zoomed-in portion of MM error during GPS outage three.

already a small margin of error between the map data and the reference solution.

5.1. Kingston Downtown Trajectory. The first trajectory examined pertained to downtown Kingston, ON. The majority of this trajectory is urban driving. There were ten intentionally introduced GPS outages of 60 seconds to examine the performance during the outages, focusing on areas like sharp turns and curves in the road which represent the most demanding scenarios for the proposed MM approach. Another challenging feature of this trajectory was its variable speed with frequent stops and sudden accelerations. The velocity of the vehicle was constantly changing due to traffic lights, pedestrians, and sharp turns. Figure 16 displays the downtown Kingston trajectory with GPS outages depicted with circles during which the results of the proposed KF/MM method were compared to the standalone KF and reference solutions.

The maximum position error (meters) for all the ten outages of the downtown Kingston trajectory are shown by the bar graph in Figure 17. Both solutions, the KF based and KF/MM based, were compared against the reference solution, which is shown in Figure 16. It may be noticed that the proposed KF/MM method showed an improvement over the standalone KF method in all the ten GPS outages. During outage 10, the proposed KF/MM showed the most

improvement (90%) which happened to be during a turn. The KF solution had a maximum position error of 88 m whereas the KF/MM solution differed only 8 m from the reference trajectory. For GPS outage seven, we observed a maximum position error of 22 m for the KF-based solution and only 11 m for the KF/MM solution which is an improvement of 50%. We will now take a closer look at three of the GPS outages and discuss how MM greatly improved the results.

GPS outage three is a good example of an outage occurring during a turn at a higher speed which is depicted in Figure 18. The trend of KF solution is similar to GPS outage one where it constantly drifts away from the reference whereas KF/MM solution limits the error growth.

It should be observed that the KF/MM solution matches the wrong road about half way through the GPS outage. This portion of Figure 18 is zoomed in and shown in Figure 19. This is a unique situation because there is a small separation between two streets until they meet to make one street. The two streets have the same azimuth angle or heading, therefore they would both pass the azimuth threshold check during the MM algorithm. When the first match occurred to the wrong street, displayed by the circle, the KF-based solution is closer to the wrong street until the two streets connect together. Therefore, the MM algorithm is actually operating correctly but in this unique situation the algorithm is matching to the

FIGURE 20: GPS outage nine during the Kingston downtown trajectory.

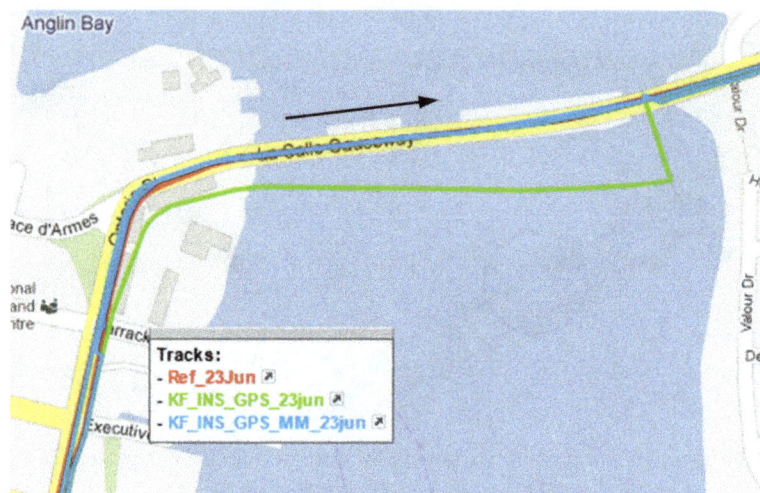

FIGURE 21: GPS outage ten during the Kingston downtown trajectory.

wrong street for approximately 2 seconds. A possible way to correct this is to use the previous map matched position and the velocity. If the distance from the new map-matched position to the previous map matched position is too large for the velocity being travelled, the algorithm would choose the second nearest line segment.

Figure 20 shows a closer look at GPS outage nine. The KF-based solution has a maximum position error of 26 m and the KF/MM-based solution has a maximum error of 6 m, which is an improvement of 77%. It is easily recognizable that the KF-based solution (green) has a constant error drift from the road or reference solution. This is due to the MEMS sensors which have a constant error drift over time. The developed method of KF/MM (blue) limited the error drift, by constantly matching the position back to the actual road. The MM is performed at every iteration during the outage.

GPS outage ten is the best example of how the MM algorithm improved the overall accuracy of the position information. Figure 21 shows the performance during GPS outage ten, which starts during a sharp turn before the LaSalle causeway. The KF-based solution has a maximum position error of 88 m and the KF/MM solution has a maximum

position error of only 8 m. This is an improvement of 90%. The KF-based solution (green) is constantly drifting from the reference and the KF/MM-based solution is limiting the error growth by restricting it to the actual road.

The KF/MM has shown promising results for the Kingston downtown trajectory where the majority of the GPS outages occurred on urban road ways. The next trajectory, which took place in Kingston suburbs, is chosen to assess the performance of the proposed algorithm in mostly rural areas with some urban portions.

5.2. Kingston Suburbs Trajectory. The second trajectory that was examined was a trajectory of the suburbs of Kingston, ON. The majority of this trajectory is rural with some urban driving. There were ten intentionally introduced GPS outages of 60 seconds to examine the performance during the outages, focusing on areas like sharp turns and curves on the road ways. Another great feature about this trajectory is that some of the GPS outages could be simulated at much higher speeds reaching 80 km/hr. Figure 22 displays the Kingston suburbs trajectory with the ten circles marking the location of the simulated GPS outages during which the positional error of KF and KF/MM is compared. The

FIGURE 22: Kingston suburbs trajectory, ten outages are shown by circles.

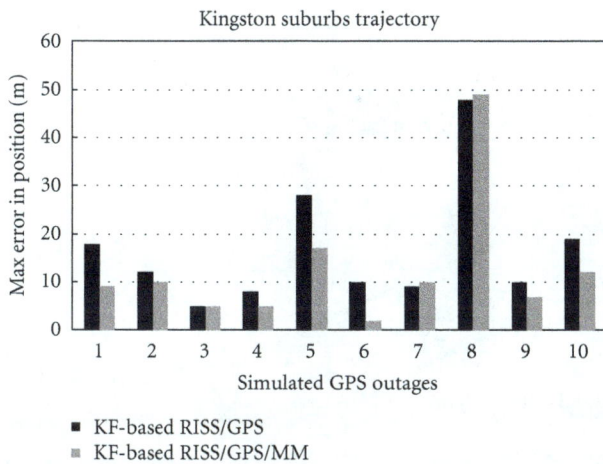

FIGURE 23: Maximum error in position (m) for the Kingston suburbs trajectory.

maximum position error (meters) for the Kingston suburbs trajectory is shown in the bar graph of Figure 23.

Both solutions, the KF based and KF/MM based, were compared against the reference solution, which is shown in Figure 22. The proposed KF/MM method showed an improvement over the KF-based solution in all GPS outages except GPS outage 8 which will be examined in detail later. It can be seen that GPS outage five, which was simulated during a sharp turn (greater than 90°), has a maximum position error of 28 m for the KF-based solution and 16 m for the KF/MM based solution. This is an improvement of 43% or 12 m in positional error. GPS outage six has a maximum position error of 10 m for the KF-based solution and only 2 m for the KF/MM-based solution, which is an improvement of 80%.

In GPS outage eight, the KF-based solution had a maximum position error of 48 m and the KF/MM-based solution also had a large maximum position error of 49 m. As shown in Figure 24, the KF-based solution (green) has a familiar drift which takes the solution away from the road or the reference solution. The developed KF/MM method

does correct this error drift throughout the outage except for one iteration as shown in Figure 25. For this one iteration the MM algorithm selects the wrong road; matched to the intersecting road (Caton Road). This is the reason why both solutions have a large maximum position error. During the next iteration, the algorithm corrected itself and matched back to the correct road. It may be visualized that KF/MM is still better then KF because it stays close to the reference except for a short time whereas KF starts to go away from the reference right from the onset of the outage. This is evident from the RMS error of the outage which is only 19 m for KF/MM-based solution as compared to 31 m for KF-based solution.

GPS outage nine is another example of how MM is restricting the positioning solution of the developed KF/MM method to the actual road. The KF-based solution has a maximum position error of 10 m and the KF/MM-based solution has a maximum error of 7 m, which is an improvement of 30%. This GPS outage occurred during a straight stretch with an average velocity of 50 km/hr.

5.3. Gananoque Trajectory. The third trajectory was conducted mostly in rural areas with slow to medium speeds in straight as well as winding patches of the road. This trajectory is called Gananoque trajectory as it passes through this town and is shown in Figure 26.

There were ten intentionally introduced GPS outages of 60 seconds to examine and compare the performance of the algorithms. The insertion of the outages was carefully planned such that they include straight portions, sharp turns, and curves on the road ways. This trajectory also included areas where outages could be simulated at higher speeds reaching up to 80 km/hr. Figure 26 displays the Gananoque trajectory with the ten GPS outages which were included for performance analysis.

As shown by the bar graph in Figure 27, the developed KF/MM method greatly improved the accuracy of the results. RISS/GPS integration for land vehicles using the developed method of KF/MM had an average maximum position error of 13.5 m and the KF-based solution had an average

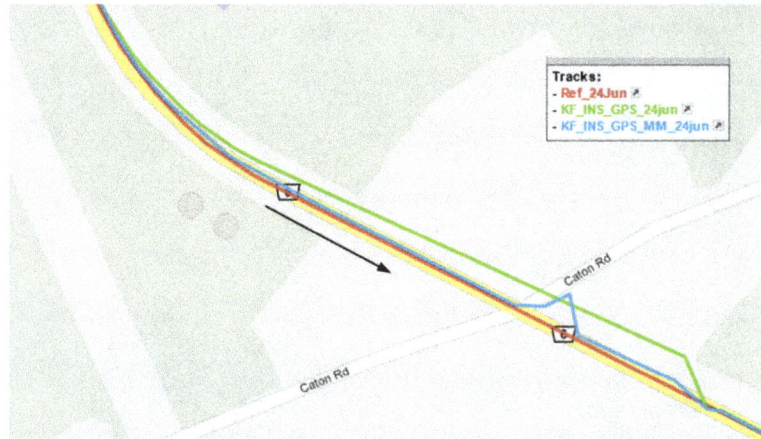

FIGURE 24: GPS outage eight during the Kingston suburbs trajectory.

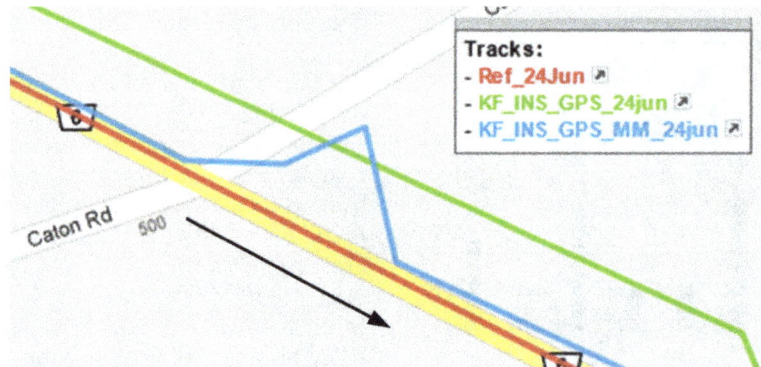

FIGURE 25: MM error during GPS outage eight.

maximum position error of 25.8 m. This is an overall average improvement of 46%.

We will now take a closer look at the GPS outage which was simulated on highway 2 during a slight turn with an average velocity of 80 km/hr. During this outage, the KF-based solution has a maximum position error of 22 m and the KF/MM-based solution has a maximum error of 11 m which is an improvement of 50%. Although there are errors in the KF/MM solution due to piecewise approximation of the road inherent in the map data, the output trajectory is mostly within the road boundaries.

GPS outage ten had the most significant improvement. The KF-based solution has a maximum position error of 52 m while the developed KF/MM based solution has a maximum position error of 12 m. This is a large improvement of 77% or 40 m in accuracy.

6. Conclusion

This paper focused on reducing the KF-based RISS/GPS integration errors by augmenting it with MM. MM limited the error growth during GPS outages by restricting the position solution to the road network. This was accomplished by using digital maps of the road networks as a constraint in the integration process. To reduce the errors contributed by

the low-cost inertial sensors, a reduced inertial sensor system was used where only one gyroscope and two accelerometers were used along with built-in odometer of the vehicle, which constitute the 3D RISS. Owing to its simplicity and comparable accuracy during good satellite visibility, loosely coupled integration was used for integration of inertial sensors and odometer with GPS. The proposed method, augmented KF/MM for 3D RISS/GPS integration, was tested on three disparate trajectories by simulating ten GPS outages in each trajectory at various locations including straight portions, slight turn as well as sharp bends. It was also ensured to include different dynamics by choosing low and high speeds, stops, and sudden accelerations. The results of the proposed method were analyzed in detail and compared with the traditional method of KF-based 3D RISS/GPS integration for land vehicles. To elucidate the comparison and clarify the exceptions to the performance of the proposed algorithm, individual outages were discussed. It was found that the proposed method outperformed the KF solution in all the three trajectories with a clear margin despite being dependent on the accuracy of the map data. For the first trajectory, the average improvement in maximum position error of the KF/MM method over KF-only method was 59%. For the second trajectory and third trajectory, it was 30% and 46%, respectively. By this account, overall, the average

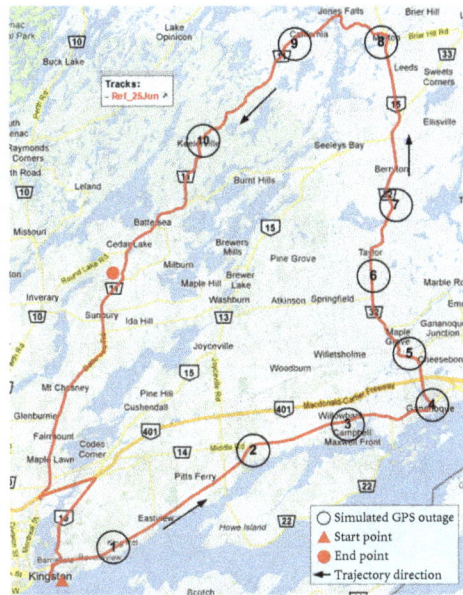

FIGURE 26: Gananoque trajectory, ten outages are shown by circles.

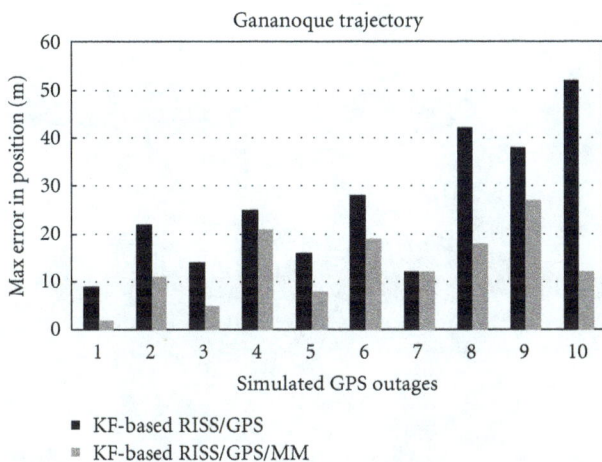

FIGURE 27: Maximum error in position (m) for the Gananoque trajectory.

improvement in maximum position error of the proposed method over traditional KF-only method was about 45%. There were few instances where the apparent performance of the proposed algorithm was poorer. These were highlighted and probable reasons and possible solutions were furnished where required.

References

[1] P. D. Groves, *Principles of GNSS, Inertial, and Multi-Sensor Integrated Navigation Systems*, Artech-House, Boston, Mass, USA, 2008.

[2] J. A. Farrell, *Aided Navigation : GPS with High Rate Sensors*, McGraw-Hill, New York, NY, USA, 2008.

[3] M. S. Grewal, L. R. Weill, and A. P. Andrews, *Global Positioning Systems, Inertial Navigation, and Integration*, John Wiley & Sons, New York, NY, USA, 2nd edition, 2007.

[4] K. R. Britting, *Inertial Navigation Systems Analysis*, John Wiley & Sons, New York, NY, USA, 1971.

[5] C. Jekeli, *Inertial Navigation Systems with Geodetic Applications*, Walter De Gruyter, Berlin, Germany, 2001.

[6] A. Noureldin, T. B. Karamat, M. D. Eberts, and A. El-Shafie, "Performance enhancement of MEMS-based INS/GPS integration for low-cost navigation applications," *IEEE Transactions on Vehicular Technology*, vol. 58, no. 3, pp. 1077–1096, 2009.

[7] J. Bernstein, "An overview of MEMS inertial sensing technology," *Sensors*, vol. 20, no. 2, pp. 14–21, 2003.

[8] C. Goodall, "Intelligent integration of a MEMS IMU with GPS using a reliable weighting scheme," in *Proceedings of the 18th International Technical Meeting of the Satellite Division of the Institute of Navigation (ION GNSS '06)*, pp. 1661–1670, Fort Worth, Tex, USA, September 2006.

[9] D. H. Titterton and J. Weston, *Strapdown Inertial Navigation Technology*, American Institute of Aeronautics and Astronautics, New York, NY, USA, 2nd edition, 2005.

[10] A. Lawrence, *Modern Inertial Technology: Navigation, Guidance, and Control*, Springer, New York, NY, USA, 2nd edition, 1998.

[11] P. Aggarwal, Z. Syed, A. Noureldin, and N. El-Sheimy, *MEMS-Based Integrated Navigation*, Artech House, Norwood, Mass, USA, 2010.

[12] J. Georgy, A. Noureldin, M. J. Korenberg, and M. M. Bayoumi, "Low-cost three-dimensional navigation solution for RISS/GPS integration using mixture particle filter," *IEEE Transactions on Vehicular Technology*, vol. 59, no. 2, pp. 599–615, 2010.

[13] P. S. Maybeck, *Stochastic Models, Estimation, and Control*, vol. 1, Academic Press, New York, NY, USA, 1979.

[14] G. Minkler and J. Minkler, *Theory and Application of Kalman Filtering*, Magellan Book, Palm Bay, Fla, USA, 1993.

[15] A. Doucet, N. De Freitas, and N. Gordon, *Sequential Monte Carlo Methods in Practice*, Springer, New York, NY, USA, 2001.

[16] B. Ristic and S. Arulampalam, *Beyond the Kalman Filter: Particle Filters for Tracking Applications*, Artech-House, Boston, Mass, USA, 2004.

[17] J. Georgy, A. Noureldin, Z. Syed, and C. Goodall, *Nonlinear Filtering for Tightly Coupled RISS/GPS Integration*, Indian Wells, Ca, USA, 2010.

[18] J. Georgy, T. Karamat, U. Iqbal, and A. Noureldin, "Enhanced MEMS-IMU/Odometer/GPS Integration Using Mixture Particle Filter," *GPS Solutions*, vol. 15, no. 3, pp. 239–252, 2011.

[19] N. El-Sheimy, K. W. Chiang, and A. Noureldin, "The utilization of artificial neural networks for multisensor system integration in navigation and positioning instruments," *IEEE Transactions on Instrumentation and Measurement*, vol. 55, no. 5, pp. 1606–1615, 2006.

[20] R. Sharaf and A. Noureldin, "Sensor integration for satellite-based vehicular navigation using neural networks," *IEEE Transactions on Neural Networks*, vol. 18, no. 2, pp. 589–594, 2007.

[21] W. Abdel-Hamid, A. Noureldin, and N. El-Sheimy, "Adaptive fuzzy prediction of low-cost inertial-based positioning errors," *IEEE Transactions on Fuzzy Systems*, vol. 15, no. 3, pp. 519–529, 2007.

[22] L. Semeniuk, *Bridging global positioning system outages using neural network forward prediction of inertial navigation position and velocity errors*, M.S. thesis, Department of Electrical and Computer Engineering, Royal Military College of Canada, Kingston, Canada, 2006.

[23] K. W. Chiang, A. Noureldin, and N. El-Sheimy, "A new weight updating method for INS/GPS integration architectures based

on neural networks," *Measurement Science and Technology*, vol. 15, no. 10, pp. 2053–2061, 2004.

[24] N. El-Sheimy, W. Abdel-Hamid, and G. Lachapelle, "An adaptive neuro-fuzzy model for bridging GPS outages in MEMS-IMU/GPS land vehicle navigation," in *Proceedings of the 17th International Technical Meeting of the Satellite Division of the Institute of Navigation (ION GNSS '04)*, pp. 1088–1095, Long Beach, Ca, USA, 2004.

[25] W. B. Zavoli and S. K. Honey, "Map matching augmented dead reckoning," in *Proceedings of the 36th IEEE Vehicular Technology Conference.*, pp. 359–362, Dallas, Tex, USA.

[26] C. Fouque, P. Bonnifait, and D. Betaille, *Enhancement of Global Vehicle Localization Using Navigable Road Maps and Dead-Reckoning*, Monterey, Ca, USA, 2008.

[27] Y. Cui and S. S. Ge, "Autonomous vehicle positioning with GPS in urban canyon environments," *IEEE Transactions on Robotics and Automation*, vol. 19, no. 1, pp. 15–25, 2003.

[28] P. Davidson, M. A. Vazquez, and R. Piche, *Uninterrupted Portable Car Navigation System Using GPS, Map and Inertial Sensors Data*, Piscataway, NJ, USA, 2009.

[29] O. Pink and B. Hummel, "A statistical approach to map matching using road network geometry, topology and vehicular motion constraints," in *Proceedings of the 11th International IEEE Conference on Intelligent Transportation Systems (ITSC '08)*, pp. 862–867, Piscataway, NJ, USA, December 2008.

[30] C. E. White, D. Bernstein, and A. L. Kornhauser, "Some map matching algorithms for personal navigation assistants," *Transportation Research Part C*, vol. 8, no. 1–6, pp. 91–108, 2000.

[31] D. Bernstein and A. Kornhauser, "An introduction to map matching for personal navigation assistants," Tech. Rep., New Jersey Institute of Technology, New Jersey, NJ, USA, 1996.

[32] J. Georgy, U. Iqbal, and A. Noureldin, "Quantitative comparison between Kalman filter and particle filter for low cost INS/GPS integration," in *Proceedings of the 6th International Symposium on Mechatronics and its Applications (ISMA '09)*, pp. 1–7, Sharjah, UAE, 2009.

[33] R. G. Brown and P. Y. C. Hwang, *Introduction to Random Signals and Applied Kalman Filtering: with MATLAB Exercises and Solutions*, John Wiley & Sons, New York, NY, USA, 3rd edition, 1997.

[34] M. S. Grewal and A. P. Andrews, *Kalman Filtering: Theory and Practice Using MATLAB*, John Wiley & Sons, New York, NY, USA, 2nd edition, 2001.

[35] P. S. Maybeck, *Stochastic Models, Estimation, and Control*, vol. 2, Academic Press, New York, NY, USA, 1982.

[36] A. Gelb, Ed., *Applied Optimal Estimation*, MIT Press, Cambridge, Mass, USA, 1974.

[37] T. Karamat, J. Georgy, U. Iqbal, and A. Noureldin, "A tightly-coupled reduced multi-sensor system for urban navigation," in *Proceedings of the 22nd International Technical Meeting of the Satellite Division of the Institute of Navigation (ION GNSS '09)*, pp. 177–187, Savannah, Ga, USA, September 2009.

[38] U. Iqbal, A. F. Okou, and A. Noureldin, "An integrated reduced inertial sensor system—RISS / GPS for land vehicle," in *Proceedings of the IEEE/ION Position, Location and Navigation Symposium (PLANS '08)*, pp. 1014–1021, Monterey, Ca, USA, May 2008.

[39] J. Georgy, U. Iqbal, M. Bayoumi, and A. Noureldin, *Reduced Inertial Sensor System (RISS)/GPS Integration Using Particle Filtering for Land Vehicles*, Savannah, Ga, USA, 2008.

[40] U. Iqbal, T. B. Karamat, A. F. Okou, and A. Noureldin, "Experimental results on an integrated GPS and multisensor system for land vehicle positioning," *International Journal of Navigation and Observation*, vol. 2009, Article ID 765010, 18 pages, 2009.

[41] IMU300—6DOF Inertial Measurement Unit: Crossbow Technology Inc., 2011, http://www.davisnet.com/product_documents/drive/spec_sheets/8211-21-25_carchip_specsB.pdf.

[42] SPAN Technology System User Manual OM-20000062: NovAtel Inc., 2011, http://www.novatel.com/Documents/Manuals/om-20000062.pdf.

[43] CarChip OBDII-Based Vehicle Data Logger and Software: Davis Instruments, 2011, http://www.davisnet.com/product_documents/drive/spec_sheets/8211-21-25_carchip_specsB.pdf.

[44] A. Schneider, Draw a Google Map from a GPS file, 2008, http://www.gpsvisualizer.com/map_input?form=google.

12

Pedestrian Tracking Solution Combining an Impulse Radio Handset Transmitter with an Ankle-Mounted Inertial Measurement Unit

Joe Youssef, Benoît Denis, Christelle Godin, and Suzanne Lesecq

CEA-Leti Minatec Campus, 17 rue des Martyrs, 38054 Grenoble Cedex 09, France

Correspondence should be addressed to Christelle Godin, christelle.godin@cea.fr

Academic Editor: Amadou Idrissa Bokoye

We address the indoor tracking problem by combining an *Impulse Radio-Ultra-Wideband* handset with an ankle-mounted *Inertial Measurement Unit* embedding an accelerometer and a gyroscope. The latter unit makes possible the detection of the stance phases to overcome velocity drifts. Regarding radiolocation, a *time-of-arrival* estimator adapted to energy-based receivers is applied to mitigate the effects of dense multipath profiles. A novel quality factor associated with this estimator is also provided as a function of the received *signal-to-noise ratio*, enabling us to identify outliers corresponding to obstructed radio links and to scale the covariance matrix of radiolocation measurements. Finally, both radio and inertial subsystems are loosely-coupled into one single navigation solution relying on a specific *extended Kalman filter*. In the proposed fusion strategy, processed inertial data control the filter state prediction whereas *Combined Time Differences Of Arrival* are formed as input observations. These combinations offer low computational complexity as well as a unique filter structure over time, even after removing outliers. Experimental results obtained in a representatively harsh indoor environment emphasize the complementarity of the two technologies and the relevance of the chosen fusion method while operating with low-cost, noncollocated, asynchronous, and heterogeneous sensors.

1. Introduction

For the last past years, new *location and tracking* (LT) needs have been gradually introduced into a wide variety of applications, such as security, health care, rescue, logistics; or house automation. A growing interest has been more particularly expressed in location-dependent indoor services, which require seamless pedestrian navigation capabilities in harsh environments where satellite-based solutions cannot operate. In this context, alternative technologies are currently under investigation, based on for example, location-enabled wireless networks [1, 2].

On their own, most of modern wireless networks can indeed retrieve the positions of mobile radio devices relative to the known position of reference anchors or *base stations* (BS). The radiolocation functionality simply relies on the measurement of radio metrics, which depend on the distance traveled in the air by transmitted signals. For instance, when a mobile radio device is synchronized with a BS (e.g., using

n-way ranging protocol transactions [3]), range information can be derived from the *time of arrival* (TOA) of the received signal. Several TOA-based range measurements collected (with respect to fixed BSs with known locations) can hence feed positioning or tracking algorithms to solve out a circular (resp., spherical) location estimation problem in 2D (resp., 3D).

Alternatively, if the surrounding BSs are strictly synchronized (i.e., independently of the clock of a mobile transmitter), the *time difference of arrival* (TDOA) can be considered, leading to a hyperbolic problem formulation.

Benefiting from unprecedented resolution and synchronization capabilities for the acquisition of such temporal radiolocation measurements, the *Impulse Radio-Ultra-Wide band* (IR-UWB) technology is today viewed as a credible solution to address LT applications through short-range and *low-data-rate* (LDR) links [4–9], *a fortiori* in the context of *ad hoc wireless sensor networks* (WSN). However, despite the claimed fine properties of IR-UWB signals, obstructed radio links notoriously introduce additional biases and high dispersion onto measurements, for example, when direct paths are blocked, or when transmitted paths are severely attenuated and shifted in time [3]. Overall, IR-UWB systems claim to provide very fine location and tracking performances in *line of sight* (LOS) (i.e., typically within submetric precision), but they can hardly guarantee such a precision level under generalized *non-line-of-sight* (NLOS) conditions (e.g., affecting several links with respect to distinct BSs simultaneously), what is a rather common situation in indoor environments. Hence, one first challenge is to derive robust TOA estimators with practical receiver architectures. One more point would be to evaluate the instantaneous quality of TOA estimates, or even to efficiently remove measurement outliers due to NLOS to assist and enhance the fed LT algorithms.

But it is also well known that radiolocation solutions operating in harsh radio environments could benefit from external means, like assisting *inertial navigation systems* (INSs) based on *inertial measurement units* (IMUs). In a navigation scenario, one INS can deliver relevant information out of raw inertial measurements for example, the pedestrian displacement amplitude, velocity, or heading [10–12]. Such information can then be used in addition to radiolocation measurements, as proposed by recent works in the field of hybrid data fusion for example, with GNSS pseudoranges in [13], with IR-UWB TOA in [14, 15], with IR-UWB TDOA and AOA in [16, 17], or even with WiFi *received signal strength indicators* (RSSI) fingerprints in [18].

Besides strict performance considerations, fusion is all the more relevant, since mobile devices (e.g., personal terminals) are expected in the near future to physically integrate multiple standards and sensors and/or to cooperate with other systems or networks, making heterogeneous modalities naturally available on the user side. Accordingly, several system architectures and configurations involving either collocated or noncollocated radio devices and IMUs can be considered for navigation purposes. As an illustrating example, smart clothes comprising distributed IMUs might form a *body area network* (BAN) using one first radio access technology and interacting with a portable handset displaying or relaying inertial information to an external access point, to the infrastructure equipping the building, or to another wearable network in the vicinity (potentially through another radio technology). Finally, coupling low-power and low-cost technologies particularly makes sense for the perennial and massive deployment of such systems. This implies the use of adequate radio technologies with energy-efficient transceiver design, reasonably simple inertial units (i.e., with a limited number of embedded sensors), and low-computational complexity for further postprocessing (including data fusion tasks).

In this context, we address herein the pedestrian navigation problem in indoor environments that present dense multipath profiles and magnetic perturbations, by loosely coupling an IR-UWB handset transmitter with a shoe-mounted IMU endowed with a 3-axis accelerometer and a 3-axis gyroscope. A specific *extended Kalman filter* (EKF) is defined to optimally hybridize the radio and inertial data

and to cope with the specific system architecture constrains (i.e., operating with noncollocated sensors). Adapting and extending previous results from [19], this formulation also combines UWB measurements (obtained through the method proposed in [20]) into new observations defined as *combined TDOA* (CTDOA). While reducing filter complexity (as a function of the number of available measurements), the proposed solution allows us to remove outlier measurements, without reconfiguring the whole filter structure and without omitting relevant measurement information. The selection of nonoutlier measurements is performed by monitoring the instantaneous quality of TOA estimates, based on a new practical SNR-dependent indicator. As for inertial data, we take benefit from the IMU place to detect stance phases and to reset the foot velocity, hence mitigating in turn the drift that usually affects the estimated INS velocity. The obtained pedestrian heading and average body velocity subsequently feed the fusion filter to control the state prediction phase. Indoor experiments were carried out to validate the proposed fusion approach, as well as to draw intermediary statistical UWB channel parameters useful to TOA estimation.

The paper is structured as follows. In Section 2, we present a selection of techniques and concerns from the recent state of the art in the fields of IR-UWB TOA-based ranging, inertial navigation systems, and the fusion of both modalities. We also try to position the main contributions of this paper in comparison with existing solutions. Then in Section 3, we briefly discuss the specificities of the statistical UWB channel models, and we estimate channel parameters from real indoor measurements. Then a robust TOA estimator adapted to energy detection receivers is recalled. We also show how to practically evaluate the quality of this estimator and to remove outlier measurements, from a filter-oriented perspective. In Section 4, we present our INS, which basically consists of an ankle/foot-mounted magnetometer-free IMU. In particular, we show how to use detected stance phases to remove the drift on INS velocity and infer the average body velocity. In Section 5 we detail and justify further our loosely-coupled fusion strategy, along with the corresponding filter structure. Then we account in Section 6 for experimental results, which were obtained in a typical indoor environment offering mixed operating conditions in terms of *signal-to-noise ratio* (SNR), radio links obstructions, *geometric dilution of precision* (GDOP), and magnetic disturbances. Finally, Section 7 concludes the paper.

2. Related State-of-the-Art and Paper Contributions

2.1. Impulse Radio-Ultra-Wideband Time-of-Arrival Estimation. Many solutions have been described in the literature to cope with IR-UWB TOA estimation, including sophisticated algorithms inspired by former high-resolution channel estimation solutions [6] requiring high sampling rates. More recently, various other techniques adapted to the low-complexity LDR context have been proposed and compared (e.g., [7]). A specific focus is usually made on noncoherent

Pedestrian Tracking Solution Combining an Impulse Radio Handset Transmitter with an Ankle-Mounted Inertial Measurement Unit

139

receivers like *energy detectors* (ED), for which one simple approach consists in comparing the energy collected in consecutive time bins with an appropriate detection threshold. The index of the first bin exceeding this threshold is then associated with the TOA estimate [7, 21]. Unfortunately, within realistic indoor channels, threshold-based ED suffers from overlapping *multipath components* (MPC) and poor *signal-to-noise ratio* (SNR) conditions, introducing significant estimation errors and biases. One weakness of these methods is that they do not take benefit from the whole MPC profile (though conveying constructive information), but they only depend on marginal and independent energy terms. Therefore, new TOA estimators have been proposed very recently in [20], assuming realistic path amplitude statistics in compliance with IEEE 802.15.4a recommendations [22], and considering the whole observed energy profile before making a decision on the estimated TOA. These estimators were shown to exhibit low-estimation dispersion around the actual channel leading edge, over a wide range of practical SNRs and channels.

But one more challenge is to allow the real-time prediction of TOA estimation uncertainty or dispersion. As regards to this measurement quality assessment, for most of the TOA estimators proposed in the literature, one could estimate offline the uncertainty based on the received signal power and conditioned on the environment category (e.g., indoor industrial, indoor residential, outdoor, etc.), on the channel configuration (e.g., LOS, NLOS, severe NLOS, etc.), and/or on the actual distance. A relationship could then be established *a priori* between the error affecting the measured TOA and the SNR through simulations, experimental campaigns, or even theoretical analysis (e.g., [23]). However, this kind of method can hardly benefit from the specificities of the received signal at each instant (e.g., of the current multipath energy profile) under mobility. As an example, under given SNR conditions, the received signal could have either sparse or dense multipath profile, which directly impacts the reliability of the estimated TOA. Moreover, offline characterization is usually mostly intended for ranging performance assessment, but still remains unexploited for online tracking purposes.

Hence a new practical method is still required to associate the instantaneous TOA estimation quality with practical estimators (e.g., [20]).

2.2. Inertial Navigation Systems. INSs based on integrated IMUs are more and more used for navigation purposes due to their low cost, low weight, and low consumption. Moreover, as stand-alone autonomous solutions, they can be used indifferently in indoor or outdoor situations, in the lack of surrounding means or infrastructure. They can also be considered for *dead reckoning navigation* (DRN) (i.e., estimation of the current position from a known starting point) when any absolute positioning system or GNSS is not available, ensuring the navigation service continuity.

A typical IMU consists of a combination of low-cost *microelectromechanical sensors* (MEMS) for example, accelerometers, magnetometers, or gyroscopes. Such units, available as commercial devices now [24], are intended for numerous applications such as motion capture, unmanned vehicular control, antenna stabilization, video gaming, and pedestrian navigation. Those IMUs provide metrics related to acceleration, orientation, magnetic field, and angular rate. The raw measurements must be processed and analyzed further in order to get relevant navigation information, such as the direction, displacement, speed, and so forth.

In the specific pedestrian navigation context, DRN systems rely on step detection and, for each detected step, on length and heading estimation [25]. Examples of step detection methods are given in [26] or [27]. For step length estimation, two kinds of approaches can be considered. One approach is based on an empirical model linking the step length with some parameters extracted from the sensors measurement during the step [28, 29]. The advantage of this approach is that the models can be adapted to any placement of the IMU. The disadvantage is that a calibration is needed for each pedestrian. A second approach uses the integration of the gyroscope and the accelerometer raw data. Nevertheless, even small errors on the measurements would lead, after integration, to a large error cumulated over time. The drift of the estimated velocity can be reduced with a foot-mounted IMU using *zero velocity update* (ZUPT) [10], which enables to reset the velocity to zero when the foot is on the floor. Other methods enable to limit the drift of foot-mounted INSs [30]. As no calibration is needed and drifts are limited, this kind of method leads to better performances than empirical methods. However, the IMU has to be placed on the foot in order to benefit from zero velocity. For heading estimation, magnetometers can be used. But they can be subject to local magnetic disturbances induced by pieces of furniture, buried or on-body metallic materials. Those perturbation are particularly present near walls and floors, making hazardous the use of foot-mounted magnetometers in typical indoor environments.

2.3. Hybrid Data Fusion Strategies. Three different fusion strategies can be applied to merge heterogeneous data in a tracking context.

The first one is the so-called *decoupled* strategy, for example [13]: the mobile node position is estimated separately and independently by each subsystem and then the set of estimated positions (delivered by independent systems) is fused into one final solution. This strategy is preferably applied when both location systems are not subject to drift, or when raw data are not available.

The second strategy is *tightly coupled*, for example [15]: some raw data are available at each subsystem, which are processed at once to track the location. Hence, it necessitates a global model that relates the heterogeneous measurements of both systems to a correction model accounting for defected measurements (i.e., orientation drift, outlier ranging measurements, magnetic disturbance). The tightly coupled fusion seems to be the best method at first sight since it enables to jointly optimize the output estimate given all the available data. However, these solutions require that the involved subsystems are physically collocated or rigidly connected, what is neither necessarily relevant nor practical for inertial-based pedestrian navigation. Finally, the synchronization issue between subsystems, which is inherent to any fusion strategy, is all the more critical within these tightly coupled solutions.

The last fusion strategy is *loosely coupled*, for example [16, 17]: the inputs to the location estimator can be raw and/or preprocessed data issued from sub-systems. This strategy enables to mitigate the previously mentioned drawbacks of the decoupled and tightly coupled fusion strategies. It tolerates that the radio part includes a controlling module to remove outlier measurements and/or to adjust the assumed quality of radio measurements. Finally, this loosely-coupled strategy enables different placements of the involved subsystems.

Besides the coupling strategy itself, the fusion of heterogeneous data also requires the use of specific estimation tools and advanced filtering techniques, such as a particle filter in [14], a combination of a particle filter and an extended Kalman filter (EKF) in [15, 18], or a backward/forward Kalman filter with a recording/smoothing unit in [31]. But another feature of navigation fusion filters concerns the way the IMU data are exploited. In a first approach, the inertial data are used in the prediction phase of the filter [17]. In the second one, they are integrated as observations in the correction phase [16].

In the following, we will provide detailed justifications for retaining a loosely-coupled scheme based on EKF and using inertial data in the prediction phase of the filter, given our system architecture constraints.

2.4. Summary of the Main Paper Contributions. Overall, the main paper contributions are as follows:

(i) application of an energy-based Bayesian TOA estimator under realistic channel parameters;

(ii) proposal of a new quality indicator for TOA estimates, feeding the tracking filter;

(iii) proposal of a specific fusion-oriented filter admitting:

    (a) IMU-based pedestrian heading and velocity measurements as control inputs into the filter state prediction;

    (b) combined temporal radiolocation measurements as filter observations, reducing computational complexity and mitigating local harmful effects due to EKF linearization;

(iv) evaluation of the previous items within one single unified tracking scheme through representative experiments in a typical indoor environment.

3. TOA Estimation with UWB Energy Detection Receivers

3.1. Received Signal Statistics. The received IR-UWB multipath signal $r(t)$ can be modeled as a sum of weighted

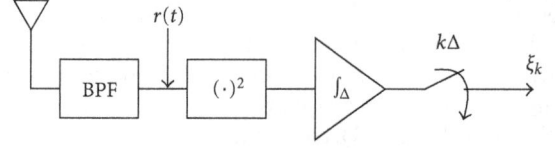

FIGURE 1: Typical block diagram of an energy detection (ED) receiver.

replicas of an energy-normalized pulse $p(t)$ of duration T_p, as follows:

$$r(t) = \sqrt{E_p} \sum_{j=1}^{\infty} \alpha_j p\left(t - \tau_j - \tau_{\text{TOA}}\right) + n(t), \qquad (1)$$

where E_p is the overall channel energy received per transmitted pulse, τ_j and $\alpha_j = \beta_j e^{i\Phi_j}$ are, respectively, the arrival time and the normalized complex gain of the jth path (with $\sum_j \mathbb{E}[\beta_j^2] = 1$ without loss of generality), if β_j and Φ_j denote the path amplitude and phase, respectively, and τ_{TOA} is the TOA of the first path. The received signal is filtered with an ideal *Bandpass filter* (BPF) of bandwidth W. Accordingly, a zero-mean *additive white Gaussian noise* (AWGN) process with double-sided power spectral density $N_0/2$ is also filtered in the band of the transmitted signal into $n(t)$, with the resulting variance $\sigma^2 = N_0 W$. Finally, we define SNR $\triangleq E_p/N_0$.

Considering a typical ED receiver [20, 21, 32] (see Figure 1), the overall observation time T is divided into K time slots of length $\Delta \approx T_p$. As we assume nonoverlapping replicas of the transmitted pulse hereafter, the noise samples taken in different slots are considered as statistically independent. The slots are numbered starting from slot 1 (i.e., for $t \in [0, \Delta]$) up to slot K (i.e., for $t \in [T - \Delta, T]$). The output sample associated with the kth time slot of the ED can be written as

$$\mu_k = \int_{(k-1)\Delta}^{k\Delta} |r(t)|^2 \, dt. \qquad (2)$$

Let k_{TOA} be the slot index to be estimated, which is associated with the bin that contains τ_{TOA}. Then the slots with indexes $\{k_{\text{TOA}}, \ldots, K\}$ correspond to the multipath region. According to [33], under the AWGN assumption the normalized energy samples $\xi_k = \mu_k/N_0$ at the integrator output of the ED follow a Chi-square distribution. For $k = \{1, \ldots, K\}$, ξ_k is either a central Chi-square distributed ($C\chi^2$) *random variable* (r.v.) with $V = 2W\Delta$ degrees of freedom or a noncentral Chi-square distributed r.v. (NCχ^2), still with V degrees of freedom, but with a noncentrality parameter $E_k = \text{SNR}\beta_k^2$ that accounts for the energy of the useful signal in slot k. β_k is usually assumed as a Nakagami-m distributed r.v. with parameters m_k and $\mathbb{E}[\beta_k^2] = \Lambda_k$. Finally, in [34] the *average power delay profile* (APDP) is modeled by a single exponential decay, with a decay constant ϵ. In [22], the APDP also follows an exponential decay, but with further multipath clustering effects.

To verify these model assumptions and characterize key channel parameters, a channel measurement campaign was carried out in the indoor environment described in

Pedestrian Tracking Solution Combining an Impulse Radio Handset Transmitter with an Ankle-Mounted Inertial Measurement Unit

141

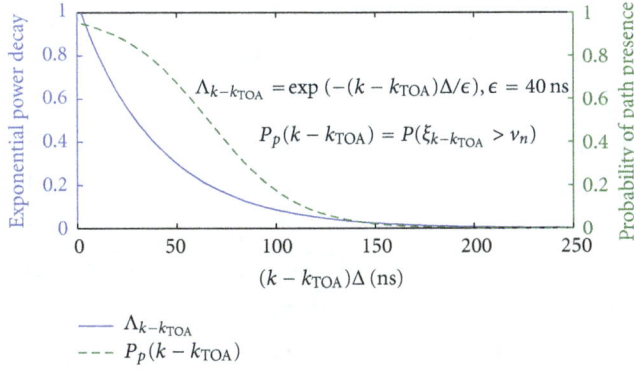

FIGURE 2: Fitted average power decay and probability of path presence out of measured channel profiles, as two functions of the excess delay.

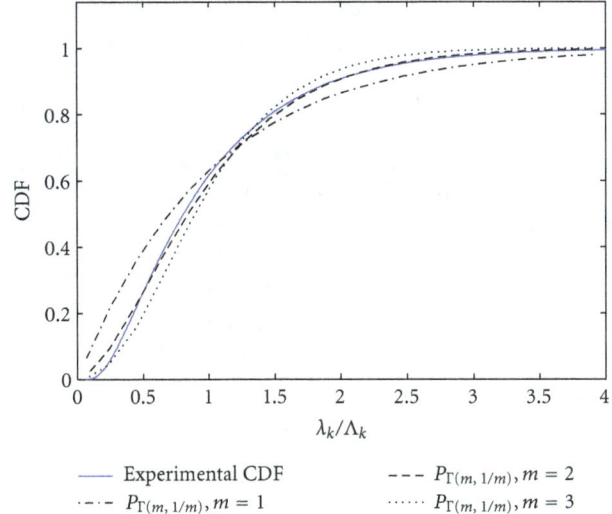

FIGURE 3: Empirical and fitted CDFs of the normalized path energy, for different Nakagami-m path parameter assumptions.

Section 6.1 for our tracking experiments. Averaging over all the measured channel profiles, an exponential model was fitted to the empirical average power decay, and an empirical function was drawn for the path presence probability, depending on the channel excess delay (or time index, equivalently). To get a better fit of the power decay to our measurements, we preliminarily conditioned received energy upon the presence of a path in each slot so that the slots with no detected path energy were not taken into account in the average. In order to decide whether ξ_k contains only noise energy, we simply set a threshold v_n equal to 10 dB above the noise power. Finally, the probability of path presence P_p was computed over all the observed channel profiles, as follows:

$$P_p(k) = \Pr(\xi_k > v_n). \tag{3}$$

According to [22], each received profile realization is affected by a distinct shadowing effect. Thus, the measured channel profiles were normalized in power before averaging. Figure 2 represents the exponential power decay model Λ_k fitted to the measured channel profiles, along with the empirical path presence probability $P_p(k)$, both of which decrease as a function of the excess slot index (i.e., the time index after channel leading edge). For the exponential power decay, the best fit was obtained for $\epsilon = 40$ ns, which is rather close to the mean value $\bar{\epsilon} = 39.8$ ns found in [34].

Since β_k is Nakagami-m distributed with a mean value Λ_k, then $\lambda_k = \beta_k^2$ is Gamma distributed with a shape parameter equal to m_k and a scale parameter equal to Λ_k/m_k. For the sake of simplicity, we assume that all the received paths have the same parameter $m_k = m$ (for all k corresponding to the part of the received signal comprising the useful multipath energy), like in [21]. As shown in Figure 3, for $m = 2$ the theoretical *cumulative density function* (CDF) of the normalized path energy satisfactorily fits to the empirical one computed out of our experiments. In the following, the exponential power decay and Nakagami-m parameters, as well as the probability of path presence as a function of the excess time index, are used as prior information to feed our TOA estimator. Similarly, in [21, 33],

such channel characteristics are assumed available *a priori* to set optimal detection thresholds.

3.2. TOA Estimation.

We consider the *minimum mean square error* (MMSE) estimator proposed in [20]. Accordingly, the sample index associated with the estimated TOA is defined as

$$\hat{k}_{\text{TOA}} \triangleq \sum_{k_{\text{TOA}}=1}^{K} k_{\text{TOA}}\, p\left(\frac{k_{\text{TOA}}}{\xi_1, \ldots, \xi_K}\right), \tag{4}$$

where $p(\cdot)$ is now a conditional probability density function (pdf). Complete analytical developments with their intermediary results can be found in [20]. The TOA estimate is finally obtained as TOA $= \Delta\, \hat{k}_{\text{TOA}}$, where Δ is the ED sampling period.

3.3. TOA Covariance Estimation.

We now intend to provide an indicator reflecting the quality of energy-based TOA estimation. One step beyond, the underlying idea is to feed a tracking filter with further information to dynamically adjust the covariance matrices used in the filter correction step, and/or even to help the detection of outlier TOA-based observations, for instance based on a filter innovation test [35].

For this purpose, we rely here on the instantaneous SNR of the received signal. First, the noise power spectral density $N_0/2$ can be rather straightforwardly estimated in the absence of transmitted signal. Thus we assume that it is available on the receiver side. One ideal solution would consist in jointly estimating the SNR and the TOA. However, for the sake of practicability and simplicity, it is chosen to estimate the SNR first out of all the collected normalized energy samples $\{\xi_k\}_{k=1,\ldots,K}$. Unfortunately, there is no close form for the *maximum likelihood estimator* (MLE) of the corresponding

$(NC\chi^2)$ noncentrality parameters $\{E_k\}_{k=1,...,K}$. However, it is at least shown in [36] that the following estimator:

$$\widetilde{E}_k = \max\{\xi_k - V, 0\} \qquad (5)$$

has a lower *mean square error* (MSE) than the MLE for a single observation and for $V \geq 0.5$, which is practically verified in our case. Now, since the TOA is *a priori* unknown before applying (4), then the mean energy value $\Lambda_k = \mathbb{E}[\lambda_k]$ of the multipath components within each time slot k is also unknown. Hence, we use $E_r \triangleq \sum_{k=1}^{K} E_k$ to determine the SNR. Here, two possibilities are available, as follows:

$$\widetilde{E}_{r1} = \sum_{k=1}^{K} \widetilde{E}_k,$$

$$\widetilde{E}_{r2} = \max\left\{ \sum_{k=1}^{K} \xi_k - KV, 0 \right\}, \qquad (6)$$

where $\sum_{k=1}^{K} \xi_k$ is an $NC\chi^2$ distributed r.v. with KV degrees of freedom and a noncentrality parameter equal to E_r. Both estimators in (6) provide the same estimation result if $\xi_k \geq V$, for all $k \in \{1 : K\}$, which is more probable at high SNR values. Otherwise, through simulations, we found out that $\widetilde{E}_{r2}$ gives lower MSE than $\widetilde{E}_{r1}$. Moreover, for $\xi_k \geq V$, for all $k \in \{1 : K\}$, $\widetilde{E}_{r2}$ is unbiased, and the standard deviation of the relative error $(\widetilde{E}_{r2} - E_r)/E_r$ is equal to $\sqrt{4E_r + 2V}/E_r$, which decreases when E_r increases (i.e., at high SNR).

Let $E_r = \text{SNR} \lambda_r$, where λ_r is the sum of K Gamma distributed r.v. $\{\lambda_k\}_{k=1,...,K}$, each with a probability $P_p(k)$, then λ_r is also a r.v. whose *characteristic function* (CF) $\phi_{\lambda_r}(t)$ is given by

$$\phi_{\lambda_r}(t) = \prod_{k=1}^{K} \left(P_p(k) \left(1 - J \frac{\Lambda_k}{m} t \right)^{-m} + 1 - P_p(k) \right), \qquad (7)$$

where $\pm J$ is the square root of -1, and still assuming $m_k = m$ (for all k corresponding to the part of the received signal comprising the useful multipath energy). Suppose now that E_r is given, then a nonbiased estimator of the SNR can be built as

$$\widehat{\text{SNR}} = \frac{E_r}{\mathbb{E}[\lambda_r]}, \qquad (8)$$

where $\mathbb{E}[\lambda_r]$ is the expected value of λ_r, which can be practically computed as $(d\phi_{\lambda_r}(t)/dt)|_{t=0} = \sum_{k=1}^{K} \Lambda_k P_p(k)$, that is to say, as the derivative of (7) with respect to t evaluated at $t = 0$.

Using the experimental Λ_k and $P_p(k)$ found in Section 3.1, it turns out that $\sigma_{\widehat{\text{SNR}}} \approx 0.1 \, \text{SNR}$ practically, which is fairly acceptable.

Finally, SNR is estimated at the real receiver using the approximation $\widetilde{E}_r$ from (6) instead of E_r in (8), as follows:

$$\widetilde{\text{SNR}} = \frac{\widetilde{E}_r}{\mathbb{E}[\lambda_r]}. \qquad (9)$$

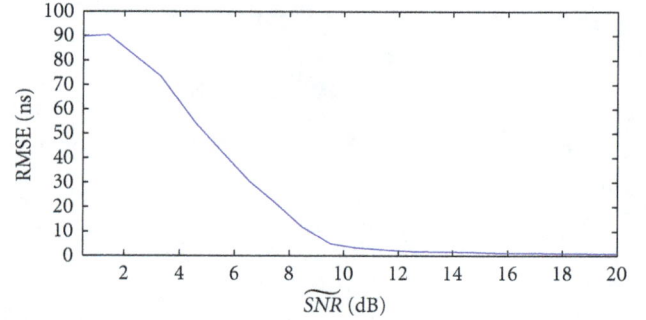

FIGURE 4: RMSE of the proposed ED-based TOA estimator, as a function of the SNR value approximated at the receiver side $\widetilde{\text{SNR}}$.

At this point, deriving the analytical expression for the *root mean square error* (RMSE) of the estimated TOA as a function of $\widetilde{\text{SNR}}$ still remains challenging. Consequently, we proceed by simulations to draw the required relationship. In these simulations, the true TOA is a uniformly distributed r.v. in the time interval $[0, T]$ (i.e., in the entire observation window), and the received signal is simulated according to the statistical parameters given in Section 3.1. Figure 4 shows the RMSE of the estimated TOA as a function of $\widetilde{\text{SNR}}$, the latter being computed on the receiver side based on (9). This monotonic evolution can be tabulated as a function of $\widetilde{\text{SNR}}$. In an online tracking context, the instantaneous standard deviation σ_{TOA_i} of the current TOA estimate with respect to the ith BS can then be approximated by the previous tabulated RMSE function, assuming unbiased TOA estimate (i.e., $\widehat{\sigma}_{\text{TOA}_i} = \text{RMSE} \approx \sigma_{\text{TOA}_i}$). In the following, $\widehat{\sigma}_{\text{TOA}_i}^2$ is used in Sections 5 and 6 to determine the instantaneous measurement covariance matrix at each update of a tracking filter (out of the collected energy samples).

3.4. Measurement Outliers Detection and Discarding. In the previous section, we have proposed a dispersion estimator $\widehat{\sigma}_{\text{TOA}}$, assuming unbiased estimated TOAs. Unfortunately, most of NLOS situations result in biased estimates.

From a tracking filter perspective, a first approach to mitigate harmful NLOS effects consists in increasing the diagonal elements of the measurement covariance matrix. Different solutions can also be used to detect NLOS outliers, for instance based on innovation tests [15, 16]. In this case, the innovation is related to the difference between the current observation and the predicted one, computed under an unbiased hypothesis. However, the complete measurement set would be discarded, even if only one single measurement was biased. Moreover, in some scenarios, severely obstructed NLOS situations might also coincide with situations where the location information is adversely conditioned, for instance due to a poor *geometrical dilution of precision* (GDOP). In this case, the values in the predicted state covariance matrix are so large (and so are the values in the covariance matrix of predicted measurements) that outliers can hardly be detected through simple innovation

Pedestrian Tracking Solution Combining an Impulse Radio Handset Transmitter with an Ankle-Mounted Inertial Measurement Unit

143

tests. Another method is to preliminarily identify the channel state LOS/NLOS based for example, on collected channel energy [37, 38]. Then, identified NLOS channel conditions are associated with TOA measurement outliers. Whatever the detection method, outliers can be discarded before the location estimation step [15] independently of the latest state estimate. In our context, as ED receivers naturally provide access to the channel energy profile, the last approach is considered.

One indicator of the channel status (and hence indirectly of potential TOA estimation biases) is conditioned on the received signal energy as follows:

$$I_\sigma = \frac{\Delta^2}{12} + \Delta^2 \sum_{k'_{\text{TOA}}=1}^{K} \left(k'_{\text{TOA}} - \hat{k}_{\text{TOA}}\right)^2 p\left(\frac{k'_{\text{TOA}}}{\xi_1, \ldots, \xi_K}\right). \tag{10}$$

Practically, this indicator reflects the variance of the estimated TOA conditioned on the energy samples $\{\xi_k\}_{k=1,\ldots,K}$ of the received signal, averaging over all the possible true k_{TOA}, supposed to be uniformly distributed in the observation window. Since the multipath energy spread of NLOS channels is notoriously larger [22], the hypothesis $k'_{\text{TOA}} = k_{\text{TOA}}$ conditioned on the observed energy samples $\{\xi_k\}_{k=1,\ldots,K}$ in (10) is likely equiprobable for all the slot indexes. Therefore, the indicator will exhibit larger values for NLOS channels. Note that computing I_σ is rather convenient, since all the required quantities are already available for the calculation of $\hat{k}_{\text{TOA}}$ (see Section 3.2).

Practically, to decide if one TOA measurement is biased (i.e., if it is an outlier) or not, $\sqrt{I_\sigma}$ is compared with a threshold, which is set at three times above the LOS $\hat{\sigma}_{\text{TOA}}$ shown in Figure 4. This threshold ensures correctly identifying 97% of the LOS channels under the default centered Gaussian assumption.

4. Inertial Navigation System

As already pointed out in Section 2.2, it is better to place the INS on the foot in order to benefit from ZUPT resetting methods [10] or other update methods [30]. Moreover, we do not consider the use of a magnetometer as magnetic disturbances are too important near the floor. Then we assume a foot/ankle-mounted IMU with a triaxis gyrometer and a triaxis accelerometer only.

First, two main frames are defined:

(i) the *body frame* (BF) is the frame in which raw data are measured. BF data are referred to their definition frame with $(\cdots)^b$;

(ii) the *navigation frame* (NF) is the reference frame, which is linked to the earth frame. The NF data are referred to this frame with $(\cdots)^n$. Since no magnetometer is used, it is not possible to estimate the orientation of the sensor with respect to the North. Hence the navigation frame includes a rotation around the vertical axis with respect to the North frame defined by the East, North, and Up orientations. This rotation is fixed at the beginning

of the walk, using the projection of the sensor axis on the horizontal plane instead of East and North axis. Vertical axis (orthogonal to the horizontal plane) is given during the stance phase by the measurement of the accelerometer.

Let $\mathbf{R}_{bn}$ be the rotation matrix related to the unitary quaternion q_{bn}, denoting the orientation of the NF in the BF using the relation given in the appendix. Initial orientation with respect to the vertical axis is given, at the beginning of the walk and during the stance phase, by the accelerometer measurement. Considering our definition of the NF, with respect to horizontal axis, initial orientation is set to zero. Then, as the gyroscope measures the angular rate, q_{bn} and then $\mathbf{R}_{bn}$ can be estimated by integrating its raw data. This rotation matrix also transforms the coordinates of a vector in the NF into its coordinates in the BF. Then the raw data issued at the accelerometer can be written as

$$\mathbf{a}^b(t) = \mathbf{R}_{bn}(t)\mathbf{a}^n(t) = \mathbf{R}_{bn}(t)\left(\mathbf{a}_p^n(t) - \mathbf{g}^n\right), \tag{11}$$

where $\mathbf{g}^n$ and $\mathbf{a}_p^n$ denote, respectively, the gravity field and the proper acceleration in the NF. The proper acceleration $\mathbf{a}_p^n$ corresponds to the derivative of the velocity and to the second derivative of the position. The velocity can hence be estimated by integrating this proper acceleration. In order to limit the drift due to noise or bias integration, it is necessary to detect correctly the stance phase and suppose the velocity is null during this phase, as proposed within the ZUPT method [10].

4.1. Stance Phase Detection. Many step detection methods are proposed in the literature. For instance, in [39] the step detection method is based on the Fourier transform through counting zero-crossing points over a threshold of the accelerometer output. In [26], the pedestrian step pattern is first detected. Then, the beginning and the end of the step can be defined according to this detected pattern. In [27] several other methods are also compared.

In the scenario considered here, the pedestrian is assumed to be walking in a building, moving from one room to another one with some stops and a direction that may change after a few steps. The pedestrian cannot walk very quickly, thus the stance phase of his foot is large enough to be easily detectable. A simple method used to detect the stance phase consists in comparing the acceleration variance to a threshold. Note that even if the ankle trajectory looks reproducible over distinct steps, the pattern of the IMU measurements will depend on the sensor orientation, see (11). To overcome this problem, as $\|\mathbf{a}^b(t)\| = \|\mathbf{a}^n(t)\|$ does not depend on the instantaneous rotation of the IMU, one can simply use the amplitude $\|\mathbf{a}^b(t)\|$ of the 3D accelerometer measurement [26]. During the stance phase, this amplitude is mostly constant and equal to the amplitude of the gravity vector. We then compute the variance of $\|\mathbf{a}^b(t)\|$ in a sliding window of temporal length equal to 0.25 s. When the variance value falls under a fixed threshold, a stance phase is detected. During our experiments presented in Section 6, this very simple method has efficiently detected all the steps.

4.2. IMU Proper Acceleration. Let $[t_1^\kappa, t_2^\kappa]$ be the swing phase time interval related to the κth detected step, and let l be the index of inertial data within this time interval (i.e., $\{l_1 : l_2\}$). Then the acceleration has to be integrated only during this time interval, whereas during the remaining time, the velocity is set to zero. Thus, the velocity of the pedestrian ankle is computed step by step, whereas the orientation of the NF in the BF at time index l is continuously determined with the quaternion $q_{bn}^{(l)} \cdots q_{bn}^{(l)}$ results from the integration of the 3D angular rate measurements $\boldsymbol{\omega}^{b(l)} = [\omega_x^{b(l)} \ \omega_y^{b(l)} \ \omega_z^{b(l)}]^T$, as follows:

$$q_{bn}^{(l+1)} = q_{-\omega}^{b(l)} \odot q_{bn}^{(l)}, \tag{12}$$

where $\odot$ denotes the quaternion product and

$$q_{-\omega}^{b(l)} = \left[\cos\left(\left\|\boldsymbol{\omega}^{b(l)} \ T_s\right\|/2\right) - \sin\left(\left\|\boldsymbol{\omega}^{b(l)} \ T_s\right\|/2\right)\mathbf{q}_\omega^{b(l)}\right]^T, \tag{13}$$

where $T_s = 1/f_s$ is the time sampling interval of the IMU and $\mathbf{q}_\omega^{b(l)} = \boldsymbol{\omega}^{b(l)}/\|\boldsymbol{\omega}^{b(l)}\|$. Once the body orientation is determined, the accelerometer measurement terms are simply rotated into the NF as follows:

$$\mathbf{a}^{n(l)} = \mathbf{R}_{nb}^{(l)}\mathbf{a}^{b(l)}, \tag{14}$$

where $\mathbf{R}_{nb}^{(l)} = \mathbf{R}_{bn}^{T(l)}$.
Then we have

$$\int_{t_1^\kappa}^{t_2^\kappa} \mathbf{a}^n(t)dt = \int_{t_1^\kappa}^{t_2^\kappa} \left(\mathbf{a}_p^n(t) - \mathbf{g}^n\right) dt = \int_{t_1^\kappa}^{t_2^\kappa} \left(\mathbf{a}_p^n(t) - \mathbf{g}^n\right) dt$$
$$= (\mathbf{v}_I^n(t_2^\kappa) - \mathbf{v}_I^n(t_1^\kappa)) - (t_2^\kappa - t_1^\kappa)\mathbf{g}^n, \tag{15}$$

where $\mathbf{v}_I(t)$ denotes the IMU velocity. Since at $t = t_1^\kappa$ and $t = t_2^\kappa$ the foot is on the floor, then $\mathbf{v}_I^n(t_1^\kappa) = \mathbf{v}_I^n(t_2^\kappa) = 0$, and consequently

$$\mathbf{g}^n = \frac{-1}{(t_2^\kappa - t_1^\kappa)} \int_{t_1^\kappa}^{t_2^\kappa} \mathbf{a}^n(t)dt. \tag{16}$$

Finally, the proper acceleration is given by

$$\mathbf{a}_p^n(t) = (\mathbf{a}^n(t) + \mathbf{g}^n), \tag{17}$$

where $\mathbf{g}^n$ in (17) compensates at the same time the gravity and the velocity bias.

4.3. Inertial Support to Pedestrian Navigation. Our aim is to continuously compute the ankle velocity of the pedestrian together with his heading. Note that the heading is defined as the angle given by the direction of the walk in the horizontal plane of the Navigation Frame. Since during the stance phase the velocity is set to zero, then the velocity is computed separately for each step with

$$\mathbf{v}_I^n(t) = \int_{t_1^\kappa}^t \mathbf{a}_p^n(t)dt. \tag{18}$$

Unlike in [10], the velocity here is continuous between different gait phases after centering the acceleration in (17) during the swing phase of each step. Thus, the computed IMU velocity in (18) starts at zero and ends up at zero.

We consider that the pedestrian walks on a flat floor. The pieces of horizontal information of the ankle velocity and heading are then given as follows:

$$v_I^{(l)} = \sqrt{\left(v_{x,I}^{n(l)}\right)^2 + \left(v_{y,I}^{n(l)}\right)^2},$$
$$\varphi_b^{(l)} = \operatorname{atan}\left(\frac{v_{y,I}^{n(l)}}{v_{x,I}^{n(l)}}\right). \tag{19}$$

During a walk with a constant speed, the ankles alternate between stance and swing phases. Hence, the pedestrian waist experiences almost a constant velocity, whereas the ankles experience high velocity variations alternating between maximum and null speed values. As the horizontal waist velocity of the pedestrian does not vary as much as the ankle velocity, then we compute the waist velocity v_W using a smoothed version of v_I. In the following, the horizontal velocity v_W of the pedestrian and the heading φ_b are the two processed inertial data simultaneously incorporated into the tracking filter, contrarily to the loosely-coupled fusion strategies in [16, 17], where the IMU is used only for heading and step detection. Furthermore, in our proposal, there is no need to have the same pedestrian heading and displacement orientation. For instance, pedestrian side walk and back walk are freely enabled since the real displacement of the foot is estimated. Moreover, there is no need to estimate the step length as in [16, 17] or to calibrate the leg length as in [40].

5. Tracking Problem Statement and Fusion Filter Formulation

In the considered scenario, we remind that inertial sensors are attached on the ankle, whereas the pedestrian holds in his hands an IR-UWB transmitter, and N_A known reference receivers are disseminated in the environment. Hence, out of $\{r_i\}_{i=1,...,N_A}$, $\{\hat{o}_i\}_{i=1,...,N_A}$, v_W, and φ_b the problem here is to track at least the unknown pedestrian position $\mathbf{p} = [x \ y]^T$, where x and y refer to the Cartesian coordinates. For this purpose, we consider using an *Extended Kalman Filter* (EKF), which is widely used in mobile tracking applications. After recalling the general EKF formulation in Section 5.1, , we will then justify our overall fusion strategy in Section 5.2. Finally, the state and observation models will be detailed in Sections 5.3 and 5.4.

5.1. Generic Extended Kalman Filter Formulation. We start formulating the tracking problem with the following generic state equation

$$\mathbf{s}^{(l)} = \mathbf{f}\left(\mathbf{s}^{(l-1)}, \mathbf{u}^{(l-1)}\right) + \mathbf{n}_s^{(l)}, \tag{20}$$

where l is the time index, $\mathbf{s}^{(l)}$, $\mathbf{u}^{(l)}$ and $\mathbf{n}_s^{(l)}$ denote respectively the state vector that contains all the parameters to be

Pedestrian Tracking Solution Combining an Impulse Radio Handset Transmitter with an Ankle-Mounted Inertial Measurement Unit

145

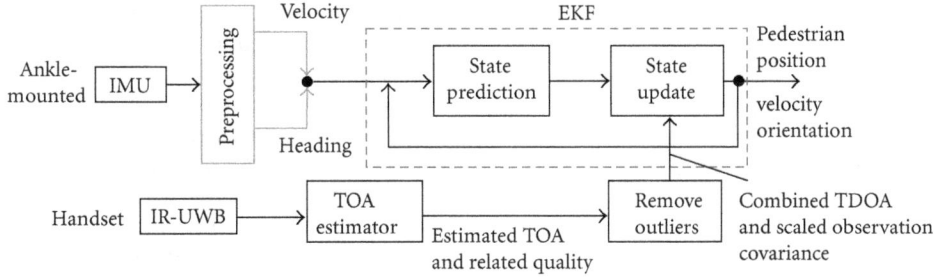

FIGURE 5: Global inputs from building subsystems and fusion filter structure.

estimated, the input control to the dynamic system, and an i.i.d. process noise sequence. $\mathbf{f}$ is the function that relates two consecutive states. As for the observation function $\mathbf{h}$, it binds the state vector to an observation vector $\mathbf{r}$ composed of measurements, as follows:

$$\mathbf{r}^{(l)} = \mathbf{h}\left(\mathbf{s}^{(l)}\right) + \mathbf{n}^{(l)}, \qquad (21)$$

where $\mathbf{n}^{(l)}$ is an i.i.d. process noise sequence.

The EKF is a popular linearized version of the Kalman filter that can handle nonlinear functions for both estimated states and observations [35]. It enables to estimate the state vector $\hat{\mathbf{s}}^{(l)}$ and the related state covariance matrix $\mathbf{P}^{(l)}$ at time increment l from the previous estimates $\hat{\mathbf{s}}^{(l-1)}$ and $\mathbf{P}^{(l-1)}$, the control input $\mathbf{u}^{(l-1)}$ obtained at $l - 1$, and the current observations $\mathbf{r}^{(l)}$.

The estimation of $\mathbf{s}^{(l)}$ with the EKF is typically split in two steps, namely the state prediction and the state update, which rely, respectively, on the dynamic state equation in (20) and the observation equation in (21).

(a) Prediction:

$$\hat{\mathbf{s}}^{(l/l-1)} = \mathbf{f}\left(\hat{\mathbf{s}}^{(l-1)}, \mathbf{u}^{(l-1)}\right)$$
$$\mathbf{P}^{(l/l-1)} = \mathbf{F}\mathbf{P}^{(l-1)}\mathbf{F}^T + \mathbf{Q}. \qquad (22)$$

(b) Update:

$$\mathbf{K}^{(l)} = \mathbf{P}^{(l/l-1)}\mathbf{H}^T\left(\mathbf{H}\mathbf{P}^{(l/l-1)}\mathbf{H}^T + \mathbf{R}\right)^{-1}$$
$$\hat{\mathbf{s}}^{(l)} = \hat{\mathbf{s}}^{(l/l-1)} + \mathbf{K}^{(l)}\left(\mathbf{r}^{(l)} - \mathbf{h}\left(\hat{\mathbf{s}}^{(l/l-1)}\right)\right) \qquad (23)$$
$$\mathbf{P}^{(l)} = \left(\mathbf{I} - \mathbf{K}^{(l)}\mathbf{H}\right)\mathbf{P}^{(l/l-1)},$$

where $\mathbf{F}$ is the Jacobian of $\mathbf{f}$ with respect to $\mathbf{s}$ computed in $\hat{\mathbf{s}}^{(l-1)}$, and $\mathbf{H}$ is the Jacobian of $\mathbf{h}$ computed in $\hat{\mathbf{s}}^{(l/l-1)}$. $\mathbf{Q}^{(l)}$ and $\mathbf{R}^{(l)}$, which are assumed to be known, are the covariance matrices of $\mathbf{n_s}^{(l)}$ and $\mathbf{n}^{(l)}$.

5.2. Overall Fusion Strategy and Filter Structure.
In Section 2.3, we presented "decoupled," "tightly coupled," and "loosely coupled" options as three different fusion strategies to merge radio and inertial subsystems into a single navigation solution.

We choose a "loosely coupled" strategy here because it authorizes different placements of the subsystems. For instance, the pedestrian can hold the UWB transmitter in his hand, avoiding many near-floor obstacles, and increasing visibility with respect to anchor nodes, whereas a foot-mounted IMU can take advantage of ZUPT to overcome the INS velocity drift. Moreover, with the "loosely coupled" strategy, each drift-free observation issued from the UWB subsystem can be used to limit the INS drifts.

In our application, we choose to integrate the IMU data in the prediction phase of the filter for the following reasons:

(i) it is hard to quantify the errors affecting the inertial data model due to for example, step missing and pedestrian walk behavior (variations of walk speed or unusual pedestrian movements);

(ii) in an alternative approach, the fusion filter update would have to run at the rate of the inertial data instead of the UWB rate (e.g., $f_s = 200\,\text{Hz}$ instead of 3-4 Hz in our experiments), while the update part of the filter includes matrix inversions/manipulations and a few matrix multiplications, which is computationally demanding for real-time applications;

(iii) the first proposed approach naturally supports different sampling rates for the prediction and the correction phases. For instance, in our application, the UWB rate is applied to the correction phase, whereas the inertial rate is applied to the prediction phase.

Figure 5 shows the global structure of the proposed EKF fusion filter. The raw inertial data are preprocessed as described in Section 4 to get pedestrian velocity and heading as inputs to the filter prediction phase. In parallel, IR-UWB TOA measurements are delivered at each receiver with their estimated variance, as computed in Section 3.3. In the update part of the filter, specific combinations of such measurements are performed, as introduced in [19], enabling us to remove outliers using the indicator presented in Section 3.4 without changing the filter structure (e.g., the size of involved matrices).

5.3. IMU-Based State Prediction.
Generally speaking, at time index l, in the absence of state noise, $\mathbf{s}^{(l)}$ is expressed as a function of the previous state $\mathbf{s}^{(l-1)}$, the pedestrian velocity

$v^{(l-1)}$, and the nonbiased or biased compensated heading, $\varphi^{(l-1)}$ as follows:

$$\mathbf{s}^{(l)} = \mathbf{f}\left(\mathbf{s}^{(l-1)}, v^{(l-1)}, \varphi^{(l-1)}\right). \qquad (24)$$

For instance if $s = \mathbf{p}$, then $\mathbf{f}$ is fully accounted by

$$x^{(l)} = x^{(l-1)} + \frac{v^{(l-1)}\cos\left(\varphi^{(l-1)}\right)}{f_s},$$
$$y^{(l)} = y^{(l-1)} + \frac{v^{(l-1)}\sin\left(\varphi^{(l-1)}\right)}{f_s}, \qquad (25)$$

with f_s the inertial refreshment rate.

As already pointed out, the INS on its own suffers from heading drift because of the integration of the biased angular rate measurement from the gyroscope. Hence, the heading angle φ_b determined in Section 4.3 is biased by b_φ. Thus, as one further variable to estimate, we incorporate the latter bias in the modeled EKF state $\mathbf{s} = [\mathbf{p}^T \ b_\varphi]^T$, whereas the control input is composed of the data issued from the INS, namely, $\mathbf{u} = [v_W \ \varphi_b]^T$ (i.e., assuming v_W is sufficiently representative for v). Therefore, in our filter the state prediction at time step l is given by

$$\hat{x}^{(l/l-1)} = \hat{x}^{(l-1)} + \frac{v_W^{(l-1)}\cos\left(\varphi_b^{(l-1)} - \hat{b}_\varphi^{(l-1)}\right)}{f_s},$$
$$\hat{y}^{(l/l-1)} = \hat{y}^{(l-1)} + \frac{v_W^{(l-1)}\sin\left(\varphi_b^{(l-1)} - \hat{b}_\varphi^{(l-1)}\right)}{f_s}, \qquad (26)$$
$$\hat{b}_\varphi^{(l/l-1)} = \hat{b}_\varphi^{(l-1)},$$

with f_s the inertial refreshment rate and hence the Jacobian $\mathbf{F}$ of $\mathbf{f}$ computed at the predicted state $\hat{\mathbf{s}}^{(l/l-1)}$ is as follows:

$$\mathbf{F} = \begin{bmatrix} 1 & 0 & \dfrac{v_W^{(l-1)}\sin\left(\varphi_b^{(l-1)} - \hat{b}_\varphi^{(l/l-1)}\right)}{f_s} \\ 0 & 1 & -\dfrac{v_W^{(l-1)}\cos\left(\varphi_b^{(l-1)} - \hat{b}_\varphi^{(l/l-1)}\right)}{f_s} \\ 0 & 0 & 1 \end{bmatrix}. \qquad (27)$$

Note that the state covariance prediction also runs at the inertial rate f_s. The state covariance matrix $\mathbf{Q} = \mathbf{Q}^{(l)}$ is supposed to be time invariant. The update part of the EKF is achieved only when new radiolocation measurements are available. Thus $\hat{b}_\varphi^{(l)}$ changes only at the update step, and holds the same value during the prediction step.

5.4. Filter Update with Combined T(D)OA Measurements. At the ith *base station* (BS) ($i = 1,\dots,N_A$), TOA-based measurements $r_i = c\,\mathrm{TOA}_i$ are performed through ED (see Section 3.2), along with an approximation $\hat{\sigma}_i = c\hat{\sigma}_{\mathrm{TOA}_i}$ for the corresponding standard deviation $\sigma_i = c\,\sigma_{\mathrm{TOA}_i}$ (see Section 3.3). It is also assumed that the N_A receivers are

synchronized, but independently of the mobile transmitter, so that all the pseudo-TOA measurements are biased by a common unknown delay.

Filter complexity may be critical with respect to both hardware and software capabilities, especially when many available measurements must be processed in realtime. It is hence worth noting that the most significant part of EKF complexity results from filter gain computations, which directly depend on the size of the observation vector. Hence we consider using the combination-based observations proposed in [19] to reduce the number of observation functions from the number N_A of anchors (or measurements) down to the dimension $\mathcal{K}$ of the location problem with no precision degradation.

For instance, let $\mathbf{W}(\mathbf{p}) = [\mathbf{w}_1 \ \mathbf{w}_2]$ be an $N_A \times \mathcal{K}$ combination matrix generating $\mathcal{K} = 2$ observation functions used in filter update (out of $N_A = 4$ measurements), as follows:

$$\mathbf{h}(\mathbf{p}) = [h_1(\mathbf{p}) \ h_2(\mathbf{p})]^T = \mathbf{W}(\mathbf{p})^T\mathbf{d}(\mathbf{p}), \qquad (28)$$

where $\mathbf{d}(\mathbf{p}) = [d_1(\mathbf{p}),\dots,d_4(\mathbf{p})]^T$ is a vector composed of the pseudodistances between the mobile node and the 4 BSs, and it is reminded $\mathbf{p} = [x \ y]^T$. Thus, the same position accuracy as that of the standard TDOA formulation can be preserved using CTDOA, as follows:

$$\mathbf{W}(\mathbf{p}) = \mathbf{R}_\mathbf{d}^{-1/2}\left[\boldsymbol{\Delta}'^\perp_x(\mathbf{p}) \ \boldsymbol{\Delta}'^\perp_y(\mathbf{p})\right], \qquad (29)$$

where

$$\boldsymbol{\Delta}'^\perp_x(\mathbf{p}) = \boldsymbol{\Delta}'_x(\mathbf{p}) - \left(\mathbf{1}'^T_{N_A,1}\boldsymbol{\Delta}'_x(\mathbf{p})\right)\mathbf{1}'_{N_A,1},$$
$$\boldsymbol{\Delta}'^\perp_y(\mathbf{p}) = \boldsymbol{\Delta}'_y(\mathbf{p}) - \left(\mathbf{1}'^T_{N_A,1}\boldsymbol{\Delta}'_y(\mathbf{p})\right)\mathbf{1}'_{N_A,1},$$
$$\boldsymbol{\Delta}'_x(\mathbf{p}) = \mathbf{R}_\mathbf{d}^{-1/2}\left.\frac{\partial\mathbf{d}(\mathbf{p})}{\partial x}\right|_\mathbf{p}, \qquad (30)$$
$$\boldsymbol{\Delta}'_y(\mathbf{p}) = \mathbf{R}_\mathbf{d}^{-1/2}\left.\frac{\partial\mathbf{d}(\mathbf{p})}{\partial y}\right|_\mathbf{p},$$

with $\mathbf{1}'_{N_A,1} = \mathbf{R}_\mathbf{d}^{-1/2}\mathbf{1}_{N_A,1}/\sqrt{\mathrm{tr}(\mathbf{R}_\mathbf{d}^{-1})}$, if $\mathbf{1}_{N_A,1}$ is an $N_A \times 1$ vector of ones and $\mathbf{R}_\mathbf{d}$ the diagonal covariance matrix of TOA-based measurements, whose diagonal is practically composed of $\hat{\sigma}_i^2$ ($i = 1,\dots,N_A$). Note that similar expressions are available in [19] for the simpler CTOA formulation.

Using such combinations, it is still possible to remove outliers with the indicator $I_{\sigma,i}$, while maintaining a constant filter architecture as a function of time. Within a more classical filter formulation, this would lead to change the size of the filter gain matrix, whereas this size is constant in our proposal and only dependent on $\mathcal{K}$. Practically, so as to discard the ith measurement, the only operation consists in replacing in $\mathbf{R}_\mathbf{d}^{-1/2}$ the ith diagonal element by 0 when computing $\mathbf{W}(\mathbf{p})$. Finally, the problem of choosing the best referential BS among the BSs providing nonoutlier measurements does not exist anymore here, unlike in the classical TDOA formulation.

Pedestrian Tracking Solution Combining an Impulse Radio Handset Transmitter with an Ankle-Mounted Inertial Measurement Unit

147

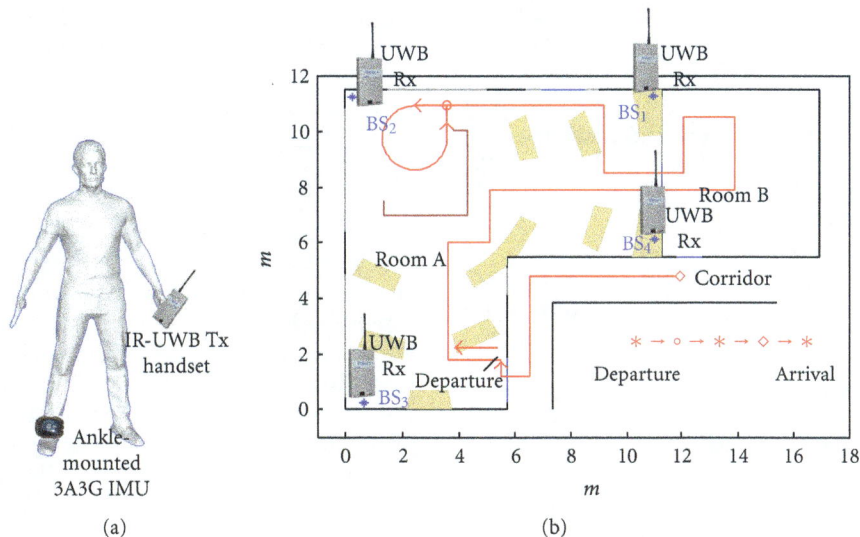

FIGURE 6: Experimental scenario: pedestrian equipped with an IR-UWB Tx handset and an ankle-mounted 3A3G IMU (a), back-forth real trajectory and layout of the indoor scene, including 4 isochronous IR-UWB Rx (BSs) in room A, with walls separating the 2 Rooms and the corridor.

6. Indoor Experiments

6.1. Experimental Setup. Real-life tracking experiments were carried out in a typical indoor environment at CEA-Leti Minatec premises, as shown on Figure 6. A pedestrian followed a 100 m-long round-trip path in two rooms and a corridor, referring to visual markers on the floor. The pedestrian was holding an IR-UWB transmitter in his hand and an IMU was attached to one of his ankles, in compliance with the fusion scenario and system architecture considered so far. We used inertial data from a 3A3G IMU at the sampling rate $f_s = 200$ Hz to estimate the heading and the waist average velocity. The latter was obtained by smoothing the online ankle velocity with a low-pass filter whose impulse response is 1.5 s-width rectangular. Without loss of generality with respect to TOA resolution or fusion concepts, the considered IR-UWB transmitter emitted in the band $[0.5, 1.1]$ GHz for implementation convenience at the *pulse repetition period* (PRP) of $1\,\mu s$. UWB antennas were connected to the four synchronous channels of a 6 GHz-bandwidth *digital storage oscilloscope* (DSO) using cables with N/SMA connectors, serving as four surrounding BSs in Room A. The DSO, enabling signal acquisition at 10 Gsps, integrated a PC for postprocessing. Relying on this setup, after averaging the incoming signal over sequences of 40 successive pulses to increase SNR before energy integration, the received signal was stored for consecutive time intervals of $T = 250$ ns (i.e., for the duration of the ED observation window). Then the ED-based TOA estimation method described in Section 3.2 was applied to the acquired signals. At the emulated receivers, the observation window was divided into $K = 125$ resolvable time bins of length $\Delta = 2$ ns (i.e., approximately the pulse width). The four DSO Rx channels were triggered synchronously but affected by the same unknown delay, which could vary from one acquisition to the next, hence requiring CTDOA as observations in the tracking filter, as

previously mentioned. Finally, the UWB acquisition rate was not constant, and reached 3 to 4 acquisitions per second, which is much lower than the inertial sampling rate.

6.2. Evaluation Procedure and Algorithms Benchmark. Five different estimator settings, depicted as so-called "scenarios" in the following, were tested and compared:

- (i) scenario 1: EKF tracking with IR-UWB only, using in the observation vector 2 CTDOA combinations of the 4 available pseudo-TOA measurements, with a time-invariant covariance matrix $\mathbf{R}_d^{(l')} = \mathbf{R}_d$ for all l' (i.e., l' being the time step whenever an UWB observation is available, with l a multiple of l');

- (ii) scenario 2: idem as scenario 1, except the covariance matrix $\mathbf{R}_d^{(l')}$ that is adjusted depending on $\{\widetilde{SNR}_i^{(l')}\}_{i=1,\dots,4}$ (using the tabulated function in Figure 4);

- (iii) scenario 3: idem as scenario 2, with additional outliers detection with $\{I_{\sigma,i}^{(l')}\}_{i=1,\dots,4}$;

- (iv) scenario 4: tracking with a stand-alone INS only;

- (v) scenario 5: EKF tracking fusing IR-UWB and INS data as in Section 5, that is, idem as scenario 3, with INS outputs controlling the filter state prediction (with predictions of the IMU frequency and updates whenever UWB measurements are available).

For scenarios 1, 2, and 3 (i.e., UWB only) the state vector of the corresponding EKF just contains the 2D cartesian

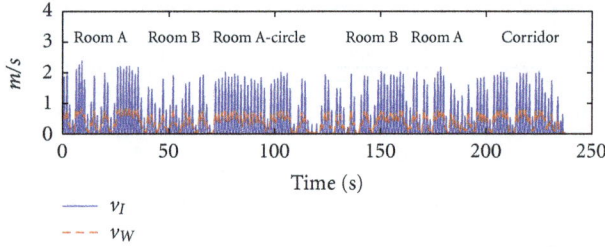

FIGURE 7: Pedestrian velocity v_W obtained by smoothing the horizontal velocity of the pedestrian ankle v_I, along the traveled trajectory.

TABLE 1: Average estimated traveled distance per actually traveled meter, for different tracking scenarios.

Scenarios	1	2	3	4	5
D (/1 m)	1.8911	2.1088	1.5712	0.8245	0.9466

coordinates and velocities $\mathbf{s} = [x\, y\, v_x\, v_y]^T$, with a classical linear transformation matrix

$$\mathbf{F} = \begin{bmatrix} 1 & 0 & dt^{(l'-1)} & 0 \\ 0 & 1 & 0 & dt^{(l'-1)} \\ 0 & 0 & 1 & 0 \\ 0 & 0 & 0 & 1 \end{bmatrix}, \tag{31}$$

where $dt^{(l'-1)}$ is the time elapsed between samples $l'-1$ and l' at the UWB subsystem. Scenario 4 being without UWB (i.e., stand-alone INS), only the state prediction step is applied, and the heading bias is systematically set to zero.

6.3. Performance Indicators. To assess and benchmark the performances of the proposed tracking schemes, two kinds of metrics are used. First, we consider an error distance between each estimated position $\hat{\mathbf{p}}^{(l)}$ issued at the filter output at time step l and its nearest orthogonal projection $\check{\mathbf{p}}^{(l)}$ onto the actual trajectory

$$\varepsilon_d^{(l)} = \left\| \hat{\mathbf{p}}^{(l)} - \check{\mathbf{p}}^{(l)} \right\|. \tag{32}$$

This instantaneous distance error is finally averaged into $\bar{\varepsilon}_d$, over specific portions of the trajectory or geographic areas (i.e., over selected but contiguous sequences of l).

As for the second performance indicator, we consider the difference between the real and estimated traveled distance (over the entire back and forth trajectory), normalized by the real traveled distance. This second relative performance indicator gives an idea about the uncertainty on the overall trajectory length, what could be interesting in several applications besides navigation (e.g., sports analysis, activity monitoring in physical rehab or as dietetics support, etc.).

6.4. Results and Discussion. Relying uniquely on the heading angle φ_b and on the pedestrian velocity v_W (i.e., after smoothing the ankle velocity v_I) shown on Figure 7 as a function of time, the tracking performance of a stand-alone dead-reckoning INS (i.e., scenario 4) is illustrated in Figure 8(a). The drifts affecting both the estimated heading and position clearly justify the use of a side IR-UWB subsystem here.

Figure 8(b) shows the estimated trajectory obtained within scenario 5, where the location drift is now significantly reduced. The dashed part of the trajectory refers to the first part of the walk (i.e., the one-way portion of the trajectory forth). Noting that BS_1 and BS_4 are in NLOS configurations in this first portion, thus the mobile could not properly correct the position drift until it gets sufficiently good pseudorange estimates. Consequently, the mobile gets closer to the real trajectory in the middle of the scene, and it even sticks to the real trajectory for the remaining part of the walk (see e.g., the trip back in straight lines).

For each scenario, Figure 9 shows the average location error $\bar{\varepsilon}_d$ over the entire trajectory and in each room separately. The error is particularly large in Room B for a single IR-UWB radiolocation system (i.e., scenarios 1 to 3). This is due to the combined harmful effects of generalized NLOS links (with respect to the four BSs simultaneously), to body shadowing, and to poor GDOP conditions. Comparing $\bar{\varepsilon}_d$ for the three first scenarios in Room B specifically, one can notice that using an adaptive observation covariance matrix $\mathbf{R}^{(l')}$ (scenario 2) and removing further outlier measurements (scenario 3) clearly help to reduce the error. In more favorable areas, for example in Room A (with at least one LOS links systematically) or in the corridor with light NLOS conditions (i.e., through plasterboard walls), even in the first scenario, the IR-UWB system alone would slightly outperform the INS system alone. As expected, fusing the two subsystems (scenario 5) reduces systematically the overall error, even if the enhancement is far more spectacular in Room B in comparison with both scenarios 3 and 4. These results open the floor to parsimonious fusion schemes, where one could switch from a stand-alone subsystem into the complete fusion-oriented system on demand, depending on the operating conditions, hence saving energy and complexity at the price of slight performance degradations.

Finally, Table 1 shows the average estimated distance D per actually traveled meter. This takes into account the distance between consecutively estimated locations. Due to NLOS situations, a tracking system only based on IR-UWB would tend to overestimate the traveled distance mostly because of occasional but strongly biased TOA-based measurements, which lead to nonstraight and more erratic estimated paths. Omitting measurement outliers, the estimated traveled distance is significantly reduced, but still rather large. The use of INS then enables us to reduce significantly this error.

The previous results illustrate the complementarity of the two subsystems and the potential of the proposed fusion scheme, under the architectural constraints of noncollocated and asynchronous sensors. The IR-UWB part of the system tends to correct the heading drift and resolves the growing error of the INS in time, whereas the INS part helps IR-UWB radiolocation in generalized NLOS situations and/or in penalizing mobile locations that would experience bad geometrical configurations with respect to the BSs

Pedestrian Tracking Solution Combining an Impulse Radio Handset Transmitter with an Ankle-Mounted Inertial Measurement Unit

149

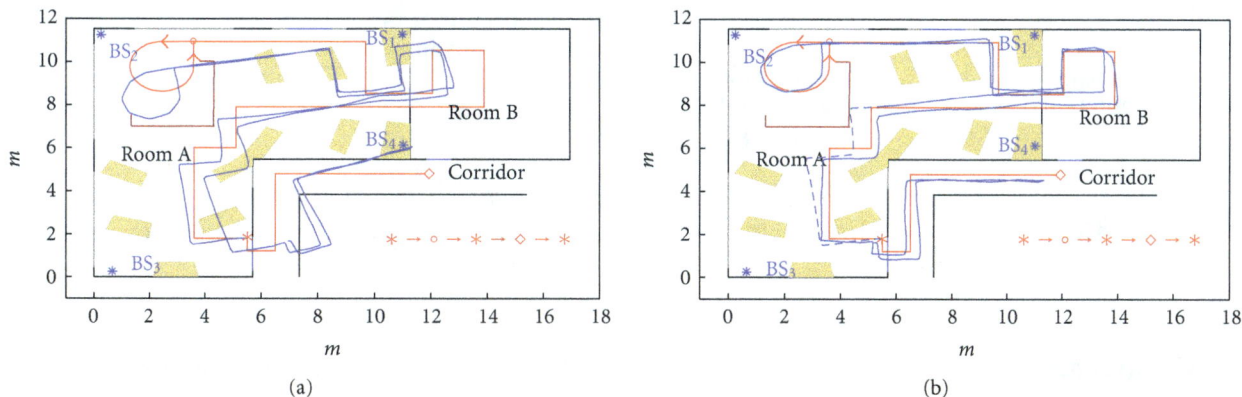

(a)

(b)

FIGURE 8: Indoor layout of tracking experiments, with 4 BSs in room A, the 2 rooms and the corridor being separated by walls. Actual (red) and estimated (blue) trajectories with an ankle-mounted 3A3G IMU only (a), or a 3A3G IMU loosely coupled with IR-UWB (b).

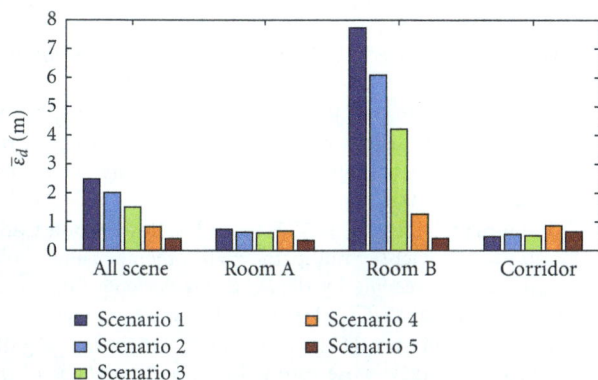

FIGURE 9: Average location errors for the whole trajectory and for each room in different tracking scenarios.

(e.g., in room B). Finally coupling both systems enables reliable and robust pedestrian tracking with an almost uniform quality all over the scene.

7. Conclusion

In this paper, we have addressed the problem of pedestrian tracking by coupling an IR-UWB transmitter handset with an ankle-mounted INS device. One motivation to couple these subsystems was to overcome their respective limitations in harmful operating indoor environments, while benefiting from their complementary capabilities.

The raw measurements of each subsystem have been carefully studied, and new preprocessing techniques have been proposed before applying hybrid data fusion techniques. Regarding IR-UWB first, TOAs are estimated at low complexity energy-based receiver following a Bayesian approach. A new practical criterion predicting the standard deviation of estimated TOAs and a methodology to identify outlier measurements have been proposed. Real channel measurements have been carried out and exploited to validate a few statistical multipath parameters. As for INS, in order to limit the drift due to noise integration and

avoid magnetic disturbances, we have considered one ankle-mounted IMU with a 3-axis gyroscope and a 3-axis accelerometer, whose measurements are processed to determine the pedestrian average velocity and biased heading. This ankle-mounted INS (when considered with a UWB Tx handset) obviously imposes further challenging constraints in terms of system architecture, while operating with non-collocated and asynchronous heterogeneous devices, hence impacting in turn the fusion strategy.

To merge the heterogeneous data from both subsystems, we have proposed a specific EKF filter formulation. Combined time difference of arrival [19] of nonoutlier TOA measurements are used as observation inputs to the filter, the observation covariance matrix being dynamically scaled thanks to the new practical TOA estimation quality indicator. The use of CTDOA enables to remove outliers without changing the low complexity structure of the tracking filter. The velocity and heading estimates issued at the IMU are taken into account into the filter during the prediction phase. This option enables us to take benefit from zero velocity information during the stance phase, as well as to operate under different sampling/refreshment rates for the IMU and IR-UWB subsystems.

Experimental results illustrate the complementarity of the two subsystems and the efficiency of the proposed fusion scheme. In particular, the IR-UWB part corrects the heading drift and resolves the growing error of the INS as a function of time, whereas the INS part advantageously assists IR-UWB radiolocation in generalized NLOS situations and/or in penalizing mobile locations that would experience bad geometrical configurations with respect to the BSs. Overall, coupling both systems enables reliable and robust tracking with uniform quality of service over the scene.

Finally, the possibility to apply parsimonious fusion schemes, hence saving energy and complexity at the price of slight performance degradations, has been pointed out and briefly discussed. Accordingly, one could switch from a stand-alone subsystem into the complete fusion-oriented system on demand, depending on the operating conditions (i.e., while experiencing generalized NLOS or generalized LOS, poor or favorable GDOP, etc.).

Appendix

Let $q_{xy} = [q_{xy,0} \quad q_{xy,1} \quad q_{xy,2} \quad q_{xy,3}]^T$ be the quaternion that denotes the orientation of the "X" frame relative to the "Y" frame, then the associated rotation matrix $\mathbf{R}_{xy}$ is given in the following equation:

$$\mathbf{R}_{xy} = \begin{bmatrix} 2\,q_{xy,0}^2 - 1 + 2\,q_{xy,1}^2 & 2\,q_{xy,1}\,q_{xy,2} - 2\,q_{xy,0}\,q_{xy,3} & 2\,q_{xy,1}\,q_{xy,3} + 2\,q_{xy,0}\,q_{xy,2} \\ 2\,q_{xy,1}\,q_{xy,2} + 2\,q_{xy,0}\,q_{xy,3} & 2\,q_{xy,0}^2 - 1 + 2\,q_{xy,2}^2 & 2\,q_{xy,2}\,q_{xy,3} - 2\,q_{xy,0}\,q_{xy,1} \\ 2\,q_{xy,1}\,q_{xy,3} - 2\,q_{xy,0}\,q_{xy,2} & 2\,q_{xy,2}\,q_{xy,3} + 2\,q_{xy,0}\,q_{xy,1} & 2\,q_{xy,0}^2 - 1 + 2\,q_{xy,3}^2 \end{bmatrix}. \tag{A.1}$$

Acknowledgments

This work has been performed in the framework of the *WHERE* (ICT-217033) and *WHERE2* (ICT-248894) projects, which are partly funded by the European Union. The authors would like to thank also the people from Movea (http://www.movea.com/) for their precious technological support as regards to inertial systems. The authors also thank the reviewers for their valuable comments and suggestions.

References

[1] A. H. Sayed, A. Tarighat, and N. Khajehnouri, "Network-based wireless location: challenges faced in developing techniques for accurate wireless location information," *IEEE Signal Processing Magazine*, vol. 22, no. 4, pp. 24–40, 2005.

[2] K. Yu, I. Sharp, and Y. J. Guo, *Ground-Based Wireless Positioning*, IEEE Series on Digital & Mobile Communication, John Wiley & Sons, 2009, 2012 DRAFT.

[3] B. Denis, J. B. Pierrot, and C. Abou-Rjeily, "Joint distributed synchronization and positioning in UWB Ad Hoc networks using TOA," *IEEE Transactions on Microwave Theory and Techniques*, vol. 54, no. 4, part 2, pp. 1896–1910, 2006.

[4] "PART 15.4: Wireless Medium Access Control (MAC) and Physical Layer (PHY) Specifications for Low-Rate Wireless Personal Area Networks (LR-WPANs)," IEEE Std 802.15.4a-2007, Amendment 1, Add Alternate PHYs, 2007.

[5] D. Lachartre, D. Morche, L. Ouvry et al., "A 1.1nJ/bit 802.15.4a-compliant fully integrated UWB transceiver in 0.13μm CMOS," in *Proceedings of the IEEE International Solid-State Circuits Conference (ISSCC '09)*, San Francisco, Calif, USA, February 2009.

[6] J.-Y. Lee and R. A. Scholtz, "Ranging in a dense multipath environment using an UWB radio link," *IEEE Journal on Selected Areas in Communications*, vol. 20, no. 9, pp. 1677–1683, 2002.

[7] I. Guvenc, S. Gezici, and Z. Sahinoglu, "Ultra-wideband range estimation: theoretical limits and practical algorithms," in *Proceedings of the IEEE International Conference on Ultra-Wideband (ICUWB '08)*, pp. 93–96, September 2008.

[8] S. Gezici, Z. Tian, G. B. Giannakis et al., "Localization via ultra-wideband radios: a look at positioning aspects of future sensor networks," *IEEE Signal Processing Magazine*, vol. 22, no. 4, pp. 70–84, 2005.

[9] I. Guvenc, S. Gezici, and Z. Sahinoglu, "Ultra-wideband range estimation: theoretical limits and practical algorithms," in *Proceedings of the IEEE International Conference on Ultra-Wideband (ICUWB '08)*, pp. 93–96, September 2008.

[10] E. Foxlin, "Pedestrian tracking with shoe-mounted inertial sensors," *IEEE Computer Graphics and Applications*, vol. 25, no. 6, pp. 38–46, 2005.

[11] C. W. Tan and S. Park, "Design of accelerometer-based inertial navigation systems," *IEEE Transactions on Instrumentation and Measurement*, vol. 54, no. 6, pp. 2520–2530, 2005.

[12] L. Fang, P. J. Antsaklis, L. A. Montestruque et al., "Design of a wireless assisted pedestrian dead reckoning system—the NavMote experience," *IEEE Transactions on Instrumentation and Measurement*, vol. 54, no. 6, pp. 2342–2358, 2005.

[13] F. Caron, E. Duflos, D. Pomorski, and P. Vanheeghe, "GPS/IMU data fusion using multisensor Kalman filtering: introduction of contextual aspects," *Information Fusion*, vol. 7, no. 2, pp. 221–230, 2006.

[14] D. B. Jourdan, J. J. Deyst, M. Z. Win, and N. Roy, "Monte Carlo localization in dense multipath environments using UWB ranging," in *Proceedings of the IEEE International Conference on Ultra-Wideband (ICU '05)*, pp. 314–319, September 2005.

[15] J. D. Hol, F. Dijkstra, H. Luinge, and T. B. Schöny, "Tightly coupled UWB/IMU pose estimation," in *Proceedings of the IEEE International Conference on Ultra-Wideband (ICUWB '09)*, pp. 688–692, September 2009.

[16] S. Pittet, V. Renaudin, B. Merminod, and M. Kasser, "UWB and MEMS based indoor navigation," *Journal of Navigation*, vol. 61, no. 3, pp. 369–384, 2008.

[17] V. Renaudin, B. Merminod, and M. Kasser, "Optimal data fusion for pedestrian navigation based on UWB and MEMS," in *Proceedings of the IEEE/ION Position, Location and Navigation Symposium (PLANS '08)*, pp. 341–349, May 2008.

[18] F. Evennou and F. Marx, "Advanced integration of WiFi and inertial navigation systems for indoor mobile positioning," *Eurasip Journal on Applied Signal Processing*, vol. 2006, Article ID 86706, 2006.

[19] J. Youssef, B. Denis, C. Godin, and S. Lesecq, "Reducing the complexity order of position estimators with combined radiolocation measurements," in *Proceedings of the 6th Workshop on Positioning, Navigation and Communication (WPNC '09)*, pp. 217–222, March 2009.

[20] J. Youssef, B. Denis, C. Godin, and S. Lesecq, "New TOA estimators within energy-based receivers under realistic UWB channel statistics," in *Proceedings of the IEEE 71st Vehicular Technology Conference (VTC '10)*, pp. 1–5, May 2010.

[21] D. Dardari, C. C. Chong, and M. Z. Win, "Threshold-based time-of-arrival estimators in UWB dense multipath channels," *IEEE Transactions on Communications*, vol. 56, no. 8, pp. 1366–1378, 2008.

[22] A. F. Molisch, D. Cassioli, C. Chong et al., "A comprehensive standardized model for ultrawideband propagation channels," *IEEE Transactions on Antennas and Propagation*, vol. 54, no. 11, part 1, pp. 3151–3166, 2006.

Pedestrian Tracking Solution Combining an Impulse Radio Handset Transmitter with an Ankle-Mounted Inertial Measurement Unit

151

[23] D. Dardari, C.-C. Chong, and M. Z. Win, "Improved lower bounds on time-of-arrival estimation error in realistic UWB channels," in *Proceedings of the IEEE International Conference on Ultra-Wideband (ICUWB '06)*, pp. 531–538, September 2006.

[24] http://www.movea.com/.

[25] A. Croci, M. De Agostino, and A. M. Manzino, "A GNSS/INS-based architecture for rescue team monitoring," in *Proceedings of the International Conference on Indoor Positioning and Indoor Navigation (IPIN '10)*, pp. 827–841, Zurich, Switzerland, September 2010.

[26] H.-J. Jang, J. W. Kim, and D.-H. Hwang, "Robust step detection method for pedestrian navigation systems," *Electronics Letters*, vol. 43, no. 14, pp. 749–751, 2007.

[27] I. Skog, J.-O. Nilsson, and P. Handel, "Evaluation of zero-velocity detectors for foot-mounted inertial navigation systems," in *Proceedings of the International Conference on Indoor Positioning and Indoor Navigation (IPIN '10)*, pp. 959–969, Zurich, Switzerland, September 2010.

[28] S. H. Shin, C. G. Park, J. W. Kim, H. S. Hong, and J. M. Lee, "Adaptive step length estimation algorithm using low-cost MEMS inertial sensors," in *Proceedings of the IEEE Sensors Applications Symposium (SAS '07)*, San Diego, Calif, USA, February 2007.

[29] S. Miyazaki, "Long-term unrestrained measurement of stride length and walking velocity utilizing a piezoelectric gyroscope," *IEEE Transactions on Biomedical Engineering*, vol. 44, no. 8, pp. 753–759, 1997.

[30] N. Castaneda and S. Lamy-Perbal, "An improved shoe-mounted inertial navigation system," in *Proceedings of the International Conference on Indoor Positioning and Indoor Navigation(IPIN '10)*, pp. 934–944, Zurich, Switzerland, September 2010.

[31] P. Coronel, S. Furrer, W. Schott, and B. Weiss, "Indoor location tracking using inertial navigation sensors and radio beacons," *Lecture Notes in Computer Science*, vol. 4952, pp. 325–340, 2008.

[32] I. Guvenc, Z. Sahinoglu, and P. V. Orlik, "TOA estimation for IR-UWB systems with different transceiver types," *IEEE Transactions on Microwave Theory and Techniques*, vol. 54, no. 4, pp. 1876–1886, 2006.

[33] I. Guvenc and Z. Sahinoglu, "Threshold-based TOA estimation for impulse radio UWB systems," in *Proceedings of the IEEE International Conference on Ultra-Wideband (ICU '05)*, pp. 420–425, Zurich, Switzerland, September 2005.

[34] D. Cassioli, M. Z. Win, and A. F. Molisch, "The ultra-wide bandwidth indoor channel: from statistical model to simulations," *IEEE Journal on Selected Areas in Communications*, vol. 20, no. 6, pp. 1247–1257, 2002.

[35] Y. Bar-Shalom and T. E. Fortmann, *Tracking and Data Association*, Academic Press Professional, 1987.

[36] K. M.-L. Saxena and K. Alam, "Estimation of the non-centrality parameter of a chi squared distribution," *The Annals of Statistics*, vol. 10, no. 3, pp. 1012–1016, 1982.

[37] S. Gezici, H. Kobayashi, and H. V. Poor, "Non-parametric non-line-of-sight identification," in *Proceedings of the IEEE 58th Vehicular Technology Conference (VTC-Fall '03)*, pp. 2544–2548, Orlando, Fla, USA, October 2003.

[38] A. Rabbachin, I. Oppermann, and B. Denis, "GML ToA estimation based on low complexity UWB energy detection," in *Proceedings of the IEEE 17th International Symposium on Personal, Indoor and Mobile Radio Communications (PIMRC 06)*, Helsinki, Finland, September 2006.

[39] Q. Ladetto and B. Merminod, "In step with INS: navigation for the blind, tracking emergency crews," *GPS World*, vol. 13, no. 10, pp. 30–38, 2002.

[40] V. Renaudin, O. Yalak, P. Tom, and B. Merminod, "Indoor navigation of emergency agents," *European Journal of Navigation*, vol. 5, no. 3, pp. 36–45, 2007.

13

Visual Flight Control of a Quadrotor Using Bioinspired Motion Detector

Lei Zhang,[1] Tianguang Zhang,[1] Haiyan Wu,[1] Alexander Borst,[2] and Kolja Kühnlenz[1,3]

[1] Institute of Automatic Control Engineering (LSR), Technische Universität München, 80290 München, Germany
[2] Department of Systems and Computational Neurobiology, Max Planck Institute of Neurobiology, Am Klopferspitz 18, D-82152 Martinsried, Germany
[3] Institute for Advanced Study (IAS), Technische Universität München, 80290 München, Germany

Correspondence should be addressed to Lei Zhang, lei.zhang0918@gmail.com

Academic Editor: Farid Melgani

Motion detection in the fly is extremely fast with low computational requirements. Inspired from the fly's vision system, we focus on a real-time flight control on a miniquadrotor with fast visual feedback. In this work, an elaborated elementary motion detector (EMD) is utilized to detect local optical flow. Combined with novel receptive field templates, the yaw rate of the quadrotor is estimated through a lookup table established with this bioinspired visual sensor. A closed-loop control system with the feedback of yaw rate estimated by EMD is designed. With the motion of the other degrees of freedom stabilized by a camera tracking system, the yaw-rate of the quadrotor during hovering is controlled based on EMD feedback under real-world scenario. The control performance of the proposed approach is compared with that of conventional approach. The experimental results demonstrate the effectiveness of utilizing EMD for quadrotor control.

1. Introduction

Flying insects have tiny brains and mostly possess compound eyes which can get panoramic scene to provide an excellent flying performance. Comparing with state-of-the-art artificial visual sensors, the optics of compound eye provide very low spatial resolution. Nevertheless, the behavior of flying insects is mainly dominated by visual control. They use visual feedback to stabilize flight [1], control flight speed, [2] and measure self-motion [3]. On the other hand, highly accurate real-time stabilization and navigation of unmanned aerial vehicles (UAVs) or microaerial vehicles (MAVs) is becoming a major research interest, as these flying systems have significant value in surveillance, security, search, and rescue missions. Thus, the implementation of a bio-plausible computation for visual systems could be an accessible method to replace the traditional image processing algorithms in controlling flying robots such as a quadrotor.

Most of early applications using insect-inspired motion detector focus on motion detection tasks rather than velocity estimation. In robotics and automation applications, EMDs

are mainly used for a qualitative interpretation of video image sequence, to provide general motion information such as orientation and infront obstacles. In [4], a microflyer with an onboard lightweight camera is developed, which is able to fly indoor while avoiding obstacles by detecting certain changes in optic flow. The recent approach for the navigation in a corridor environment on an autonomous quadrotor by using optical flow integration is shown in [5]. Another example is a tethered optic flow-based helicopter that mimics insect behaviors such as taking off, cruise, and landing [6, 7], and in [8] the EMDs visual sensors were tested and characterized in field experiments under various lighting conditions. Numerous authors have pointed out that the Reichardt model, while sensitive to motion, does not measure velocity [9–11]. However, some efforts have been made, examining the possibility of velocity estimation tasks by introducing elaborated models [12, 13]. In [14], yaw rate estimates on a coaxial helicopter testbed are obtained using a matched filter approach which yet incorporates a virtual 3D environment in the control loop. Although a lot of work has been done on a simulation level or involving simulation tools, further robotic applications with EMDs considering closed-loop velocity control are still to be investigated under real-world scenarios.

In this paper, a quadrotor system with bioinspired visual sensor is described. The novel image processing methods and the control laws are implemented in real-time experiments. The yaw rate control is totally based on the visual feedback of the on-board camera. The reference velocity value is provided by the on-board inertial measurement unit (IMU). For a 6-DOF (degrees of freedom) flying robot control, a tracking system of multicamera configuration is also utilized to achieve the altitude and attitude stabilization near hover. Due to the noise in real-world, the velocity estimation tasks would be more challenging. In this work, the approach of building an empirical lookup table from open-loop test results is introduced for this task. The Reichardt motion detector is modified which describes, at an algorithmic level, the process of local motion detection in flying behaviors. Certain patterns of receptive fields, which respond to particular optic flow, are utilized to estimate the global ego-motion through the environment. Another main issue which needs to be tackled carefully in this work is that the flying robot should be well stabilized during hovering. By multicamera tracking, the absolute position as well as the pose is determined from the positions of four on-board markers.

The remainder of this paper is organized as follows: in Section 2, we firstly introduce the bioinspired visual image processing methods used in this work. In Section 3, the 3D pose estimation using visual tracking system is described. The control strategy of the system as well as the software structure of algorithms is presented in Section 4. Then in Section 5, an overview of the whole experimental platform is illustrated. The control performance is also evaluated based on the experimental results in this section. Conclusions are given in Section 6, with directions on future works.

2. Bioinspired Image Processing

In this section, we introduce the essential part of this work: using biological models for yaw rate estimation of a quadrotor. The EMDs are utilized for this task. The whole methodology is introduced in detail. To achieve the yaw rate control, the system also requires accurate visual tracking for pose stabilization (Section 3) and efficient controllers (Section 4).

In an insect's perspective, motion information has to be computed from the changing retinal images by the nervous system [15]. For engineering applications, some properties of the biological visual system are converted into computational algorithms.

The elaborated EMD model used in this work is a modified model of the famous Reichardt motion detector [16]. The original Reichardt motion detector (Figure 1(a)) has only low-pass filters and two correlations. In this work, a temporal high-pass filter is added before the low-pass filter to obtain a simple response to step edges [17] (Figure 1(b)). The high-pass filters and low-pass filters in this model are all designed to be of first order.

In [12], a mathematical analysis of the original Reichardt motion detector is given regarding the response to different images (sinusoidal gratings as well as natural images). Without loss of generality, we firstly consider the response of this modified model to a moving natural image (which possesses energy at all spatial frequencies). Similar to the response of the simplified model [12], for this modified model, the output is

$$R = \int_0^\infty \frac{2\pi\tau_L f_s v \sin(2\pi f_s \Delta\phi)}{1 + (\tau_L/\tau_H)^2 + (1/2\pi f_s v \tau_H)^2 + (2\pi f_s v \tau_L)^2} P(f_s) df_s, \tag{1}$$

where $\Delta\phi$ is the angular displacement between the two vision sensors, f_s is the spatial frequency of the image input to the detector, v stands for the velocity of the moving image, τ_L and τ_H are the time constants of the low- and high-pass filters, respectively, and $P(f_s)$ represents the power spectral density. So according to (1), the local motion information is calculated.

To obtain a global ego-motion estimation, certain receptive fields of the motion-sensitive widefield neurons in the fly brain are applied. Considering the specified experimental scenario in this work, two novel templates of receptive fields for rotation detection are utilized (Figure 2), which are proposed in [17].

The algorithms for calculating the rotation global response can be described as (image size: length × width; R_H: response of local horizontal motion; R_V: response of local vertical motion):

$$\text{Horizontal rotation} = \sum_{i=1}^{\text{width/2}} \sum_{j=1}^{\text{length}} R_H - \sum_{i=\text{width/2}+1}^{\text{width}} \sum_{j=1}^{\text{length}} R_H,$$

$$\text{Vertical rotation} = \sum_{i=1}^{\text{width}} \sum_{j=1}^{\text{length/2}} R_V - \sum_{i=1}^{\text{width}} \sum_{j=\text{length/2}+1}^{\text{length}} R_V. \tag{2}$$

Now we examine the feasibility of using this model for velocity estimation tasks. In [12], two criteria are quantified for an accurate velocity estimation system: (1) image motion at a fixed velocity should always have approximately response; (2) at a given velocity, the response to motion should be unambiguous over certain range. In simulation, we find that by introducing this modified model, the response to a specific velocity of image motion can meet the two basic requirements at low velocities (above which the response output is ambiguous). Thus, for velocity estimation tasks, the motion velocity should be limited in a certain range due to essential property (bell-shaped response) of the Reichardt model. In order to reduce the brightness sensitivity, logarithmic transformation could be also applied (as the modified model in [17]). However, by doing this the discrimination of response is also highly reduced. That means, the response at a given velocity cannot differ significantly from the response at other velocities, which is not appropriate for quantifying velocity. Moreover,

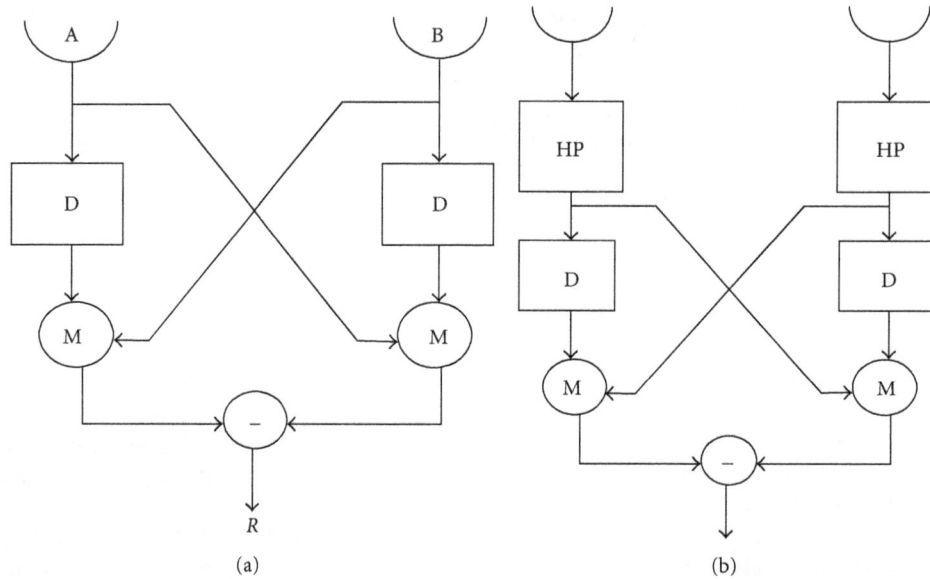

FIGURE 1: The simple Reichardt motion detector (a) and the elaborated EMD model (b). A and B are two visual signal inputs. The moving luminance signal is observed by this pair of visual sensors (such as the ommatidia of a fruit fly). The detector compares visual sensor inputs and generates a direction sensitive response R corresponding to the visual motion. D: delay block. Here it refers to low-pass filter; HP: High-pass filter; M: multiplication.

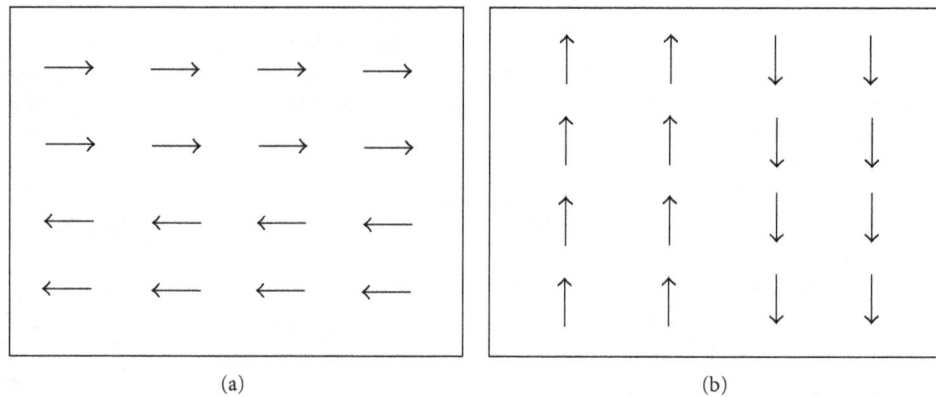

FIGURE 2: Rotation templates: horizontal (a) and vertical (b). The response of horizontal rotation is calculated by subtracting the upper part and lower part of the template of receptive fields, while the vertical rotation is detected by the difference between left part and right part of the global optical flow.

regarding the brightness sensitivity problem, a stable lighting condition is demanded in real-time experiments.

We firstly examine the open-loop characteristics of the system only for yaw rate estimation. The quadrotor is tethered in the air, with an on-board camera looking directly to the ground texture (the complete system is further introduced in Section 5). This scenario in the indoor environment involves a black-white chessboard ground texture. It is considered to be the most suitable scenario for detecting rotation motion of a flying robot. Compared to other forms of textures, the high image contrast can also help to improve the discrimination for quantifying velocity (due to the characteristics of the biological model itself). The quadrotor is rotated on horizontal level without control, and we can get the relationship between the response output and

the rotation velocity (yaw rate). A lookup table is then built. Due to the system noise and discrimination limitations of the experiments, the curve has some nonmonotonic regions. The polynomial minimum quadric method is used to fit the curve, (where d_i is a residual which is defined as the difference between the predicted value y_i and the actual value $f(x_i)$):

$$\sum_{i=1}^{n} d_i^2 = \sum_{i=1}^{n} [y_i - f(x_i)]^2 = \text{minimum}. \qquad (3)$$

The experimental results of the open-loop characteristics are shown and further discussed in Section 5.

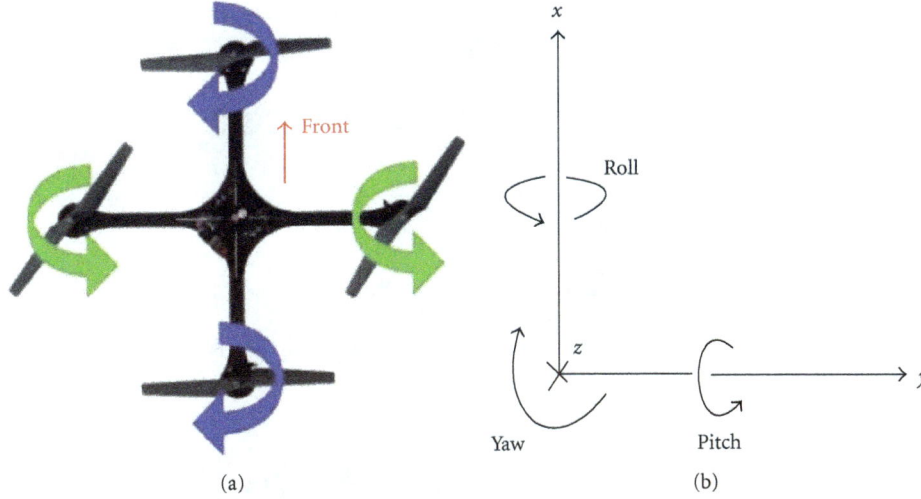

FIGURE 3: Quadrotor dynamics. (a) the overhead view of a quadrotor; (b) six degrees of freedom of the flying robot.

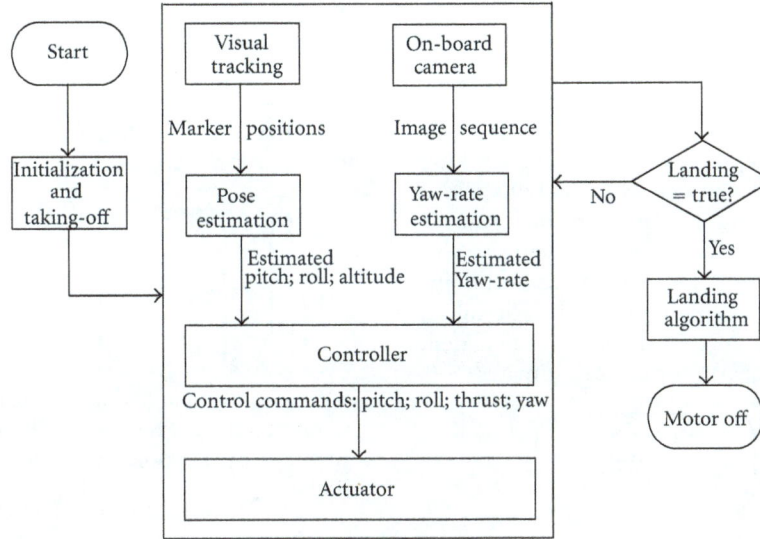

FIGURE 4: Software architecture.

3. Multicamera 3D Pose Estimation

Quadrotor is an underactuated vehicle. The 6 DOFs of the quadrotor are controlled by four motor inputs (pitch, roll, thrust, and yaw) by varying the lift forces and the torque balance through changing the rotating speed of the rotors (see Figure 3(a)). Yaw control is realized by tuning the differential speed between the two counterrotating rotor pairs. Increasing the rotating speed of all the four motors at the same amount will cause an upward movement of the quadrotor. When tuning the differential speed between the two motors of either single rotor pair, the quadrotor will fly sideways.

The work in this section is based on the former related work in [18]. In this work, we set up an indoor GPS system by using multicamera tracking instead of the former two-camera tracking. By tracking the four markers installed on the axis of the quadrotor, the 3D position as well as the pose of the flying robot can be estimated. The experimental setup of 3D tracking is further introduced in Section 5. The frame of quadrotor dynamics is the same as the in Figure 3(b). We have the following definition:

Marker i position vector: $S_i = (x_i, y_i, z_i)^T$ $(i = 1, 2, 3, 4)$.

Central point vector between two nonadjacent markers: $M_j = (x_{M_j}, y_{M_j}, z_{M_j})^T$ $(j = 1, 2)$;

Estimated central point vector of the quadrotor: $M_q = (x_M, y_M, z_M)^T$;

Orientation of marker i: $V_i = (x_{V_i}, y_{V_i}, z_{V_i})^T$ $(i = 1, 2, 3, 4)$;

The counting of the markers is clockwise, while the first marker is on the main axis. For 3D pose control, the central point should be used as the reference position of the quadrotor. The central points of the distance between marker 1 and marker 3 as well as between marker 2 and

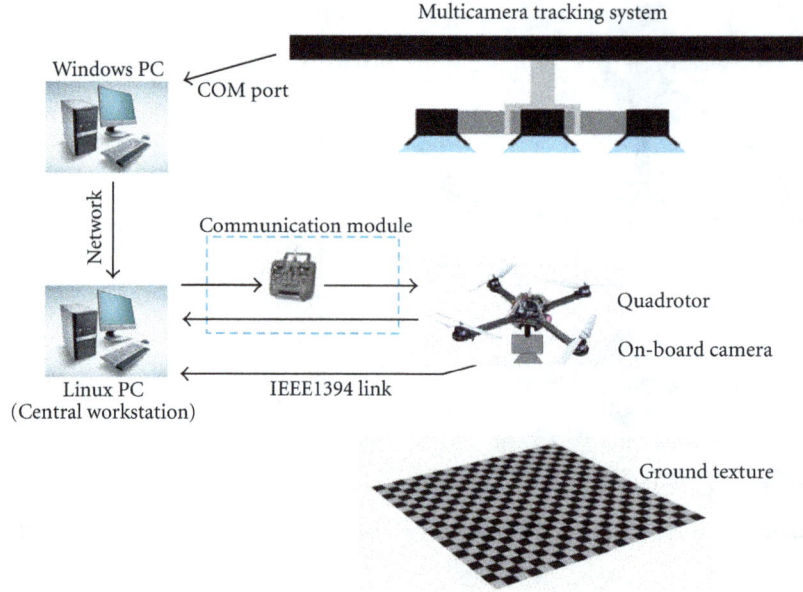

FIGURE 5: Experimental setup.

marker 4 are $M_1 = (1/2)(S_1 + S_3)$ and $M_2 = (1/2)(S_2 + S_4)$. In consequence of the marker's noise through the tracking system, the two central points in the two equations above are not identical. Thus, the central point of the quadrotor is $M_q = (1/2)(M_1 + M_2)$. The vectors between the central point and marker 1 for pitch as well as between the central point and marker 2 for roll are $V_i = S_i - M_q$ $(i = 1, 2)$. The values of pitch θ, roll ϕ, and yaw ψ angles can be then calculated, and thus the 3D pose of the quadrotor can be estimated:

$$\theta = \text{sgn}(z_{v_1}) \arccos\left(\frac{\sqrt{x_{v_1}^2 + y_{v_1}^2}}{\sqrt{x_{v_1}^2 + y_{v_1}^2 + z_{v_1}^2}}\right), \qquad (4)$$

$$\phi = \text{sgn}(z_{v_2}) \arccos\left(\frac{\sqrt{x_{v_2}^2 + y_{v_2}^2}}{\sqrt{x_{v_2}^2 + y_{v_2}^2 + z_{v_2}^2}}\right), \qquad (5)$$

$$\psi = -\text{sgn}(y_{v_1}) \arccos\left(\frac{x_{v_1}}{\sqrt{x_{v_1}^2 + y_{v_2}^2}}\right). \qquad (6)$$

4. Controller

At first the quadrotor should be regulated to hover in the air on horizontal plane with little shaking. That means, the stable state commands (u_p^s for pitch, u_r^s for roll, u_y^s for yaw, and u_t^s for thrust) should be adjusted firstly. Basing on these parameters, the control commands can be calculated next. For each controller, we have an output value (u_p^q for pitch, u_r^q for roll, u_y^q for yaw, and u_t^q for thrust) between -1 and 1, which is then added with the corresponding stable state command. Since we only consider the rotation movement in this experiment, which means, the quadrotor is not always heading with the main axis towards X direction, the pitch

and roll commands should be adjusted in the ψ direction with a rotation matrix:

$$\begin{pmatrix} u_p \\ u_r \\ u_t \\ u_y \end{pmatrix} = \begin{bmatrix} \cos\psi & -\sin\psi & 0 & 0 \\ \sin\psi & \cos\psi & 0 & 0 \\ 0 & 0 & 1 & 0 \\ 0 & 0 & 0 & 1 \end{bmatrix} \begin{pmatrix} u_p^q \\ u_r^q \\ u_t^q \\ u_y^q \end{pmatrix} + \begin{pmatrix} u_p^s \\ u_r^s \\ u_t^s \\ u_y^s \end{pmatrix}. \qquad (7)$$

We choose proportional-integral (PI) controller for the yaw rate control. The pitch, roll, and thrust commands are controlled by proportional-integral-derivative (PID) controllers:

$$u_y^q = k_y^P(\dot{\psi}_0 - \dot{\psi}) + k_y^I \int (\dot{\psi}_0 - \dot{\psi}) dt,$$

$$u_p^q = k_p^P(\theta_0 - \theta) + k_p^I \int (\theta_0 - \theta) dt + k_p^D \frac{d}{dt}(\theta_0 - \theta),$$

$$u_r^q = k_r^P(\phi_0 - \phi) + k_r^I \int (\phi_0 - \phi) dt + k_r^D \frac{d}{dt}(\phi_0 - \phi),$$

$$u_t^q = k_t^P(z_0 - z) + k_t^I \int (z_0 - z) dt + k_t^D \frac{d}{dt}(z_0 - z). \qquad (8)$$

In this experiment, the reference values θ_0 and ϕ_0 are set to zero. The desired altitude z_0 is 0.35 m near hover. The measured values θ and ϕ are calculated from the received data of the visual multicamera tracking system using (4) and (5), whereas ψ is searched out from a certain empirical lookup table using the response value, which is calculated by insect-inspired motion detectors. The yaw velocity can also be obtained from (6) by time derivative, which is, for the heading stabilization, used as a reference (ground truth). The closed-loop results will be shown and further discussed in Section 5.

(a)

(b)

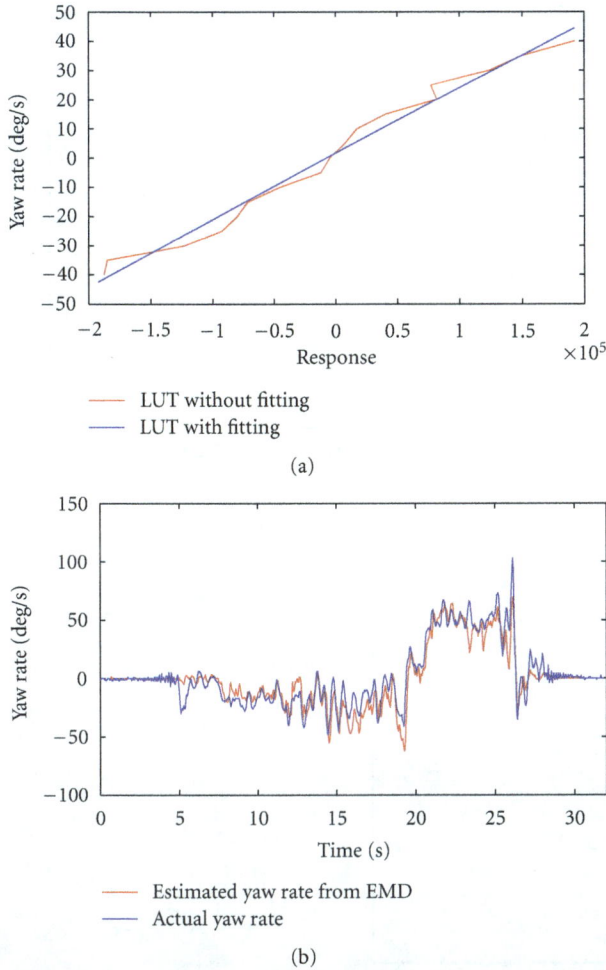

FIGURE 6: (a) lookup Table (LUT) in this work. This empirical LUT shows the relationship between yaw rate (in low speed) and visual response under the experimental environment; (b) open-loop characteristics for rotation motion near hover (without control).

The main loop in the whole software architecture (Figure 4) consists of two simultaneous processes: yaw rate control using on-board camera visual feedback and X, Y, and Z position/poses control using visual tracking system. A graphical user interface (GUI) is developed basing on Qt cross-platform application, which provides data visualization (e.g., 3D trajectory of the quadrotor, battery voltage and sensory data information), commands input and real-time online configuration of control parameters.

5. Experiments and Results

5.1. Experimental Setup. The whole experimental platform is shown in Figure 5. It mainly consists of a quadrotor testbed, off-board workstation, and video camera tracking system.

(1) Quadrotor. The miniquadrotor used in this work is a "Hummingbird" with an "AutoPilot" central control board from ascending technologies. It offers a 1 kHz control frequency and motor update rate, which guarantees fast

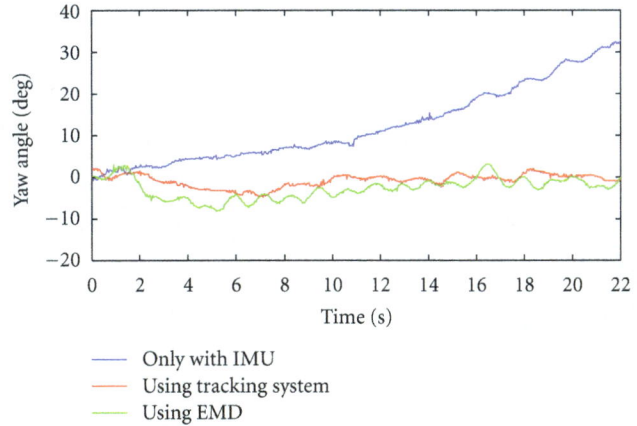

FIGURE 7: Heading stabilization.

response to changes in the environment. The size of the whole quadrotor testbed is 36.5 cm in diameter, and the four rotors (each with a propeller size of 19.8 cm) are directly driven by four high-torque DC brushless motors respectively. Powered by a state-of-the-art 3-cell 2100mAh lithium-polymer battery, the vehicle is able to hover up to 15 minutes (with about 120 g of payload in this work).

(2) On-Board Camera. Considering the limited payloads of the quadrotor, the PointGrey Firefly MV CMOS camera which has a light weight (14 g) and a tiny size (25 mm × 40 mm) is selected as the on-board camera. We choose a standard resolution of 640 × 480 (pixels), and the frame rate is 60 Hz. The camera is equipped with a 6 mm microlens, providing a viewing angle of 56 deg and 38 deg in the length and width directions, respectively. It uses a 5-pin USB 2.0 digital interface with a 480 Mb/s transfer rate and 8-bit raw Bayer data format (connected through IEEE 1394 to workstation). The camera is mounted under the base board of the quadrotor, looking directly down to the ground texture.

(3) Workstation and Communication Module. An off-board Linux PC (AMD Athlon 5200+; 2 GB RAM) is used for image data processing, 3D pose estimation and control law execution in this case. The quadrotor is equipped with XBeePro wireless communication module from MaxStream/Digi, which enables the data transmission from the on-board inertial measurement unit (IMU) and the control command reception (with R/C transmitter enabled) from workstation at a rate of 100 Hz.

(4) Visual Tracking System and Marker Placement. The tracking system VisualeyezII VZ4000 from Phoenix Technologies Incorporated is used to get the absolute position of the quadrotor. It has three cameras inside which can capture the certain markers installed on the four axes of the quadrotor in an accuracy of millimeter level. In this work, the tracking system is installed on the ceiling of the lab (Figure 5). The software VZSoft is installed in another Windows PC

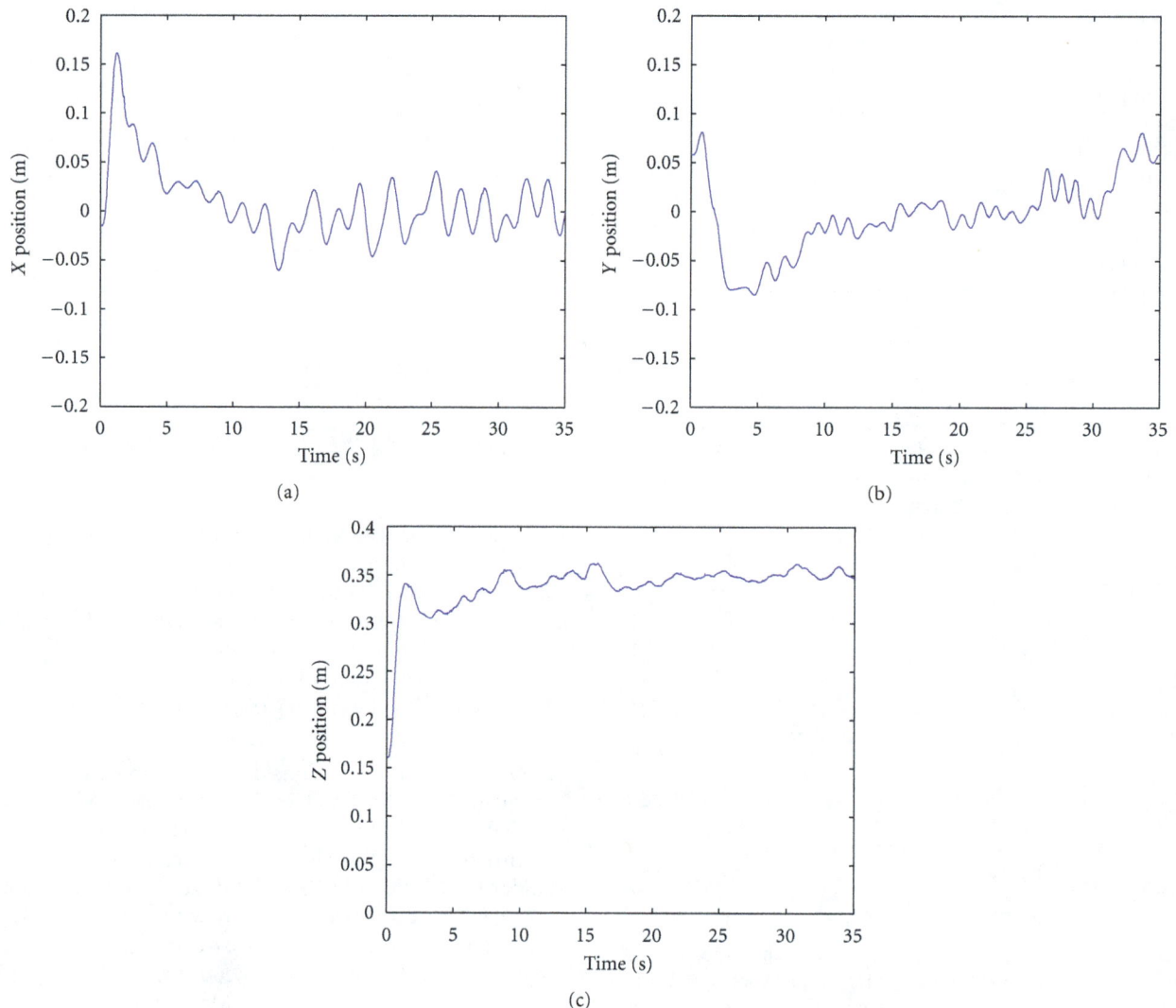

FIGURE 8: 3D position measurements when the quadrotor is in stabilized hover ((a), (b), and (c) for X, Y, and Z positions, resp.). The 3D position is stabilized by multicamera tracking system exclusively. The X and Y positions are set to zero, which means the on-board camera is heading directly to the central point of the ground texture as well as the origin of world reference coordinate. The altitude (Z position) is set to 0.35 m.

(AMD Athlon XP 3000+; 2.1 GHz). It gets the data from the tracking system through a COM interface. The data will be then sent to the workstation with the interface Babelfish which is developed by the Institute of Autonomic Control Engineering, Technical University Munich, using Internet Communications Engine (ICE).

5.2. Velocity Estimation.
To validate the designed templates of receptive fields for rotation detection, the bioinspired image processing algorithm is implemented with C++ language using Open CV.

Under low velocities and within certain altitude range, the response can be regarded as monotonic and near linear from the test results. In this work, the yaw rate is under 100 deg/s and the altitude value is set to 0.35 m. The lookup table is shown in Figure 6(a), with a polynomial curve fitting. From the comparison in Figure 6(b), this approach provides

a fairly accurate yaw rate estimation (the mean error is 1.85 deg/s and the standard deviation of error is 10.22 deg/s). This lookup table could be then used in the closed-loop control under the same light condition.

5.3. Heading Stabilization.
In this experiment, we compare the heading control performance using EMD with those using IMU or tracking system respectively. At first the stable commands (u_p^s, u_r^s, u_y^s, and u_t^s) should be determined experimentally, so that without any controllers off board, the quadrotor can be hovering in the air nearly on a horizontal level and rotating as little as possible, with all the payloads mounted (in this experiment, with on-board bread board for tracking system using TCM8 mode, and with cable power supply instead of battery). The X, Y, and Z positions should be further controlled using the feedback from the tracking system, while the yaw position has no controllers

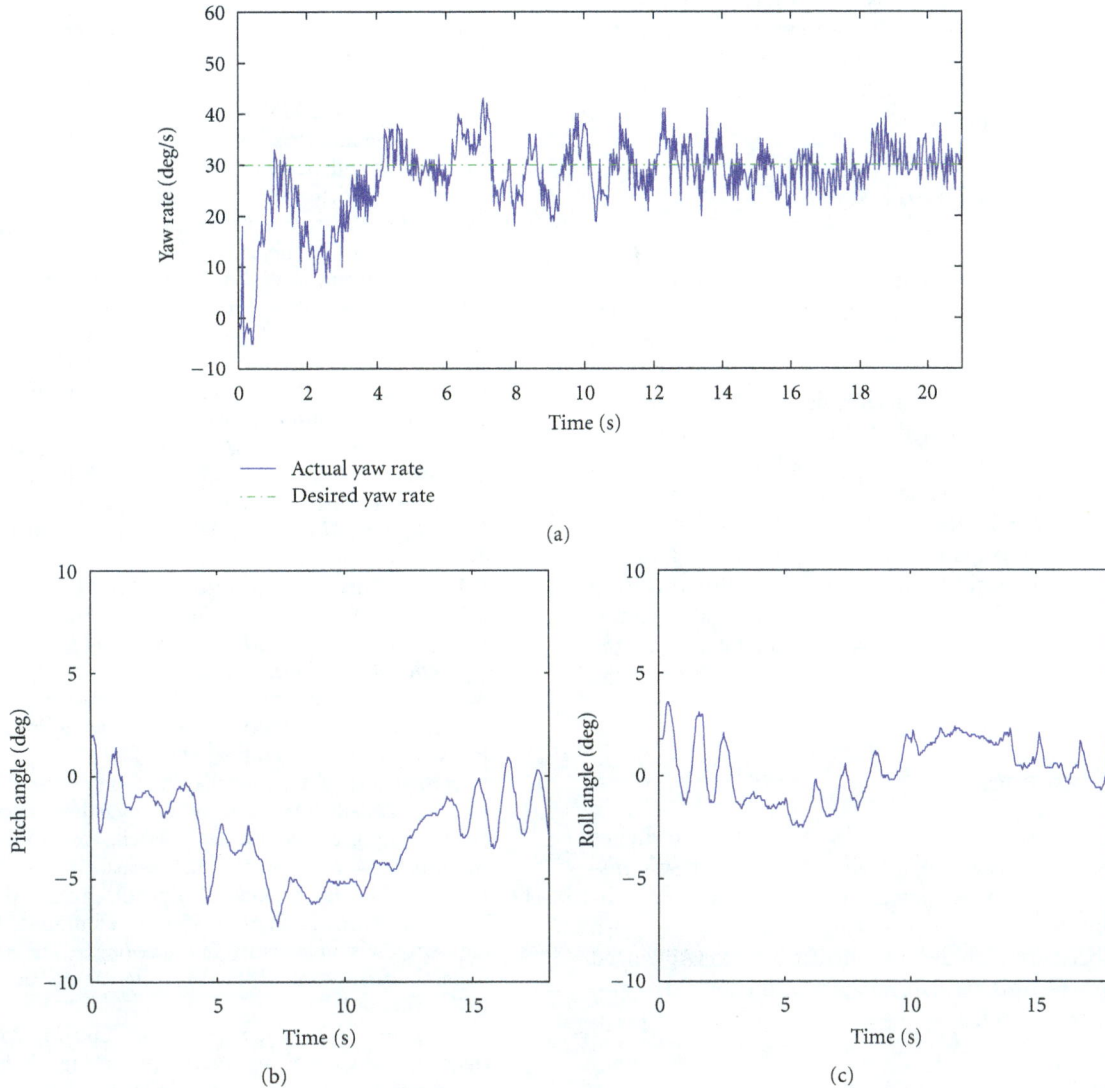

FIGURE 9: Attitude measurements of the flying robot near hover with yaw-rotation (a): yaw rate control. After takeoff, the quadrotor is switched to a desired yaw rate of 30 deg/s; (b) and (c): pitch and roll angles of the quadrotor.

except the on-board IMU at the first attempt. The IMU controller is already integrated on the base board so that no other off-board controller is needed for IMU controlling. A major disadvantage of using IMUs for navigation is that they typically suffer from accumulated error (see Figure 7 blue curve). In this case, in about 20 seconds the yaw position will deviate by 30 degrees if only IMU is used for the heading stabilization. The second reference is the yaw position with the control using tracking system, but without using EMD (the red curve in Figure 7). In Figure 8, the 3D position when using EMD for heading stabilization is shown. By using tracking system and EMDs (the green curve in Figure 7) a satisfying performance could be both achieved. Despite some deviation (for tracking system maximal ±5 degrees and for EMD maximal ±7 degrees), the quadrotor can hover very well with straight heading direction.

5.4. Yaw Rate Control.
The next step is to achieve the velocity control using EMDs. The yaw rate should be set in a low-speed area considering a monotonic relationship between response and velocity. In this case, the desired velocity is 30 degrees/s. The results are shown in Figure 9(a). The settling time is about 4 seconds and the maximal error is about ±10 degrees/s. The inflight performance of 3D pose is shown in Figures 9(b) and 9(c). Since the IMU provides only the angular velocity values, the angle positions are integrated by the base control board on the quadrotor in order to get the angle positions, which are sent to the central workstation.

For velocity estimation, although the EMD is not a pure velocity detector, a closed-loop control of yaw rate is achieved with restrictions of the structured environment and the limitation of velocity in low-speed area. Including image translation delay through IEEE1394/USB cable, image processing by CPU costs 10–20 ms (the program takes nearly 10 ms to wait for images from the camera, while the actual computing time is only several milliseconds). So the EMD computing is extremely fast, which provides an evidence of the efficiency when using biological models.

6. Conclusions and Future Works

In this work, the closed-loop control of a flying robot is achieved by using bioinspired image processing method, which proves to be an effective approach with low computational cost. For real-time implementation, the experimental results of heading stabilization show that, by using EMD response as a feedback, the accumulating drift from the on-board IMU is compensated. Another trial regarding the EMDs as a velocity sensor has realized a low-speed control of the yaw rate on the quadrotor in real-world scenario. In the future, all the 6 DOFs should be controlled by using bioinspired image processing exclusively, without relying on any off-board visual sensors or GPS. The absolute position should be determined by certain advanced algorithms. For the future works, some efforts should be put in the development of novel approaches for highly robust flying performance.

Acknowledgments

This work is supported in part by the DFG excellence initiative research cluster Cognition for Technical Systems-CoTeSys, see also www.cotesys.org, the Bernstein Center for Computational Neuroscience Munich, see also http://www.bccn-munich.de/, and the Institute for Advanced Study (IAS), Technische Universität München, see also http://www.tum-ias.de/.

References

[1] M. Egelhaaf and A. Borst, "Motion computation and visual orientation in flies," *Comparative Biochemistry and Physiology A*, vol. 104, no. 4, pp. 659–673, 1993.

[2] M. V. Srinivasan and S. W. Zhang, "Visual navigation in flying insects," *International Review of Neurobiology*, vol. 44, pp. 67–92, 2000.

[3] H. G. Krapp and R. Hengstenberg, "Estimation of self-motion by optic flow processing in single visual interneurons," *Nature*, vol. 384, no. 6608, pp. 463–466, 1996.

[4] A. Beyeler, J.-C. Zufferey, and D. Floreano, "3D vision-based navigation for indoor microflyers," in *Proceedings of the IEEE International Conference on Robotics and Automation (ICRA '07)*, pp. 1336–1341, April 2007.

[5] J. Conroy, G. Gremillion, B. Ranganathan, and J. S. Humbert, "Implementation of wide-field integration of optic flow for autonomous quadrotor navigation," *Autonomous Robots*, vol. 27, no. 3, pp. 199–200, 2009.

[6] F. Ruffier and N. Franceschini, "Optic flow regulation: the key to aircraft automatic guidance," *Robotics and Autonomous Systems*, vol. 50, no. 4, pp. 177–194, 2005.

[7] N. Franceschini, F. Ruffier, and J. Serres, "A bio-inspired flying robot sheds light on insect piloting abilities," *Current Biology*, vol. 17, no. 4, pp. 329–335, 2007.

[8] F. Expert, S. Viollet, and F. Ruffier, "Outdoor field performances of insect-based visual motion sensors," *Journal of Field Robotics*, vol. 28, no. 4, pp. 529–541, 2011.

[9] E. Buchner, "Behavioural analysis of spatial vision in insects," in *Photoreception and Vision in Invertebrates*, M. A. Ali, Ed., pp. 561–621, Plenum Press, New York, NY, USA, 1984.

[10] M. Egelhaaf and W. Reichardt, "Dynamic response properties of movement detectors: theoretical analysis and electrophysiological investigation in the visual system of the fly," *Biological Cybernetics*, vol. 56, no. 2-3, pp. 69–87, 1987.

[11] M. V. Srinivasan, S. W. Zhang, M. Lehrer, and T. S. Collett, "Honeybee navigation en route to the goal: visual flight control and odometry," *Journal of Experimental Biology*, vol. 199, no. 1, pp. 237–244, 1996.

[12] R. O. Dror, D. C. O'Carroll, and S. B. Laughlin, "Accuracy of velocity estimation by Reichardt correlators," *Journal of the Optical Society of America A*, vol. 18, no. 2, pp. 241–252, 2001.

[13] S. Rajesh, D. O'Carroll, and D. Abbott, "Elaborated Reichardt correlator for velocity estimation tasks," in *Proceedings of the Biomedical Applications of Micro- and Nanoengineering*, pp. 241–253, December 2002.

[14] S. Han, A. Straw, M. Dickinson, and R. Murray, "A real-time helicopter testbed for insect-inspired visual flight control," in *Proceedings of the IEEE International Conference on Robotics and Automation*, Kobe, Japan, 2009.

[15] A. Borst, C. Reisenman, and J. Haag, "Adaptation of response transients in fly motion vision. II: model studies," *Vision Research*, vol. 43, no. 11, pp. 1311–1324, 2003.

[16] B. Hassenstein and W. Reichardt, "Systemtheoretische Analyse der Zeit-Reihenfolgen, und Vorzeihenauswertung bei der Bewegungsperzeption des Ruesselkaefers Chlorophanus," *Naturforsch*, vol. 11, pp. 513–524, 1956.

[17] T. Zhang, H. Wu, A. Borst, K. Kuehnlenz, and M. Buss, "An FPGA implementation of insect-inspired motion detector for high-speed vision systems," in *Proceedings of the IEEE International Conference on Robotics and Automation*, Pasadena, Calif, USA, May 2008.

[18] M. Achtelik, T. Zhang, K. Kuhnlenz, and M. Buss, "Visual tracking and control of a quadcopter using a stereo camera system and inertial sensors," in *Proceedings of the IEEE International Conference on Mechatronics and Automation*, Changchun, China, 2009.

14

Optimized Carrier Tracking Loop Design for Real-Time High-Dynamics GNSS Receivers

Pedro A. Roncagliolo, Javier G. García, and Carlos H. Muravchik

LEICI, Facultad Ingeniería, UNLP, B1900TAG La Plata, Argentina

Correspondence should be addressed to Pedro A. Roncagliolo, agustinr@ing.unlp.edu.ar

Academic Editor: Carles Fernández-Prades

Carrier phase estimation in real-time Global Navigation Satellite System (GNSS) receivers is usually performed by tracking loops due to their very low computational complexity. We show that a careful design of these loops allows them to operate properly in high-dynamics environments, that is, accelerations up to 40 g or more. Their phase and frequency discriminators and loop filter are derived considering the digital nature of the loop inputs. Based on these ideas, we propose a new loop structure named Unambiguous Frequency-Aided Phase-Locked Loop (UFA-PLL). In terms of tracking capacity and noise resistance UFA-PLL has the same advantages of frequently used coupled-loop schemes, but it is simpler to design and to implement. Moreover, it can keep phase lock in situations where other loops cannot. The loop design is completed selecting the correlation time and loop bandwidth that minimize the pull-out probability, without relying on typical rules of thumb. Optimal and efficient ways to smooth the phase estimates are also presented. Hence, high-quality phase measurements—usually exploited in offline and quasistatic applications—become practical for real-time and high-dynamics receivers. Experiments with fixed-point implementations of the proposed loops and actual radio signals are also shown.

1. Introduction

A fundamental task of every Global Navigation Satellite System receiver is to synchronize with the visible satellite signals. Since Direct Sequence Spread Spectrum (DS-SS) signals are utilized, code and carrier synchronization is required, but a correlation stage is necessary to despread the signals before the synchronization algorithms can be applied. In real-time receivers the required economy of operations usually precludes the use of complex estimation schemes and tracking loops are preferred. Due to the correlation process these loops are necessarily discrete. The typical trade-off in tracking loop design is bandwidth versus dynamic performance: output noise increases with a larger loop bandwidth, while dynamic tracking error decreases with it [1]. Thus, the loop design becomes particularly challenging when the receivers are subject to high dynamics. To overcome this limitation other receiver structures have been proposed in [1], claiming tracking capability up to 150 g of acceleration, in contrast with the 5 g regularly assigned to tracking loops. However, the required computational burden

is large since several simultaneous correlations and Fast Fourier Transform (FFT) computations are needed. In this paper we show a careful design of the digital loops that can expand their tracking ability to acceleration steps up to 40 g or even more, keeping a low computational load and reasonable tracking threshold values at the same time.

The loop structure known as FLL-assisted PLL [2] is very often adopted for GNSS receivers. It consists of a Phase-Locked Loop (PLL) and a Frequency-Locked Loop (FLL) in a coupled mode, with the advantage of reducing locking times and avoiding false locks. This solution is also a legacy of analog loops since the FLL or Automatic Frequency Control (AFC) has been used to reduce frequency errors as a previous stage to phase lock for analog PLL [3]. The advantages of adding the FLL to track spread spectrum signals in dynamic environments were already studied in [4]. For high-dynamics GNSS receivers, the focus is on carrier loops because the carrier shares the same dynamics as the code. Then, the estimation of the carrier frequency can be used to aid the estimation of the code frequency, and a first-order code loop is enough [5]. Usually, implementations of FLL-assisted

PLL are not based on optimal digital loop solutions, with each loop designed separately, leaving the analysis of their interactions and possible modifications to the simulation stage [2, 5, 6]. Moreover, schemes adopted to discriminate phase or frequency errors are often justified because of their similarity with well-known analog solutions rather than with an optimality versus implementation complexity criterion. We will show that digital implementations of optimal discriminators are not necessarily more complex and allow designing the FLL-assisted PLL in a coupled way.

Nevertheless, the FLL-assisted PLL leads to a more complex design and a computationally more expensive implementation than a single PLL. Moreover, when coupled-loops lose phase lock for a moment, they present cycle slips introducing a phase ambiguity. We will show how to use the same frequency information as that of an FLL to build a nonambiguous phase detector, the Unambiguous Frequency-Aided (UFA) phase discriminator. A PLL with this new phase discriminator, that is, a UFA-PLL, keeps the desirable properties of an FLL without demanding an extra loop and avoiding cycle slips. Other nonambiguous phase discriminators are known for analog PLLs, that is, with analog loop filter, such as the sequential discriminators built with flip-flops presented in [7, 8] or the nonsequential discriminator of [9]. While their goals are quite similar to ours, they increase the PLL implementation complexity, demanding some digital circuitry and a digital-to-analog converter to get the analog phase error. On the contrary, the UFA phase discriminator is easily implemented and naturally suited for a software-based PLL, leading to a less complex implementation than a FLL-assisted PLL. Section 2 introduces the UFA-PLL structure for GNSS tracking loops.

The optimum loop filter structure for analog PLLs was introduced in [10], solving the mentioned bandwidth trade-off by minimizing a quadratic functional. A widespread technique for designing digital loops is discretizing an analog loop with a sample rate $1/T$ at least ten times faster than loop bandwidth B_N [5, 11, 12]. As $B_N T$ increases above the rule-of-thumb value of 0.1, the resulting loop deviates from optimal and may become unstable [5], especially when accounting for the delays of a digital implementation. This limit imposed to the loop bandwidth is not fundamental and an attempt to avoid it has been presented in [13]. They introduced a digital loop design based on pole placement that allows somewhat larger $B_N T$ values. However, the pole location is assigned with standard second-order analog-system rules. Our approach is to consider a completely digital loop model and pose the bandwidth trade-off directly in the digital domain, building upon the early and often overlooked work of [14] for hybrid loops. We include two delays in the loop to consider the effect of the correlation stage, similar to the inclusion of an accumulator before the loop error discriminator for signals without spreading codes [15]. Our method [16] allows the design of stable loops with $B_N T > 0.1$, a particularly useful feature for high-dynamics receivers. Specifically, we will focus on dynamics modeled as acceleration steps, that is, unbounded jerk, as in the case of launching vehicles when the engine turns on or off. In Section 3 we first derive the optimal loop filter for arbitrary

phase inputs and then for the case of acceleration steps that produce quadratic ramps of input phase or a linear ramp of the input frequency. Simulations comparing the different loop structures are also shown.

Optimization gives the structure for the loop filter, leaving the choice of T and B_N unsolved. Usually, these parameters are selected based on some rule of thumb [2, 5], and the ultimate loop performance, as measured by the pull-out probability and/or tracking threshold, is obtained later by simulation. An optimal choice of these fundamental parameters demands an analysis of the nonlinear aspects of the tracking loop with noise. This is quite difficult, although some results are known for analog loops with relatively simple loop filters, by solving a Fokker-Planck equation [17]. They can be extended to digital loops when an analog approximation is valid [18]. Our approach is to get a reasonable approximation for the pull-out probability and its relationship to the loop parameters. This new approach introduced in [19] allows us considering dynamics modeled as acceleration steps and digital loop filters with zero stationary error response to these inputs. Previous analyses are based on stationary loop responses or sinusoidal acceleration profiles [2, 5]. For these cases, we derive approximate expressions for the probability of starting a nonlinear behavior of the mentioned loops. These expressions quantify the role of B_N and T and let us choose them in order to obtain lower tracking thresholds for different dynamic scenarios, as presented in Section 4.

Our optimized digital carrier tracking loops also allow smoothing of the phase estimates incorporating more measurements, at the expense of some delay. In general, an output delay of a few samples should not be a limitation since the navigation task in a GNSS receiver is usually slow compared with the loop sample rate. This update of the phase estimates can significantly reduce the noise variance and the transient responses in high-dynamics environments. This strategy is suitable for real-time receivers because it can be efficiently calculated. Hence, some of the precise positioning techniques would be applicable in real-time and for high-dynamics receivers. Consider, for instance, smoothing of code delay measurements with carrier phase estimates in stand-alone receivers [12, 20], or differential positioning applications [12, 21], or even attitude estimation with GNSS signals [22]. In all these cases, an improvement in the phase estimation has a direct impact on the positioning performance. The expressions for optimal smoothing filters are derived in Section 5, and their efficient implementation is also discussed there. In addition, we present experimental results obtained with actual RF signals and a fixed point implementation of our loops tracking acceleration steps of up to 40 g. Finally, the conclusions of this work are given in Section 6.

2. Digital Loops Models

Correlations of the received signal with the locally generated replicas for each visible satellite are the inputs to the discriminator of the carrier tracking loops in a GNSS receiver. The complex correlation for a given satellite with

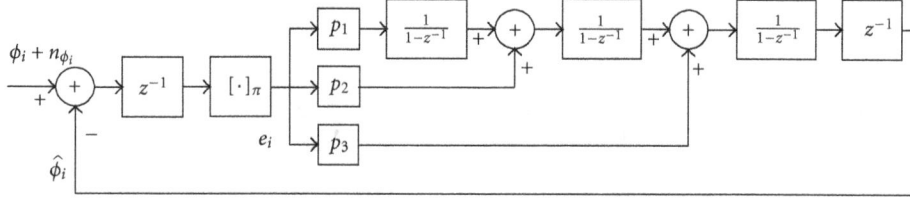

FIGURE 1: Block diagram of classical PLL structure.

carrier power to noise power spectral density C/N_0 and for the ith correlation interval of duration T can be written as [12]

$$C_i = D_i \sqrt{\frac{TC}{N_0}} \sin c(\Delta f_i) R(\Delta \tau_i) e^{j(\pi \Delta f_i + \Delta \theta_i)} + n_i, \quad (1)$$

where $\Delta \tau_i = \tau_i - \hat{\tau}_i$ is the code delay estimation error, $\Delta f_i = f_i - \hat{f}_i$ the frequency estimation error, and $\Delta \theta_i = \theta_i - \hat{\theta}_i$ the phase estimation error, all assumed constant during the integration time. The sequence n_i is a complex white Gaussian noise process with unit variance, $R(\cdot)$ is the code correlation function, and $\sin c(x) = \sin(\pi x)/(\pi x)$. It is also assumed that the signal has binary data bits $D_i = \pm 1$ and that correlations are computed within a single bit period. This Binary Phase Shift Keying (BPSK) modulation is present in many GNSS signals like the GPS civil signals or in the data components of composite modernized GNSS signals [23].

In tracking conditions (i.e., after the acquisition process has been completed [12]), estimation errors are small and then the functions $\sin c(\cdot)$ and $R(\cdot)$ can be approximated by 1. In this case the expression (1) reduces to

$$C_i = I_i + jQ_i \approx D_i \sqrt{\frac{TC}{N_0}} e^{j\Delta\phi_i} + n_{I_i} + jn_{Q_i}, \quad (2)$$

where I_i and Q_i are the so-called in-phase and in-quadrature correlations respectively, $n_{I_i} = \Re\{n_i\}$, $n_{Q_i} = \Im\{n_i\}$, and we have defined $\Delta\phi_i = \phi_i - \hat{\phi}_i$, with $\phi_i = \pi f_i + \theta_i$ and $\hat{\phi}_i = \pi \hat{f}_i + \hat{\theta}_i$. These sequences allow to model the carrier tracking loop as a purely digital single-input single-output (SISO) system. When the frequency is changing according to a constant acceleration error of a m/s^2, we verified—by numerical integration—that expression (2) is a good approximation if $aT^2/\lambda \ll 1$, where λ is the wavelength of the signal. For L1 GPS $\lambda = 0.19$ m and with $T = 5$ ms, this implies that $a \ll 7600 \equiv 775$ g. In this case, the terms Δf_i and $\Delta\phi_i$ have to be reinterpreted as the average frequency error and average phase error during the correlation interval, respectively.

In the following we briefly review the basic concepts of PLL and FLL-assisted PLL from our digital point of view, and later we introduce the UFA-PLL.

2.1. PLL Model. The phase estimation error is typically obtained using one of several possible discriminators [5],

which give the desired phase modified by different memoryless nonlinearities. The optimal one—maximum likelihood estimator—is given by

$$e_i = \tan^{-1}\left(\frac{Q_i}{I_i}\right) = \left[\Delta\phi_i + n_{\phi_i}\right]_\pi, \quad (3)$$

where the notation $[\cdot]_\pi$ indicates that its argument is kept within the interval $(-\pi/2, \pi/2]$ by adding or subtracting π as many times as needed. The zero-mean noise term n_{ϕ_i} has a rather complicated probability distribution [24], but in high C/N_0 it can be approximated by a Gaussian distribution with zero mean and variance $\sigma_{\phi_i}^2 \approx 1/(2TC/N_0)$.

A four-quadrant $\tan^{-1}(\cdot)$ is not appropriate if there is BPSK data modulation because the discriminator becomes sensitive to the data phase changes. On the contrary, for signals without data the range of the discriminator can be doubled with a four-quadrant $\tan^{-1}(\cdot)$. We chose this discriminator because it is not amplitude dependent and the calculation of $\tan^{-1}(\cdot)$ can be easily implemented with a lookup table, since in practice I_i and Q_i are frequently quantized to a few bits.

In order to close the loop in our model, it is of crucial importance to consider the delays present in a real implementation. Failure to account for a delay may turn unstable an optimal loop design. Since ours loops are digital, a single sample delay is expected but in fact there are two. One of them is due to the time spent in I_i and Q_i calculations. The other delay appears because the estimated values used in the present correlations have to be known before the calculations begin. That is, the value $\hat{\phi}_i$ is obtained with the loop filter output of the $(i-1)$th correlation interval, which in turn is calculated with the estimation errors of $\hat{\phi}_{i-2}$. Then, with these considerations, the model of a PLL using the classical loop filter structure of type 3, that is, with three accumulators, is shown in Figure 1.

2.2. FLL-Assisted PLL Model. To add an FLL to our previous PLL, a frequency discriminator is needed. In a digital loop a frequency error estimate may be obtained as the difference of two successive phase errors, and in fact this is often correct. A problem appears when the discontinuities caused by $[\cdot]_\pi$ make that the difference to be wrong in $\pm\pi$. However, our discrete system cannot distinguish frequencies greater than half of the sample rate, that is, phase changes of π between consecutive samples, and so the measured frequency errors must be bounded. In fact, if the phase discriminator is insensitive to BPSK data, the phase changes caused by

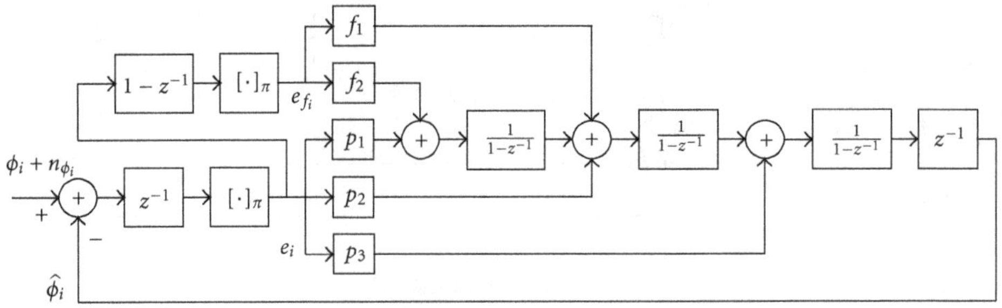

FIGURE 2: Block diagram of the typical FLL-assisted PLL structure.

frequency errors must lie in the interval $(-\pi/2, \pi/2]$ [25]. Thus, the difference of two consecutive outputs of the phase discriminator can be corrected just using the operation $[\cdot]_\pi$. Therefore, the frequency discriminator for the FLL can be obtained by

$$e_{f_i} = [e_i - e_{i-1}]_\pi. \tag{4}$$

Figure 2 shows a diagram of the FLL-assisted PLL presented in [2], where the second-order loop filter of the FLL shares the same cascade of accumulators used by the PLL filter.

In the locked condition $e_i = \Delta\phi_i$ and $e_{f_i} = \Delta\phi_i - \Delta\phi_{i-1}$ are small enough to justify a linear analysis of the loop. The complete loop is seen as an equivalent PLL with filter coefficients p_3, $p_2 + f_2$, and $p_1 + f_1$, instead of p_3, p_2, and p_1. Thus, the FLL is inserted into the model of the PLL at a design stage. This eliminates the constrain of using a narrow bandwidth FLL to not significantly perturb the PLL behavior, as in [2, 6]. A wide bandwidth FLL allows the loop to have two regions of operation: "phase-locked" as it was described before, and "frequency-locked" when the dynamics unlocks the PLL but the FLL keeps the frequency error within the linear range of its discriminator. In the latter region the loop is governed by the FLL (coefficients f_1 and f_2) and the phase error input acts like a zero-mean perturbation [25]. As soon as the dynamics let the loop reduce its frequency error close to zero, the phase lock can be restored.

2.3. The UFA-PLL Model. As we have seen so far, due to the cyclic nature of phase a memory-less discriminator is unable to distinguish changes of an integer number of cycles—or half cycles if there is BPSK data—, that is, its output is ambiguous. However, it is possible to obtain a frequency error estimate from these ambiguous phase error estimates correcting their difference with the nonlinear operation $[\cdot]_\pi$. This is the reason why an FLL can cope with carrier tracking in situations when a single PLL cannot. Assume that there is BPSK data and the PLL phase error is rising and crosses the value $\pi/2$. The output of the phase discriminator abruptly changes to a value close to $-\pi/2$, reversing the evolution of the PLL phase. Hence, the phase error will increase since the PLL is now moving in the wrong direction. We should instruct the phase discriminator with information of the phase derivative to keep moving in the right direction, that is, we should feed it with proper frequency information

available at the FLL. Therefore, the idea of the Unambiguous Frequency-Aided (UFA) phase discriminator is to use the same frequency information used by the FLL to get a better phase discriminator. It works correcting the ambiguous values of e_i by adding or subtracting an integer number of π so that the difference of successive values of the corrected phase error, u_i, gives the right frequency error. Then, the equations that define our new phase error estimate, for $i \in \mathbb{N}$, are

$$u_i = k_i\pi + e_i, \quad k_i \text{ such that } I_\pi(u_i - u_{i-1}) = 0, \tag{5}$$

where we define $I_\pi(x) = x - [x]_\pi$, an operation similar to the function integer part but with steps at the multiples of π. A practical formula to compute k_i can be derived noting that $I_\pi(x + l\pi) = I_\pi(x) + l\pi$, $l \in \mathbb{Z}$ since

$$I_\pi(k_i\pi + e_i - u_{i-1}) = k_i\pi + I_\pi(e_i - u_{i-1}) = 0 \tag{6}$$

and then $k_i\pi = -I_\pi(e_i - u_{i-1})$. Substituting this in (5), we can recursively calculate the UFA phase error from the ambiguous e_i:

$$u_i = e_i - I_\pi(e_i - u_{i-1}) \tag{7}$$

with starting value $u_0 = e_0$. Then, the PLL structure in Figure 1 transforms into a UFA-PLL just adding a block that implements (7) immediately after the phase discriminator output, as shown in Figure 3.

It is interesting to note that the UFA scheme acts like the phase unwrapping algorithm proposed in [18] for correcting cycle slips in the phase estimates of feed-forward synchronizers. In this case, the phase correction does not affect the phase estimation process since it is done once the estimation stage is finished. On the contrary, the UFA phase discriminator modifies the behavior of our feedback estimator, the PLL, changing its nonlinear characteristics. As a result, cycle slips and the rather complex transient responses induced by them are avoided as long as the frequency error is compatible with the loop sample rate.

2.4. Equivalence between UFA-PLL and FLL. We saw that the frequency error estimate can be obtained as the difference of two successive phase errors if the result is kept in range by adding or subtracting an integer number of π. Thus, the frequency discriminator for the FLL can be obtained

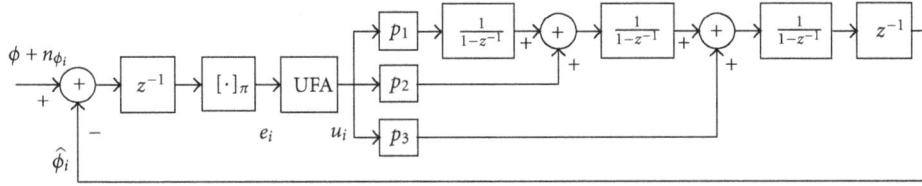

FIGURE 3: Block diagram of the new UFA-PLL structure.

correcting the difference of two consecutive outputs of the phase discriminator just using the operation $[\cdot]_\pi$. Figure 4(a) shows a block diagram of a digital FLL, with loop filter transfer function $F(z)$. Notice that the two delays and the accumulator that convert the frequency estimation to phase before the feedback are not included in $F(z)$.

An alternative way to obtain the same frequency error discriminator is to use the UFA algorithm previously described. Indeed, the output phase sequence u_i is built in such a way that the difference of consecutive values produces the right frequency error, as seen from (4) and (5). Therefore, the schemes of Figures 4(a) and 4(b) are equivalent. The interesting fact in Figure 4(b) is that most linear blocks are adjacent. Thus, the differentiator cancels with the accumulator without changes in the dynamic loop response, except for the mean value of the phase error, leading to the equivalent UFA-PLL model of Figure 4(c). In fact, this zero-pole cancellation shows why the FLL is insensitive to constant phase errors whereas the equivalent UFA-PLL is not. More importantly, the equivalence reveals that the nonlinear behavior of the UFA-PLL is equal to that of a FLL with the same $F(z)$, and then their tracking capacity and noise resistance are the same.

3. Optimal Loop Filter Design

We propose to design the digital loop filter minimizing a two-term quadratic functional to handle the bandwidth trade-off mentioned in the Introduction. The input signal is assumed to have a part related to phase evolution ϕ_i plus additive, zero mean, and noise $n_{\phi i}$. The functional to be optimized is

$$J = \sigma_N^2 + \alpha^2 E_T(\phi_i), \tag{8}$$

where α^2 is a weighting factor that controls the trade-off between noise and transient response, that is, the loop bandwidth, σ_N^2 is the noise variance at the loop output, and $E_T(\phi_i)$ represents the energy of the tracking error $\Delta\phi_i$ transient response. Since the functional uses the energy of the transient response, the optimum filter must produce a zero stationary response for the given input.

Suppose $F(z)$ is the loop-filter transfer function to be found, and consider that the linear model hypothesis holds for a PLL or FLL. The closed loop transfer function including the delays is

$$T(z) = \frac{F(z)z^{-2}}{1 + F(z)z^{-2}} = Y(z)z^{-2}, \tag{9}$$

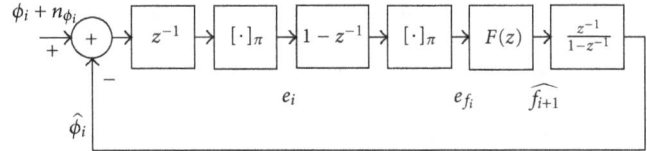

(a) Equivalent scheme with UFA

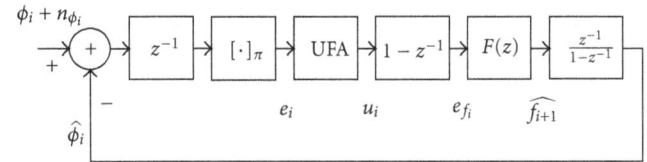

(b) FLL scheme

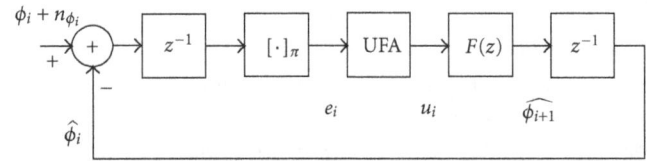

(c) UFA-PLL equivalent scheme

FIGURE 4: Equivalence between UFA-PLL and FLL.

where $Y(z)$ must be a causal and stable rational transfer function. The minimization of the functional J written in terms of $Y(z)$ is shown in the Appendix A. The optimum transfer function is given by

$$Y(z) = \frac{X(z)z}{\Psi(z)}, \tag{10}$$

where $\Psi(z)$ and $X(z)$ can be obtained from the spectral factorization of (A.6) and from the partial fraction expansion of (A.8), respectively. We repeat them here for completeness,

$$\Psi(z)\Psi(z^{-1}) = 1 + \gamma^2 \Phi(z)\Phi(z^{-1}), \tag{11}$$

$$G(z) = \frac{\gamma^2 \Phi(z)\Phi(z^{-1})z}{\Psi(z^{-1})} = X(z) + W(z^{-1}), \tag{12}$$

where $\Phi(z)$ is the z-transform of ϕ_i. The relation between minimizing (8) to Wiener filtering [26] is sketched in Appendix B. Observe that whereas the former is a mixed criterion with a term depending on the stochastic part and a term depending on the deterministic part of the phase signal;

the latter criterion stems from a purely stochastic formulation. The connection between both approaches arises when modeling the input phase as white noise passing through a rational transfer function. The Wiener filtering approach offers other possibilities such as keeping the optimality for a wide range of admissible transfer functions via a robust approach as in [27] or considering continuous models for the phase as in [28].

Optimum loop filters for an input phase step, frequency step, and frequency ramp were derived in [16]. In the following, only the last result is presented for the sake of brevity. Analog loop filters optimized for these kind of inputs are the origin of the classical methods of filter design for type one, two, and three loops, respectively. As it will be seen, our purely discrete design for each case has one extra pole, due to the loop delays. This additional pole does not appear when discretizing analog designs, but it has a decisive influence on the stability or the range of achievable product $B_N T$.

3.1. Optimum Filter for a Frequency Ramp.

The ramp is modeled as

$$\Phi(z) = \frac{\dot{\Delta\omega} T^2}{(1 - z^{-1})^3}, \tag{13}$$

where $\dot{\Delta\omega}$ is the rate of frequency change. Denoting $\nu = \dot{\Delta\omega}^2 T^4 \gamma^2$, from (11) it is necessary to solve $(z - 1)^6 - \nu z^3 = 0$. The six roots of this polynomial are obtained using the fact that three of them are the inverses of the other three. This allows us to express the following equations:

$$z_{1,2} + z_{1,2}^{-1} = 2 - \frac{1 \pm j\sqrt{3}}{2} \sqrt[3]{\nu},$$
$$z_3 + z_3^{-1} = 2 + \sqrt[3]{\nu} \tag{14}$$

that determine the values of z_1, z_2, complex conjugates and a real z_3. Using these values and (11), we get

$$\Psi(z) = \frac{(1 - z_1 z^{-1})(1 - z_2 z^{-1})(1 - z_3 z^{-1})}{(1 - z^{-1})^3 (z_1 z_2 z_3)^{1/2}}, \tag{15}$$

and replacing in (12)

$$G(z) = \frac{-(z_1 z_2 z_3)^{-1/2} \nu z^4}{(z - z_1^{-1})(z - z_2^{-1})(z - z_3^{-1})(z - 1)^3}. \tag{16}$$

Then, the corresponding $X(z)$ has only three poles in $z = 1$, and the closed-loop transfer function of (10) is

$$Y(z) = \frac{A - Bz^{-1} + Cz^{-2}}{(1 - z_1 z^{-1})(1 - z_2 z^{-1})(1 - z_3 z^{-1})}, \tag{17}$$

where $A = (6 - 3z_s + z_d)$, $B = (8 - 3z_s + z_p)$, and $C = 3 - z_s$, with $z_s = z_1 + z_2 + z_3$, $z_p = z_1 z_2 z_3$ and $z_d = z_1 z_2 + z_1 z_3 + z_2 z_3$. Then, the optimum loop filter with four poles, three of the input and the extra one, is

$$F(z) = \frac{A - Bz^{-1} + Cz^{-2}}{(1 - z^{-1})^3 (1 + Cz^{-1})}. \tag{18}$$

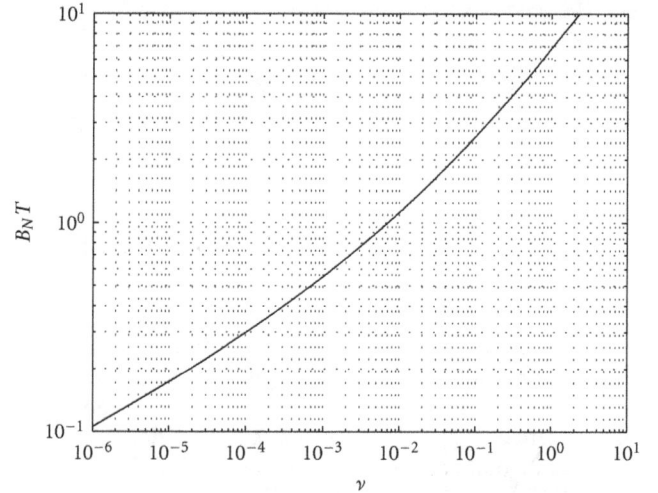

FIGURE 5: Type 3 loop noise equivalent bandwidth.

For the purpose of implementation, it is desirable to use a cascade of accumulators. Then, (18) can be rewritten as

$$F(z) = \frac{p_3 + p_2(1 - z^{-1}) + p_1(1 - z^{-1})^2}{(1 - z^{-1})^3 (1 + p_1 z^{-1})}, \tag{19}$$

where $p_1 = C$, $p_2 = B - 2C$, and $p_3 = A - B + C$. The closed loop noise equivalent bandwidth is shown in Figure 5. This curve allows choosing the appropriate value of ν for a given normalized noise bandwidth. Observe also from this figure that the larger is the step or the parameter ν, the more emphasis is given to the transient energy, causing the normalized noise bandwidth to increase. The product $B_N T$ levels out to a value of approximately 54.5. This part is not included in the figure since it is of minor importance for most designs of practical interest.

3.2. Design Example: Loops for Launching Vehicles.

We simulate and compare the loop models presented in Section 2 taking as an example a GPS carrier tracking loop for launching vehicles [29]. In this case the dynamic input can be modeled as an acceleration step, which becomes a quadratic ramp in terms of phase and a linear one, in terms of frequency. For these inputs the optimal loop filter for a PLL was obtained in Section 3.1. The case of an FLL-assisted PLL, taking results from [29], leads to a type 2 FLL and a type 3 PLL.

The FLL-assisted PLL in [29] was designed to operate in "phase-locked" mode with steps up to 10 g and in "frequency-locked" mode up to 20 g of acceleration. These requirements were too demanding for the commonly adopted correlation time of 10 ms, and then it was lowered to 5 ms (at the cost of almost doubling the processor load and an increase in the tracking threshold). A typical rule of thumb for keeping a reasonable distance from the pull-out values of the loop is that the peak of the error transient has a maximum value given by half the linear range of the phase discriminator, an eighth of cycle [2]. As it will be shown in the simulations this condition was obtained with a value of $\nu = 0.00025$ for the PLL. However, as we will explain

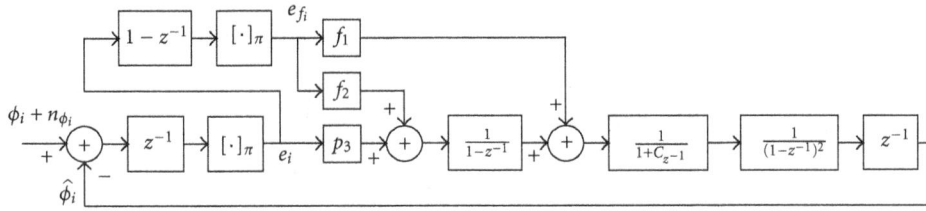

FIGURE 6: Block diagram of the FLL-assisted PLL in Section 3.

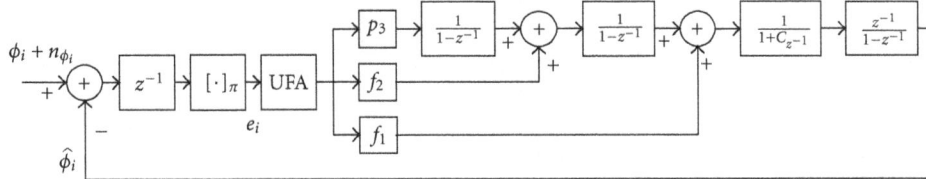

FIGURE 7: Block diagram of the UFA-PLL designed in Section 3.

in Section 4, this rule is not that useful. The resulting filter transfer function for the PLL was

$$F(z) = \frac{A - Bz^{-1} + Cz^{-2}}{(1 - z^{-1})^3(1 + Cz^{-1})}, \tag{20}$$

where $A = 0.6173$, $B = 1.105$, and $C = 0.5$. Then, in the structure of Figure 1 this implies $p_1 + f_1 = C = 0.5$, $p_2 + f_2 = B - 2C = 0.105$, and $p_3 = A - B + C = 0.0123$, plus a block that implements the extra pole in $z = -C$. The resulting PLL equivalent noise bandwidth is $B_N = 75.6$ Hz.

Since the FLL design does not affect the previous results, it was designed wider than strictly necessary in order to facilitate the posterior implementation. The selected transfer function is

$$F(z) = \frac{D - Ez^{-1}}{(1 - z^{-1})^2(1 + Ez^{-1})}, \tag{21}$$

where $D = 0.6$ and $E = 0.5$, resulting in that the extra pole needed for the FLL and the PLL is the same. Then, $f_1 = E = 0.5$ and $f_2 = D - E = 0.1$. This implis $p_1 = 0$ and $p_2 \approx 0$. With these simplifications the complete loop design reduces to the diagram showed in Figure 6. This FLL loop can track steps up to 40 g with transient error peaks smaller than 25 Hz, half of the linear range of the frequency discriminator, with an equivalent noise bandwidth of $B_N = 61.3$ Hz.

We will use the previous loop filter as a basis for the comparison of different loop configurations. We consider a PLL and a UFA-PLL with the same loop filter as before, that is $p_1 = C = 0.5$, $p_2 = 0.105$, $p_3 = 0.0123$. This structure is showed in Figure 7. The phase error response for a step of 10 g of acceleration, common to the three loops as expected, is depicted in Figure 8. In this case the phase error detected by the discriminators is equal to the actual phase error since its magnitude is always less than a quarter of cycle.

In Figure 9 the discriminated phase error during a 40 g step in the three loops is illustrated. It can be seen that

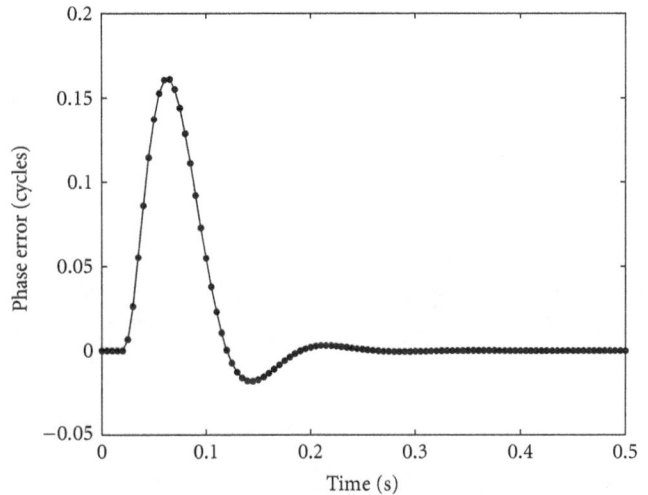

FIGURE 8: Phase error during a 10 g step.

the PLL cannot track this step, whereas the others can. The response of the UFA-PLL is a scaled version of the response for 10 g, showing the effect of linearization of the discriminator characteristic achieved by the UFA algorithm. In the case of the FLL-assisted PLL the loop lost phase lock for a moment, but could still track the dynamics because the FLL remained locked. This nonlinear behavior could correspond to a cycle slip. To verify this analysis the actual phase error for each configuration is shown in Figure 10. Clearly, there is one cycle slip in the tracked phase of the FLL-assisted PLL, whereas the UFA-PLL is able to track this step of acceleration without any cycle slip.

The limit of the tracking capability of the FLL-assisted PLL and the UFA-PLL is the frequency error—the phase change between samples—, since both are frequency-aided loops. If this error becomes greater than 50 Hz in magnitude, the frequency estimation will be ambiguous, in the same way

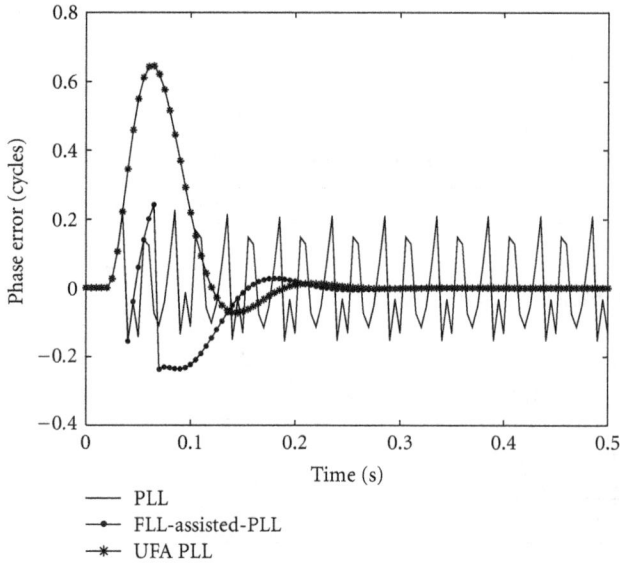

FIGURE 9: Discriminated phase error during a 40 g step.

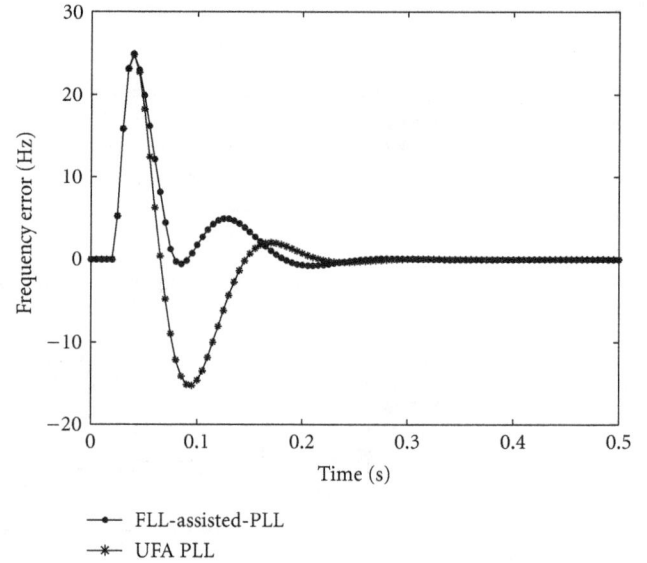

FIGURE 10: Actual phase error during a 40 g step.

as the phase of the PLL does, and the loops will lose their lock. This can be caused by excessively high dynamics or noise power or a combination of both. Notice that the noise power considered in this case is twice the input phase noise power due to the differencing. In fact, we can use the UFA algorithm applied to the frequency discrimination to further extend the dynamics resistance of the loops. However, it will be of little practical importance due to the noise power increase caused by a new differentiation. The frequency error of both loops during a 40 g step is shown in Figure 11. The peak error is 25 Hz—half the limit—and thus, using the same rule of thumb that we used for the phase error of the PLL, it can be argued that 40 g is the level of acceleration steps that can be tracked with a reasonable safety margin for noise effects. In Section 4 we will give a totally different approach for the consideration of the noise in the UFA-PLL.

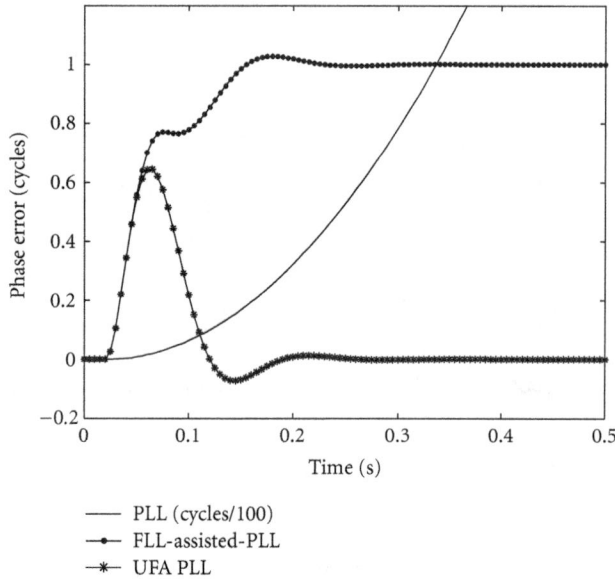

FIGURE 11: Frequency error during a 40 g step.

Another shared feature of the FLL-assisted PLL and the UFA-PLL is the resistance to false locks. For the sake of brevity this analysis is not included here but it can be seen in [16].

4. Pull-Out Probability Analysis

In this section we will compute an approximation to the pull-out probability for a loop in a given operating condition. Since the pull-out is necessarily a consequence of nonlinear behavior, a simple way to bound the pull-out probability is with the probability of entering nonlinear behavior, that is, the limits of the $\tan^{-1}(\cdot)$ range. We are interested in tracking acceleration steps. Then, it is clear that for a given noise level the instants when the loop is closer to these boundaries approximately correspond to the peaks of the loop transient response. Therefore, we focus on calculating the probability of entering nonlinear behavior at the instant of the transient response peak, given that the loop behavior has been linear up to that time.

4.1. PLL Analysis. The PLL enters nonlinear behavior when the phase error becomes larger than $\pi/2$. This is equivalent to a sign reversal of the in-phase component, with respect to the sign of the data bit. Then, the probability of nonlinear behavior at the transient peak $i = p$ is [19]

$$
\begin{aligned}
P_p &= P\left\{ \left| \Delta\phi_p + n_{\phi_p} \right| > \frac{\pi}{2} \right\} \\
&= P\left\{ \cos\left(\phi_p - \hat{\phi}_p \right) + n'_{I_p} < 0 \right\},
\end{aligned}
\tag{22}
$$

where $n'_{I_p} = D_p n_{I_p}/\sqrt{TC/N_0}$ has variance $\sigma^2 = 1/(2TC/N_0)$. Assuming that the PLL has had a linear behavior up to the analyzed instant, $\hat{\phi}_p$, can be thought of as a deterministic value plus output noise $n_{\hat{\phi}_p}$. Subtracting this deterministic value from ϕ_p we find the peak value of the loop transient

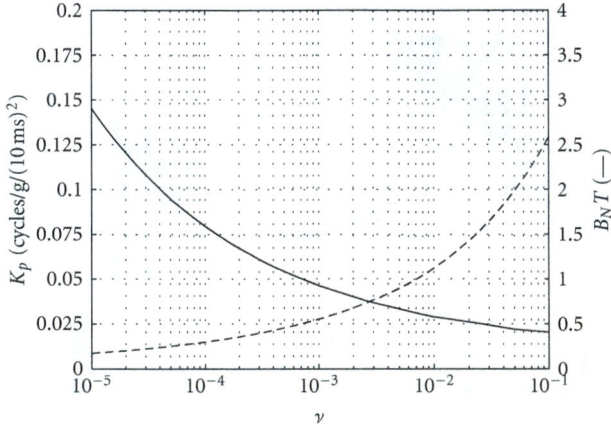

FIGURE 12: Noise Bandwidth and Peak Error Response for type 3 PLL.

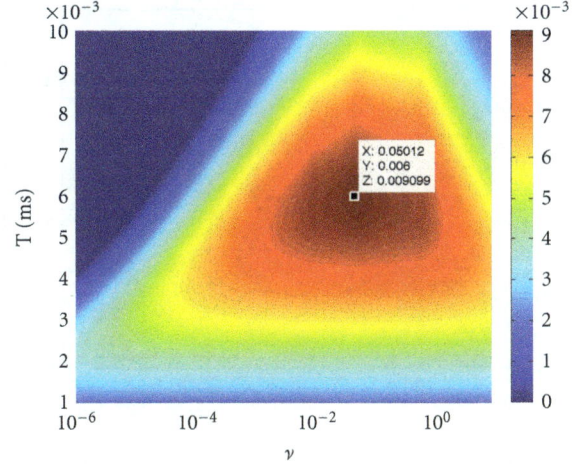

FIGURE 13: Function $f_{\text{PLL}}(\nu, T)$ for type 3 PLL and 5 g.

response to an acceleration step, which is shown in Figure 12 as a function of ν for an integration time of 10 ms. Note that with acceleration we refer to the Doppler rate the loop has to track, scaled in units of g = 9.8 m/s^2 instead of Hz/s to keep an easy physical interpretation. This peak value is proportional to the square of the integration time because an acceleration step is a parabolic ramp of phase. The values of the normalized closed-loop equivalent noise bandwidth are also plotted in Figure 12 for completeness.

The noise $n_{\hat{\phi}_p}$ depends on the filtered past input noise—from instants $p - 2$, $p - 3$, and so on—and is statistically independent of n'_{I_p}. Then, we can write $\phi_p - \hat{\phi}_p = K_p a T^2 + n_{\hat{\phi}_p}$, where a is the acceleration amplitude in g's. In (22), we obtain

$$P_P = P\left\{\cos\left(K_p a T^2 + n_{\hat{\phi}_p}\right) + n'_{I_p} < 0\right\}. \tag{23}$$

To further simplify (23), we use a "low noise" approximation $\cos(n_{\hat{\phi}_p}) \approx 1$ and $\sin(n_{\hat{\phi}_p}) \approx n_{\hat{\phi}_p}$ and obtain

$$P_P \approx P\left\{\cos\left(K_p a T^2\right) - \sin\left(K_p a T^2\right) n_{\hat{\phi}_p} + n'_{I_p} < 0\right\}. \tag{24}$$

Using central limit theorem arguments, it is reasonable to approximate the distribution of $n_{\hat{\phi}_p}$ as a zero mean Gaussian with variance approximated by $2B_N T \sigma^2_{\phi_i} \approx B_N/(C/N_0)$, using the high C/N_0 variance expression for the input noise. In this case, both random terms in (24) become a single Gaussian random variable with zero mean and variance $(1 + 2B_N T \sin^2(K_s a T^2))\sigma^2$. Therefore,

$$P_P \approx Q\left(\sqrt{\frac{2T C/N_0 \cos^2\left(K_p a T^2\right)}{1 + 2B_N T \sin^2\left(K_p a T^2\right)}}\right), \tag{25}$$

where $Q(x)$ is the cumulative Gaussian distribution from x to ∞. Since the function $Q(\sqrt{x})$ is monotonically decreasing, we can define a function

$$f_{\text{PLL}}(\nu, T, a) = \frac{2T \cos^2\left(K_p a T^2\right)}{1 + 2B_N(\nu) T \sin^2\left(K_p a T^2\right)} \tag{26}$$

such that the larger is f_{PLL}, the smaller is P_P for a given C/N_0, and then a low tracking threshold is attained. For example, $f_{\text{PLL}}(\nu, T, a)$ for a 5 g acceleration step is plotted in Figure 13. It can clearly be seen that the larger values of f_{PLL} are found in the region 5 ms < T < 7 ms and $10^{-3} < \nu < 10^0$, and the maximum is approximately at $\nu = 0.05$ and $T = 6$ ms. The value of f_{PLL} in this region is about 0.009. Larger values of ν are not preferred because they lead to loops that produce larger phase estimation error and only a slightly lower P_P. However, it must be emphasized that for values of $\nu > 10^{-3}$ the loop bandwidth is $B_N T > 0.5$. These values of bandwidth show that a filter loop design based on discretization of analog solutions, only valid for $B_N T < 0.1$, is not appropriate to design loops with better tolerance to nonlinear behavior. If $\nu = 0.02$, then $B_N T = 1.5$ and, with $T = 5$ ms, leads to an optimum loop bandwidth $B_N = 300$ Hz, which is too large from the point of view of output phase error variance. This shows that the main cause of pull-out in PLLs with narrow bandwidths is the transient error response, due to the input phase and the input noise, rather than output noise. In other words, 5 g of acceleration is very demanding for a single PLL and then a large bandwidth is required to track them with small pull-out probability.

4.2. UFA-PLL Analysis.
An equivalent description of the UFA algorithm presented in Section 2 is to consider it as a modified $\tan^{-1}(\cdot)$ function that produces output values in the range $(-\pi/2 + u_{i-1}, \pi/2 + u_{i-1}]$ instead of $(-\pi/2, \pi/2]$. Hence, we conclude that nonlinear behavior will occur if the actual phase error differs from the previously discriminated one by more than $\pi/2$. Assuming that the behavior before the analysis time has been linear, $u_{i-1} = \Delta\phi_{i-1} + n_{\phi_{i-1}}$ and then

$$P_U = P\left\{\left|\Delta\phi_i + n_{\phi_i} - \Delta\phi_{i-1} - n_{\phi_{i-1}}\right| > \frac{\pi}{2}\right\}. \tag{27}$$

Writing the phase noise terms as a function of the corresponding in-phase and in-quadrature components, it can be shown that this condition is equivalent to a sign reversal of

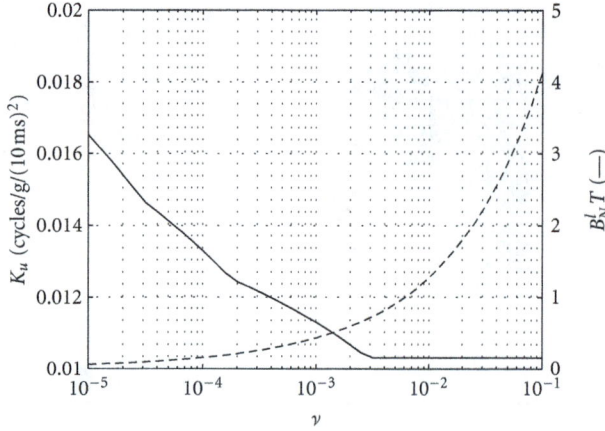

FIGURE 14: Noise bandwidth and peak error response for type 3 UFA-PLL.

the noise component in phase with $\Delta\phi_{i-1}$, rather than in phase with $\hat{\phi}_{i-1}$. Therefore,

$$P_U = P\Big\{C + n''_{I_i} + n''_{I_{i-1}} + n'_{I_i}n'_{I_{i-1}}$$

$$+ n'_{Q_i}n'_{Q_{i-1}} < 0\Big\}, \tag{28}$$

where

$$n''_{I_i} = n'_{I_i}\cos(\Delta\phi_{i-1}) - n'_{Q_i}\sin(\Delta\phi_{i-1}),$$

$$n''_{I_{i-1}} = n'_{I_{i-1}}\cos(\Delta\phi_i) - n'_{Q_{i-1}}\sin(\Delta\phi_i) \tag{29}$$

with $n'_{I_i} = D_i n_{I_i}\delta$, $n'_{Q_i} = D_i n_{Q_i}\delta$, $n'_{I_{i-1}} = D_{i-1}n_{I_{i-1}}\delta$, $n'_{Q_{i-1}} = D_{i-1}n_{Q_{i-1}}\delta$, $\delta = 1/\sqrt{TC/N_0}$ and $C = \cos(\Delta\phi_i - \Delta\phi_{i-1})$. The deterministic part of the argument of this cosine is a differenced version of the phase error transient, and the random part due to the output noise in the estimates is a differenced version of $n_{\hat{\phi}_i}$. Therefore, for the analysis of UFA-PLL two additional loop parameters are needed: the maximum difference of the error transient response, denoted by K_u, and the equivalent noise bandwidth of the linear model of the loop plus a differentiator, denoted by $B'_N T$. These quantities calculated by means of residues are plotted in Figure 14. The value K_u is constant for values of $\nu > 0.003$ because for this region the largest difference occurs between the two first samples of the transient response, which in turn are equal to the corresponding input samples due to the delays of the loop. Then, if the peak of the differenced transient occurs at $i = d$, we can write $\Delta\phi_d - \Delta\phi_{d-1} = K_u a T^2 + n_{\hat{\phi}_d} - n_{\hat{\phi}_{d-1}}$.

The quadratic terms in (28) have zero mean, and σ^4 variance and are uncorrelated between them and with the linear ones. For practical values of σ^2 their variance is much smaller than the variance of the linear terms. Even in this case, they cause the probability distribution of the sum in (28) to differ considerably from Gaussian. A more accurate calculation will require numerical computations of the actual distribution. On the contrary, our aim is to get reasonable and easy-to-handle approximation and, then, we will discard

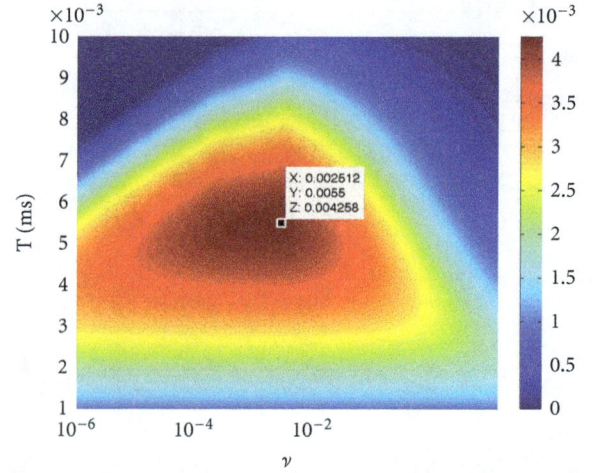

FIGURE 15: Function $f_{\mathrm{UFA}}(\nu, T)$ for type 3 UFA-PLL and 20 g.

them, but being aware that the Gaussian assumption is only a coarse approximation. Then,

$$P_U \approx P\Big\{\cos\big(K_u a T^2 + \Delta n_{\hat{\phi}_d}\big) + n''_{I_i} + n''_{I_{i-1}} < 0\Big\} \tag{30}$$

where $\Delta n_{\hat{\phi}_d} = n_{\hat{\phi}_d} - n_{\hat{\phi}_{d-1}}$. The input noise terms are independent of each other and of the output ones because of both loop delays. Therefore, comparing (30) with (23) we replicate the reasoning for the PLL, but doubling the input noise contribution, and changing K_p by K_u and $B_N T$ by $B'_N T$. Then, using the same approximations made for (25), we get

$$P_U \approx Q\left(\sqrt{\frac{2TC/N_0 \cos^2\left(K_u a T^2\right)}{2 + 2B'_N T \sin^2\left(K_u a T^2\right)}}\right), \tag{31}$$

and then

$$f_{\mathrm{UFA}}(\nu, T, a) = \frac{T\cos^2\left(K_u a T^2\right)}{1 + B'_N(\nu)T \sin^2\left(K_u a T^2\right)} \tag{32}$$

is the function to analyze which values of ν and T are better for the design of low tracking threshold loops. For accelerations of 20 g $f_{\mathrm{UFA}}(\nu, T, a)$ is plotted in Figure 15 as an example. It can be seen that the larger values of f_{UFA} are found in the region of T near 5 ms and $10^{-4} < \nu < 10^{-2}$, and the maximum is approximately at $\nu = 0.0025$ and $T = 5.5$ ms. The value of f_{UFA} in this region is about 0.004. In this case, the optimum loop bandwidth is about $B_N = 70$ Hz, which is a more reasonable value than in the case of the PLL. For the UFA-PLL, the minimum pull-out probability and minimum output variance seem not to be as contradictory criteria as for the PLL. This can be understood noticing that the ability of the UFA-PLL for tracking in high dynamics depends on the smoothness of the transient error response rather than its absolute value, and then the output noise contribution becomes more relevant. Another important fact that must be emphasized is that even for 20 g accelerations it is not advisable to use correlation times T lower than 5 ms.

Notice that, as it was explained in Section 2, this probability analysis applies also for an FLL as long as the right $F(z)$ is used in the computations of K_u and $B'_N T$.

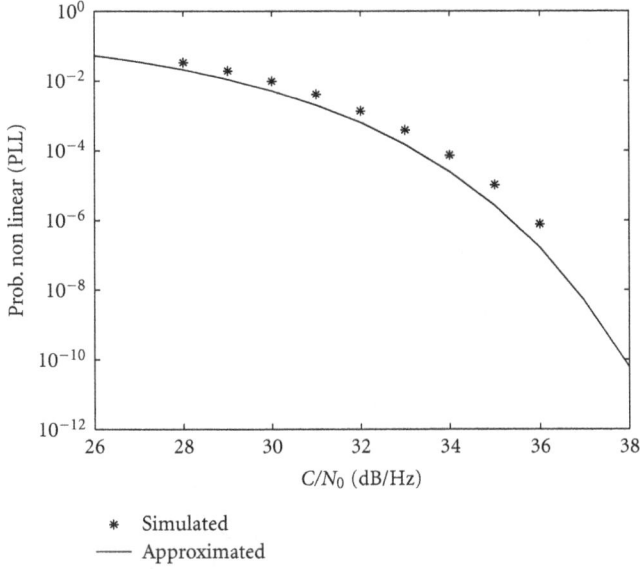

FIGURE 16: Probability of NL behavior for type 3 PLL and 5 g.

* Simulated
—— Approximated

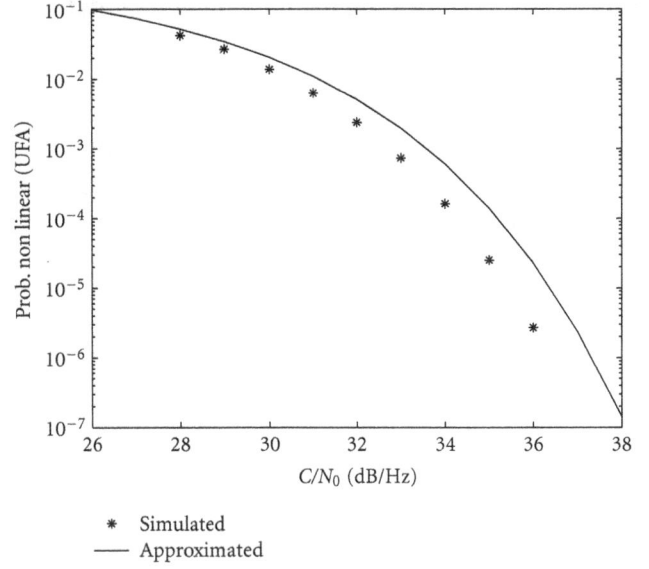

FIGURE 17: Probability of NL behavior for type 3 UFA-PLL and 20 g.

* Simulated
—— Approximated

4.3. Simulations. In this section we assess the accuracy of the approximations made in the previous analysis. According to them the filter design of Section 3 is almost optimum when used in a UFA-PLL for tracking steps of 20 g, that is, it produces the minimum tracking threshold. The loop evolution with input noise according to a given C/N_0 was simulated until the transient peak instant. Runs entering the nonlinear region before this peak were discarded. A variable number of runs were used in order to reduce simulation time as much as possible but keeping statistical significance of the results. Specifically, 100,000 runs were enough for the lowest C/N_0 values, whereas 100 million had to be used for the highest.

A step of 5 g is considered in the simulation of the PLL, which produces a transient peak of $K_p 5\,\mathrm{g}\,(5\,\mathrm{ms})^2 = 1$ rad. Using (26) we found $f_{\mathrm{PLL}} = 0.0033$. The results of the simulations compared with expression (25) are presented in Figure 16. It can be seen that the approximation is slightly optimistic and that the error is almost constant in the simulated range of C/N_0. For the UFA-PLL we adopt a step of 20 g that produces a transient peak of $K_p 20\,\mathrm{g}\,(5\,\mathrm{ms})^2 = 2$ rad and $K_u 20\,\mathrm{g}\,(5\,\mathrm{ms})^2 = 0.38$ rad of peak difference. Using (25), we get $f_{\mathrm{UFA}} = 0.004$. The results of the simulations compared with expression (31) are shown in Figure 17. In this case the approximation is still acceptable for tracking threshold determination, but now it is pessimistic and the error grows for increasing values of C/N_0. This behavior is caused by the Gaussian approximation that neglects the quadratic noise terms in the probability expression (28).

4.4. Tracking Threshold Analysis. To illustrate how this analysis can lead to loop designs with lower tracking thresholds we consider the design of a type 3 UFA-PLL for 20 g acceleration steps. The tracking threshold is determined by a given probability of pull-out, and then we will define it by a given level of probability of starting a nonlinear behavior.

Therefore, if P is the admissible pull-out probability and $C/N_0|_{\mathrm{TH}}$ is the tracking threshold, then

$$P_U \approx Q\left(\sqrt{\left.\frac{C}{N_0}\right|_{\mathrm{TH}}} f_{\mathrm{UFA}}(\nu, T)\right) = P, \qquad (33)$$

or equivalently,

$$\left.\frac{C}{N_0}\right|_{\mathrm{TH}} \approx \frac{(Q^{-1}(P))^2}{f_{\mathrm{UFA}}(\nu, T)}. \qquad (34)$$

Clearly, minimum tracking threshold values will be achieved when $f_{\mathrm{UFA}}(\nu, T)$ is near its maximum, that is, using $\nu \approx 0.00025$ and $T \approx 5.5$ ms. Considering $\nu = 0.0003$ and $T = 5$ ms, for example, the loop parameters are $B_N = 80$ Hz, $B_N' = 51$ Hz $K_p a T^2 = 2$, $K_u a T^2 = 0.38$, and $f_{\mathrm{UFA}} = 0.004$. Taking for instance a value of $P = 0.001$ and replacing in (34), we find that $C/N_0|_{\mathrm{TH}} \approx 34$ dB/Hz.

Another design, similar to those based on analog prototypes, can use $\nu = 0.000001$ and $T = 2$ ms and then $B_N T = 0.1$. The loop parameters become $B_N = 50$ Hz, $B_N' = 13$ Hz $K_p a T^2 = 1.4$, $K_u a T^2 = 0.1$, and $f_{\mathrm{UFA}} = 0.002$. Hence, for the same $P = 0.001$ if we replace in (34) the tracking threshold results $C/N_0|_{\mathrm{TH}} \approx 37$ dB/Hz. Therefore, the use of the digital design method together with the proposed pull-out probability analysis can lower 3 dB the tracking threshold compared with traditional analog-based designs. An additional advantage is the use of longer values of T, requiring less computational load than analog designs.

It has to be mentioned that, even though the actual probability distribution can be different because of the Gaussian approximation, only the arguments of the $Q(\cdot)$ are used in the comparison of both designs. Therefore, the comparison is not affected by the Gaussian approximation, that is, a modified $Q(\cdot)$ function could be used for a more accurate probability calculation but the 3 dB threshold gain would remain.

5. Optimal Smoothing of the Phase Estimates

Due to the presence of two delays in the loop, the phase estimate obtained at a given instant is not computed with measurements up to this instant, but with measurements up to two previous instants. Thus, in the notation of [30] the loop phase estimate at the feedback branch is actually $\hat{\phi}[i \mid i-2] = \hat{\phi}_i$. Naturally, the use of "closer" measurements would produce a smoothing effect, that is, a better estimation. The real-time constraint does not allow taking advantage of this for the loop itself, but it is possible for other purposes as data detection and raw data generation for the navigation processes of the GNSS receivers. In this case, the optimal phase estimate can be obtained in the same way as before, but without forcing the two delays as in (9).

5.1. One Sample Smoothing. In this case the problem is equivalent to obtaining the optimal loop filter with only one delay. Then, if $F'(z)$ is the new loop-filter transfer function to be found, the corresponding closed loop transfer function is

$$T'(z) = \frac{F'(z)z^{-1}}{1 + F'(z)z^{-1}} = Y'(z)z^{-1}, \qquad (35)$$

where $Y'(z)$ is the rational and stable transfer function to be found minimizing $J(Y'(z))$ in (8). Following the same optimization process done in the Appendix A, the result is

$$Y'(z) = \frac{X'(z)z}{\Psi(z)}, \qquad (36)$$

where $\Psi(z)$ is the same minimum phase rational function of (11) and $X'(z)$ is the rational and stable transfer function obtained from

$$G'(z) = \frac{\gamma^2 \phi(z)\,\phi(z^{-1})}{\psi(z^{-1})} = X'(z) + W'(z^{-1}). \qquad (37)$$

Noting that $G'(z) = G(z)/z$, it is simple to relate $X'(z)$ with $X(z)$ since the only change needed is to extract the possible pole in $z = 0$ of $W(z^{-1})/z$ to obtain $W'(z^{-1})$. Hence,

$$X'(z) = \frac{X(z)}{z} + \frac{(G(0) - X(0))}{z} \qquad (38)$$

since $W(z^{-1}) = G(z) - X(z)$. For the case we are interested in, which is tracking of acceleration steps, according to (16) $G(0) = 0$. Then,

$$X'(z) = \frac{X(z) - X(0)}{z}, \qquad (39)$$

Therefore, the new optimum close-loop transfer function of (36) is

$$Y'(z) = \frac{C + (A - 3C)z^{-1} + (3C - B)z^{-2}}{(1 - z_1 z^{-1})(1 - z_2 z^{-1})(1 - z_3 z^{-1})}, \qquad (40)$$

and the corresponding optimum loop filter is now only the three poles of the input,

$$F'(z) = \frac{C + (A - 3C)z^{-1} + (3C - B)z^{-2}}{(1 - z^{-1})^3}. \qquad (41)$$

Even more interesting is the expression of (36) in terms of the cascade of accumulators,

$$Y'(z) = \frac{p_3 + (p_2 - p_3)(1 - z^{-1}) + (p_1 - p_2)(1 - z^{-1})^2}{(1 - z_1 z^{-1})(1 - z_2 z^{-1})(1 - z_3 z^{-1})}, \qquad (42)$$

If it could be possible to implement this loop filter, the feedback of the complete loop would be $\hat{\phi}[i \mid i-1]$. Of course $T'(z) = Y'(z)z^{-1}$ cannot be implemented with a real-time loop, which has two delays, but $Y'(z)z^{-2}$ can. Indeed, if the loop filter structure of the UFA-PLL is slightly modified as shown in Figure 18, it can be shown that amazingly $\hat{\phi}[i \mid i-1] = x_i^{(1)} - x_i^{(2)}$. Clearly, the delay on the feedback branch precludes the use of this value for the loop, but not for the rest of the GNSS receiver.

5.2. Two Samples Smoothing. The previous process can be applied again. Now, the problem is equivalent to obtaining the optimal loop filter without delay. Then, if $F''(z)$ is the loop-filter transfer function and the corresponding closed loop transfer function is

$$T''(z) = \frac{F''(z)z^{-1}}{1 + F''(z)} = Y''(z), \qquad (43)$$

where $Y''(z)$ is the rational and stable transfer function to be found minimizing $J(Y''(z))$ in (8) and replicating (36)–(39), we obtain

$$X''(z) = \frac{X'(z) - X'(0)}{z}. \qquad (44)$$

Therefore, the optimum transfer function of (43) is

$$Y''(z) = \frac{p_3 + P_2(1 - z^{-1}) + P_1(1 - z^{-1})^2}{(1 - z_1 z^{-1})(1 - z_2 z^{-1})(1 - z_3 z^{-1})}, \qquad (45)$$

where $P_2 = (p_2 - 2p_3)$ and $P_1 = (p_1 - 2p_2 - p_3)$. Now $T''(z) = Y''(z)$ cannot be implemented with a real-time loop, but $Y''(z)z^{-2}$ can. Again, based on the loop structure of Figure 18 it can be shown that $\hat{\phi}[i \mid i] = x_{i+1}^{(1)} - 2x_{i+1}^{(2)} + x_{i+1}^{(3)}$.

5.3. More Samples Smoothing. As it was previously mentioned, since $G(z)$ in (16) has four zeros at $z = 0$ for inputs modeled as accelerations steps, the previous smoothing procedure can be done two more times. In this way, if some latency is allowed, the following phase estimates can be obtained based only on the real-time tracking loop of Figure 18—the first three equations are repeated for clarity:

$$\hat{\phi}[i \mid i - 2] = x_{i-1}^{(1)},$$

$$\hat{\phi}[i \mid i - 1] = x_i^{(1)} - x_i^{(2)},$$

$$\hat{\phi}[i \mid i] = x_{i+1}^{(1)} - 2x_{i+1}^{(2)} + x_{i+1}^{(3)}, \qquad (46)$$

$$\hat{\phi}[i \mid i + 1] = x_{i+2}^{(1)} - 3x_{i+2}^{(2)} + 3x_{i+2}^{(3)},$$

$$\hat{\phi}[i \mid i + 2] = x_{i+3}^{(1)} - 4x_{i+3}^{(2)} + 6x_{i+3}^{(3)}.$$

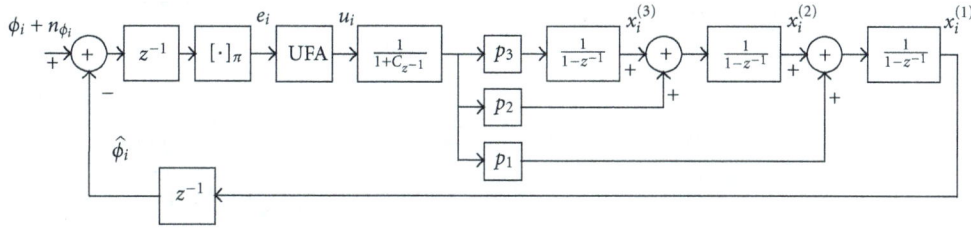

FIGURE 18: Block diagram of the UFA-PLL model with loop filter structure modified for phase smoothing.

These phase estimates can be interpreted as a prediction in the context of estimation theory [30]. In fact, $\hat{\phi}[i \mid i-2] = x_{i-1}^{(1)}$ can be obtained by propagating $\hat{\phi}[i-2 \mid i-2]$ with the signal dynamic model adopted for the phase estimation. If we consider $\mathbf{x}_i = [\mathbf{x}_i^{(1)} \, \mathbf{x}_i^{(2)} \, \mathbf{x}_i^{(3)}]$ as the state of the loop input phase model, the transition matrix must be

$$\mathbf{A} = \begin{bmatrix} 1 & 1 & 1 \\ 0 & 1 & 1 \\ 0 & 0 & 1 \end{bmatrix}. \tag{47}$$

Since x_{i-1} include measurements up to the instant $(i-2)$, $\hat{\phi}[i-2 \mid i-2]$ can be found propagating backwards this state as $\{\mathbf{A}^{-2}x_{i-1}\}^{(1)}$. Notice that using this backward propagation process with the matrix $\mathbf{A}^{-1}$, all the estimators of (46) can be obtained. Actually, more smoothed estimates can be built. For example, the equation

$$\hat{\phi}[i \mid i+3] = \mathbf{x}_{i+4}^{(1)} - 5\mathbf{x}_{i+4}^{(2)} + 10\mathbf{x}_{i+4}^{(3)} \tag{48}$$

can be used, but it is not optimal. As it will be shown in the simulations it can be considered useful because of its extremely simple implementation. In some way, the quantity of zeros in (16) at $z = 0$ gives a measure of the backward propagation capacity of the states estimated by the loop filter.

5.4. Simulations. The simulated loop model is a UFA-PLL as shown in Figure 18 with the same filter coefficients of Section 3. The phase estimation error for an acceleration step of 50 g (starting at $i = 5$ and without noise) for the different estimators is plotted in Figure 19. In this situation the loop error grows up to almost one cycle and therefore the data detection during this transient will not be possible. However, applying the smoothing process described by (46) the transient error is consistently reduced each time that a new input sample is used for the estimation. The response of the suboptimal estimator of (48) is also shown. It can be seen that its transient response is slightly worse than the obtained with $\hat{\phi}[i \mid i+2]$.

The smoothing process also produces a decrease of the estimation noise variance. In Figure 20 the standard deviation of the six previous phase estimators is plotted for three different signal levels. These results were obtained simulating a linearized loop fed with Gaussian noise of variance $1/(2TC/N_0)$. As expected, an increase of 3 dB in the signal corresponds to a reduction of approximately $\sqrt{2}$ in the standard deviation. It is also possible to verify that

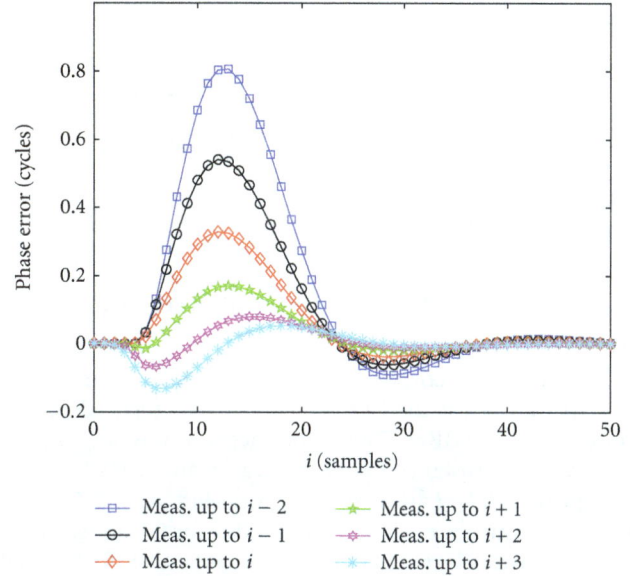

FIGURE 19: Phase estimation error during a step of 50 g.

the standard deviation for the loop output, that is, without smoothing, is equal to $\sigma_{\hat{\phi}_i} = \sqrt{B_N/(C/N_0)}$. This expression gives values 15.75°, 11.14°, and 7.88° when the values of C/N_0 are 30, 33, and 36 dB/Hz, respectively.

6. RF Test Experiments

The FLL-assisted PLL and the UFA-PLL designed in Section 3 were implemented in a System Developer Kit (SDK) for GPS receivers from SiRF [31]. Due to the real-time nature of this task all calculations for the loop had to be done in a fraction of 5 ms. They were programmed in fixed-point arithmetic, using some scaling and approximating coefficient values by powers of 2. Details of these implementations were given in [32]. To verify the tracking capability of the loops with real signals and without relying on expensive equipment like a GPS signal simulator, we used an RF signal generator to produce a frequency modulated carrier at 1575.42 MHz (the L1 GPS frequency). The signal was not spread with the code of a particular satellite, and thus the code generators of the GPS receiver were turned off during the test. This is not a limitation since the focus of this analysis is on the carrier loop, rather than the code loop. A triangular waveform was used as a frequency modulation to simulate steps in acceleration. The frequency deviation was selected according to the magnitude of the step (an instantaneous frequency

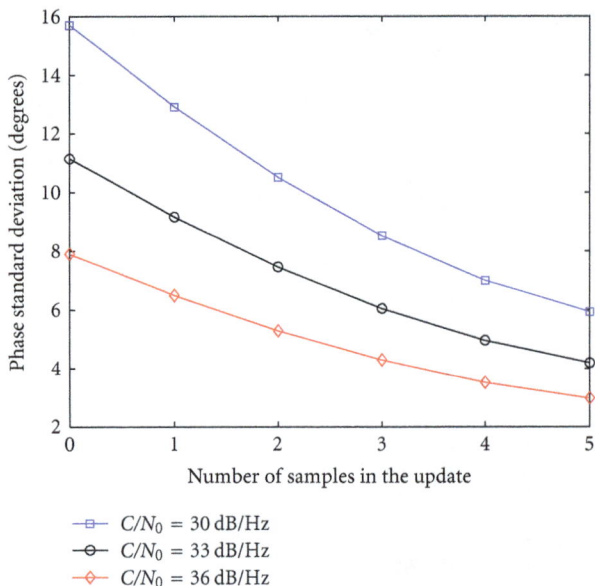

FIGURE 20: Phase estimation standard deviation.

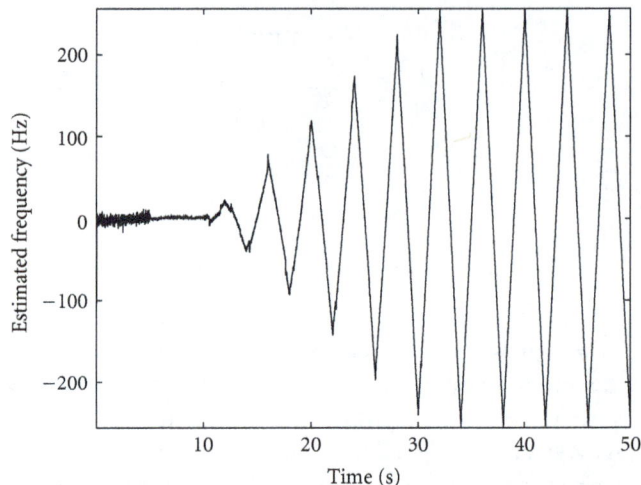

FIGURE 21: Tracked frequency during a 10 g test.

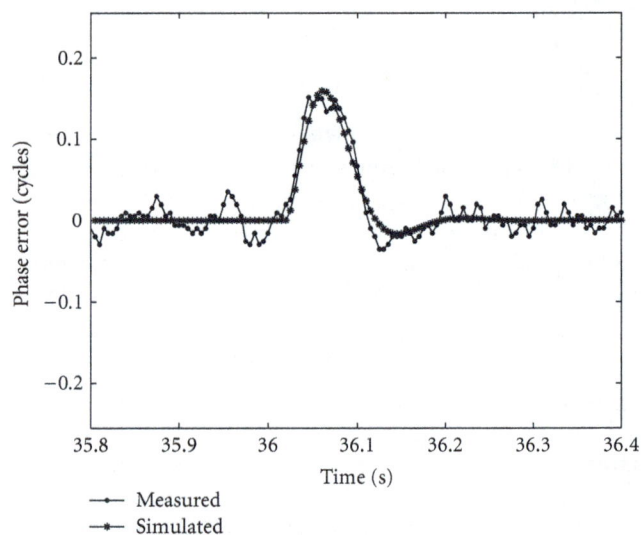

FIGURE 22: Measured phase error during a 10 g step.

deviation of Δf corresponds to a $\lambda \cdot \Delta f$ instantaneous velocity, where λ is the L1 wavelength). The selected carrier power was -113 dBm. Taking into account the noise of the $50\,\Omega$ output resistance of the generator and a noise figure of approximately 8 dB of the RF front-end gives a $C/N_0 = 53$ dB/Hz, which is a relatively high value for GNSS receivers. This value was selected to obtain low noise curves for a visual comparison with the simulated responses since the noise performance has been already characterized.

A test of 10 g acceleration steps is shown in Figure 21, depicting the tracked frequency detrended by a linear fit that accounts for the local clock drift. The amplitude of the triangular waveform was increased gradually up to the desired value to avoid large frequency steps. The measured phase error response of the FLL-assisted PLL at the output of the phase discriminator in one of the steps is shown in Figure 22, the response of the UFA-PLL is the same. The simulated response of the loop (as in Figure 8) is also displayed to appreciate that the implemented loop is properly characterized by our model.

The same experiment was performed for acceleration steps of 40 g. The phase error responses of the FLL-assisted PLL and the UFA-PLL are presented in Figures 23 and 24, respectively. In the case of Figure 23, the step presented is negative and therefore the phase response is upside down with respect to the simulation in Figure 9. Again, a fine agreement between the measurements with real laboratory generated signals and the simulations can be appreciated.

7. Conclusions

A new carrier tracking loop design method for real-time GNSS receivers has been presented, which is completely optimized from the perspective of the digital nature of the correlation measurements. An analysis of the phase and frequency discrimination ideas from this point of

view allowed us to choose optimum discriminators often discarded because of the complexity of their analog counterparts. Also, the known structure of FLL-assisted PLL has been considerably improved leading to a carrier loop that operates normally in phase locked condition and in frequency locked condition if the dynamics become severe enough. The effect of coupling the FLL to the PLL is considered at the design stage allowing a fine control of the effective loop bandwidth. Moreover, this approach allowed us to develop the UFA algorithm that corrects the cycle ambiguities of measured phase errors using the frequency information exploited by an FLL. With this algorithm it was possible to conceive a PLL that has the same advantages of an FLL-assisted PLL but avoids cycle slips and yet is easy to implement.

Regarding loop filters, their optimization was achieved directly in the digital domain. Our procedure solves the bandwidth trade-off considering the discrete nature of the filter and the unavoidable computational delays in the loops and makes possible to design loops extending beyond

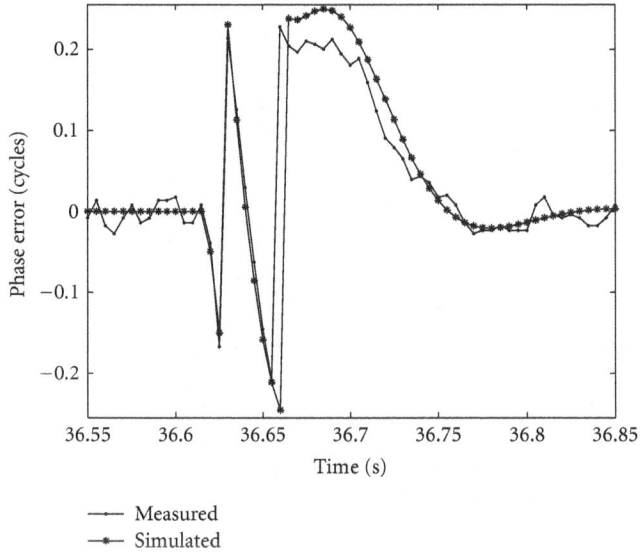

FIGURE 23: Phase error of FLL-assisted PLL during a 40 g step.

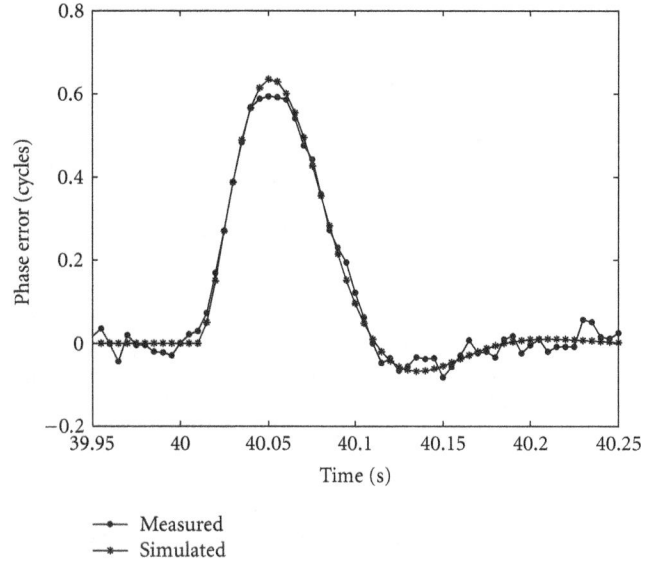

FIGURE 24: Phase error of UFA-PLL during a 40 g step.

the restrictive constraint $B_N T < 0.1$. This limitation is caused by the standard technique of discretizing analog loop filters, which only gives acceptable results for narrow loop designs, not suitable for high-dynamics receivers. Even more important, with our method the designed loop bandwidth is the same as the actually implemented one, since there are no approximations involved. This is of great importance considering the difficulties found in ([5] pp. 183) with respect to the stability of designed loops.

An approximate analysis of the nonlinear behavior of digital loops has been presented. The pull-out probability is approximated as an efficient tool for selecting the correlation time and the loop bandwidth so that the tracking threshold is minimized. Contrary to classical pull-out probability studies, our approach considers nonstationary scenarios as is often found in high-dynamics applications. The UFA-PLL examples presented show that 3 dB of improvement in the tracking threshold can be attained by properly selecting the integration time and loop bandwidth. It is worthwhile to emphasize that even for accelerations of 20 g, a loop sample time shorter than 5 ms is not appropriate.

It was also shown that using some state variables of the loop, smoothed phase estimates can be efficiently built with a latency of only a few samples. This can be very useful in many GNSS applications that use phase measurements, such as code measurement smoothing, differential positioning, and attitude estimation. It can also extend the use of these techniques to real-time and high-dynamics applications, where more complex phase estimation schemes are not practical and the usual tracking loop estimates do not provide enough phase accuracy. The simulations presented show that using the proposed smoothing scheme the transient responses of the loop to an acceleration step of 50 g can be almost eliminated and the estimation noise reduced by half.

Finally, our technique was used to design a loop that can optimally track steps of 20 g with a tracking threshold

of $C/N_0 \approx 34$ dB/Hz. It was implemented in a GPS receiver using fixed-point arithmetic and tested with frequency modulated RF signals. Simulations and experimental results confirm that our new loop designs can track the input phase in these severe conditions, with the same implementation complexity as the usual loops.

This work has not considered adaptive schemes because they tend to increase the computational requirements and transient behavior. An adaptive version of the UFA-PLL could be formulated for situations where time-varying filters are affordable. Our assumptions do not deal with heavily unmodeled dynamics or disturbances on the tracked signals that justify a robust H_∞ approach. Nevertheless, given the Wiener filtering relation established for the present hypotheses, the loops obtained are equivalent to a steady-state version of Kalman-filter based-tracking loops [33]. Including robustness as one of the loop design considerations can also be addressed with the Wiener filtering formulation [27, 34], and a balance between robustness and adaptiveness will be pursued elsewhere.

Appendices

A. Minimizing $J(Y)$

Expression (8) for J can be minimized applying a standard procedure from variational calculus.

Assuming that the input noise is white with power spectral density $\eta/2$, the variance of the output noise can be calculated as

$$\sigma_N^2 = \frac{\eta}{2} \int_{-\pi}^{\pi} \left| Y\left(e^{j\omega}\right) \right|^2 \frac{d\omega}{2\pi} = \frac{\eta}{4\pi j} \oint Y(z) Y(z^{-1}) \frac{dz}{z},$$

$$(A.1)$$

where $Y(e^{j\omega})$ is the frequency response of $Y(z)$ and the last integral extends over the unit circle of the complex plane.

Using Parseval's theorem, we find that

$$E_T(\phi_i) = \sum_{i=0}^{\infty} \Delta\phi_i^2 = \int_{-\pi}^{\pi} \left| \Delta\Phi\left(e^{j\omega}\right) \right|^2 \frac{d\omega}{2\pi}$$

$$= \frac{1}{2\pi j} \oint \Phi(z)\Phi(z^{-1})[1 - Y(z)z^{-2}] \qquad \text{(A.2)}$$

$$\times [1 - Y(z^{-1})z^2]\frac{dz}{z},$$

where $\Delta\Phi(z) = \Phi(z)[1 - Y(z)z^{-2}]$ is the z transform of $\Delta\phi_i$ and $\Phi(z)$ the z transform of ϕ_i. Then, replacing (A.1) and (A.2) in (8), we get the expression of the functional J with explicit dependence of $Y(z)$:

$$J(Y(z)) = \frac{\eta}{4\pi j} \oint \{ Y(z)Y(z^{-1}) + \gamma^2\Phi(z)\Phi(z^{-1})$$

$$\times [1 - Y(z)z^{-2}][1 - Y(z^{-1})z^2] \}$$

$$\times \frac{dz}{z},$$

$$\text{(A.3)}$$

where $\gamma^2 = 2\alpha^2/\eta$.

Let $Y(z)$ be the argument optimizing J and $y(z)$ any other stable causal rational transfer function:

$$\frac{\partial J(Y(z) + \varepsilon y(z))}{\partial\varepsilon}\bigg|_{\varepsilon=0}$$

$$= \frac{-\eta}{2\pi j} \oint \{\gamma^2\Phi(z)\Phi(z^{-1})z \qquad \text{(A.4)}$$

$$- Y(z)[1 + \gamma^2\Phi(z)\Phi(z^{-1})]z^{-1}\}$$

$$\times y(z^{-1})dz,$$

where we used the following identity that holds for all f : $\mathbb{C} \to \mathbb{C}$ such that the integrals exist,

$$\oint f(z^{-1})\frac{dz}{z} = \oint f(z)\frac{dz}{z}. \qquad \text{(A.5)}$$

The optimizing value $Y(z)$ has to produce a zero derivative (A.4) for every $y(z)$. To make the expression between braces zero would not assure to get a stable $Y(z)$. It is first necessary to make the following spectral decomposition:

$$\Psi(z)\Psi(z^{-1}) = 1 + \gamma^2\phi(z)\phi(z^{-1}). \qquad \text{(A.6)}$$

This is always possible because the term on the right has an even number of poles and zeros. And besides, poles as well as zeros can be separated in two sets. If z_i belongs to a set, $1/z_i$ belongs to the other. Hence, if z_i is within the unit circle, there is also $1/z_i$ outside the the unit circle. Therefore, a rational, minimum phase (all zeros inside the unit circle) and stable $\Psi(z)$ can be found. Then, the required $Y(z)$ has to make zero the following integral:

$$\oint \left\{ \frac{\gamma^2\Phi(z)\Phi(z^{-1})z}{\Psi(z^{-1})} - \frac{G(z)\Psi(z)}{z} \right\} g(z^{-1})\Psi(z^{-1})dz.$$

$$\text{(A.7)}$$

Finally, we have to split the first term between braces in (A.7) in two, separating its partial fraction expansion in a part with all poles inside the unit circle $X(z)$ and a part with all poles outside it $W(z^{-1})$, that is,

$$G(z) = \frac{\gamma^2\Phi(z)\Phi(z^{-1})z}{\Psi(z^{-1})} = X(z) + W(z^{-1}), \qquad \text{(A.8)}$$

and note that

$$\oint W(z^{-1})y(z^{-1})\Psi(z^{-1})dz = 0 \qquad \text{(A.9)}$$

because the integrand is analytic inside the unit circle.

Therefore, the optimum transfer function is given by (10), where $\Psi(z)$ and $X(z)$ can be obtained from (11) and (12), respectively. Notice the important role of asking $\Psi(z)$ to be of minimum phase: since it will be inverted in (10) its zeros become poles of $Y(z)$ that has to be causal and stable.

B. Relation to Wiener Filtering

The Wiener filtering is usually posed as minimizing the variance of the estimation error when a measured signal ω_i related to another one ϕ_i is used to produce a linear prediction $\hat{\phi}[i + m|i]$ of ϕ_i. In our case, $\omega_i = \phi_i + n_{\phi i}$. The general solution in terms of spectral densities is given for instance in [26],

$$Y(z) = \left\{ \frac{S_{\phi\omega}(z)z^m}{H(z^{-1})} \right\}_+ H(z), \qquad \text{(B.1)}$$

where $\{\cdot\}_+$ means the part of the partial fraction expansion with poles inside the unit circle and H is the whitening filter that produces a white sequence at its output when ω_i is the input. This solution looks simple, but it does not give yet explicit equations for the coefficients of the causal and stable rational transfer functions needed.

The so-called polynomial framework [34] permits a systematization of this issue. Assume the signal part of the phase is modeled as $\phi_i = (C/D)b_i$ where C and D are polynomials, with C being an abbreviated notation for $C(q) = c_0 + c_1q^{-1} + c_2q^{-2} + \ldots$, and the operator q^{-1} represents the delay operator such that $q^{-1}\phi_i = \phi_{i-1}$. The signal b is a white noise sequence of variance σ_b^2. The noise contaminating the measurements is also modeled linearly as $w_i = (M/N)v_i$ where M and N are polynomials and v is a white noise sequence of variance σ_v^2, independent of b.

The polynomial solution is given by

$$Y = \frac{Q_1 N}{\beta}, \qquad \text{(B.2)}$$

where $\beta(q) = \beta_0 + \beta_1q^{-1} + \beta_2q^{-1} + \cdots$ is a monic polynomial ($\beta_0 = 1$) obtained from the following factorization problem:

$$\sigma_\varepsilon^2\beta\beta_* = \sigma_b^2CC_*NN_* + \sigma_v^2MM_*DD_*. \qquad \text{(B.3)}$$

The notation β_* means $\beta_*(q^{-1}) = \beta_0^* + \beta_1^*q + \beta_2^*q^2 + \cdots$, with β_i^* being the complex conjugate of β_i. In order to obtain Q_1, the Diophantine equation, with $r^2 = \sigma_\varepsilon^2/\sigma_b^2$,

$$q^mCC_*N_* = r^2\beta_*Q_1 + qDL_* \qquad \text{(B.4)}$$

must be solved for Q_1 and L_*. Equivalently, (B.4) is the same as doing the partial fraction expansion of

$$q^{m-1}\frac{CC_*N_*}{r^2 D\beta_*} = \frac{Q_1}{qD} + \frac{L_*}{r^2\beta_*}. \tag{B.5}$$

Notice that if we enforce $\Phi(z) = C(z)/D(z)$, $m = 2$, $w[n] = v[n]$, that is, $M = 1 = N$, $\sigma_v^2 = \eta/(2T)$, and $\sigma_b^2 = \gamma^2\sigma_v^2$; then (11) is the same as (B.3) and $\Psi(z) = \gamma r\beta(z)/D(z)$. Moreover, (12) is similar to (B.5), and thus $\gamma r Q_1(z)/zD(z) = X(z)$. We can see how both solutions match.

Acknowledgment

This work was funded by ANPCyT PICT 00535, UNLP 11-I-127, and CIC-PBA.

References

[1] W. J. Hurd, J. I. Statman, and V. A. Vilnrotter, "High dynamic GPS receiver using maximum likelihood estimation and frequency tracking," *IEEE Transactions on Aerospace and Electronic Systems*, vol. 23, no. 4, pp. 425–437, 1987.

[2] P.W. Ward, "Performance comparisons between FLL, PLL and a novel FLL-assisted-PLL carrier tracking loop under RF interference conditions," in *Proceedings of the 11th International Technical Meeting of The Satellite Division of The Institute of Navigation (ION GPS '98)*, Nashville, Tenn, USA, September 1998.

[3] F. M. Gardner, *Phaselock Techniques*, John Wiley and Sons, New York, NY, USA, 1979.

[4] C. R. Cahn, D. K. Leimer, C. L. Marsh, F. J. Huntowski, and G. D. Larue, "Software implementation of a PN spread spectrum receiver to accommodate dynamics," *IEEE Transactions on Communications*, vol. 25, no. 8, pp. 832–840, 1977.

[5] E. D. Kaplan and C. J. Hegarty, *Understanding GPS: Principles and Applications*, Artech House, Boston, Mass, USA, 2nd edition, 2006.

[6] S. B. Son, I. K. Kim, S. H. Oh, S. H. Kim, and Y. B. Kim, "Commercial GPS receiver design for high dymanic launching vehicles," in *Proceedings of the International Symposium on GNSS/GPS*, Sydney, Australia, December 2004.

[7] P. V. Brennan, "Performance of phase-locked loop frequency synthesiser using accumulative phase detector," *IEE Proceedings of Circuits, Devices and Systems*, vol. 143, no. 5, pp. 249–254, 1996.

[8] R. C. Den Dulk, "Digital fast acquisition method for phase-lock loops," *Electronics Letters*, vol. 24, no. 17, pp. 1079–1080, 1988.

[9] D. Weinfeld and I. Bar-David, "Phase locked loop with extended range phase detector," in *Proceedings of the 18th Convention of Electrical and Electronics Engineers in Israel*, Tel Aviv, Israel, March 1995.

[10] R. Jaffe and E. Rechtin, "Design and performance of phase-lock circuits capable of near-optimun performance over a wide range of inputs signal and noise levels," *IEEE Transaction on Information Theory*, vol. 76, no. 1, pp. 66–76, 1955.

[11] W. C. Lindsey and C. M. Chie, "Survey of digital phase-locked loops," *Proceedings of the IEEE*, vol. 69, no. 4, pp. 410–431, 1981.

[12] B. W. Parkinson and J. J. Spilker, Eds., *Global Positioning System: Theory and Applications*, American Institute of Aeronautics and Astronautics (AIAA), Washington. DC, USA, 1996.

[13] S. A. Stephens and J. B. Thomas, "Controlled-root formulation for digital phase-locked loops," *IEEE Transactions on Aerospace and Electronic Systems*, vol. 31, no. 1, pp. 78–95, 1995.

[14] S.C. Gupta, "On digital phase-locked loops," *IEEE Transactions on Communication Technology*, vol. 16, no. 2, pp. 340–344, 1968.

[15] S. Hinedi and J. I. Statman, "Digital accumulators in phase and frequency tracking loops," *IEEE Transactions on Aerospace and Electronic Systems*, vol. 26, no. 1, pp. 169–180, 1990.

[16] P. A. Roncagliolo and J. G. Garcia, "High dynamics and false lock resistant GNSS Carrier Tracking Loops," in *Proceedings of the 20th International Technical Meeting of The Satellite Division of The Institute of Navigation (ION GNSS '07)*, Fort Worth, Tex, USA, September 2007.

[17] W. C. Lindsey and M. K. Simon, *Telecommunication Systems Engineering*, Prentice-Hall, Englewood Cliffs, NJ, USA, 1973.

[18] H. Meyr and G. Ascheid, *Phase-, Frequency-Locked Loops, and Amplitude Control*, vol. 1 of *Synchronization in Digital Communication*, John Wiley and Sons, New York, NY, USA, 1990.

[19] P.A. Roncagliolo, J.G. Garcia, and C.H. Muravchik, "Pull-out probability considerations in high dynamics GNSS tracking loops design," in *Proceedings of the 10th International Symposium on Spread Spectrum Techniques and Applications (ISSSTA '08)*, Bologna, Italy, August 2008.

[20] F. D. Nunes, J. M. N. Leitão, and F. M. G. Sousa, "Nonlinear filtering in GNSS pseudorange dynamics estimation combining code delay and carrier phase," *IEEE Journal on Selected Topics in Signal Processing*, vol. 3, no. 4, pp. 639–650, 2009.

[21] H. K. Lee and C. Rizos, "Position-domain hatch filter for kinematic differential GPS/GNSS," *IEEE Transactions on Aerospace and Electronic Systems*, vol. 44, no. 1, pp. 30–40, 2008.

[22] P. A. Roncagliolo, J. G. García, P. I. Mercader, D. R. Fuhrmann, and C. H. Muravchik, "Maximum-likelihood attitude estimation using GPS signals," *Digital Signal Processing: A Review Journal*, vol. 17, no. 6, pp. 1089–1100, 2007.

[23] S. Lo, A. Chen, P. Enge et al., "GNSS album. Image and spectral signatures of the new GNSS signals," *Inside GNSS*, pp. 46–56, 2006.

[24] V. I. Tikhonov, "The effect of noise on phase-lock oscillator operation," *Automat. i Tekmekh.*, vol. 22, no. 9, 1959.

[25] J. A. Areta, P. A. Roncagliolo, and C. H. Muravchik, "Sincronización digital de señales de espectro expandido con perturbaciones de alta dinámica," in *Proceedings of the 10th Reunión de trabajo en Procesamiento de la Información y Control*, pp. 762–767, Buenos Aires, Argentina, September 2003.

[26] T. Kailath, *Lectures on Wiener and Kalman Filterings*, Springer, New York, NY, USA, 1981.

[27] M. Sternad and A. Ahlén, "Robust filtering and feedforward control based on probabilistic descriptions of model errors," *Automatica*, vol. 29, no. 3, pp. 661–679, 1993.

[28] R. H. Milocco and C. H. Muravchik, "H2 optimal linear robust sampled-data filtering design using polynomial approach," *IEEE Transactions on Signal Processing*, vol. 51, no. 7, pp. 1816–1824, 2003.

[29] P. A. Roncagliolo, C. E. DeBlasis, and C. H. Muravchik, "GPS digital tracking loops design for high dynamic launching vehicles," in *Proceedings of the 9th International Symposium on Spread Spectrum Techniques and Applications (ISSSTA '06)*, Manaos, Brazil, August 2006.

[30] S. M. Kay, *Fundamentals of Statistical Signal Processing: Estimation Theory*, Prentice Hall, NJ, USA, 1993.

[31] "SiRF technology," http://www.sirf.com/.

[32] C. E. De Blasis, "Seguimiento de señales de GPS," Electronic Engineering Graduation Project, UNLP, Argentina, 2005.

[33] D. Simon, *Optimal State Estimation: Kalman, H-infinity, and Nonlinear Approaches*, John Wiley and Sons, Hoboken, NJ, USA, 2006.

[34] A. Ahlen and M. Sternad, "Wiener filter design using polynomial equations," *IEEE Transactions on Signal Processing*, vol. 39, no. 11, pp. 2387–2399, 1991.

15

INS/GPS for High-Dynamic UAV-Based Applications

Junchuan Zhou,[1] Stefan Knedlik,[2] and Otmar Loffeld[1]

[1] *Center for Sensor Systems (ZESS), University of Siegen, 57076 Siegen, Germany*
[2] *iMAR Navigation GmbH, 66386 St. Ingbert, Germany*

Correspondence should be addressed to Junchuan Zhou, zhou@zess.uni-siegen.de

Academic Editor: Farid Melgani

The carrier-phase-derived delta pseudorange measurements are often used for velocity determination. However, it is a type of integrated measurements with errors strongly related to pseudorange errors at the start and end of the integration interval. Conventional methods circumvent these errors with approximations, which may lead to large velocity estimation errors in high-dynamic applications. In this paper, we employ the extra states to "remember" the pseudorange errors at the start point of the integration interval. Sequential processing is employed for reducing the processing load. Simulations are performed based on a field-collected UAV trajectory. Numerical results show that the correct handling of errors involved in the delta pseudorange measurements is critical for high-dynamic applications. Besides, sequential processing can update different types of measurements without degrading the system estimation accuracy, if certain conditions are met.

1. Introduction

GPS receivers are widely used in navigation. However, the system performance largely depends on the signal environment, and the measurement update rate is low. This raises the need to integrate GPS with the inertial navigation system (INS) to have a robust continuous navigation solution regardless of the environment. Among several integration architectures, the tightly coupled INS/GPS integration is one of the most promising methods to fuse the GPS and INS data, where the code-derived pseudorange and carrier-phase-derived delta pseudorange measurements are often exploited. However, the delta pseudorange is an integrated measurement with errors strictly related to the pseudorange errors at the endpoints of the integration interval. In practice, various approximations are made to handle these errors. The weakest but often used approach is to simply consider the integral of velocity divided by time (an average value) as the instantaneous velocity measurement at the endpoint of the integration interval [1]. It may fulfill the accuracy requirements in static or low dynamic applications. Nevertheless, if the vehicle is maneuvering under high dynamics with low GPS data update rate, large velocity estimation

errors will appear. That is, the velocity errors will be strongly correlated to the accelerations and jerks involved in the trajectory [2]. In order to tackle this problem, we use delay states to "remember" the pseudorange errors at the start of the integration interval. Besides, for reducing the processing load, sequential processing is utilized to avoid the time consuming computation of matrix inversion when deriving the Kalman gain. Different types of measurements (e.g., integrated and nonintegrated measurements) are sequentially processed with no compromise made on their errors.

In the remainder of this paper, the content is organized as follows. In Section 2, the often used velocity determination method based on the delta pseudorange measurements is overviewed, and its problem is indicated. In Section 3, the approach based on the augmented state vector is proposed for tackling this problem. Moreover, sequential processing is utilized and the computational burden is analyzed. In Section 4, the INS/GPS tightly coupled system state space models are introduced. In Section 5, simulation tests are conducted. Numerical results are compared and analyzed. In the Appendix, the number of numerical operations involved in the calculation of the matrix inversion using the Gauss-Jordan elimination method is counted.

2. Velocity Determination

The delta pseudorange is formed by integrating the pseudorange rate over the filter measurement update interval. It is equal to the difference of pseudoranges measured at the endpoints of the filter update interval. Accordingly, the errors involved in delta pseudorange are strictly related to the corresponding pseudorange errors at the start and end of the integration interval, which are given in (1):

$$\Delta \boldsymbol{\rho}_k = \int_{k-1}^{k} \delta \dot{\boldsymbol{\rho}} \, dt = \delta \boldsymbol{\rho}_k - \delta \boldsymbol{\rho}_{k-1}, \tag{1}$$

where $\Delta \boldsymbol{\rho}_k$ represents the delta pseudorange error measured at time instance k; $\delta \dot{\boldsymbol{\rho}}$ denotes the pseudorange rate error; $\delta \boldsymbol{\rho}_k$ and $\delta \boldsymbol{\rho}_{k-1}$ are the pseudorange errors at the two endpoints of the integration interval, which can be related to error states (i.e., position errors and receiver clock bias error) as

$$\begin{aligned} \delta \boldsymbol{\rho}_k &= \hat{\boldsymbol{\rho}}_k - \boldsymbol{\rho}_k = \mathbf{H}_k \delta \mathbf{x}_k + \mathbf{v}_k, \\ \delta \boldsymbol{\rho}_{k-1} &= \hat{\boldsymbol{\rho}}_{k-1} - \boldsymbol{\rho}_{k-1} = \mathbf{H}_{k-1} \delta \mathbf{x}_{k-1} + \mathbf{v}_{k-1}, \end{aligned} \tag{2}$$

where the error states are defined as $\delta \mathbf{x} = \hat{\mathbf{x}}_{\text{estimated}} - \mathbf{x}_{\text{true}}$; the $\mathbf{H}_k$ and $\mathbf{H}_{k-1}$ are the Jacobian matrices from the first-order linearization of the nonlinear pseudorange observation model at time instants of the start and end of the integration interval; $\mathbf{v}_k$ and $\mathbf{v}_{k-1}$ represent the zero-mean Gaussian white noises.

Substituting (2) into (1) yields

$$\delta \boldsymbol{\rho}_k - \delta \boldsymbol{\rho}_{k-1} = \mathbf{H}_k \delta \mathbf{x}_k - \mathbf{H}_{k-1} \delta \mathbf{x}_{k-1} + \mathbf{v}_{k,k-1}. \tag{3}$$

For clarity purpose, we will denote $\mathbf{v}_k^{\phi}$ to represent the cumulative effect of the carrier-phase-derived delta pseudorange measurement noises $\mathbf{v}_{k,k-1}$ from time instants $k-1$ to k. We divide (3) by the time interval $\Delta t = t_k - t_{k-1}$ and assume that this time interval is small. Thus, the changes in the Jacobian matrix $\mathbf{H}$ are negligible (i.e., $\mathbf{H}_{k-1} \approx \mathbf{H}_k$), and we arrive at

$$\begin{aligned} \frac{\delta \boldsymbol{\rho}_k - \delta \boldsymbol{\rho}_{k-1}}{\Delta t} &= \mathbf{H}_k \frac{(\delta \mathbf{x}_k - \delta \mathbf{x}_{k-1})}{\Delta t} + \mathbf{v}_k^{\phi} \\ &\Longrightarrow \delta \dot{\boldsymbol{\rho}}_k = \mathbf{H}_k \delta \dot{\mathbf{x}}_k + \mathbf{v}_k^{\phi}, \end{aligned} \tag{4}$$

where $\delta \dot{\mathbf{x}}_k$ contains the system velocity errors and receiver clock drift error.

This equation is often employed for velocity determination using delta pseudorange measurements. However, it relates the average information (i.e., $(\delta \boldsymbol{\rho}_k - \delta \boldsymbol{\rho}_{k-1})/\Delta t$) to the instantaneous velocity estimation errors at time instant k (i.e., $\delta \dot{\mathbf{x}}_k$). If the vehicle is maneuvering under high dynamics, this approximation will be very poor.

Therefore, we lay (4) aside. And from (2), we have

$$\begin{aligned} \delta \boldsymbol{\rho}_k - \delta \boldsymbol{\rho}_{k-1} &= \left(\hat{\boldsymbol{\rho}}_k - \boldsymbol{\rho}_k \right) - \left(\hat{\boldsymbol{\rho}}_{k-1} - \boldsymbol{\rho}_{k-1} \right) \\ &= \left(\hat{\boldsymbol{\rho}}_k - \hat{\boldsymbol{\rho}}_{k-1} \right) - \left(\boldsymbol{\rho}_k - \boldsymbol{\rho}_{k-1} \right). \end{aligned} \tag{5}$$

We denote $\hat{\mathbf{y}}_k^{\phi} = \hat{\boldsymbol{\rho}}_k - \hat{\boldsymbol{\rho}}_{k-1}$, $\mathbf{y}_k^{\phi} = \boldsymbol{\rho}_k - \boldsymbol{\rho}_{k-1}$ and $\delta \mathbf{y}_k^{\phi} = \hat{\mathbf{y}}_k^{\phi} - \mathbf{y}_k^{\phi}$. Thus, (3) can be reformulated as

$$\delta \mathbf{y}_k^{\phi} = \mathbf{H}_k \delta \mathbf{x}_k - \mathbf{H}_{k-1} \delta \mathbf{x}_{k-1} + \mathbf{v}_k^{\phi}. \tag{6}$$

Equation (6) relates the system current and delay error states to the delta pseudorange innovation vector. It can be used as the system observation model. However, this equation is not in the general format for applying a Kalman filter (KF). That is, the delay state term $\mathbf{H}_{k-1} \delta \mathbf{x}_{k-1}$ appears on the right side of the equation. In order to solve this problem, modifications are required. In this contribution, we augment the state vector to include the delay states. Details will be given in the next section.

3. Augmentation of the System State Vector

We augment the system state vector to include the delay states to "remember" the system estimation errors at the start of the integration interval. These delay states are explicitly estimated from the previous measurement update and do not evolve over time (e.g., [3]). Therefore, we should model them as random constants in the current measurement update. The augmented system model is

$$\begin{bmatrix} \delta \mathbf{x}_k \\ \hline \delta \mathbf{x}_{d,k} \end{bmatrix} = \begin{bmatrix} \boldsymbol{\Phi}_{k,k-1} & \mathbf{O} \\ \hline \mathbf{O} & \mathbf{I} \end{bmatrix} \begin{bmatrix} \delta \mathbf{x}_{k-1} \\ \hline \delta \mathbf{x}_{d,k-1} \end{bmatrix} + \begin{bmatrix} \mathbf{w}_{k-1} \\ \hline \mathbf{0} \end{bmatrix}, \tag{7}$$

where $\delta \mathbf{x}_{d,k}$ denotes the delay error state vector; $\boldsymbol{\Phi}_{k,k-1}$ is the transition matrix for current states; $\mathbf{I}$ is an identity matrix; $\mathbf{w}_{k-1}$ represents the zero-mean Gaussian white noises.

The state estimation error covariance and system process noise error covariance matrices are given in their partitioned forms as:

$$\mathbf{P}_k' = \begin{bmatrix} \mathbf{P}_k & \mathbf{P}_{cd,k} \\ \hline \mathbf{P}_{dc,k} & \mathbf{P}_{d,k} \end{bmatrix}, \qquad \mathbf{Q}_k' = \begin{bmatrix} \mathbf{Q}_k & \mathbf{O} \\ \hline \mathbf{O} & \mathbf{O} \end{bmatrix}, \tag{8}$$

where $\mathbf{P}_k$ is the error covariance of $\delta \mathbf{x}_k$; $\mathbf{P}_{d,k}$ is the error covariance of $\delta \mathbf{x}_{d,k}$, which is explicitly estimated from the previous measurement update; $\mathbf{P}_{cd,k}$ and $\mathbf{P}_{dc,k}$ are the cross-error covariance matrices between the current error state vector $\delta \mathbf{x}_k$ and the delay error state vector $\delta \mathbf{x}_{d,k}$; $\mathbf{Q}_k$ is the process noise error covariance of $\mathbf{w}_k$; $\mathbf{O}$ is a zero matrix.

For the delta pseudorange observation model, it is given as

$$\delta \mathbf{y}_k^{\phi} = \begin{bmatrix} \mathbf{H}_k & -\mathbf{H}_{d,k} \end{bmatrix} \begin{bmatrix} \delta \mathbf{x}_k \\ \delta \mathbf{x}_{d,k} \end{bmatrix} + \mathbf{v}_k^{\phi}, \tag{9}$$

where $\mathbf{H}_k$ and $\mathbf{H}_{d,k}$ are the Jacobian matrices related to the time instants of the start and end of the integration interval; $\delta \mathbf{y}_k^{\phi}$ is the delta pseudorange innovation vector; $\mathbf{v}_k^{\phi}$ represents the measurement noise.

As compared with (6), in this arrangement, we actually consider the "delay states" as a part of the current states. Thus, all the terms are assumed to be related with the current time instant k.

Besides the delta pseudorange, the code-derived pseudorange measurements are also considered in the filter. For the pseudorange measurements, the observation model is

$$\hat{\boldsymbol{\rho}}_k - \boldsymbol{\rho}_k = \mathbf{H}_k \delta \mathbf{x}_k + \mathbf{v}_k^\rho. \tag{10}$$

Substituting $\delta \mathbf{y}_k^\rho = \hat{\boldsymbol{\rho}}_k - \boldsymbol{\rho}_k$ into (10), the overall system observation model can be derived in its partitioned form as

$$\begin{bmatrix} \delta \mathbf{y}_k^\rho \\ \delta \mathbf{y}_k^\phi \end{bmatrix} = \begin{bmatrix} \mathbf{H}_k & \mathbf{O} \\ \mathbf{H}_k & -\mathbf{H}_{d,k} \end{bmatrix} \begin{bmatrix} \delta \mathbf{x}_k \\ \delta \mathbf{x}_{d,k} \end{bmatrix} + \begin{bmatrix} \mathbf{v}_k^\rho \\ \mathbf{v}_k^\phi \end{bmatrix},$$

$$\text{with } \mathbf{R}_k' = \begin{bmatrix} \mathbf{R}_k^\rho & \mathbf{O} \\ \mathbf{O} & \mathbf{R}_k^\phi \end{bmatrix}, \tag{11}$$

where $\mathbf{R}_k^\rho$ and $\mathbf{R}_k^\phi$ are the pseudorange and delta pseudorange measurement error covariance matrices, which are assumed to be uncorrelated with each other.

Thus, (7) and (11) comprise the new system propagation and observation models, which will be used in the INS/GPS integration system. In the following subsections, we will firstly apply the sequential processing to update different types of measurements based on the new developed system models. And then, its computational burden is analyzed with respect to the conventional batch processing.

3.1. Measurement Updates. In this contribution, we process the pseudorange and delta pseudorange measurements as two batches of data. In this way, the dimension of observation vector will be half of the case, if we process the data in one batch.

For sequential processing, if the measurement errors are uncorrelated, they can be processed one after another with zero-width time interval until all measurements are sequentially updated. Otherwise, decoupling of the correlated measurement errors must be made, and linear combinations of measurements should be conducted to yield a new set of measurements whose errors are uncorrelated. After sequentially updating the measurements, the final state estimates and the associated error covariance matrix will be the same as if the measurements are processed in one batch. In this way, the time-consuming calculation of the matrix inversion can be prevented (e.g., [4, 5]). The exact number of numerical operations involved in the matrix inversion is calculated in the Appendix using the Gauss-Jordan Elimination method (e.g., [6]).

3.1.1. Sequential Updates of Delta Pseudorange Measurements. The conventional batch measurement update of delta pseudorange measurements is performed as follows:

$$\begin{bmatrix} \mathbf{K}_k \\ \mathbf{K}_{d,k} \end{bmatrix} = \begin{bmatrix} \mathbf{P}_k^- & \mathbf{P}_{cd,k}^- \\ \mathbf{P}_{dc,k}^- & \mathbf{P}_{d,k}^- \end{bmatrix} \begin{bmatrix} \mathbf{H}_k^T \\ -\mathbf{H}_{d,k}^T \end{bmatrix}$$

$$\times \left[\begin{bmatrix} \mathbf{H}_k & -\mathbf{H}_{d,k} \end{bmatrix} \begin{bmatrix} \mathbf{P}_k^- & \mathbf{P}_{cd,k}^- \\ \mathbf{P}_{dc,k}^- & \mathbf{P}_{d,k}^- \end{bmatrix} \right.$$

$$\left. \times \begin{bmatrix} \mathbf{H}_k^T \\ -\mathbf{H}_{d,k}^T \end{bmatrix} + \mathbf{R}_k^\phi \right]^{-1},$$

$$\begin{bmatrix} \delta \hat{\mathbf{x}}_k^+ \\ \delta \hat{\mathbf{x}}_{d,k}^+ \end{bmatrix} = \begin{bmatrix} \delta \hat{\mathbf{x}}_k^- \\ \delta \hat{\mathbf{x}}_{d,k}^- \end{bmatrix} + \begin{bmatrix} \mathbf{K}_k \\ \mathbf{K}_{d,k} \end{bmatrix}$$

$$\times \left[\delta \mathbf{y}_k^\phi - \begin{bmatrix} \mathbf{H}_k & -\mathbf{H}_{d,k} \end{bmatrix} \begin{bmatrix} \delta \hat{\mathbf{x}}_k^- \\ \delta \hat{\mathbf{x}}_{d,k}^- \end{bmatrix} \right],$$

$$\begin{bmatrix} \mathbf{P}_k^+ & \mathbf{P}_{cd,k}^+ \\ \mathbf{P}_{dc,k}^+ & \mathbf{P}_{d,k}^+ \end{bmatrix} = \left[\begin{bmatrix} \mathbf{I} & \mathbf{O} \\ \mathbf{O} & \mathbf{I} \end{bmatrix} - \begin{bmatrix} \mathbf{K}_k \\ \mathbf{K}_{d,k} \end{bmatrix} \begin{bmatrix} \mathbf{H}_k & -\mathbf{H}_{d,k} \end{bmatrix} \right]$$

$$\times \begin{bmatrix} \mathbf{P}_k^- & \mathbf{P}_{cd,k}^- \\ \mathbf{P}_{dc,k}^- & \mathbf{P}_{d,k}^- \end{bmatrix}, \tag{12}$$

where $\mathbf{K}_k$ is the Kalman gain for the current error state vector $\delta \mathbf{x}_k$, while $\mathbf{K}_{d,k}$ is the Kalman gain for the delay error state vector $\delta \mathbf{x}_{d,k}$.

As stated before, the delay error states will not be updated from current measurement update. Thus, we can set $\mathbf{K}_{d,k}$ to be a zero matrix. Hereby, (12) can be simplified as

$$\mathbf{K}_k = \begin{bmatrix} \mathbf{P}_k^- \mathbf{H}_k^T - \mathbf{P}_{cd,k}^- \mathbf{H}_{d,k}^T \end{bmatrix}$$

$$\times \begin{bmatrix} \mathbf{H}_k \mathbf{P}_k^- \mathbf{H}_k^T - \mathbf{H}_{d,k} \mathbf{P}_{dc,k}^- \mathbf{H}_k^T \\ -\mathbf{H}_k \mathbf{P}_{cd,k}^- \mathbf{H}_{d,k}^T + \mathbf{H}_{d,k} \mathbf{P}_{d,k}^- \mathbf{H}_{d,k}^T + \mathbf{R}_k^\phi \end{bmatrix}^{-1},$$

$$\delta \hat{\mathbf{x}}_k^+ = \delta \hat{\mathbf{x}}_k^- + \mathbf{K}_k \begin{bmatrix} \delta \mathbf{y}_k^\phi - \mathbf{H}_k \delta \hat{\mathbf{x}}_k^- + \mathbf{H}_{d,k} \delta \hat{\mathbf{x}}_{d,k}^- \end{bmatrix}, \tag{13}$$

$$\mathbf{P}_k^+ = (\mathbf{I} - \mathbf{K}_k \mathbf{H}_k) \mathbf{P}_k^- + \mathbf{K}_k \mathbf{H}_{d,k} \mathbf{P}_{dc,k}^-,$$

$$\mathbf{P}_{cd,k}^+ = (\mathbf{I} - \mathbf{K}_k \mathbf{H}_k) \mathbf{P}_{cd,k}^- + \mathbf{K}_k \mathbf{H}_{d,k} \mathbf{P}_{d,k}^-,$$

$$\mathbf{P}_{dc,k}^+ = \left(\mathbf{P}_{cd,k}^+ \right)^T.$$

It is worth mentioning that, in (13), we have not considered the measurement update of $\delta \hat{\mathbf{x}}_{d,k}$, but the uncertainty of $\delta \hat{\mathbf{x}}_{d,k}$ (i.e., $\mathbf{P}_{d,k}^-$, $\mathbf{P}_{dc,k}^-$, $\mathbf{P}_{cd,k}^-$) is accounted for in the derivation of the Kalman gain $\mathbf{K}_k$.

In sequential update of delta pseudorange measurements, the following parameters need to be firstly initialized. We denote $\mathbf{H}_k(i)$ and $\mathbf{H}_{d,k}(i)$ as the ith row of $\mathbf{H}_k$ and $\mathbf{H}_{d,k}$, respectively, and use $\mathbf{R}_k^\phi(i,i)$ to represent the ith main diagonal element of $\mathbf{R}_k^\phi$.

$$\mathbf{P}_k(0) = \mathbf{P}_k^-, \mathbf{P}_{cd,k}(0) = \mathbf{P}_{cd,k}^-, \mathbf{P}_{dc,k}(0) = \mathbf{P}_{dc,k}^-, \delta \hat{\mathbf{x}}_k(0) = \delta \hat{\mathbf{x}}_k^-,$$

$$\mathbf{H}_k = \begin{bmatrix} \mathbf{H}_k(1) \\ \vdots \\ \mathbf{H}_k(m) \end{bmatrix}, \qquad \mathbf{H}_{d,k} = \begin{bmatrix} \mathbf{H}_{d,k}(1) \\ \vdots \\ \mathbf{H}_{d,k}(m) \end{bmatrix},$$

$$\mathbf{R}_k^\phi = \begin{bmatrix} \mathbf{R}_k^\phi(1,1) & \cdots & 0 \\ \vdots & \ddots & \vdots \\ 0 & \cdots & \mathbf{R}_k^\phi(m,m) \end{bmatrix}. \tag{14}$$

From $i = 1$ to m, the measurements are updated sequentially as follows:

$$\mathbf{K}_k(i) = \left[\mathbf{P}_k(i-1)\mathbf{H}_k(i)^T - \mathbf{P}_{cd,k}(i-1)\mathbf{H}_{d,k}(i)^T\right]$$
$$\times\left[\mathbf{H}_k(i)\mathbf{P}_k(i-1)\mathbf{H}_k(i)^T\right.$$
$$-\mathbf{H}_{d,k}(i)\mathbf{P}_{dc,k}(i-1)\mathbf{H}_k(i)^T$$
$$-\mathbf{H}_k(i)\mathbf{P}_{cd,k}(i-1)\mathbf{H}_{d,k}(i)^T$$
$$\left.+\mathbf{H}_{d,k}(i)\mathbf{P}_{d,k}\mathbf{H}_{d,k}(i)^T + \mathbf{R}_k^\phi(i,i)\right]^{-1},$$

$$\delta\hat{\mathbf{x}}_k(i) = \delta\hat{\mathbf{x}}_k(i-1) + \mathbf{K}_k(i)$$
$$\times\left[\delta\mathbf{y}_k^\phi(i) - \mathbf{H}_k(i)\delta\hat{\mathbf{x}}_k(i-1) + \mathbf{H}_{d,k}(i)\delta\hat{\mathbf{x}}_{d,k}\right],$$

$$\mathbf{P}_k(i) = [1 - \mathbf{K}_k(i)\mathbf{H}_k(i)]\mathbf{P}_k(i-1) + \mathbf{K}_k(i)\mathbf{H}_{d,k}(i)\mathbf{P}_{dc,k}(i-1),$$

$$\mathbf{P}_{cd,k}(i) = [1 - \mathbf{K}_k(i)\mathbf{H}_k(i)]\mathbf{P}_{cd,k}(i-1) + \mathbf{K}_k(i)\mathbf{H}_{d,k}(i)\mathbf{P}_{d,k},$$

$$\mathbf{P}_{dc,k}(i) = \mathbf{P}_{cd,k}(i)^T. \tag{15}$$

3.1.2. Sequential Updates of Pseudorange Measurements. The batch processing of pseudorange measurements is handled after we process the delta pseudorange measurements. The *a posteriori* estimates from the delta pseudorange measurement update (i.e., $\delta\hat{\mathbf{x}}_k^+$ and $\mathbf{P}_k^+$) are used here as the *a priori* estimates (i.e., $\delta\hat{\mathbf{x}}_k^-$ and $\mathbf{P}_k^-$).

$$\mathbf{K}_k = \mathbf{P}_k^-\mathbf{H}_k^T\left(\mathbf{H}_k\mathbf{P}_k^-\mathbf{H}_k^T + \mathbf{R}_k^\rho\right)^{-1},$$
$$\delta\hat{\mathbf{x}}_k^+ = \delta\hat{\mathbf{x}}_k^- + \mathbf{K}_k\left[\delta\mathbf{y}_k^\rho - \mathbf{H}_k\delta\hat{\mathbf{x}}_k^-\right], \tag{16}$$
$$\mathbf{P}_k^+ = (\mathbf{I} - \mathbf{K}_k\mathbf{H}_k)\mathbf{P}_k^-.$$

For sequential processing, we define

$$\mathbf{H}_k = \begin{bmatrix}\mathbf{H}_k(1)\\\vdots\\\mathbf{H}_k(m)\end{bmatrix}, \qquad \mathbf{R}_k^\rho = \begin{bmatrix}\mathbf{R}_k^\rho(1,1) & \cdots & 0\\\vdots & \ddots & \vdots\\0 & \cdots & \mathbf{R}_k^\rho(m,m)\end{bmatrix}. \tag{17}$$

From $i = 1$ to m, the measurements are sequentially updated as

$$\mathbf{K}_k(m+i) = \mathbf{P}_k(m+i-1)\mathbf{H}_k(i)^T$$
$$\times\left[\mathbf{H}_k(i)\mathbf{P}_k(m+i-1)\mathbf{H}_k(i)^T + \mathbf{R}_k^\rho(i,i)\right]^{-1},$$
$$\delta\hat{\mathbf{x}}_k(m+i) = \delta\hat{\mathbf{x}}_k(m+i-1) + \mathbf{K}_k(m+i)$$
$$\times\left[\delta\mathbf{y}_k^\rho(i) - \mathbf{H}_k(i)\delta\hat{\mathbf{x}}_k(m+i-1)\right],$$
$$\mathbf{P}_k(m+i) = [1 - \mathbf{K}_k(m+i)\mathbf{H}_k(i)]\mathbf{P}_k(m+i-1). \tag{18}$$

After all the measurements are updated, the *a posteriori* estimates are calculated as

$$\delta\hat{\mathbf{x}}_k^+ = \delta\hat{\mathbf{x}}_k(m+m), \qquad \mathbf{P}_k^+ = \mathbf{P}_k(m+m). \tag{19}$$

They are used as the delay states and covariance parameters in the next recursion

$$\delta\hat{\mathbf{x}}_{d,k+1} = \delta\hat{\mathbf{x}}_k(m+m), \qquad \mathbf{P}_{d,k+1} = \mathbf{P}_k(m+m). \tag{20}$$

The whole process repeats until the end of the trajectory.

3.2. Computational Burden Comparisons. It is known that the sequential processing can save computational time. But how much can be exactly saved? In order to answer this question, the number of numerical operations involved in addition and multiplication should be counted separately. The method utilized in [7] has been employed in this paper, where subtraction is counted as addition and the division over scalar is considered as multiplication by the inverse of the scalar. The "n" represents the dimension of state vector and "m" is the dimension of the measurement vector. We denote the "nn" as an n-by-n square matrix, and "$nn \times nn$" stands for an n-by-n matrix multiplies another n-by-n matrix. "$n1$" means an n-dimensional vector, "1" means a scalar value. The numbers in the column of "Num." represent the amount of operations involved in the measurement update of both the pseudorange and delta pseudorange measurements, while the numbers in the parentheses denote the operations encountered in the delta pseudorange measurement update only. For batch measurement update (i.e., (13) and (16)), the involved numerical operations are given in the left side of Table 1. For measurement update in sequence (i.e., (15) and (18)), they are given in the right side of Table 1. Table 2 summaries the overall conducted numerical operations involved in sequential and batch measurement updates.

The correctness of the calculations in Table 2 can be easily verified when we consider the case that there is only one satellite in view ($m = 1$). In this case, there should be no difference between the sequential and batch processing. We can prove this by assigning m to be 1. The batch processing returns $18n^2 - 2n$ for addition, and $18n^2 + 12n + 2$ for multiplication, which are obviously equal to the corresponding calculations in sequential processing.

Based on the calculations in Table 2, the relationships between the number of numerical operations and the dimension of the observation vector are illustrated in Figures 1 and 2, where the dimension of the state vector is 17 (i.e., $n = 17$).

Figure 1 shows that if more measurements are processed, more computational time can be saved from sequential processing. However, if the number of measurements is smaller than 4, no big difference can be observed. In this paper, we update the delta pseudorange and pseudorange as two batches of data. And hence, in the calculation, the dimension of observation vector is always smaller than 11 (i.e., $m \leq 11$), which saves a large number of numerical operations.

Figure 2 illustrates that the measurement update of the delta pseudorange requires much more numerical operations than the pseudorange measurement update due to the increased computational complexity in (15) as compared with (18). However, when compared with the batch measurement update (i.e., red curves in Figure 2), they are both saved.

4. Tightly Coupled INS/GPS State Space Models

The INS propagation model we use is often seen for low-cost microelectromechanical- (MEMS-) based IMU, which

TABLE 1: Numerical operations involved in sequential and batch measurement updates.

Operation	Num.	Add.	Multi.	Operation	Num.	Add.	Multi.
$nn \times nm$	3 (2)	$(n-1)nm$	mn^2	$nn \times n1$	3 (2) $\times m$	$n(n-1)$	n^2
$nm \times mn$	5 (4)	$(m-1)n^2$	mn^2	$n1 + n1$	3 (2) $\times m$	n	—
$nm + nm$	1 (1)	nm	—	$1n \times nn$	10 (8) $\times m$	$n(n-1)$	n^2
$nm \times mm$	2 (1)	$(m-1)mn$	nm^2	$1n \times n1$	8 (6) $\times m$	$n-1$	n
$mn \times nn$	10 (8)	$mn(n-1)$	mn^2	$1 + 1$	8 (6) $\times m$	1	—
$mn \times nm$	5 (4)	$(n-1)m^2$	nm^2	1×1	2 (1) $\times m$	—	1
$mm + mm$	5 (4)	m^2	—	$n1 \times 1$	4 (2) $\times m$	—	n
$nn + nn$	5 (4)	n^2	—	$n1 \times 1n$	5 (4) $\times m$	—	n^2
$mn \times n1$	3 (2)	$m(n-1)$	mn	$nn + nn$	5 (4) $\times m$	n^2	—
$m1 + m1$	3 (2)	m	—				
$nm \times m1$	2 (1)	$(m-1)n$	nm				
$n1 + n1$	2 (1)	n	—				
$\text{inv}(mm)$	2 (1)	$(3/2)m^3 - 2m^2 + (m/2)$	$(3/2)m^3 - (1/2)m$				

TABLE 2: The sum of additions and multiplications involved in batch and sequential processing.

Operations	Additions	Operations	Multiplications
Batch	$3m^3 + (7n-4)m^2 + (18n^2 - 9n + 1)m$	Batch	$3m^3 + 7nm^2 + (18n^2 + 5n - 1)m$
Sequential	$(18n^2 - 2n)m$	Sequential	$(18n^2 + 12n + 2)m$

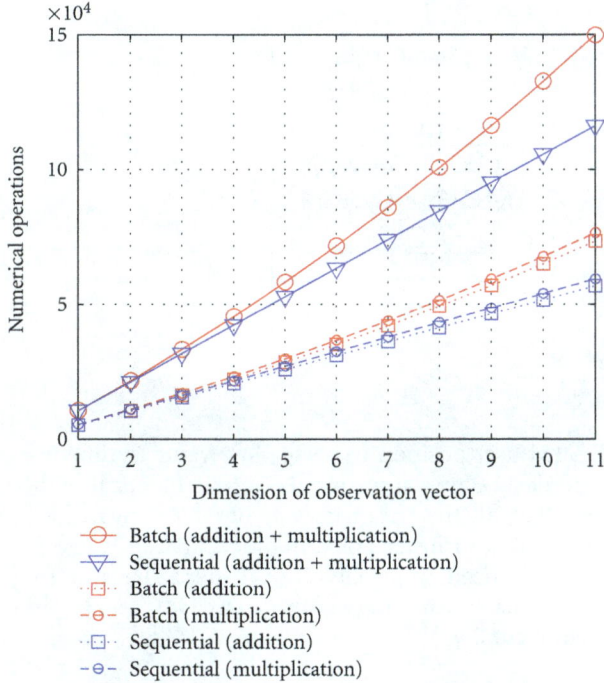

FIGURE 1: Computational burden analysis (addition versus multiplication) with varying number of measurements.

is also used as the system propagation model in INS/GPS integrated system. In terms of a low-cost IMU, the effects from the earth rotation cannot be observed. Hence, they are not considered in the model. Moreover, the transport rate and Coriolis terms have been neglected for simplicity. The simplified propagation model in discrete time domain can be expressed in navigation frame as:

$$\mathbf{p}_{n,k+1} = \mathbf{p}_{n,k} + \mathbf{v}_{n,k} \cdot \Delta t,$$
$$\mathbf{v}_{n,k+1} = \mathbf{v}_{n,k} + \left[\mathbf{R}_{b2n,k} \cdot \left(\widetilde{\mathbf{f}}_{b,k} - \hat{\mathbf{f}}_{b,k}^{\text{error}} - \mathbf{w}_f \right) + \mathbf{g}_n \right] \cdot \Delta t,$$
$$\boldsymbol{\psi}_{k+1} = \boldsymbol{\psi}_k + \mathbf{E}_{b2n,k} \cdot \left(\widetilde{\boldsymbol{\omega}}_{b,k} - \hat{\boldsymbol{\omega}}_{b,k}^{\text{error}} - \mathbf{w}_\omega \right) \cdot \Delta t,$$

$$(21)$$

where k is the time instant; $\widetilde{\mathbf{f}}_b$ is the measurement vector of the specific force; $\widetilde{\boldsymbol{\omega}}_b$ represents the measurement vector of angular rate; $\mathbf{g}_n$ is the local gravity vector; $\boldsymbol{\psi}$ is the Euler angles; $\hat{\mathbf{f}}_{b,k}^{\text{error}}$ and $\hat{\boldsymbol{\omega}}_{b,k}^{\text{error}}$ are the estimated specific force and angular rate errors; $\mathbf{w}_f$ and $\mathbf{w}_\omega$ are the remaining noise errors, which are assumed to be zero-mean Gaussian white noises; $\mathbf{R}_{b2n}$ is the frame rotation matrix from body frame to NED navigation frame; $\mathbf{E}_{b2n}$ is the rotation rate transformation matrix from body to navigation frame,

$$\mathbf{R}_{b2n} = \begin{pmatrix} C\gamma C\beta & C\gamma S\beta S\alpha - S\gamma C\alpha & C\gamma S\beta C\alpha + S\gamma S\alpha \\ S\gamma C\beta & S\gamma S\beta S\alpha + C\gamma C\alpha & S\gamma S\beta C\alpha - C\gamma S\alpha \\ -S\beta & C\beta S\alpha & C\beta C\alpha \end{pmatrix},$$
$$\mathbf{E}_{b2n} = \begin{pmatrix} 1 & S\alpha T\beta & C\alpha T\beta \\ 0 & C\alpha & -S\alpha \\ 0 & S\alpha/C\beta & C\alpha/C\beta \end{pmatrix},$$

$$(22)$$

where $CX, SX,$ and TX represent the trigonometric operations of cosine, sine, and tangent of X, respectively. The α, β, γ denote the roll, pitch, and yaw, respectively.

Regarding the INS/GPS tightly coupled integration, the extended Kalman filter (EKF) is considered as state of the art for fusing the INS and GPS data, and the error states are often used [8]. In this paper, the error states are defined as

$\delta\mathbf{x} = \hat{\mathbf{x}}_{\text{estimated}} - \mathbf{x}_{\text{true}}$. The system transition model $\mathbf{\Phi}_{k,k-1}$ (i.e., in (7)) for current error state vector is shown in (23), where the first three rows come from the first order linearization of (21) with respect to the corresponding states:

$$
\begin{bmatrix} \delta\mathbf{p}_{n,k+1} \\ \delta\mathbf{v}_{n,k+1} \\ \delta\boldsymbol{\psi}_{k+1} \\ \delta\mathbf{f}_{b,k+1}^{\text{error}} \\ \delta\boldsymbol{\omega}_{b,k+1}^{\text{error}} \\ c\delta t_{k+1} \\ c\delta\dot{t}_{k+1} \end{bmatrix} = \begin{bmatrix} \mathbf{I}_{3\times3} & \mathbf{I}_{3\times3}\Delta t & \mathbf{O}_{3\times3} & \mathbf{O}_{3\times3} & \mathbf{O}_{3\times3} & \mathbf{0}_{3\times1} & \mathbf{0}_{3\times1} \\ \mathbf{O}_{3\times3} & \mathbf{I}_{3\times3} & \mathbf{F}_{23} & -\mathbf{R}_{b2n}\Delta t & \mathbf{O}_{3\times3} & \mathbf{0}_{3\times1} & \mathbf{0}_{3\times1} \\ \mathbf{O}_{3\times3} & \mathbf{O}_{3\times3} & \mathbf{F}_{33} & \mathbf{O}_{3\times3} & -\mathbf{E}_{b2n}\Delta t & \mathbf{0}_{3\times1} & \mathbf{0}_{3\times1} \\ \mathbf{O}_{3\times3} & \mathbf{O}_{3\times3} & \mathbf{O}_{3\times3} & \mathbf{I}_{3\times3} & \mathbf{O}_{3\times3} & \mathbf{0}_{3\times1} & \mathbf{0}_{3\times1} \\ \mathbf{O}_{3\times3} & \mathbf{O}_{3\times3} & \mathbf{O}_{3\times3} & \mathbf{O}_{3\times3} & \mathbf{I}_{3\times3} & \mathbf{0}_{3\times1} & \mathbf{0}_{3\times1} \\ \mathbf{0}_{3\times1}^T & \mathbf{0}_{3\times1}^T & \mathbf{0}_{3\times1}^T & \mathbf{0}_{3\times1}^T & \mathbf{0}_{3\times1}^T & 1 & \Delta t \\ \mathbf{0}_{3\times1}^T & \mathbf{0}_{3\times1}^T & \mathbf{0}_{3\times1}^T & \mathbf{0}_{3\times1}^T & \mathbf{0}_{3\times1}^T & 0 & 1 \end{bmatrix} \cdot \begin{bmatrix} \delta\mathbf{p}_{n,k} \\ \delta\mathbf{v}_{n,k} \\ \delta\boldsymbol{\psi}_{k} \\ \delta\mathbf{f}_{b,k}^{\text{error}} \\ \delta\boldsymbol{\omega}_{b,k}^{\text{error}} \\ c\delta t_{k} \\ c\delta\dot{t}_{k} \end{bmatrix} + \mathbf{w}_k. \tag{23}
$$

In (23), the gyroscope and accelerometer sensor errors $\delta\mathbf{f}_b^{\text{error}}$ and $\delta\boldsymbol{\omega}_b^{\text{error}}$ are modeled as random-walk plus constant (e.g., [8]). The receiver clock drift error $c\delta\dot{t}$ is modeled as constant plus random walk process, while the receiver clock bias $c\delta t$ is the integral of the clock drift error. The submatrices $\mathbf{F}_{33}$ and $\mathbf{F}_{23}$ are given as:

$$
\mathbf{F}_{33} = \begin{bmatrix} 1 + \begin{pmatrix} \hat{\omega}_y C\alpha T\beta \\ +\hat{\omega}_z T\beta S\alpha \end{pmatrix}\Delta t & \begin{bmatrix} \hat{\omega}_z C\alpha(T^2\beta+1) \\ +\hat{\omega}_y S\alpha(T^2\beta+1) \end{bmatrix}\Delta t & 0 \\ -\begin{bmatrix} \hat{\omega}_z C\alpha \\ +\hat{\omega}_y S\alpha \end{bmatrix}\Delta t & 1 & 0 \\ \begin{bmatrix} \hat{\omega}_y C\alpha - \hat{\omega}_z S\alpha \\ \overline{C\beta} \end{bmatrix}\Delta t & \dfrac{C\alpha S\beta\hat{\omega}_z + S\beta S\alpha\hat{\omega}_y}{C^2\beta}\Delta t & 1 \end{bmatrix},
$$

$$
\mathbf{F}_{23} = \begin{bmatrix} \begin{matrix} \hat{f}_y(S\alpha S\gamma + C\alpha C\gamma S\beta) \\ +\hat{f}_z(C\alpha S\gamma + C\gamma S\beta S\alpha) \end{matrix} & \begin{matrix} \hat{f}_z C\beta C\alpha C\gamma \\ -\hat{f}_x C\gamma S\beta \\ +\hat{f}_y C\beta C\gamma S\alpha \end{matrix} & \begin{matrix} \hat{f}_z(C\gamma S\alpha + C\alpha S\beta S\gamma) \\ -\hat{f}_y(C\alpha C\gamma - S\beta S\alpha S\gamma) \\ -\hat{f}_x C\beta S\gamma \end{matrix} \\ \begin{matrix} -\hat{f}_z(S\beta S\alpha S\gamma + C\alpha C\gamma) \\ -\hat{f}_y(C\gamma S\alpha - C\alpha S\beta S\gamma) \end{matrix} & \begin{matrix} \hat{f}_z C\beta C\alpha S\gamma \\ -\hat{f}_x S\beta S\gamma \\ +\hat{f}_y C\beta S\alpha S\gamma \end{matrix} & \begin{matrix} \hat{f}_z(S\alpha S\gamma + C\alpha C\gamma S\beta) \\ -\hat{f}_y(C\alpha S\gamma - C\gamma S\beta S\alpha) \\ +\hat{f}_x C\beta C\gamma \end{matrix} \\ \hat{f}_y C\beta C\alpha - \hat{f}_z C\beta S\alpha & \begin{matrix} -\hat{f}_x C\beta \\ -\hat{f}_z C\alpha S\beta \\ -\hat{f}_y S\beta S\alpha \end{matrix} & 0 \end{bmatrix} \cdot \Delta t, \tag{24}
$$

where $C^2 X, T^2 X$ represent the square of cosine and tangent trigonometric operations of X, and $\hat{f}_x = \tilde{f}_x - \hat{f}_x^{\text{error}}$, $\hat{f}_y = \tilde{f}_y - \hat{f}_y^{\text{error}}$, $\hat{f}_z = \tilde{f}_z - \hat{f}_z^{\text{error}}$, $\hat{\omega}_y = \tilde{\omega}_y - \hat{\omega}_y^{\text{error}}$, $\hat{\omega}_z = \tilde{\omega}_z - \hat{\omega}_z^{\text{error}}$ are the IMU raw data compensated with the estimated sensor errors expressed in body frame. The linearization can, for example, be accomplished using the "Jacobian" function in MATLAB.

The transition matrix for the delay state vector is an identity matrix. Thus, according to (7), we have already specified the system propagation model.

For the system observation model (i.e., in (11)), $\mathbf{H}_{d,k}$ and $\mathbf{H}_k$ are the Jacobian matrices from the first order linearization of the pseudorange nonlinear observation equations around the INS-estimated position at the time instants of the start and end of the integration interval. The Jacobian matrix related with the current error state vector is shown as:

$$
\mathbf{H}_k = \begin{bmatrix} l_{n,k}^{sv(i)} & l_{e,k}^{sv(i)} & l_{d,k}^{sv(i)} & 0 & \cdots & 0 & 1 & 0 \end{bmatrix}_{i\times 17}, \tag{25}
$$

where $l_{n,k}^{sv(i)}, l_{e,k}^{sv(i)}, l_{d,k}^{sv(i)}$ ($i = 1$ to m) denote the unit direction vectors pointing from the ith satellite position to the INS

estimated user position in navigation frame at time instant k. The delay observation matrix $\mathbf{H}_{d,k}$ is identical with $\mathbf{H}_k$, except that all the parameters involved are related to the time instant $k-1$ instead of k. In the recursive KF algorithm, we do not need to specify $\mathbf{H}_{d,k}$. Instead, we can simply use previous epoch observation matrix $\mathbf{H}_{k-1}$ as $\mathbf{H}_{d,k}$ in the current recursion.

5. Simulation Experiments

In the following simulation tests, a field-collected UAV trajectory is used. The position and velocity dynamic profiles are depicted in Figures 3 and 4. On the purpose of verifying the correctness of the proposed algorithm in the processing of integrated measurements sequentially, we assume that the GPS receiver is operating at the DGPS mode, in which majority of GPS measurement errors are corrected or minimized to small values, that is, the ionospheric and tropospheric delays. Hereby, they are neglected in this simulation.

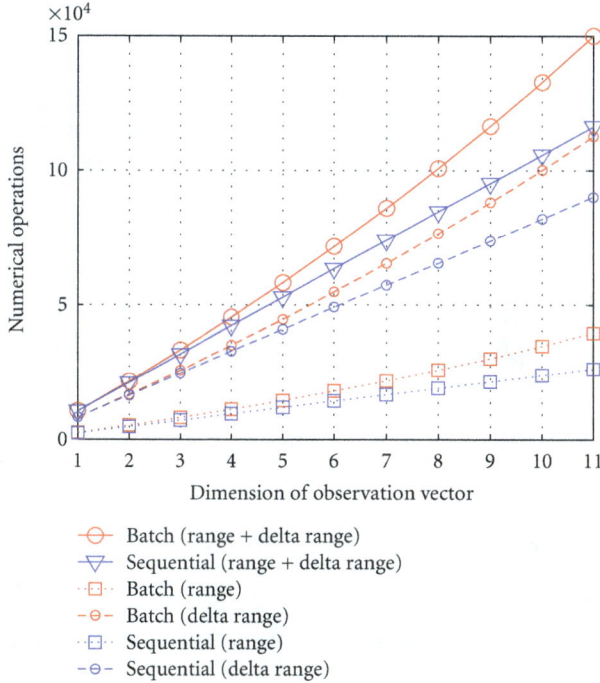

FIGURE 2: Computational burden analysis (pseudorange versus delta pseudorange) with varying number of measurements.

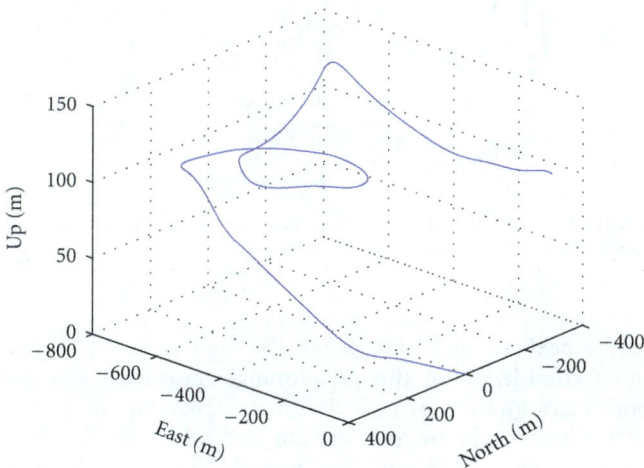

FIGURE 3: UAV trajectory.

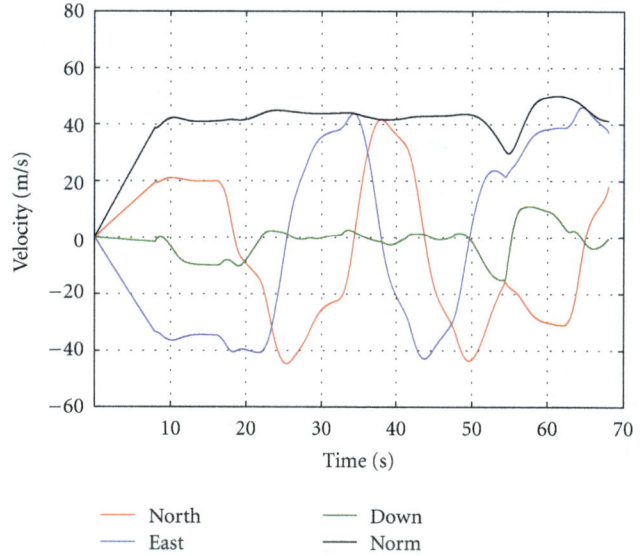

FIGURE 4: Trajectory dynamics.

TABLE 3: Parameters used for generating the receiver-related measurement errors (1σ).

	Receiver thermal noise	Multipath errors
Pseudorange (DLL)	0.5 m	0.4 m
Doppler (FLL)	0.05 m/s	0.04 m/s
Carrier phase (PLL)	0.001 m	0.004 m

For the GPS receiver, we assume that we use a NovAtel DL-4plus GPS receiver. The parameters shown in Table 3 are used to simulate the receiver-related measurement errors. The multipath errors are generated using SATNAV toolbox from GPSoft LLC. The basic idea is to form the zero-elevation angle equivalent pseudorange multipath errors using a linear autoregressive model (i.e., "filter" function in MATLAB) and scale these errors by the cosine of their elevation angles before they are applied to the range measurements. Regarding the IMU, an automotive level LandMark 20 eXT MEMS-IMU from Gladiator Technologies Inc. is simulated. The main sensor errors are generated according to its specification as

shown in Table 4. The IMU raw data (simulated errors plus the field collected high quality IMU raw data) are plotted in Figure 5. For the integrated KF, the system time-update happens at the IMU measurements update rate, which is 50 Hz, while the measurement-update happens at the GPS measurement update rate, which is 5 Hz. The simulation is conducted with 8 satellites in view.

With this simulation setup, two tests are made. In the first test, we verify the advantage of correctly handling the errors in the delta pseudorange measurements. And in the second one, we prove that the sequential measurement update of different types of measurements presents identical system estimation accuracy with respect to the batch processing, if certain conditions are met.

5.1. System Estimation Accuracy Using "Augmented" System Model.
We initialize the filter to be 1 m away to the north, east and down directions from their true values. The errors analyzed here, represent the norm of position, velocity and attitude errors, respectively. For instance, the norm of position errors will be calculated as $\|\Delta \mathbf{x}\| = \sqrt{\Delta x_n^2 + \Delta x_e^2 + \Delta x_d^2}$. The comparisons are made between conditions listed in Table 5, and the results are depicted in Figures 6 and 7.

For the "augmented" system (red curves), both code-derived pseudorange and carrier-phase-derived delta pseudorange measurements are used in the position determination through the system observation model (i.e., (11)).

TABLE 4: Landmark 20 eXT MEMS-based IMU performance specification.

Gyroscope (angular rates)	Bias in-run stability 20 [°/h] (1σ)	Noise (ARW) 0.035 [°/s/$\sqrt{\text{Hz}}$] (1σ)	Scale factor error $\leq$1000 [ppm]
Accelerometer (specific forces)	Bias in-run stability 20 [μg] (1σ)	Noise (VRW) 40 [μg/$\sqrt{\text{Hz}}$] (1σ)	Scale factor error $\leq$1000 [ppm]

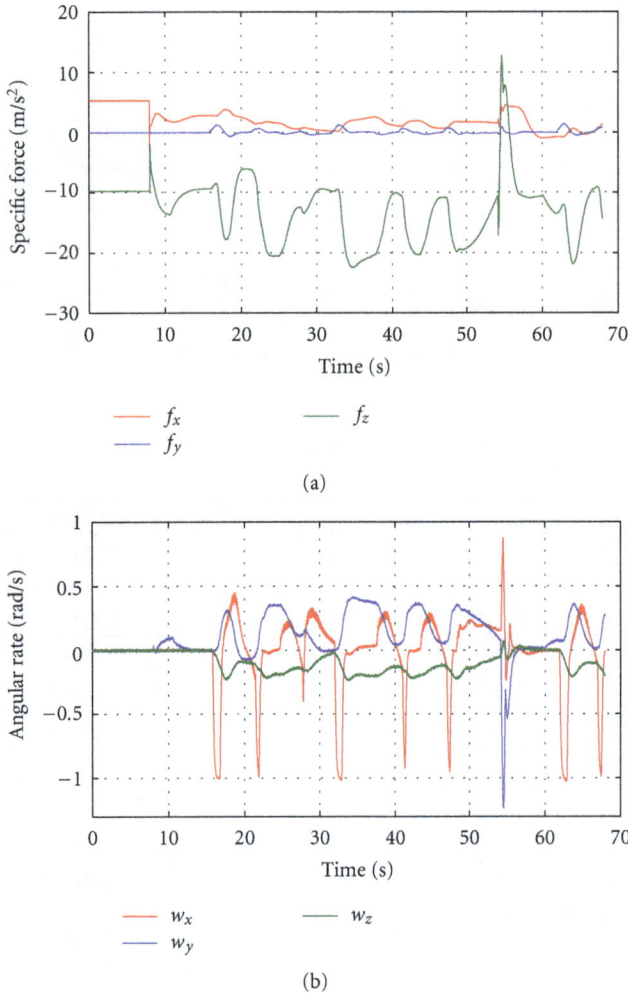

(a)

(b)

FIGURE 5: Simulated LandMark 20 eXT MEMS IMU raw data.

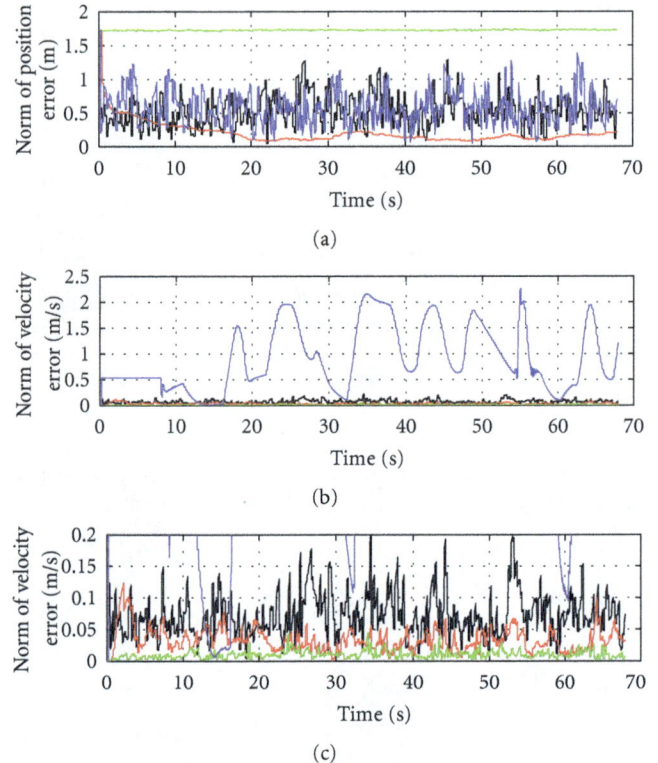

(a)

(b)

(c)

FIGURE 6: Comparison of position and velocity estimates (the 3rd subplot is the zoom-in of the 2nd subplot).

TABLE 5: Conditions for comparison.

Red	"Augmented" system using pseudorange and delta pseudorange measurements
Green	"Augmented" system using delta pseudorange measurements alone
Blue	System using pseudorange for position and delta pseudorange for velocity
Black	System using pseudorange for position and Doppler shift for velocity

The pseudorange provides the absolute positioning information, while the delta pseudorange provides the relative displacement of the system position from one time instant to the next. In such case, when the filter is tuned to give more confidence on the pseudorange measurements, the position estimation errors will converge faster, but the system performance will be noisier. On the other hand, when the filter gives more confidence on the delta pseudorange measurements, the estimation errors will converge slower, but system presents much smoothed estimation results. By giving the weights on the pseudorange and delta pseudorange measurements based on the errors introduced onto the GPS raw data, the red curves are obtained. In Figure 6, the third subplot is the "zoom-in" version of the second subplot, where the red curve shows the velocity estimation errors using the "augmented" system model. It demonstrates that with correctly handling the errors involved in the delta pseudorange measurement, the filter can correctly track the accelerations and jerks involved in the trajectory.

The green curves in Figure 6 show an extreme case. That is, if there are no pseudorange measurements (absolute positioning information) available in the filter, with only the delta pseudorange measurements (relative positioning information), the system presents the best velocity estimation

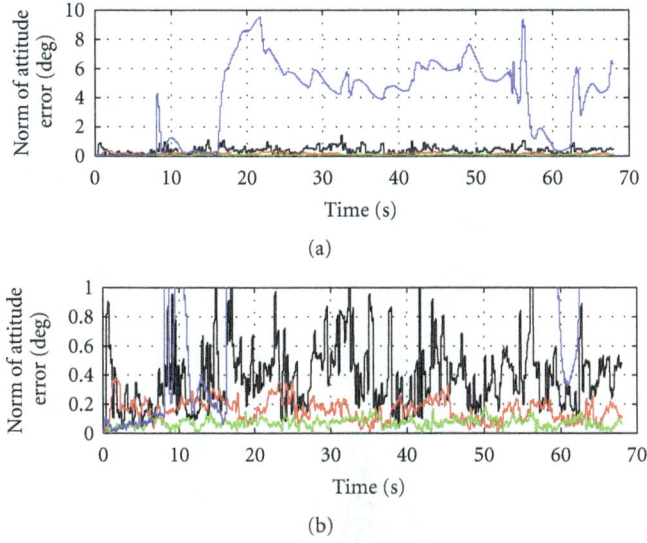

FIGURE 7: Comparison of attitude estimates (the 2nd subplot is the zoom-in of the 1st subplot).

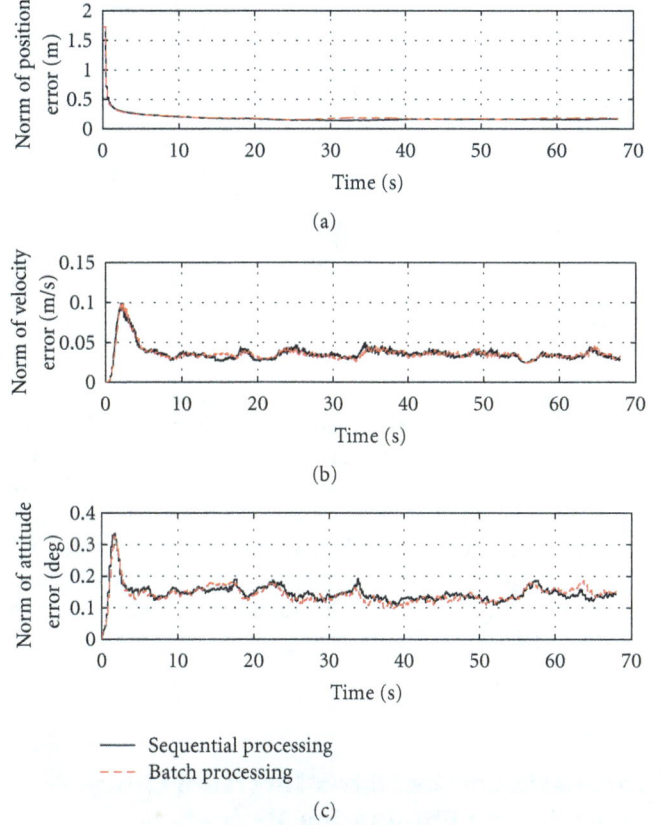

— Sequential processing
--- Batch processing

(c)

FIGURE 8: System performance comparisons.

accuracy. However, for position, the errors from the incorrect initialization remain through the whole trajectory, as shown in the first subplot of Figure 6.

The blue curves in Figure 6 represent the estimation results using the conventional approach. That is, to relate the pseudorange for position estimation, and the delta pseudorange for velocity estimation. In this test, the dynamics encountered in the trajectory is large. Therefore, large velocity estimation errors appear. The same problem can be observed from the attitude estimation results (i.e., Figure 7).

In this paper, we integrate the measurements obtained from a single-GPS receiver antenna with measurements from an INS without redundant attitude information (e.g., from magnetometers or a multiantenna GPS system). Therefore, no attitude fixes are available. In such case, the attitude information is contained in the position and velocity information, and its errors are related to the position and velocity estimation errors through the off-diagonal parameters in the error covariance matrices. Hence, the greater the accuracy of the position and velocity estimates, the greater the dampening on the attitude errors.

As an alternate approach, the Doppler shift measurements are often used for velocity determination (black curves in Figures 6 and 7). However, they are the raw measurements from the receiver frequency lock loop (FLL), which is known to be noisier than the carrier-phase-derived measurements.

5.2. Comparison between Sequential and Batch Measurement Updates. In this test, the simulation setup remains unchanged as the aforementioned one. Figure 8 plots the position, velocity, and attitude estimation errors from 50 Monte Carlo runs (with 50 simulated GPS and IMU data sets based on parameters in Tables 3 and 4) using the "augmented" system with sequential and batch measurement updates.

As depicted in the figure, the sequential measurement update presents equivalent performance as compared with batch measurement update. Nevertheless, two rules must be obeyed.

(1) If we process a group of measurements, which involve both integrated and nonintegrated measurements, we must always process the integrated measurements (delta pseudorange) before the nonintegrated measurements (pseudorange) to guarantee that there is no measurement update before we process the integrated measurements.

(2) The cross-covariance matrices (i.e., $\mathbf{P}_{cd,k}$ and $\mathbf{P}_{dc,k}$) need to be carefully treated. In sequential processing, among the measurement scalar updates, these cross covariance matrices, that is, $\mathbf{P}_{cd,k}$ and $\mathbf{P}_{dc,k}$, are updated from one measurement to the other with zero-width time interval, as shown in (15).

6. Conclusion

The errors involved in delta pseudorange measurements are related to the pseudorange errors at the start and end of the integration interval. The correct handling of these errors is essential in cases that the vehicle is maneuvering under high dynamics with low GPS update rate. In this contribution, the method with the "augmented" system state vector is considered, and is employed in the INS/GPS tightly coupled

TABLE 6: Computational cost of matrix inverse calculation.

Steps	Left side		Right side		
	Multi.	Add.	Multi.	Add.	Rows processed
1	n	$n-1$	n	n	$n-1$
2	$n-1$	$n-2$	n	n	$n-1$
$\vdots$	$\vdots$	$\vdots$	$\vdots$	$\vdots$	$\vdots$
$n-1$	2	1	n	n	$n-1$
n	1	0	n	n	$n-1$
Division	0	0	n	0	n

integration. Simulation is conducted using a field-collected UAV trajectory. Numerical results show that the filter can correctly track the dynamics in the trajectory. Moreover, sequential processing is used for measurement update in the KF to prevent the matrix inverse calculation. Monte Carlo runs are conducted and verify that the sequential processing can update different types of measurements (i.e., integrated and nonintegrated) without degrading the system estimation accuracy, if certain rules are met.

Appendix

Calculation of the Matrix Inversion Using Gauss-Jordan Elimination Method

The Gauss-Jordan Elimination method is one of the methods which are widely used to calculate the inverse of a square matrix (e.g., [6]). It is done by augmenting the square matrix with an identity matrix of the same dimensions, and passing through the following matrix operations

multiply
$$\begin{bmatrix} a_{11} & a_{12} & \cdots & a_{1n} & 1 & 0 & \cdots & 0 \\ a_{21} & a_{22} & \cdots & a_{2n} & 0 & 1 & \cdots & 0 \\ \vdots & \vdots & \ddots & \vdots & \vdots & \vdots & \ddots & \vdots \\ a_{n1} & a_{n2} & \cdots & a_{nn} & 0 & 0 & \cdots & 1 \end{bmatrix},$$

by $\mathbf{A}^{-1}$ to have
$$\begin{bmatrix} 1 & 0 & \cdots & 0 & b_{11} & b_{12} & \cdots & b_{1n} \\ 0 & 1 & \cdots & 0 & b_{21} & b_{22} & \cdots & b_{2n} \\ \vdots & \vdots & \ddots & \vdots & \vdots & \vdots & \ddots & \vdots \\ 0 & 0 & \cdots & 1 & b_{n1} & b_{n2} & \cdots & b_{nn} \end{bmatrix}.$$
(A.1)

After the Gauss-Jordan elimination, the right side of the augmented matrix is the inverse of the original square matrix. In this contribution, we firstly transform the left side of the augmented matrix (original square matrix) to be a diagonal matrix and divide the augmented matrix by its diagonal elements on each row to have an identity matrix on the left side and the inverse of original matrix on the right side. The computation is shown as follows.

In the first step, we use the elementary row operations to introduce zeros in the first column beginning from the second row, which involves $2n$ multiplications and $2n-1$ additions:

$$\begin{bmatrix} a_{11} & a_{12} & \cdots & a_{1n} & b_{11} & b_{12} & \cdots & b_{1n} \\ 0 & a_{22}^{(1)} & \cdots & a_{2n}^{(1)} & b_{21}^{(1)} & b_{22}^{(1)} & \cdots & b_{2n}^{(1)} \\ \vdots & \vdots & \ddots & \vdots & \vdots & \vdots & \ddots & \vdots \\ 0 & a_{n2}^{(1)} & \cdots & a_{nn}^{(1)} & b_{n1}^{(1)} & b_{n2}^{(1)} & \cdots & b_{nn}^{(1)} \end{bmatrix}_{n \times 2n}.$$
(A.2)

After the first step, the elements in the rows below the first will change. We indicate this by denoting the elements with a superscript, which stands for the number of arithmetic operations involved.

In the second step, we introduce zeros in the second column except the second row, which involves $2n-1$ multiplications and $2n-2$ additions.

In the n-th step, we introduce zeros in the n-th column except the n-th row, which involves $n+1$ multiplications and n additions:

$$\begin{bmatrix} a_{11} & 0 & \cdots & 0 & b_{11}^{(n-1)} & b_{12}^{(n-1)} & \cdots & b_{1n}^{(n-1)} \\ 0 & a_{22}^{(1)} & \cdots & 0 & b_{21}^{(n-1)} & b_{22}^{(n-1)} & \cdots & b_{2n}^{(n-1)} \\ \vdots & \vdots & \ddots & \vdots & \vdots & \vdots & \ddots & \vdots \\ 0 & 0 & \cdots & a_{nn}^{(n-1)} & b_{n1}^{(n-1)} & b_{n2}^{(n-1)} & \cdots & b_{nn}^{(n-1)} \end{bmatrix}_{n \times 2n}.$$
(A.3)

In the last step, we divide the augmented matrix by its diagonal elements on each row to have the matrix as shown in (A.1). The involved numerical operations are given in Table 6.

Using the summation formula

$$\sum_{i=1}^{n} i = \frac{n(n+1)}{2}.$$
(A.4)

According to Table 6, the number of multiplications involved in Gauss-Jordan Elimination is

$$(n-1)\sum_{i=1}^{n} i + n^2(n-1) + n^2$$
$$= (n-1)\frac{n(n+1)}{2} + n^2(n-1) + n^2 = \frac{3}{2}n^3 - \frac{1}{2}n.$$
(A.5)

And the required number of additions is computed as

$$(n-1)\sum_{i=1}^{n-1} i + n^2(n-1) = \frac{3}{2}n^3 - 2n^2 + \frac{n}{2}.$$
(A.6)

We can verify our calculation by inverting a scalar number as an extreme example for square matrix inversion (i.e., $n = 1$). The arithmetic operations involved contain obviously only one multiplication (division), which is the same as the results computed from (A.5) and (A.6).

Acknowledgment

Part of the work reported herein has been funded by the German Research Foundation (DFG), Grant no. KN 876/1-2, which is gratefully acknowledged.

References

[1] R. G. Brown and P. Y. C. Hwang, *Introduction to Random signals and Applied Kalman Filtering*, John Wiley and Sons, New York, NY, USA, 1997.

[2] P. Misra and P. Enge, *Global Positioning System: Signals, Measurements, and Performance*, Ganga-Jamuna Press, 2007.

[3] D. Hartman and D. B. Tyler, "New Kalman filter formulation for GPS delta pseudorange processing," in *ION Proceedings of GPS*, pp. 1395–1400, Nashville, Tenn, USA, 1998.

[4] P. S. Maybeck, *Stochastic Models, Estimation, and Control*, vol. 1, Academic Press, New York, NY, USA, 1982.

[5] D. Simon, *Optimal State Estimation: Kalman, H Infinity, and Nonlinear Approaches*, Wiley-Interscience, Hoboken, NJ, USA, 2006.

[6] G. Strang and K. Borre, *Linear Algebra, Geodesy, and GPS*, Wellesley-Cambridge Press, 1997.

[7] Y. Li, C. Rizos, J. L. Wang, P. Mumford, and W. D. Ding, "Sigma-Point Kalman filtering for tightly coupled INS/GPS integration," *Journal of the Institute of Navigation*, vol. 55, no. 3, pp. 167–179, 2008.

[8] J. Farrell, *Aided Navigation: GPS with High Rate Sensors*, McGraw-Hill Professional, New York, NY, USA, 2008.

16

Performance Analysis of Alignment Process of MEMS IMU

Vadim Bistrov

Faculty of Electronics and Telecommunication, Riga Technical University, 1019 Riga, Latvia

Correspondence should be addressed to Vadim Bistrov, bistrov@inbox.lv

Academic Editor: Jinling Wang

The procedure of determining the initial values of the attitude angles (pitch, roll, and heading) is known as the alignment. Also, it is essential to align an inertial system before the start of navigation. Unless the inertial system is not aligned with the vehicle, the information provided by MEMS (microelectromechanical system) sensors is not useful for navigating the vehicle. At the moment MEMS gyroscopes have poor characteristics and it's necessary to develop specific algorithms in order to obtain the attitude information of the object. Most of the standard algorithms for the attitude estimation are not suitable when using MEMS inertial sensors. The wavelet technique, the Kalman filter, and the quaternion are not new in navigation data processing. But the joint use of those techniques for MEMS sensor data processing can give some new results. In this paper the performance of a developed algorithm for the attitude estimation using MEMS IMU (inertial measurement unit) is tested. The obtained results are compared with the attitude output of another commercial GPS/IMU device by Xsens. The impact of MEMS sensor measurement noises on an alignment process is analysed. Some recommendations for the Kalman filter algorithm tuning to decrease standard deviation of the attitude estimation are given.

1. Introduction

Navigation can be defined as the process of determining the position, orientation, and velocity of an object. A GPS-based navigation is quick and drift free and is readily available most of the time. However, as the GPS requires direct line of sight signals from at least four GPS satellites, the navigation can be frequently interrupted in the land based applications. The GPS signal gets lost due to various factors such as the blockage by buildings, trees, and other natural and nonnatural obstructions. This affects both the amplitude and phase of the received satellite signals and causes the receiver to lose lock on the blocked satellite, meaning that it needs both to reacquire the signal and to resolve the ambiguities in the phase measurements. Both these processes take time, and if there are several satellites affected, the receiver cannot provide a position solution for a significant period of time. Also it is worth mentioning that the data rate for the GPS can be too low for the particular application. In such situations when the GPS signals are not available, a relative navigation can be performed using the inertial sensors (accelerometers and gyroscopes) and magnetometers.

The strapdown inertial navigation system (SINS) has been widely used in numerous fields such as the positioning and navigation of ships, aeroplanes, vehicles, and missiles. To initialize the navigation Kalman filter (KF), all three attitude angles including the roll, pitch, and heading (azimuth) are required. The initial misalignment is one of the major error sources of the SINS. So, it is crucial to have an accurate initial alignment in order to implement the integrated navigation system. The tactical and navigation grade sensors are limited to commercial and military applications and very expensive, but, with the introduction of the compact, low-power, and cost-efficient MEMS sensors, it is possible to have portable integrated INS/GPS navigation modules. For low-cost INS, the initial alignment is still a challenging issue because of the high noises from the low-cost inertial sensors. The SINS must be preferably aligned before positioning and navigation. The aim of the initial alignment of the SINS is to get a coordinate transformation matrix from the body frame to the navigation frame and conduct the misalignment angles to zero or as small as possible. In many applications, it is essential to achieve an accurate alignment of the SINS within a very short period of time.

Accordingly, a number of filtering algorithms have been developed for integrating the output of the gyros and the accelerometers, to estimate the attitude. Previous studies of the attitude calculation methods, using the inertial sensors only, have been researched for use in the robotics [1], aircrafts [2], and human motion tracking [3]. The initial alignments of the high precision INS and low-cost INS should be treated using different methods, because the noise levels of the inertial sensors are quite different for these two types of systems.

The low-cost IMUs usually use gyroscopes with noise levels larger than the Earth's rotation rate, and, hence, they cannot be aligned in the static mode. In this case, the external heading measurements using, for instance, the magnetic compass, are usually used to provide the alignment information [4]. Another possibility is to transfer the obtained attitudes of another, statically aligned, better quality IMU through the master-slave initialization process [5]. In addition, the dynamic alignment can be performed through the velocity matching techniques by using the velocity updates from an aiding system such as the Differential Global Positioning System (DGPS) or the Doppler radar [6–8].

This paper proposes the attitude estimation of the low-cost IMU using the linear Kalman filter (standard and modified) in the stationary and dynamic mode. The linear accelerations and angular rates are measured by MEMS IMU. The gyro and accelerometer data is denoised by wavelet algorithm. Next, the Kalman filter algorithm was used to estimate the pitch and roll. The magnetometer data was used for estimation of the heading. Finally, the test results are presented to show the performance of the proposed algorithm for attitude estimation of the low-cost IMU in the stationary mode and attitude estimation of the moving vehicle.

2. Alignment

Here we are discussing the practical aspects of alignment mainly regarding the vehicle navigation. For land vehicles, it can be assumed that the direction of travel is identical to the direction of the b-frame x-axis and the roll angle is near to zero. There are two types of alignment that are required before the navigation parameters can be estimated for the portable navigation module in a vehicle. The first alignment is the alignment of the IMU axes with respect to the vehicle axes (i.e., making the b-frame coincide with the v-frame), which is referred to as the relative alignment. Once the relative alignment is achieved, the next step is to align the b-frame with the l-frame, which is referred to as the absolute alignment [6, 8].

In a broad sense, the initial absolute alignment of the SINS can be divided into two categories, that is, the stationary based and the moving based alignment. The moving based alignment is used mainly when the good quality GPS signals are available. Here we are discussing the stationary based alignment using the inertial sensor signals only. The requirements of the initial alignment of the SINS are high accuracy and short time. An accurate

alignment is crucial; however, this is based on the alignment over a long period of time. A compromise of accuracy and time consumption of the initial alignment should be made. During the process of absolute alignment the pitch, roll, and heading are estimated. The misaligned portable navigation system (PNS) with respect to the vehicle frame is shown in Figure 1. If the relative alignment is not done properly, the navigation solution will be erroneous. The relative alignment means alignment of the b-frame with respect to the v-frame. For the vehicle frame, the axes are generally defined as x (direction of travel), z (direction of gravity), and y—completing the orthogonal set. In case of misalignment, the x-axis accelerometer will not measure the true vehicle acceleration in the travel direction. This is due to an additional accelerometer error—not correctly compensated gravity, which is caused by the IMU misalignment. Also the initial attitude angles of the vehicle will be incorrectly estimated if the relative misalignment is not taken into account. And this, in its turn, will cause additional velocity and position drift.

The orientation accuracy of the PNS in this case will totally depend on the user and it is quite easy to assume that even a careful user cannot align the system properly at every use. An easy way to solve this problem is the introduction of a holder inside the vehicle that is aligned with the vehicle and provides the user an easy way to align the system at every use [6].

Inertial sensors do not need external sources for the initial attitude measurement. It can align itself by using the measurements of the local gravity and earth rate. The gyroscope compassing method is used for estimation of the heading [8]. However, the MEMS sensors have significantly high drift rates and noise characteristics, and, therefore, the gyroscope outputs cannot be used to estimate the azimuth or heading of the vehicle. The main reason is that the gyroscope compassing uses the rotation rate of the Earth. The Earth rate is about 15 deg/h, so the noise levels of the low cost gyros are near or higher than the Earth rate. This means that the Earth rate cannot be monitored with the MEMS gyroscopes for the moment. The heading is important in the initialization of the navigation algorithm. Magnetometers are used for the heading estimation. When the vehicle is moving the heading or the azimuth of a vehicle can be determined by incorporating the north and east velocity components from the GPS receiver.

The gravity is a relatively large quantity; even the low-cost accelerometers can measure it properly. So, for the MEMS sensors, the strong gravity signals from accelerometers can be measured and these measurements are used for estimation of the roll and pitch of the inertial system with respect to the l-frame.

3. MTi-G Device

The calibrated data (the rate of turn, acceleration, and the magnetic field) is expressed in the sensor fixed coordinate system [9]. All calibrated vector sensor readings are in the right-handed Cartesian coordinate system. This coordinate

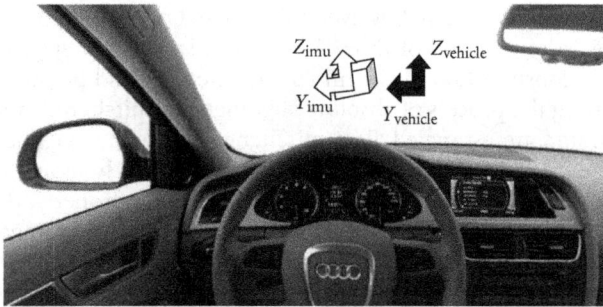

FIGURE 1: The misaligned IMU with respect to the vehicle frame.

system is body fixed to the device and is defined as the sensor co-ordinate system or the body frame. The coordinate system is aligned to the external housing of the MTi-G. The aluminium base plate of the MTi-G is carefully aligned with the output coordinate system.

The MTi-G default local tangent plane of the North-West-Up is defined in the Figure 2 above; it has X pointing to the North and it is tangent to an arbitrary reference point. The third component (Z) is chosen Up which is common for many applications (according to the ISO/IEC 18026). The vertical vector Z is perpendicular to the tangent of the ellipsoid as defined in Figure 2. The alignment of the bottom plane and the sides of the aluminium base plate of the MTi-G with respect to the sensor-fixed output coordinate system is within 0.1 deg.

The orientation output of the MTi-G is the orientation between the sensor-fixed coordinate system (the body frame) and the local tangent plane coordinate system, as the reference coordinate system. The output orientation can be presented in different parameterizations such as the quaternion, direction cosine matrix, or the Euler angles (roll, pitch, and yaw). The Euler angles have more meaningful attitude expression than the quaternion method or the direction cosine matrix method and the user recognizes the attitude of the object directly. According to the definition of the Euler angles, the coordinate system rotates in the angle ψ (yaw) on the Z-axis; the coordinate system rotates in the angle θ (pitch) on the Y-axis. Finally this coordinate system can rotate in the angle ϕ (roll) on its X-axis (see Figure 3).

4. Sensor Data Denoising

It is advisable to reduce the noise level of the inertial sensor signals before using the measurements in the alignment procedure. The objective is to provide more accurate computation of the initial attitude angles and speed up the convergence of the Kalman algorithm.

Most of the standard inertial denoising methods such as the low pass filtering are not efficient enough for reducing the high noise levels of the MEMS inertial measurement unit without corrupting some useful information in the sensors' signal.

In [10] it was demonstrated that the wavelet tool can be useful for the analysis of the inertial measurements in order to decrease the noise level.

Since the low frequency fraction ($\Delta F < 0.5$ Hz) of the inertial measurement reading contains the majority of the inertial sensor dynamics during the static alignment phase, these inertial measurement readings can be denoised using the wavelet multilevel of decomposition to separate the low and high frequencies.

Consequently, five levels of decomposition will limit the frequency band to 0.5 Hz. Five levels of decomposition (LOD) are adequate to reduce the high frequency noise from the real inertial sensor measurement [10]. Increasing the level of decomposition results in the undesired features of the navigation solution since the original features of the IMU data will be lost after wavelet filtering with too high LOD. Denoising of the inertial sensor signals by wavelet decomposition has proven its success in reducing errors of the estimated attitude angles for the integrated navigation systems [10]. The analysis of the gyroscope signal is made in order to demonstrate the effect of the wavelet denoising. The results are shown in Figure 4. As we can see, there is a considerable decrease of the noise level. The standard deviation of the gyroscope signal before the wavelet filtering was $\sigma_1 = 0.0063$ rad/s, but, after it, it decreases till $\sigma_2 = 0.0011$ rad/s. The same effect of the noise reducing can be demonstrated for the accelerometer signal too.

5. Heading Estimation Algorithm

When the GPS signal is available and the vehicle has a nonzero velocity, it's possible to calculate a heading of the vehicle using the GPS-derived velocity. When the GPS signal is not available, the magnetometers (which sense the Earth's magnetic field strength) can be used for the determination of the absolute heading with reference to the local magnetic North. The deviation between the true north and the local magnetic north is known as the magnetic declination. This declination can be calculated as a function of latitude, longitude, and time using a global model such as the World Magnetic Model (WMM). The global model is typically accurate to about 0.5 degree. The declination angle according to the WMM 2010 for latitude 55° 52′ 19″N, longitude 26° 31′ 4″E, and attitude 100 m is 7.284°E. The calculation was done using a software of the open access from the website of the British Geological Survey.

The short-term temporal variations of the Earth's magnetic field can be caused by the magnetic storms [5]. The heading with reference to the local magnetic north is derived from the horizontal force of the magnetic field. If the magnetometer was aligned with the local horizontal the heading, ψ, would be calculated as [11]

$$\psi = \arctan 2\left(M_x^{\text{level}}, M_y^{\text{level}}\right), \qquad (1)$$

where M_x^{level}, M_y^{level} represent the Earth's magnetic field along the x- and y-axes, aligned with the horizontal plane.

Different types of magnetic disturbances distort the measured magnetic field. The magnetic fields are produced not only by the earth but also by the man-made objects such as vehicles, buildings, bridges, and power lines.

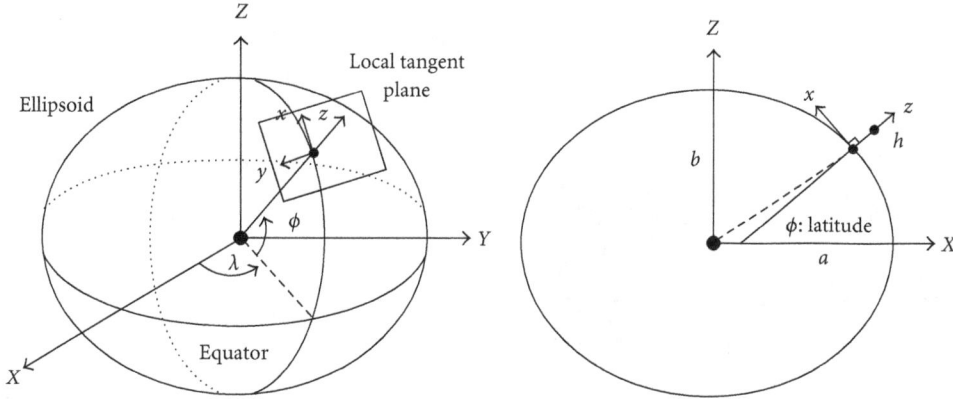

FIGURE 2: Definition of the default MTi-G local tangent plane [9].

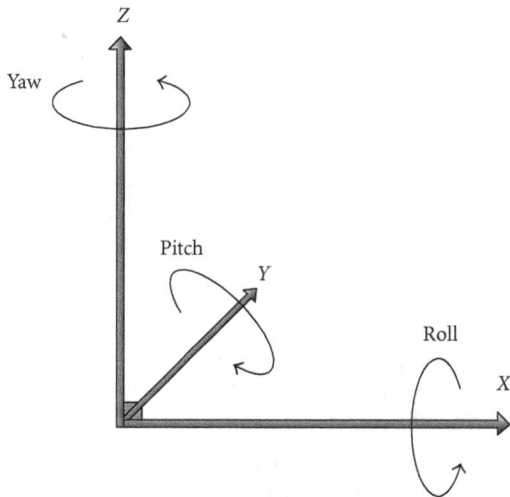

FIGURE 3: Definition of the Euler angles.

Hence, the quality of the magnetometer heading strongly depends on the tilt compensation (if not horizontally located) and the calibration procedure, including identifying the possible error sources and removing them from the measurements.

The traditional autocalibration method [4] is based on the fact that the locus of the error free magnetometer measurements is a circle if the sensor moves around the circle. The impact of various magnetometer errors would distort the shape of this circle; however, the circular constraint eventually can be used to partially estimate the local variations of the Earth's magnetic field.

Practical implementation of the auto-calibration does not require any reference headings, but this method strongly depends on the position and the effect of ferromagnetic materials in the proximity of the sensor. If the sensor's location is changed or it is placed in a new environment, a new calibration is needed for identifying and removing the effects of the new magnetic environment.

The magnetometer calibrated measurements M_x, M_y for horizontal components can be represented as follows:

$$M_x = S_x M_x^{\text{level}} + B_x,$$
$$M_y = S_y M_y^{\text{level}} + B_y,$$

(2)

where S_x, S_y are two scale factors, B_x, B_y are two biases along the horizontal axes of the magnetic field and M_x^{level}, M_y^{level} are the levelled magnetic components of the Earth. The scale factors and biases can be found using the following equations:

$$S_x = \text{Max}\left(1, \frac{M_y^{\text{Max}} - M_y^{\text{Min}}}{M_x^{\text{Max}} - M_x^{\text{Min}}}\right),$$

$$S_y = \text{Max}\left(1, \frac{M_x^{\text{Max}} - M_x^{\text{Min}}}{M_y^{\text{Max}} - M_y^{\text{Min}}}\right),$$

(3)

$$B_x = \left(\frac{M_x^{\text{Max}} - M_x^{\text{Min}}}{2} - M_x^{\text{Max}}\right)S_x,$$

$$B_y = \left(\frac{M_y^{\text{Max}} - M_y^{\text{Min}}}{2} - M_x^{\text{Max}}\right)S_y,$$

where S_x, S_y are two scale factors, B_x, B_y are two biases along the horizontal axes of the magnetic field, M_y^{Max}, M_x^{Min}, M_y^{Min}, M_x^{Max}—the maximum and minimum of the measured magnetic field along the x- and y-axes.

To calibrate the magnetometer using the above equation, it's necessary to rotate the levelled magnetometer several times by 360 degrees and then determine the scale factors as the ratio of the major and minor axes, changing the circle to an ellipse, and bias parameters as the offset center of the ellipse, after it's possible to calculate the magnetometer calibrated horizontal components using (2). And the heading can be estimated using (1).

The above-mentioned auto-calibration approach has been implemented in MATLAB and applied to the MTi-G magnetometers' signals in order to test the implementation

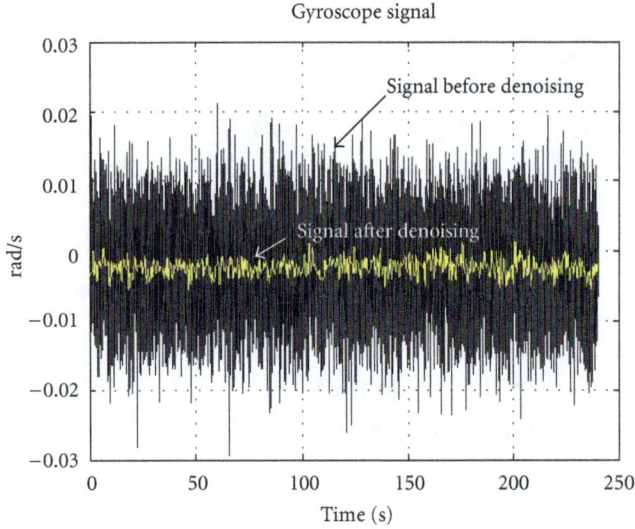

FIGURE 4: Low-cost gyroscope signal analysis using the wavelet (LOD = 5).

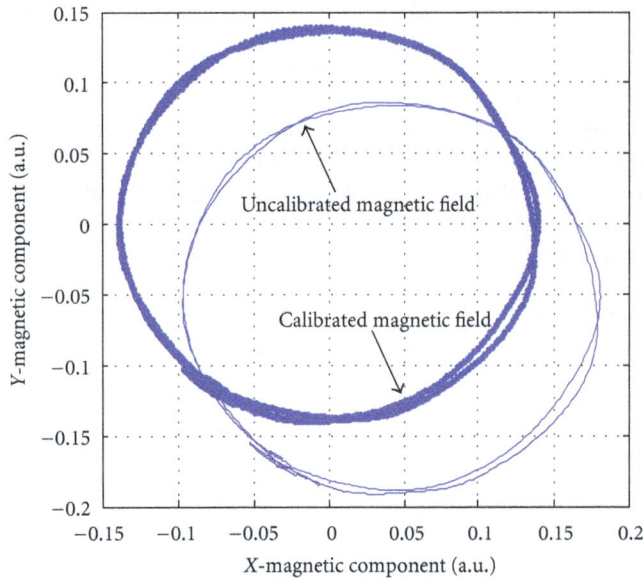

FIGURE 5: Performance of the implemented auto-calibration procedure.

of the auto-calibration procedure. The results are shown in Figure 5. Here the magnetic field strength of the Earth is defined in the arbitrary units (a.u.), normalized to the Earth's field strength.

6. Kalman Filter Algorithm for Roll and Pitch Estimation

6.1. Process Model. In this paper quaternions were used as state variables for the Kalman filter. The equations for the linear Kalman filter algorithm used here can be found in [11]. Quaternion implementation of the attitude update

is computationally efficient [5]. There's no trigonometric function during the quaternion based attitude updating.

Quaternion is hypercomplex numbers with four components:

$$\mathbf{q} = \begin{pmatrix} q_1 & q_2 & q_3 & q_4 \end{pmatrix}, \qquad (4)$$

where q_4 represents the magnitude of rotation, and other three components represent the axis about which that rotation takes place.

The differential equations for the process model have the following form [12]:

$$\begin{pmatrix} \dot{q}_1 \\ \dot{q}_2 \\ \dot{q}_3 \\ \dot{q}_4 \end{pmatrix} = \frac{1}{2} \begin{pmatrix} 0 & \omega_z & -\omega_y & \omega_x \\ -\omega_z & 0 & \omega_x & \omega_y \\ \omega_y & -\omega_x & 0 & \omega_z \\ -\omega_x & -\omega_y & -\omega_z & 0 \end{pmatrix} \cdot \begin{pmatrix} q_1 \\ q_2 \\ q_3 \\ q_4 \end{pmatrix} + \begin{pmatrix} \eta_1 \\ \eta_2 \\ \eta_3 \\ \eta_4 \end{pmatrix}, \quad (5)$$

where ω_x, ω_y, ω_z are the output signal of gyroscopes and η_1, η_2, η_3, η_4 are zero mean Gaussian noise with the corresponding PSD $\sigma_{P,1}^2, \sigma_{P,2}^2, \sigma_{P,3}^2, \sigma_{P,4}^2$. Allan variance method can be used in order to make coarse estimation of σ_P^2. Also sensor noise specification given by the manufacturer can be used for σ_P^2 estimation. Then system noise matrix for the Kalman filter can be determined taking into account σ_P^2 value. And the elements of this matrix can be precised in order to achieve better performance of the filter. For example, elements of the system noise matrix can be adjusted to minimize standard deviation of state vector values. Usually it's preferable to increase values of the noise matrix elements in order to take into account other unmodelled noise of the system.

The defined system model (5) is linear and can be used in the linear Kalman algorithm (LKF).

6.2. Measurement Model. The pitch and roll angles calculated using the accelerometer data do not diverge with time because they are not calculated via integration, but rely on the earth's gravity measured on each axis. Also, the low cost accelerometers have better characteristics (noise level, and stability) comparing with the low cost gyroscopes. The pitch and roll [5] may be determined using

$$\theta = \arctan\left(\frac{-a_x}{\sqrt{\left(a_y\right)^2 + (a_z)^2}} \right), \qquad (6)$$

$$\phi = \arctan 2\left(-a_y, a_z\right),$$

where a_x, a_y, a_z are accelerometer signal and arctan 2 is the four-quadrant arctangent function.

Transforming the pitch, yaw, and roll angles to quaternion, we get the measurement vector $\mathbf{z}$ for updating a step of the Kalman filter. The value of yaw angle for the Kalman filter update step is calculated based on magnetometer data.

The measurement equation for the LKF has the following form:

$$\hat{\mathbf{z}} = \mathbf{H} \cdot \mathbf{q} + \mathbf{u}, \qquad (7)$$

where $\mathbf{H}$ is the measurement matrix, $\mathbf{u}$ is vector of the zero mean Gaussian measurement noises and $\mathbf{q} = [q_1 \ q_2 \ q_3 \ q_4]^T$ is the state vector.

The defined measurement model is linear and can be directly used in the linear Kalman algorithm.

At any time of the LKF algorithm processing attitude angles can be calculated using [5]

$$\phi = \arctan 2(2(q_2 q_3 - q_1 q_4), \ (q_4^2 - q_1^2 - q_2^2 + q_3^2)),$$

$$\theta = -\arcsin(2(q_1 q_3 + q_2 q_4)), \qquad (8)$$

$$\psi = \arctan 2(2(q_1 q_2 - q_3 q_4), \ (q_4^2 + q_1^2 - q_2^2 - q_3^2)),$$

where $\mathbf{q} = [q_1 \ q_2 \ q_3 \ q_4]$ is the quaternion.

7. Experimental Results

Experiments have been carried out with the MTi-G manufactured by Xsens Technologies. MTi-G is an integrated GPS and Inertial Measurement Unit (IMU) with the navigation and attitude and heading reference system (AHRS) processor. The MTi-G is based on the MEMS inertial sensors and the GPS receiver and also includes a 3D magnetometer and a static pressure sensor. In this work accelerometer, the gyroscope and magnetometer signals from the MTi-G will be processed by the algorithms described in the previous sections. The sensor data postprocessing algorithms were implemented using the MATLAB software.

7.1. Pitch and Roll Estimation for Stationary Object. To test the proposed attitude calculation method, the attitude result is compared with the attitude output of the MTi-G. The attitude output from the MTi-G and the proposed method with the calibrated data from the MTi-G use the same data; thus the attitude result comparison means only the comparison of differences in the attitude calculation algorithm. The GPS signal for the MTi-G was unavailable during the tests. Experiments for attitude estimation were conducted through simulating the certain value of pitch and roll angles of the MTi-G using a tilt table. The accelerometer and gyroscope signals were denoised using the wavelet filtering prior to its processing with the LKF.

In order to demonstrate the effect of the signal denoising using the wavelet filtering for attitude angle estimation (e.g., for the pitch), the inertial sensor signals were processed with and without their initial denoising. The result of such experiment is shown in Figure 6 for the case when the MTi-G was perfectly levelled (with ±0.1° precision) relative to the local horizontal. The standard deviation of pitch estimation using the LKF and without the wavelet filtering of accelerometer signals was $\sigma_1 = 0.0746°$, but with denoising (prior to the data processing by the LKF) decreases till $\sigma_2 = 0.0105°$.

As the pitch and roll estimation for a stationary object is based on the sensed accelerometer signals, it is worth mentioning that any sensor error such as bias drift, for example, due to the temperature variations will result in the estimation mistake of attitude angles. It's impossible to differentiate

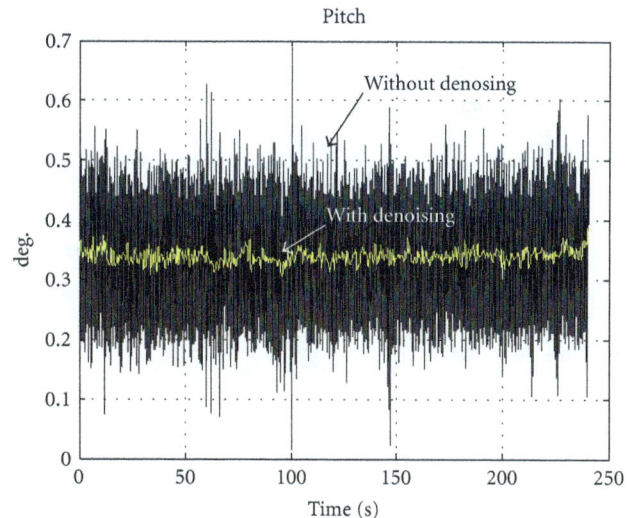

FIGURE 6: Pitch estimation.

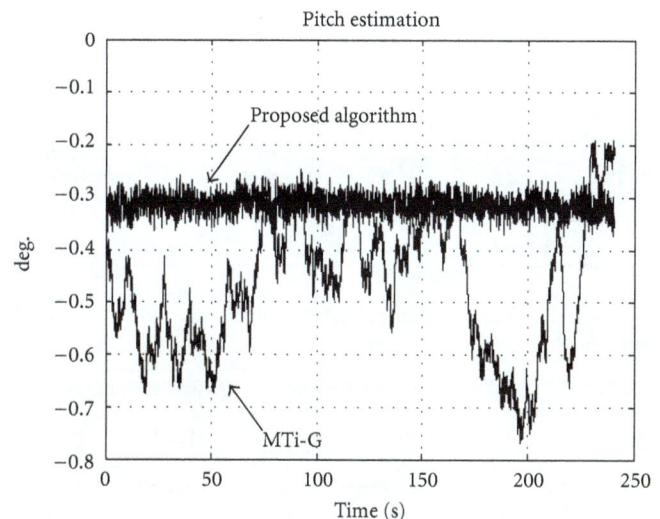

FIGURE 7: Pitch estimation, when the IMU is aligned with the local horizontal.

the accelerometer bias due to the misalignment or sensor measurement error. In order to minimize the negative effect from random (stochastic) part of the accelerometer bias, it is recommended to switch on the low cost IMU some time before (5–15 min depending on the IMU technology) the real navigation. The internal characteristics of the low-cost accelerometer will stabilize during this time.

During the first test the tilt table was aligned in order to obtain the pitch $\theta = 0°$ and roll $\phi = 0°$ with a precision ±0.1°. The results of the tilt angle estimation are shown in Figures 7 and 8. The mean value and standard deviation of the estimated pitch and roll are given in Table 1.

Then the tilt table was adjusted in order to obtain the roll $\phi = 39.5°$ with a precision ±0.5°, but pitch was kept the same ($\theta = 0°$). The roll estimation for this test is shown in Figure 9.

Then the tilt table was adjusted in order to obtain the pitch $\theta = 39.5°$ with a precision ±0.5° ; the roll value was

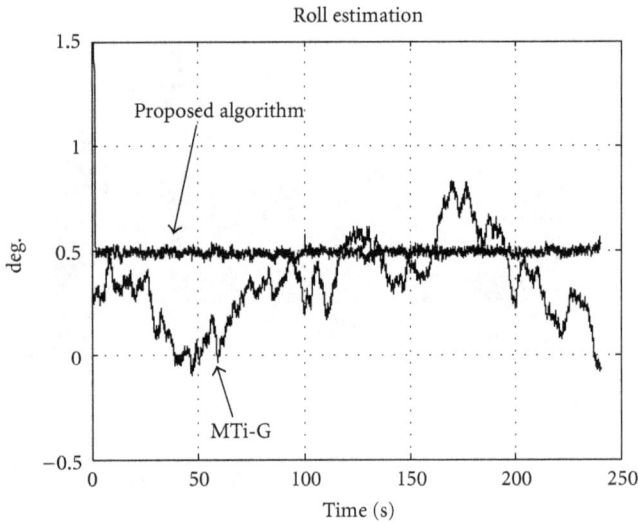

FIGURE 8: Roll estimation, when the IMU is aligned with the local horizontal.

TABLE 1: The statistical characteristics of the estimated pitch and roll.

Statistical characteristics, degree (°)	Pitch estimation by		Roll estimation by	
	MTi-G	Proposed algorithm	MTi-G	Proposed algorithm
Mean value	−0.4699	−0.3142	0.3433	0.4896
Standard deviation	0.1270	0.0168	0.1949	0.0150

TABLE 2: The statistical characteristics of the estimated pitch and roll.

Statistical characteristics, degree (°)	Pitch estimation by		Roll estimation by	
	MTi-G	Proposed algorithm	MTi-G	Proposed algorithm
Mean value	39.7758	39.4934	39.3042	39.6591
Standard deviation	0.1336	0.0307	0.3483	0.0318

TABLE 3: The statistical characteristics of the estimated heading.

Statistical characteristics, degree (°)	Heading estimation by	
	MTi-G	Proposed algorithm
Mean value	119.0114	118.5133
Standard deviation	0.8005	0.4083

TABLE 4: The statistical characteristics of the estimated pitch and roll ($\mathbf{R} = 0.01 \cdot \mathbf{I}$).

Statistical characteristics, degree (°)	Pitch estimation by		Roll estimation by	
	KF	Modified KF	KF	Modified KF
Mean value	0.2648	0.2646	0.4512	0.4510
Standard deviation	0.0096	0.0040	0.0128	0.0055

zero ($\phi = 0°$). The pitch estimation is shown in Figure 10. The statistical characteristics of the pitch and roll for the last two tests are provided in Table 2.

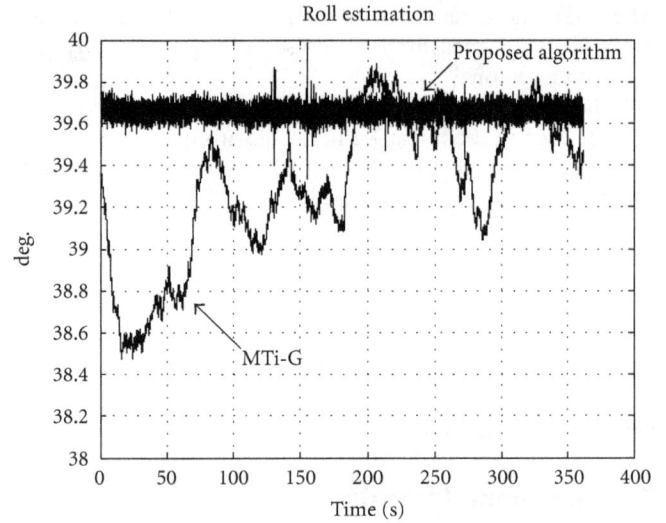

FIGURE 9: Roll estimation, when the IMU misaligned by $\phi = 39.5°$.

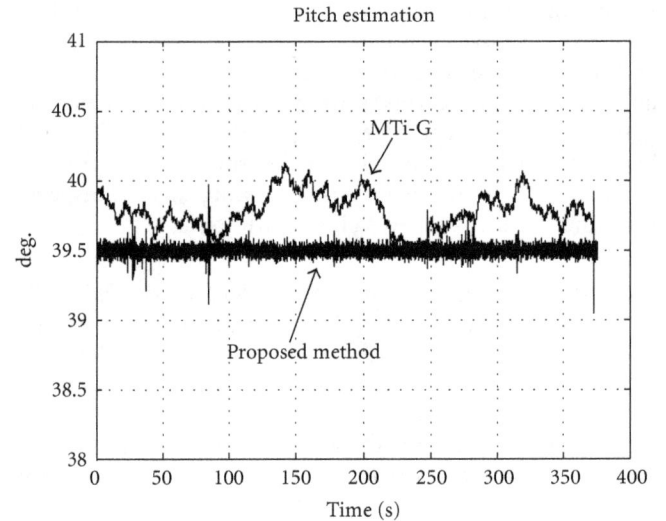

FIGURE 10: Pitch estimation, when the IMU misaligned by $\theta = 39.5°$.

7.2. Heading Estimation for Stationary Object. During this test the tilt table was aligned in order to make the pitch and roll equal to zero with a precision ±0.1°. The heading estimation for this test is shown in Figure 11. As it was not possible to obtain a true value of the heading with the necessary precision by external means, it can be assumed that the true heading value is approximately an average of two mean values, obtained by the proposed method (see Section 5) and the MTi-G output. In such a case the true heading value is $\psi \approx 118.76°$. The statistical characteristics of the estimated heading are given in Table 3.

7.3. Minimisation of the Standard Deviation of the Pitch/Roll Estimation. The functional curves of the standard deviation of the pitch/roll estimation depending on the system noise level are shown in Figures 12 and 13.

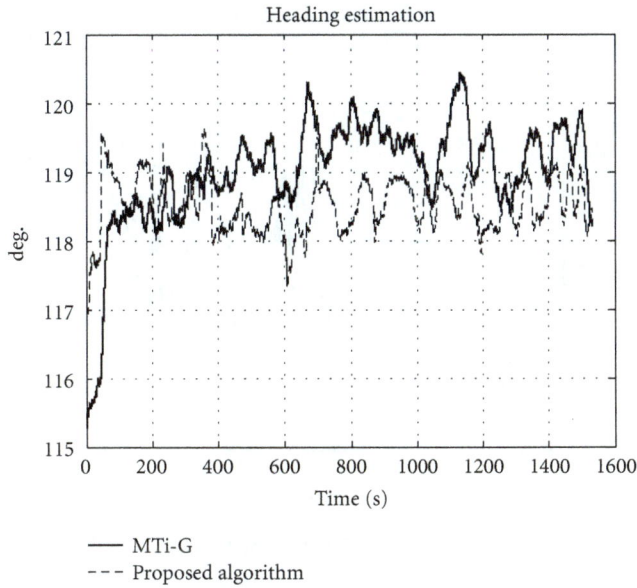

FIGURE 11: Heading estimation using the magnetometers' signals.

The standard deviation of the pitch/roll estimation has a greater minimal value, when the KF algorithm is used without signal preprocessing by a wavelet algorithm. And there is only one optimal value of the system noise deviation σ that minimizes the standard deviation of the pitch/roll estimation, when only KF algorithm is used for data processing.

When wavelet algorithm is used for sensor data preprocessing, the standard deviation of the roll/pitch estimation decreases insignificantly, when $\sigma > 0.03$.

The standard deviation of the pitch/roll estimation is high, when the system noise value is small ($\sigma < 0.01$) in the KF algorithm.

The functional curve type remains the same for different values of the measurement noise level, which is defined in the measurement noise matrix **R**.

It is possible to decrease the standard deviation of the pitch/roll estimation for a stationary object even more. For this purpose, the state transition matrix $\mathbf{\Phi}$ in the KF algorithm should be replaced by the corresponding identity matrix ($\mathbf{\Phi} = \mathbf{I}$). The system noise level should be decreased for at least 1000–10000 times comparing to the value defined according to the MEMS gyroscope specification. These changes in the algorithm are possible as the MEMS gyroscope in a stationary mode has no informative output signal. And it is assumed that such a model is perfectly suitable for the system description and hence requires very small level of system noise for setting in the KF algorithm. The results of the data processing by KF with a modified state transition matrix (modified KF) are presented in Table 4. For comparison, in the same table the results of the data processing by KF without modification of the state transition matrix are presented as well. The sensor data preprocessing by a wavelet algorithm (LOD = 5) was applied in both cases. The standard deviation of the pitch/roll estimation decreases twofold, when using the modified KF.

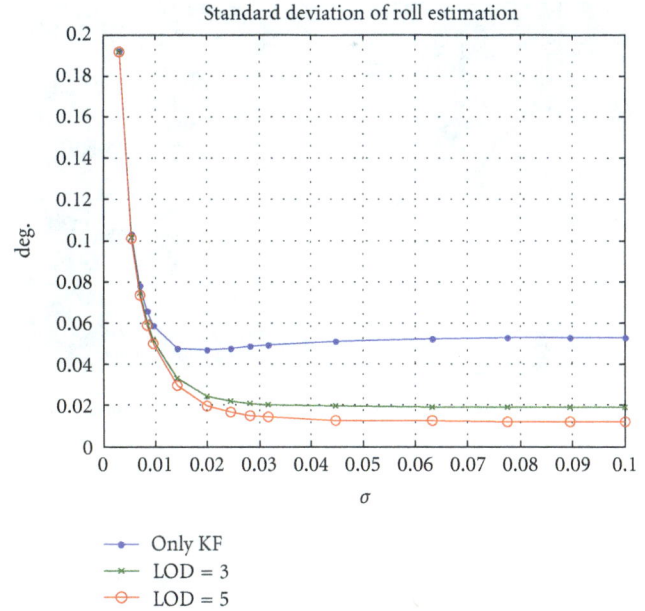

FIGURE 12: Impact of the system noise value σ for the standard deviation of the roll estimation (LOD: Level of Decomposition and KF: Kalman Filter).

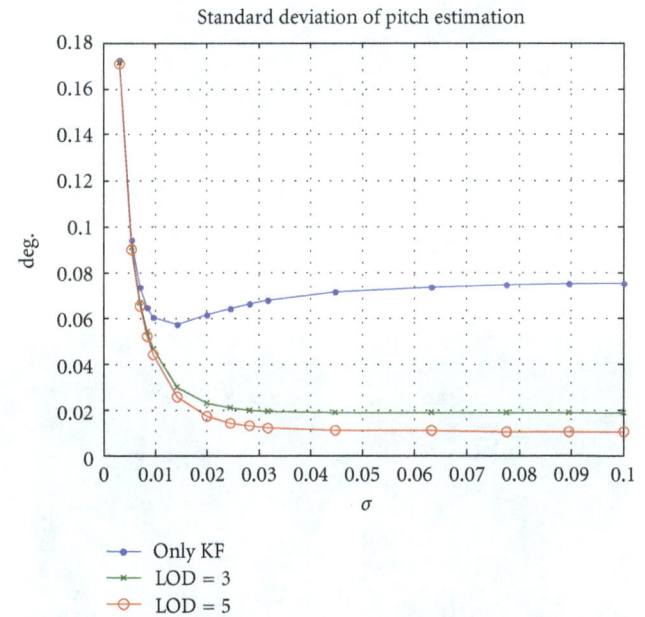

FIGURE 13: Impact of the system noise value σ for the standard deviation of the pitch estimation (LOD: Level of Decomposition and KF: Kalman Filter).

7.4. Attitude Estimation for a Vehicle. In most cases, it is sufficient to have accelerometer data for pitch and roll estimation, when a vehicle is in stationary mode. When the vehicle is moving, it is not possible to obtain a reliable solution for the vehicle pitch and roll attitude angles using only accelerometer signals. It is related to the fact that the accelerometer signal contains the information not

FIGURE 14: Relative alignment of IMU (orange unit) inside vehicle.

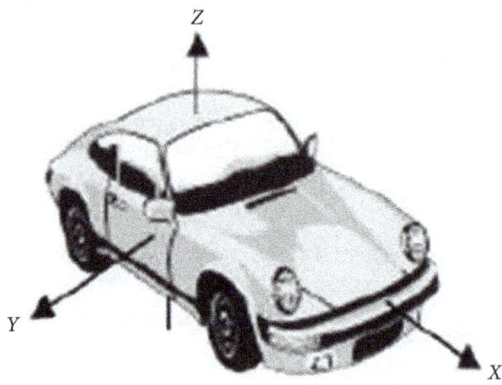

FIGURE 15: V-frame definition.

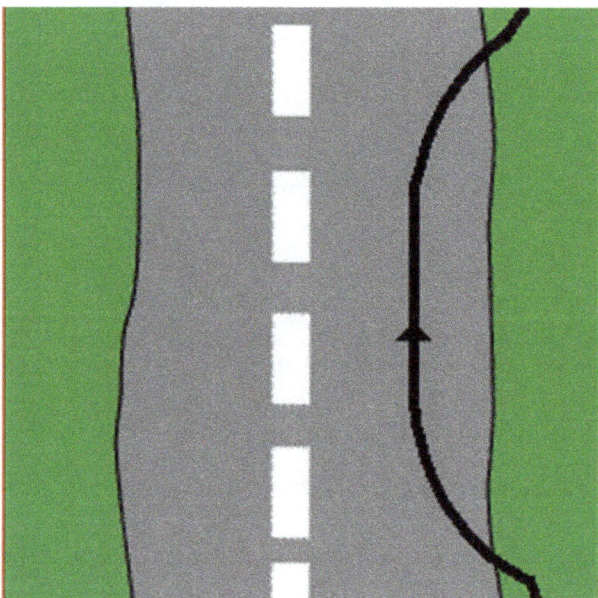

FIGURE 16: The trajectory of the vehicle.

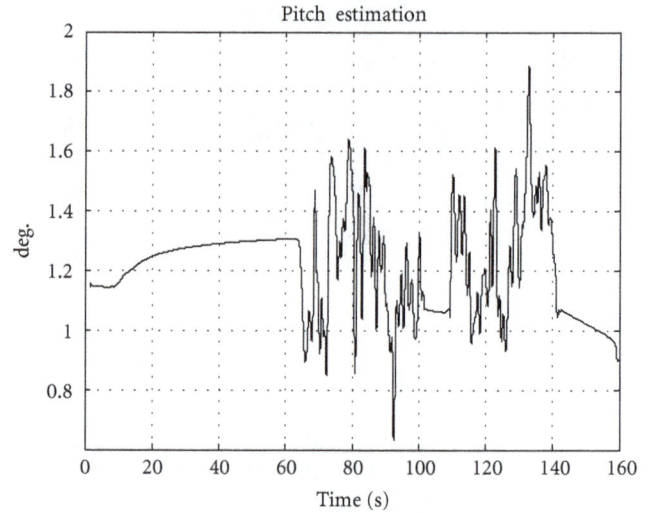

FIGURE 17: Pitch estimation.

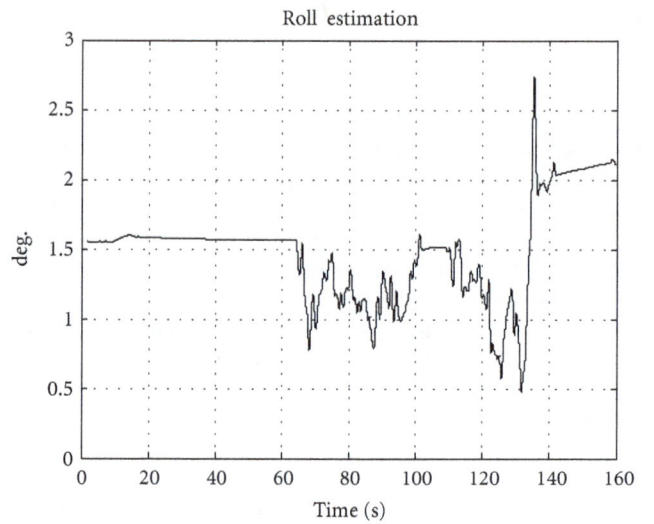

FIGURE 18: Roll estimation.

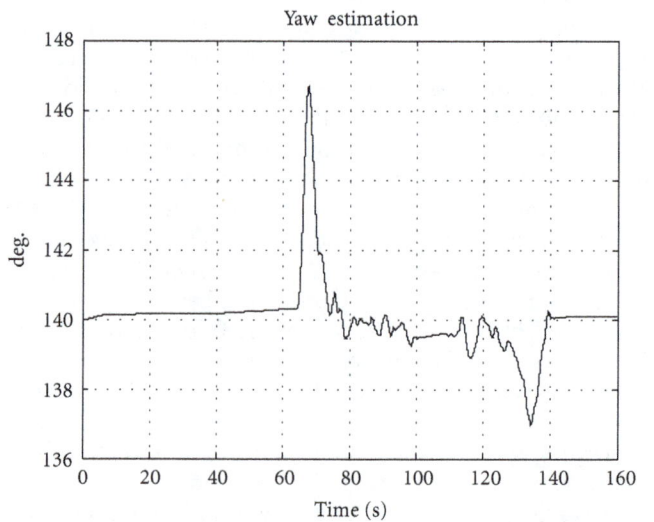

FIGURE 19: Yaw estimation.

only about the object misalignment but also additional signal components due to vehicle acceleration. Thus, the information from the gyroscope signal should be used for the attitude estimation of a moving vehicle. Taking this into account, it is necessary to adjust the algorithm (described in Section 6) in such a way that accelerometer measurements would be nearly ignored, when the vehicle is accelerating. This can be implemented by setting the Kalman gain matrix values to zero in the algorithm, when the vehicle has velocity change more than certain value during 1 second. This value of velocity change was determined empirically. This value was set to 0.15 m/s for the considered algorithm here.

The following additional modification of the algorithm was implemented taking in account results from Section 7.3—to replace state transition matrix Φ by the corresponding identity matrix ($\Phi = I$), when the vehicle has velocity change less than 0.15 m/s during 1 second.

The experiment was conducted on the straight good quality asphalt road. The IMU was fixed rigidly on a board (Figure 14) inside the vehicle. The IMU is placed on the board in order for that the IMU x-axis would coincide with the vehicle longitudinal axis. This can be achieved by a simple geometrical calculation and defining the corresponding axes orientation. (Then the y-axis takes correct orientation automatically). The z-axis of the b-frame was aligned with the gravity vector orientation. This was implemented by the verification of the tilt angles of the board. The tilt angles of the board were verified by an inclinometer in order to guarantee the minimal difference of the z-axes orientation for the b and vehicle coordinate frame. This board was fixed inside the car.

The vehicle coordinate system defines the longitudinal axis of the vehicle as x, the lateral axis to be y, and the vertical axis z-points downwards so that it is aligned with the gravity (Figure 15).

In the beginning of the test, the vehicle was in stationary mode. Then the vehicle accelerates till certain velocity, after that the velocity of the vehicle remains nearly the same for 1-2 minutes, and after that the vehicle slowed down, and the vehicle was again in the stationary mode. The trajectory of the vehicle moving is shown schematically in Figure 16.

The results of the vehicle attitude estimation are shown in Figures 17, 18, and 19. The vehicle was stationary for the first 60 seconds and the last 10 seconds of the experiment. As expected, the estimation of the pitch and roll based on accelerometer data (when vehicle is stationary) has less fluctuations comparing with the attitude estimation, when the vehicle is moving.

The standard deviation of the pitch/roll estimation (when the car is stationary) is bigger comparing to the results shown in Section 7.1. One of the reasons for this is the additional disturbances of the accelerometer signal caused by the closing car door, engine switching, and passenger moving inside the car.

The estimation of the pitch angle (when the car is moving) has a bit more of fluctuations when comparing to the roll estimation as the x-accelerometer (measures acceleration of the vehicle) signal influences the pitch estimation according to (6).

The roll value was bigger, when the vehicle was stationary since the vehicle was on the verge (Figure 16), where the road had greater inclination. The yaw of the vehicle had significant change in the beginning and in the end of the movement (Figure 19). This is due to the fact that the car made a maneuver, moving from the verge to the traffic line. The initial estimate of the yaw was made using magnetometer data.

8. Conclusion

The MEMS IMU is capable of estimating the initial attitude angles (pitch and roll) in the stationary mode without correction from an external sensor such as the GPS. The attitude estimation precision ($\pm 0.5°$) is sufficient for a vehicle navigation application. The convergence rate of the proposed algorithm is very fast: less than 1 second is necessary to obtain the estimation values of pitch and roll. The results of the comparison show that the attitude (pitch and roll) estimations by the proposed algorithm are less noisy comparing with the MTi-G output. Moreover, the practice showed that it is necessary to wait for about 1-2 minutes until the values of pitch and roll from the MTi-G output stabilize. The heading was estimated using the magnetometer data. The heading estimation was noisier and less stabile comparing with the pitch and roll estimation, even taking into account that the wavelet denoising of the signal was used.

The estimation of the vehicle attitude is quite noisy, because the inertial sensors have additional, not enough reduced measurement errors (mainly sensor biases). It's advisable to use specific methods such as stochastic modelling for reducing measurement errors of the inertial sensors or use an aiding sensor such as the GPS for the estimation of the measurement errors of the inertial sensors through fusion with the inertial sensors data. The particle filter [13] or extended Kalman filter [14] can be used as the data fusion algorithms.

References

[1] K. C. Woo and P. C. Gook, "Attitude estimation with accelerometers and gyros using fuzzy tuned kalman filter," in *Proceedings of the European Control Conference*, pp. 23–26, Budapest, Hungary, August 2009.

[2] R. Stirling, K. Fyfe, and G. Lachapelle, "Evaluation of a new method of heading estimation for pedestrian dead reckoning using shoe mounted sensors," *Journal of Navigation*, vol. 58, no. 1, pp. 31–45, 2005.

[3] H. Chao, C. Coopmans, L. Di, and Y. Q. Chen, "A comparative evaluation of low-cost IMUs for unmanned autonomous systems," in *Proceedings of the IEEE Conference on Multisensor Fusion and Integration for Intelligent Systems (MFI '10)*, pp. 211–216, University of Utah, Salt Lake City, Utah, USA, September 2010.

[4] C. Cristina, B. Ludovico, and C. Alessandro, "Design of a low-cost GPS/magnetometer system for land-based navigation: integration and autocalibration algorithms," in *TS07F—Mobile and Asset Mapping Systems*, Marrakech, Morocco, 2011.

[5] D. Groves Paul, *Principles of GNSS, Inertial, and Multisensor Integrated Navigation Systems*, Artech House, London, UK, 2008.

[6] A. Priyanka, S. Zainab, N. Aboelmagd, and E. Naser, *MEMS-Based Integrated Navigation*, Artech House, London, UK, 2010.

[7] M. S. Grewal, L. R. Weill, and A. P. Andrews, *Global Positioning Systems, Inertial Navigation, and Integration*, John Wiley & Sons, New York, NY, USA, 2001.

[8] H. Titterton David and L. Weston John, *Strapdown Inertial Navigation Technology*, The Institution of Electrical Engineers, London, UK, 2nd edition, 2004.

[9] "MTi-G User Manual and Technical Documentation," Document MT0137P, Revision G, Xsens Technologies B.V., May 2009.

[10] A. M. Hasan, K. Samsudm, A. R. Ramli, and R. S. Azmir, "Wavelet-based pre-filtering for low cost inertial sensors," *Journal of Applied Sciences*, vol. 10, no. 19, pp. 2217–2230, 2010.

[11] V. Bistrovs, "Analyse of Kalman algorithm for different movement modes of land mobile object," *Elektronika ir Elektrotechnika*, no. 6, pp. 89–92, 2008.

[12] Y. Li, D. Dusha, W. Kellar, and A. Dempster, "Calibrated MEMS inertial sensors with GPS for a precise attitude heading reference system on autonomous farming tractors," in *Proceedings of the 22nd International Technical Meeting of the Satellite Division of the Institute of Navigation (ION GNSS '09)*, pp. 2138–2145, Savannah, Ga, USA, September 2009.

[13] V. Bistrovs and A. Kluga, "MEMS INS/GPS data fusion using particle filter," *Elektronika ir Elektrotechnika*, no. 6, pp. 77–80, 2011.

[14] V. Bistrovs and A. Kluga, "Adaptive extended Kalman filter for aided inertial navigation system," *Elektronika ir Elektrotechnika*, vol. 122, no. 6, pp. 37–40, 2012.

Permissions

The contributors of this book come from diverse backgrounds, making this book a truly international effort. This book will bring forth new frontiers with its revolutionizing research information and detailed analysis of the nascent developments around the world.

We would like to thank all the contributing authors for lending their expertise to make the book truly unique. They have played a crucial role in the development of this book. Without their invaluable contributions this book wouldn't have been possible. They have made vital efforts to compile up to date information on the varied aspects of this subject to make this book a valuable addition to the collection of many professionals and students.

This book was conceptualized with the vision of imparting up-to-date information and advanced data in this field. To ensure the same, a matchless editorial board was set up. Every individual on the board went through rigorous rounds of assessment to prove their worth. After which they invested a large part of their time researching and compiling the most relevant data for our readers. Conferences and sessions were held from time to time between the editorial board and the contributing authors to present the data in the most comprehensible form. The editorial team has worked tirelessly to provide valuable and valid information to help people across the globe.

Every chapter published in this book has been scrutinized by our experts. Their significance has been extensively debated. The topics covered herein carry significant findings which will fuel the growth of the discipline. They may even be implemented as practical applications or may be referred to as a beginning point for another development. Chapters in this book were first published by Hindawi Publishing Corporation; hereby published with permission under the Creative Commons Attribution License or equivalent.

The editorial board has been involved in producing this book since its inception. They have spent rigorous hours researching and exploring the diverse topics which have resulted in the successful publishing of this book. They have passed on their knowledge of decades through this book. To expedite this challenging task, the publisher supported the team at every step. A small team of assistant editors was also appointed to further simplify the editing procedure and attain best results for the readers.

Our editorial team has been hand-picked from every corner of the world. Their multi-ethnicity adds dynamic inputs to the discussions which result in innovative outcomes. These outcomes are then further discussed with the researchers and contributors who give their valuable feedback and opinion regarding the same. The feedback is then collaborated with the researches and they are edited in a comprehensive manner to aid the understanding of the subject.

Apart from the editorial board, the designing team has also invested a significant amount of their time in understanding the subject and creating the most relevant covers. They scrutinized every image to scout for the most suitable representation of the subject and create an appropriate cover for the book.

The publishing team has been involved in this book since its early stages. They were actively engaged in every process, be it collecting the data, connecting with the contributors or procuring relevant information. The team has been an ardent support to the editorial, designing and production team. Their endless efforts to recruit the best for this project, has resulted in the accomplishment of this book. They are a veteran in the field of academics and their pool of knowledge is as vast as their experience in printing. Their expertise and guidance has proved useful at every step. Their uncompromising quality standards have made this book an exceptional effort. Their encouragement from time to time has been an inspiration for everyone.

The publisher and the editorial board hope that this book will prove to be a valuable piece of knowledge for researchers, students, practitioners and scholars across the globe.

List of Contributors

Ali Jafarnia-Jahromi, Ali Broumandan, John Nielsen and Gerard Lachapelle
Position Location and Navigation (PLAN) Group, Schulich School of Engineering, University of Calgary, 2500 University Drive, NW, Calgary, AB, Canada

Davide Margaria and Beatrice Motella
Istituto Superiore Mario Boella, Via P.C. Boggio 61, 10138 Torino, Italy

Fabio Dovis
Department of Electronics and Telecommunications, Politecnico di Torino, Corso Duca degli Abruzzi 24, 10129 Torino, Italy

Aleksandar Jovanovic, Youssef Tawk, Cyril Botteron and Pierre-Andre Farine
Electronics and Signal Processing Laboratory (ESPLAB), E cole Polytechnique Federale de Lausanne (EPFL), Rue A.-L. Breguet 2, 2000 Neuchatel, Switzerland

Cecile Mongredien
Fraunhofer Institute for Integrated Circuits IIS, Nordostpark 93, 90411 Nuernberg, Germany

Jaime Lien
Department of Electrical Engineering, Stanford University, Stanford, CA 94305, USA

Ulric J. Ferner
Research Laboratory of Electronics, Massachusetts Institute of Technology (MIT), Cambridge, MA 02139, USA

Warakorn Srichavengsup
Department of Computer Engineering, Thai-Nichi Institute of Technology, Bangkok 10250, Thailand

Henk Wymeersch
Department of Signals and Systems, Chalmers University of Technology, Gothenburg 412 96, Sweden

Moe Z. Win
Laboratory for Information and Decision Systems, Massachusetts Institute of Technology (MIT), Cambridge, MA 02139, USA

Montserrat Ros, Joshua Boom and Gavin de Hosson
School of Electrical, Computer and Telecommunications Engineering, University of Wollongong, Wollongong, NSW 2522, Australia

Matthew D Souza
CSIRO ICT Centre, Brisbane, QLD 4069, Australia

V. Dehghanian
Position Location And Navigation (PLAN) Group, University of Calgary, Calgary, AB, Canada
Department of Math Physics and Engineering, Calgary, Mount Royal University, AB, Canada

J. Nielsen and G. Lachapelle
Position Location And Navigation (PLAN) Group, University of Calgary, Calgary, AB, Canada

Muhammad Tahir and Letizia Lo Presti
Dipartimento di Elettronica e Telecomunicazioni (DET), Politecnico di Torino, Corso Duca degli Abruzzi 24, 10129 Torino, Italy

Maurizio Fantino
Navigation Laboratory, Istituto Superiore Mario Boella, Via P.C. Boggio 61, 10138 Torino, Italy

Z. Jiang and W. Lewandowski
Time Department, Bureau International des Poids et Mesures (BIPM), Pavillon de Breteuil, 92312 Sevres Cedex, France

Pau Closas and Carles Fernandez-Prades
Centre Tecnologic de Telecomunicacions de Catalunya (CTTC), Parc Mediterrani de la Tecnologia, Avenida Carl Friedrich Gauss 7, Barcelona, 08860 Castelldefels, Spain

Jose Diez and David de Castro
DEIMOS Space S.L.U, Ronda de Poniente 19, Tres Cantos 28760, Madrid, Spain

J. Nielsen, V. Dehghanian and G. Lachapelle
Position Location and Navigation Group, University of Calgary, Calgary, AB, Canada T2N 1N4

Matthew Cossaboom
Navigation and Instrumentation Research Group (NavINST), Electrical and Computer Engineering Department, Royal Military College of Canada, Kingston, ON, Canada

Jacques Georgy
Trusted Positioning Inc., Calgary, AB, Canada

Tashfeen Karamat and Aboelmagd Noureldin
Navigation and Instrumentation Research Group (NavINST), Electrical and Computer Engineering Department, Queen's University, Kingston, ON, Canada

Joe Youssef, Benoıt Denis, Christelle Godin and Suzanne Lesecq
CEA-Leti Minatec Campus, 17 rue des Martyrs, 38054 Grenoble Cedex 09, France

Lei Zhang, Tianguang Zhang and Haiyan Wu
Institute of Automatic Control Engineering (LSR), Technische Universitat Munchen, 80290 Munchen, Germany

Alexander Borst
Department of Systems and Computational Neurobiology, Max Planck Institute of Neurobiology, Am Klopferspitz 18, D-82152 Martinsried, Germany

Kolja Kuhnlenz
Institute of Automatic Control Engineering (LSR), Technische Universitat Munchen, 80290 Munchen, Germany
Institute for Advanced Study (IAS), Technische Universitat Munchen, 80290 Munchen, Germany

Pedro A. Roncagliolo, Javier G. Garcıa and Carlos H. Muravchik
LEICI, Facultad Ingenierıa, UNLP, B1900TAG La Plata, Argentina

Junchuan Zhou and Otmar Loffeld
Center for Sensor Systems (ZESS), University of Siegen, 57076 Siegen, Germany

Stefan Knedlik
iMAR Navigation GmbH, 66386 St. Ingbert, Germany

Vadim Bistrov
Faculty of Electronics and Telecommunication, Riga Technical University, 1019 Riga, Latvia